Lecture Notes in Computer Science 7092

Commenced Publication in 1973
Founding and Former Series Editors:
Gerhard Goos, Juris Hartmanis, and Jan van Leeuwen

W0246068

Editorial Board

David Hutchison
Lancaster University, UK
Takeo Kanade
Carnegie Mellon University, Pittsburgh, PA, USA
Josef Kittler
University of Surrey, Guildford, UK
Jon M. Kleinberg
Cornell University, Ithaca, NY, USA
Alfred Kobsa
University of California, Irvine, CA, USA
Friedemann Mattern
ETH Zurich, Switzerland
John C. Mitchell
Stanford University, CA, USA
Moni Naor
Weizmann Institute of Science, Rehovot, Israel
Oscar Nierstrasz
University of Bern, Switzerland
C. Pandu Rangan
Indian Institute of Technology, Madras, India
Bernhard Steffen
TU Dortmund University, Germany
Madhu Sudan
Microsoft Research, Cambridge, MA, USA
Demetri Terzopoulos
University of California, Los Angeles, CA, USA
Doug Tygar
University of California, Berkeley, CA, USA
Gerhard Weikum
Max Planck Institute for Informatics, Saarbruecken, Germany

Dongdai Lin Gene Tsudik
Xiaoyun Wang (Eds.)

Cryptology and Network Security

10th International Conference, CANS 2011
Sanya, China, December 10-12, 2011
Proceedings

Springer

Volume Editors

Dongdai Lin
Chinese Academy of Sciences
State Key Laboratory of Information Security (SKLOIS)
Beijing 100190, China
E-mail: ddlin@is.iscas.ac.cn

Gene Tsudik
University of California, Computer Science Department
Irvine, CA 92697-3435, USA
E-mail: gts@ics.uci.edu

Xiaoyun Wang
Tsinghua University, Institute of Advanced Study
Beijing 100084, China
E-mail: xywang@sdu.edu.cn

ISSN 0302-9743
e-ISSN 1611-3349
ISBN 978-3-642-25512-0
ISBN 978-3-642-25513-7 (eBook)
DOI 10.1007/978-3-642-25513-7
Springer Heidelberg Dordrecht London New York

Library of Congress Control Number: 2011941497

CR Subject Classification (1998): E.3, C.2, K.6.5, D.4.6, G.2.1, E.4

LNCS Sublibrary: SL 4 – Security and Cryptology

Typesetting: Camera-ready by author, data conversion by Scientific Publishing Services, Chennai, India

Printed on acid-free paper

Springer is part of Springer Science+Business Media (www.springer.com)

Preface

It was a real pleasure to have taken part in organizing the 10th International Conference on Cryptography and Network Security (CANS 2011). It was held during December 10–12, 2011, at the International Asia Pacific Convention Center in Sanya, on the subtropical island of Hainan (China). CANS 2011 was sponsored by the National Natural Science Foundation of China (NSFC) and Shandong University (SDU). It was also held in cooperation with the International Association for Cryptologic Research (IACR).

The CANS 2011 Program Committee (PC) consisted of 46 internationally recognized researchers with combined expertise covering the entire scope of the conference.

The recent growth in the number of cryptography venues prompted stiff competition for high-quality papers. Nonetheless, CANS has consistently attracted a number of strong submissions. This year, we received a total of 65 submissions. A few were incomplete and/or were rejected without review. Each remaining paper was reviewed by at least four reviewers. After intensive deliberations by the Program Committee, 18 submissions were accepted for presentation.

In addition to regular papers, the conference program included four excellent invited talks, by Colin Boyd (QUT), Xavier Boyen (PARC), Phong Nguyen (ENS) and Joan Daemen (STMicroelectronics).

A number of people selflessly contributed to the success of CANS 2011. First and foremost, we thank the authors of all submissions. They are the backbone of this conference and their confidence and support are highly appreciated. We are similarly grateful to the dedicated, knowledgeable and hard-working Program Committee members who provided excellent and timely reviews as well as took part in post-review discussions. Their altruistic dedication and community service spirit are commendable. Special thanks to "shepherds" for their extra efforts.

We gratefully acknowledge the organizational contributions by CANS 2011 General Chair, Dongdai Lin, without whom the conference would have been impossible. We wish to thank Meiqin Wang as well as all the members of Organizing Committee for the local arrangements, and Claudio Soriente for publicity. We are also indebted to the CANS Steering Committee members for their guidance. Last, but not least, we thank CANS 2011 sponsors: NSFC and SDU.

December 2011

Gene Tsudik
Xiaoyun Wang
Program Co-chairs

CANS 2011

The 10th International Conference on Cryptography and Network Security (In Cooperation with IACR)

Sanya, China December 10–12, 2011

Sponsored and organized by

Shandong University
and
National Natural Science Foundation of China

Steering Committee

Yvo Desmedt	University College London, UK
Matt Franklin	University of California, Davis, USA
Juan Garay	AT&T Labs - Research, USA
Yi Mu	University of Wollongong, Australia
David Pointcheval	CNRS and ENS, France
Huaxiong Wang	NTU, Singapore

General Chair

Dongdai Lin	SKLOIS, Chinese Academy of Sciences, China

Program Co-chairs

Gene Tsudik	University of California, Irvine, USA
Xiaoyun Wang	Shandong University, China

Program Committee

Jean-Philippe Aumasson	Nagravision, Switzerland
Feng Bao	I2R, Singapore
Jean-Luc Beuchat	University of Tsukuba, Japan
Mike Burmester	Florida State University, USA
Sherman S.M. Chow	University of Waterloo, Canada
Giovanni Di Crescenzo	Telcordia, USA
Emiliano De Cristofaro	University of California, Irvine, USA
Reza Curtmola	NJIT, USA

Xuhua Ding	SMU, Singapore
David Galindo	University of Luxembourg, Luxembourg
Amir Herzberg	Bar-Ilan University, Israel
Nick Hopper	University of Minnesota, USA
Tetsu Iwata	Nagoya University, Japan
Charanjit Jutla	IBM Research, USA
Khoongming Khoo	DSO National Laboratories, Singapore
Loukas Lazos	University of Arizona, USA
Gaëtan Leurent	University of Luxembourg, Luxembourg
Helger Lipmaa	Cybernetica AS and Tallinn University, Estonia
Feifei Li	Florida State University, USA
Di Ma	University of Michigan, USA
Mark Manulis	TU Darmstadt, Germany
Florian Mendel	Graz University of Technology, Austria
Atsuko Miyaji	JAIST, Japan
Jorge Nakahara Jr.	Independent Researcher
Melek Önen	EURECOM, France
Bryan Parno	Microsoft Research, USA
Andreas Pashalidis	KU Leuven, Belgium
Thomas Peyrin	NTU, Singapore
Raphael C.-W. Phan	Loughborough University, UK
Josef Pieprzyk	Macquarie University, Australia
Rei Safavi-Naini	University of Calgary, Canada
Yu Sasaki	NTT Research, Japan
Martin Schläffer	Graz University of Technology, Austria
Michael Scott	Dublin City University, Ireland
Elaine Shi	Berkeley / PARC, USA
Claudio Soriente	Universidad Politecnica de Madrid, Spain
Damien Stehlé	CNRS and ENS de Lyon, France
Ron Steinfeld	Macquarie University, Australia
Søren Steffen Thomsen	Technical University of Denmark, Denmark
Ersin Uzun	PARC, USA
Ivan Visconti	University of Salerno, Italy / UCLA, USA
Huaxiong Wang	NTU, Singapore
Meiqin Wang	Shandong University, China
Wenling Wu	Institute of Software, CAS, China
Shouhuai Xu	University of Texas, San Antonio, USA
Jianying Zhou	I2R, Singapore

Local Arrangements Chair

Meiqin Wang	Shandong University, China

Local Organizing Committee

Hongbo Yu	Tsinghua University, China
Keting Jia	Tsinghua University, China
Puwen Wei	Shandong University, China
Lidong Han	Tsinghua University, China

Publicity Chair

Claudio Soriente	Polytechnic University of Madrid, Spain

Secretary and Treasurer

Puwen Wei	Shandong University, China

WEB/Registration

Jiazhe Chen	Shandong University, China

External Reviewers

Filipe Beato
Xavier Boyen
Aldar C-F. Chan
Kai-Yuen Cheong
Kim-Kwang Raymond Choo
Cheng-Kang Chu
Yi Deng
Orr Dunkelman
Nicolas Estibals
Kazuhide Fukushima
David Gardner
Paolo Gasti
Pierrick Gaudry
Zahra Aghazadeh Gaven Watson
Moti Geva
Jian Guo
Francis Hsu
Xinyi Huang
Choy Jia Li Valerie
Markulf Kohlweiss
Simon Kramer
Mario Lamberger
Gregor Leander
Zi Lin
Weiliang Luo
Avradip Mandal
Tomislav Nad
Kris Narayan
María Naya-Plasencia
Lan Nguyen
Ryo Nojima
Wakaha Ogata
Kazumasa Omote
Andreas Peter
Natacha Portier
Axel Poschmann
Francisco Rodrguez-Henrquez
Yasuhide Sakai
Alessandra Scafuro
Jun Shao
Haya Shulman
Ashraful Tuhin
Frederik Vercauteren
Cong Wang
Wei Wu
Brecht Wyseur
Huihui Yap
Shucheng Yu
Tsz Hon Yuen
Kai Zeng
Lei Zhang
Liangfeng Zhang
Xin Zhang
Yun Zhang
Qingji Zheng

Table of Contents

Invited Talks

Symmetric Cryptanalysis

Symmetric Ciphers

Public Key Cryptography

Protocol Attacks

Privacy Techniques

Varia

Expressive Encryption Systems from Lattices
(Abstract from the Invited Lecture)

Xavier Boyen

Palo Alto Research Center

Abstract. In this survey, we review a number of the many "expressive" encryption systems that have recently appeared from lattices, and explore the innovative techniques that underpin them.

1 Introduction

Lattice-based cryptosystems are becoming an increasingly popular in the research community, owing to a unique combination of factors. On the one hand, lattice systems are often conceptually simple to understand and thus easy to implement by non-specialists, at least if one makes abstraction of the finer mathematical intricacies surrounding their security analysis. On the other, their soundness is backed by strong complexity-theoretic evidence that the underlying problems are suitably "hard", of which the most often repeated are the existence of certain average-case to worst-case equivalences [7,19] and their conjectured resistance to quantum attacks. All those factors conspire to make lattices a prime choice, if not the primary one yet, for mathematical crypto design looking out into the future.

Although empirical uses of lattices have been made in commercial cryptography, they have had a rather slow start in research circles. For more than a decade, indeed, signature schemes and basic public-key encryption have essentially remained their sole confine [7,19]. In the past few years, however, lattices have flourished into a theoretically solid, comprehensive framework, owing to the discovery of a few key concepts and techniques. This ushered the way to the construction of ever more powerful and expressive public-key encryption systems, writ large — a whole new world of cryptographic constructions waiting to be explored and conquered.

The search for encryption systems with complex functionalities arguably originates with the field of modern cryptography itself; but it is the arrival of bilinear maps, or pairings, that truly jumpstarted it, by providing such spectacular solutions to long-standing open problems as identity-based encryption [10]. Lattices are late to this game, and currently still lag in functionality and practicality with respect to pairing-based constructions. Nevertheless, an unmistakable shift from pairings to lattices is presently occurring in the research community, driven as much as the looming threat of quantum attacks that lattices seek to alleviate, as the sheer scientific draw of tackling tough problems from wholly new directions.

D. Lin, G. Tsudik, and X. Wang (Eds.): CANS 2011, LNCS 7092, pp. 1–12, 2011.

In this lecture, we set out to explore some of the recent advances in that search, and distill the essential new ideas that made them possible.[1]

2 Background

A lattice is an additive subgroup of $\mathbb{R}^n$; it is therefore generated by a *basis* of n (linearly independent) vectors in $\mathbb{R}^n$. In high dimensions, many computational problems on lattices are intractable, and in some cases are even known to be NP-hard. What makes lattices useful in cryptography, is that, though all bases are equivalent from a linear algebraic point of view, bases whose vectors have *low norm* can provide easy solutions to otherwise intractable lattice problems. For instance, the "closest vector problem" (which consists of finding a lattice point within a prescribed radius from a given reference in $\mathbb{R}^n$) becomes soluble if avails a low-norm lattice basis. Without such a *good basis*, this problem and many related ones remain intractable.

Whereas this asymmetry is, of course, central to lattices' use in asymmetric cryptography, general lattices as defined above are somewhat unwieldy to work with. One often prefers to restrict oneself to a restricted class of lattices with special properties; be it for reasons of convenience or efficiency, or both.

To wit, many of the recently developed expressive cryptosystems make use of Ajtai's lattices [6]. Those are sets of vectors $\boldsymbol{x} \in \mathbb{Z}^m$ that lie in the kernel of some $\mathbf{A} \in \mathbb{Z}^{n \times m}$ modulo some prime q, *i.e.*, defined by an equation $\mathbf{A} \cdot \mathbf{x} = 0 \pmod q$. Aside from their definitional convenience, Ajtai's lattices are appealing for two reasons: one of security, the other of flexibility. First, they induce rich and usable cryptographic key spaces, owing to the Regev's result that random instances are just as hard as worst-case ones [19]. Second, they are closely related to error-correction codes, and in particular the matrix $\mathbf{A}$ defines a "public" computational operator that can be effectively reverted with knowledge of a "private" trapdoor, as first shown by Gentry *et al.* [15]: the map $x \mapsto \mathbf{A} \cdot \boldsymbol{x}$, restricted to for *low-norm* inputs $x \in \mathbb{Z}^m$, can be reverted, in the sense of finding a colliding pre-image $x' \in \mathbb{Z}^m$, if one knows a good basis for the implied lattice. This combination of features — easy-to-sample key spaces and a kind of invertibility — are sought for in asymmetric cryptographic constructions.

By way of comparison, we mention that Gentry's fully homomorphic encryption scheme made extensive use of a different kind of lattices, constructed from polynomial rings, whose ring structure was crucial to realize full homomorphism in his original system.

[1] Around the same time, also appeared the first realization of "fully homomorphic encryption" [14], a hugely significant breakthrough of both theoretical and (one hopes) eventual practical significance. FHE undoubtedly contributed greatly to the general surge in lattice popularity, notwithstanding the quite different flavors of problems involved. FHE has since taken a life of its own, with the most recent performance and conceptual improvements seemingly taking it away from its lattice roots, and squarely into the realm of pure number theory. We refer the interested reader to the rapidly growing literature on the subject; see [16] for pointers.

Note. Due to space contraints, we do not give formal statements of the various notions and schemes in this abstract, but refer the reader to the original papers.

2.1 Lattice Notions

We let parameters q, m, n be polynomial functions of a security parameter λ.

Lattices. Let $\mathbf{B} = [\ \boldsymbol{b}_1 \mid \ldots \mid \boldsymbol{b}_m\ ]$ be an $m \times m$ real matrix with linearly independent column vectors. It generates an m-dimensional full-rank lattice Λ,

$$\Lambda = \mathcal{L}(\mathbf{B}) = \Big\{\ \boldsymbol{y} \in \mathbb{R}^m \quad \text{s.t.} \quad \exists \boldsymbol{s} = (s_1, \ldots, s_m) \in \mathbb{Z}^m\ ,\quad \boldsymbol{y} = \mathbf{B}\,\boldsymbol{s} = \sum_{i=1}^{m} s_i\,\boldsymbol{b}_i\ \Big\}$$

Of interest to us is the case of integer lattices that are invariant under translation by multiples of some integer q in each of the coordinates, or Ajtai lattices.

Ajtai lattices (and their shifts). For q prime, $\mathbf{A} \in \mathbb{Z}_q^{n\times m}$ and $\boldsymbol{u} \in \mathbb{Z}_q^n$, define:

$$\Lambda_q^{\perp}(\mathbf{A}) = \{\ \boldsymbol{e} \in \mathbb{Z}^m \quad \text{s.t.} \quad \mathbf{A}\,\boldsymbol{e} = 0 \pmod q\ \}$$
$$\Lambda_q^{\boldsymbol{u}}(\mathbf{A}) = \{\ \boldsymbol{e} \in \mathbb{Z}^m \quad \text{s.t.} \quad \mathbf{A}\,\boldsymbol{e} = \boldsymbol{u} \pmod q\ \}$$

Ajtai [6] first showed how to sample an essentially uniform matrix $\mathbf{A} \in \mathbb{Z}_q^{n\times m}$, along with a full-rank set $\mathbf{T_A} \subset \Lambda^{\perp}(\mathbf{A})$ of *low-norm* vectors or points on the lattice. We state an improved version of Ajtai's basis generator, from [8].

Trapdoors for lattices. Let $n = n(\lambda), q = q(\lambda), m = m(\lambda)$ be positive integers with $q \geq 2$ and $m \geq 5n \log q$. There exists a probabilistic polynomial-time algorithm $\mathsf{TrapGen}$ that outputs a pair of $\mathbf{A} \in \mathbb{Z}_q^{n\times m}$ and $\mathbf{T_A} \in \mathbb{Z}_q^{m\times m}$ such that $\mathbf{A}$ is statistically close to uniform and $\mathbf{T_A}$ is a basis for $\Lambda^{\perp}(\mathbf{A})$ with "Gram-Schmidt" length $L = \|\widetilde{\mathbf{T_A}}\| \leq m \cdot \omega(\sqrt{\log m})$, with all but $n^{-\omega(1)}$ probability.

2.2 Discrete Gaussians

Central to all cryptosystems based on Ajtai lattices, is the study of the distribution of various vectors of interest (e.g., preimages to the operation $\mathbf{A}$). Multidimensional discrete Gaussian distributions are particularly useful.

Discrete Gaussians. Let m be a positive integer and Λ an m-dimensional lattice over $\mathbb{R}$. For any vector $\boldsymbol{c} \in \mathbb{R}^m$ and any positive spread parameter $\sigma \in \mathbb{R}_{>0}$, let:

$\rho_{\sigma,c}(\boldsymbol{x}) = \exp\big(-\pi \frac{\|\boldsymbol{x}-\mathbf{c}\|^2}{\sigma^2}\big)$: a Gaussian function of center $\mathbf{c}$ and parameter σ;
$\rho_{\sigma,\mathbf{c}}(\Lambda) = \sum_{\boldsymbol{x}\in\Lambda} \rho_{\sigma,\mathbf{c}}(\boldsymbol{x})$: the infinite discrete sum of $\rho_{\sigma,\mathbf{c}}$ over the lattice Λ;
$\mathcal{D}_{\Lambda,\sigma,\mathbf{c}}$: the discrete Gaussian distribution on Λ of center $\mathbf{c}$ and parameter σ:

$$\forall \boldsymbol{y} \in \Lambda\ ,\ \mathcal{D}_{\Lambda,\sigma,\mathbf{c}}(\boldsymbol{y}) = \frac{\rho_{\sigma,\mathbf{c}}(\boldsymbol{y})}{\rho_{\sigma,\mathbf{c}}(\Lambda)}$$

For convenience, we abbreviate $\rho_{\sigma,0}$ and $\mathcal{D}_{\Lambda,\sigma,0}$ respectively as ρ_σ and $\mathcal{D}_{\Lambda,\sigma}$.

2.3 Sampling and Preimage Sampling

The public-key and secret-key functions we need for asymmetric cryptography arise from the previous notions. Specifically, while anyone can sample a discrete Gaussian preimage with no prescription on its image under $\mathbf{A}$, only with a trapdoor or short basis $\mathbf{B}$ can one sample a preimage hitting a specific target image $\boldsymbol{u}$ with the same conditional distribution. The following results are due to Gentry, Peikert, and Vaikuntanathan [15]. They first construct an algorithm for sampling from the discrete Gaussian $\mathcal{D}_{\Lambda,\sigma,\boldsymbol{c}}$, given a basis $\mathbf{B}$ for the m-dimensional lattice Λ with $\sigma \geq \|\widetilde{\mathbf{B}}\| \cdot \omega(\sqrt{\log m})$. Next they give an algorithm that given an trapdoor and a target, can sample a preimage with the same (conditional) discrete Gaussian distribution.

Sampling a discrete Gaussian. There exists a probabilistic polynomial-time algorithm, denoted SampleGaussian, that, on input an arbitrary basis $\mathbf{B}$ of an m-dimensional full-rank lattice $\Lambda = \mathcal{L}(\mathbf{B})$, a parameter $\sigma \geq \|\widetilde{\mathbf{B}}\| \cdot \omega(\sqrt{\log m})$, and a center $\boldsymbol{c} \in \mathbb{R}^m$, outputs a sample from a distribution statistically close to $\mathcal{D}_{\Lambda,\sigma,\boldsymbol{c}}$.

Preimage sampling from trapdoor. There exists a probabilistic polynomial-time algorithm, denoted SamplePre, that, on input a matrix $\mathbf{A} \in \mathbb{Z}_q^{n\times m}$, a short trapdoor basis $\mathbf{T_A}$ for $\Lambda_q^{\perp}(\mathbf{A})$, a target image $\boldsymbol{u} \in \mathbb{Z}_q^n$, and a Gaussian parameter $\sigma \geq \|\widetilde{\mathbf{T_A}}\| \cdot \omega(\sqrt{\log m})$, outputs a sample $\boldsymbol{e} \in \mathbb{Z}_q^m$ from a distribution within negligible statistical distance of $\mathcal{D}_{\Lambda_q^{\boldsymbol{u}}(\mathbf{A}),\sigma}$.

Micciancio and Regev [17] show that the norm of vectors sampled from discrete Gaussians is small with high probability. We omit the full statement.

2.4 Hardness Assumption

One of the classic hardness assumptions associated with Ajtai lattices, refers to the LWE — Learning With Errors — problem, first stated by [19], and since extensively studied and used. For polynomially bounded modulus q, the computational and decisional versions of the problems are polynomially reducible to each other. We give the following statement of the decisional version.

The decisional LWE problem. Consider a prime q, a positive integer n, and a distribution χ over $\mathbb{Z}_q$, all public. An $(\mathbb{Z}_q, n, \chi)$-LWE problem instance consists of access to an unspecified challenge oracle $\mathcal{O}$, being, either, a noisy pseudo-random sampler $\mathcal{O}_s$ carrying some constant random secret key $\boldsymbol{s} \in \mathbb{Z}_q^n$, or, a truly random sampler $\mathcal{O}_\$$, whose behaviors are respectively as follows:

$\mathcal{O}_s$: outputs noisy pseudo-random samples of the form $(\boldsymbol{w}_i, v_i) = \big(\boldsymbol{w}_i,\ \boldsymbol{w}_i^T\, \boldsymbol{s} + x_i\big) \in \mathbb{Z}_q^n \times \mathbb{Z}_q$, where, $\boldsymbol{s} \in \mathbb{Z}_q^n$ is a uniformly distributed persistent secret key that is invariant across invocations, $x_i \in \mathbb{Z}_q$ is a freshly generated ephemeral additive noise component with distribution χ, and $\boldsymbol{w}_i \in \mathbb{Z}_q^n$ is a fresh uniformly distributed vector revealed as part of the output.

$\mathcal{O}_\$$: outputs truly random samples $(\boldsymbol{w}_i, v_i) \in \mathbb{Z}_q^n \times \mathbb{Z}_q$, drawn independently uniformly at random in the entire domain $\mathbb{Z}_q^n \times \mathbb{Z}_q$.

The $(\mathbb{Z}_q, n, \chi)$-LWE problem statement, or LWE for short, allows an unspecified number of queries to be made to the challenge oracle $\mathcal{O}$, with no stated prior bound. We say that an algorithm $\mathcal{A}$ decides the $(\mathbb{Z}_q, n, \chi)$-LWE problem if $\big|\Pr[\mathcal{A}^{\mathcal{O}_s} = 1] - \Pr[\mathcal{A}^{\mathcal{O}_\$} = 1]\big|$ is non-negligible for a random $s \in \mathbb{Z}_q^n$.

Average to worst case. The confidence in the hardness of the LWE problem stems in part from a result of Regev [19] which shows that the for certain noise distributions χ, the LWE problem is as hard as (other) classic lattice problems (such as SIVP and GapSVP) in the worst case, under a quantum reduction. A non-quantum reduction with different parameters was later given by Peikert [18]. We state Regev's result for reference below.

The Regev reduction theorem. Consider a real parameter $\alpha = \alpha(n) \in (0, 1)$ and a prime $q = q(n) > 2\sqrt{n}/\alpha$. Denote by $\mathbb{T} = \mathbb{R}/\mathbb{Z}$ the group of reals $[0, 1)$ with addition modulo 1. Denote by Ψ_α the distribution over $\mathbb{T}$ of a normal variable with mean 0 and standard deviation $\alpha/\sqrt{2\pi}$ then reduced modulo 1. Denote by $\lfloor x \rceil = \lfloor x + \frac{1}{2} \rfloor$ the nearest integer to the real $x \in \mathbb{R}$. Denote by $\bar{\Psi}_\alpha$ the discrete distribution over $\mathbb{Z}_q$ of the random variable $\lfloor q X \rceil \mod q$ where the random variable $X \in \mathbb{T}$ has distribution Ψ_α. Then, if there exists an efficient, possibly quantum, algorithm for deciding the $(\mathbb{Z}_q, n, \bar{\Psi}_\alpha)$-LWE problem, there exists a quantum $q \cdot poly(n)$-time algorithm for approximating the SIVP and GapSVP problems, to within $\tilde{O}(n/\alpha)$ factors in the ℓ_2 norm, in the worst case.

Since the best known algorithms for 2^k-approximations of GapSVP and SIVP run in time $2^{\tilde{O}(n/k)}$, it follows that the LWE problem with noise ratio $\alpha = 2^{-n^\epsilon}$ ought to be hard for some constant $\epsilon < 1$.

3 Classic Constructions

We start this presentation with the systems from which all recent developments are based, starting with Regev's minimalistic public-key cryptosystem.

3.1 Regev Public-Key Encryption

The basic principle of Regev's original public-key cryptosystem is deceptively simple, as long as one does not delve too deep in its analysis. Paradoxically, Regev's system predated the GPV trapdoor preimage sampling, and required no other machinery than a basic random Ajtai lattice, not even a short basis.

The algorithms defining the system are as follows:

Key Generation. Pick a suitable modulus q, a random Ajtai matrix $\mathbf{A} \in \mathbb{Z}_q^{n \times m}$, and a short random vector $\boldsymbol{d} \in \mathbb{Z}^m$; and let $\boldsymbol{u} = \mathbf{A} \cdot \boldsymbol{d} \mod q \in \mathbb{Z}_q^n$. The public and secret keys are:

$$\mathsf{PK} = (q, \mathbf{A}, \boldsymbol{u}) \qquad \mathsf{SK} = \boldsymbol{e}$$

Encryption. To encrypt a bit $m \in \{0,1\}$, pick a random vector $\boldsymbol{s} \in \mathbb{Z}_q^n$, a noise scalar $y_0 \sim \psi$, and a noise vector $\boldsymbol{y}_1 \sim \psi^m$, and output:

$$\mathsf{CT} = \Big(c_0 = \boldsymbol{s}^\top \boldsymbol{u} + m \cdot \lfloor \frac{q}{2} \rfloor + y_0, \quad \boldsymbol{c}_1 = \mathbf{A}^\top \boldsymbol{s} + \boldsymbol{y}_1 \Big)$$

Decryption. The bit m is deemed to be 0 or 1, if the following quantity is respectively closer to 0 or $\frac{q}{2}$, modulo q:

$$c_0 - \boldsymbol{c}_1^\top \boldsymbol{d} \pmod q$$

It is easy to see that all terms cancel in the decryption operation, but for the noise contributions due to y_0 and $\boldsymbol{y}_1$ and the term $m \cdot \lfloor \frac{q}{2} \rfloor$ which redundantly encodes m. The noise is chosen sufficiently small so that, even after taking the inner product of $\boldsymbol{y}_1$ with the secret key vector $\boldsymbol{d}$, the message m remains recognizable. However, for an attacker who can only find *large* preimages of $\boldsymbol{u}$, decoding will fail as the noise will completely mask the message. Technically, the noise distribution ψ is chosen according to Regev's reduction theorem, so that semantic security of the system can be reduced to a worst-case lattice hardness assumption. We refer to Regev's paper for details.

Remark. We note that in Regev's original paper [19], the roles of $\boldsymbol{d}$ and $\boldsymbol{s}$ were reversed. The above is Regev's *dual*, more conveniently extended as we now describe.

3.2 GPV Identity-Based Encryption

Gentry *et al.* [15] first showed how to realize identity-based encryption from lattices. In IBE, the public key is arbitrary, and the corresponding secret key can be "extracted" from it by a central authority that holds a special trapdoor.

The GPV system can be viewed as an instantiation of the Regev system, where instead of having a single fixed "syndrome" vector $\boldsymbol{u}$ (see the description above), said vector is made to depend on the recipient's identity using a hash function, as in $\boldsymbol{u}_{id} = H(id)$. Since no predetermined $\boldsymbol{d}$ can serve to deduce $\boldsymbol{u}$, a central authority will need the preimage sampling trapdoor to compute a short preimage $\boldsymbol{d}_{id}$ for any desired target $\boldsymbol{u}_{id}$; the trapdoor is thus the IBE master key.

Their system is described as follows:

System Setup. Pick a suitable modulus q, and sample a random Ajtai matrix $\mathbf{A} \in \mathbb{Z}_q^{n\times m}$ with associated trapdoor $\mathbf{B} \in \mathbb{Z}^{m\times m}$. The public parameters and master secret key are:

$$\mathsf{PP} = (q, \mathbf{A}) \qquad \mathsf{MK} = \mathbf{B}$$

Private Key Extraction. To extract a private key corresponding to a public identity id, first compute $\boldsymbol{u}_{id} = H(id) \in Z_q^n$, and then, using the trapdoor $\mathbf{B}$, find a short preimage $\boldsymbol{d}_{id}$, i.e., a low-norm vector such that $\mathbf{A} \cdot \boldsymbol{d}_{id} = \boldsymbol{u}_{id}$ (mod q). Output the private key as:

$$\mathsf{SK}_{\mathsf{id}} = d_{id}$$

Encryption. To encrypt a bit $m \in \{0,1\}$ for an identity *id*, compute $\boldsymbol{u}_{id} = H(id) \in Z_q^n$ and then encrypt as in the Regev system; i.e., picking a random vector $\boldsymbol{s} \in \mathbb{Z}_q^n$, a noise scalar $y_0 \sim \psi$, and a noise vector $\boldsymbol{y}_1 \sim \psi^m$, output:

$$\mathsf{CT} = \left(c_0 = \boldsymbol{s}^\top \boldsymbol{u}_{id} + m \cdot \lfloor \frac{q}{2} \rfloor + y_0, \quad \boldsymbol{c}_1 = \mathbf{A}^\top \boldsymbol{s} + \boldsymbol{y}_1\right)$$

Decryption. Proceed as in the Regev system, using the private key $\boldsymbol{d}_{id}$; i.e., decrypt as 0 or 1 depending on whether the following is closer to 0 or $\frac{q}{2}$, modulo q:

$$c_0 - \boldsymbol{c}_1^\top \boldsymbol{d}_{id} \pmod q$$

The proof of security follows readily from that of Regev's system, given the properties of trapdoor preimage sampling, in the random-oracle model.

4 Techniques and Refinements

Building upon those earlier results, a number of significant refinements were quick to appear, showing that full security reductions were possible and practical, even in the standard model.

4.1 Bit-by-Bit Standard-Model IBE

The first step was taken concurrently by several teams [3,13], that quickly figured out a way to realize IBE from lattices *in the standard model*, albeit with a stiff efficiency penalty over the GPV system.

The idea was to encode the identity not in Regev's vector $\boldsymbol{u}$ as in GPV (which required a random oracle), but in the matrix $\mathbf{A}$ itself, in a binary fashion reminiscent of the pairing-based from [12]. Specifically, for an ℓ-bit identity $id = (b_1, \ldots, b_\ell) \in \{0,1\}^\ell$, the matrix $A_{id} \in \mathbb{Z}_q^{n \times (\ell+1)m}$ is defined as the following concatenation of $\ell + 1$ constant matrices of dimension $n \times m$:

$$\mathbf{A}_{id} = \left[\mathbf{A}_0 \middle| \mathbf{A}_{1,b_1} \middle| \mathbf{A}_{2,b_2} \middle| \ldots \middle| \mathbf{A}_{\ell,b_\ell}\right]$$

from which the following relationship between (a user's) public and private key will be enforced:

$$\mathbf{A}_{id} \cdot \boldsymbol{d}_{id} = \boldsymbol{u} \pmod q$$

It is easy to see (but harder to prove) that all that is needed to find a short solution d_{id} in the above equation, is a preimage sampling trapdoor for *any* of the matrices $A_{i,\cdot}$ intervening in $\mathbf{A}_{id}$. Accordingly, all the submatrices $\mathbf{A}_{i,b_i}$ for $i \geq 1$ can be picked at random, as merely a trapdoor $\mathbf{B}_0$ for $\mathbf{A}_0$ suffices to find short preimages under the whole of $\mathbf{A}_{id}$. Hence, such shall be the IBE master key in the real system.

The full system is described as follows:

System Setup. Pick a suitable modulus q, and sample a random Ajtai matrix $\mathbf{A}_0 \in \mathbb{Z}_q^{n\times m}$ with associated trapdoor $\mathbf{B}_0 \in \mathbb{Z}^{m\times m}$. Also sample 2ℓ random matrices $\mathbf{A}_{i,b} \in \mathbb{Z}_q^{n\times m}$ for $i \in [\ell]$ and $b \in \{0,1\}$, and a random vector $\boldsymbol{u} \in \mathbb{Z}_q^n$. The public parameters and master secret key are:

$$\mathsf{PP} = (q, \mathbf{A}_0, \{\mathbf{A}_{i,b}\}, \boldsymbol{u}) \qquad \mathsf{MK} = \mathbf{B}_0$$

Private Key Extraction. To extract a private key corresponding to a public identity *id*, using the trapdoor $\mathbf{B}_0$, find a low-norm vector $\boldsymbol{d}_{id}$ such that $\mathbf{A}_{id} \cdot \boldsymbol{d}_{id} = \boldsymbol{u} \pmod q$. The private key is: $\mathsf{SK}_{\mathsf{id}} = d_{id}$.

Encryption. Proceed as in the Regev system substituting $\mathbf{A}_{id}$ for $\mathbf{A}$.

Decryption. Proceed as in the Regev system, substituting $\boldsymbol{d}_{id}$ for $\boldsymbol{d}$.

The large matrix $\mathbf{A}_{id}$ renders the system rather inefficient, but enables a security proof against "selective-identity" attacks (where the attacker reveals the target id^* in advance) *in the standard model.* One builds a simulator that can extract private keys for all identities but the pre-announced target id^*. The simulator shall set itself up with a trapdoor for every submatrix $A_{i,(1-b_i^*)}$ where b_i^* is the i-th bit of the target identity — but not $\mathbf{A}_0$ (which shall be assembled from an LWE challenge to show a reduction). This way, the resulting concatenation $\mathbf{A}_{id}$ will have one or more known trapdoors, unless $id = id^*$.

4.2 All-at-Once Standard-Model IBE

Just like the "bit-by-bit" construction of [3,13], above, is a lattice analogue to the pairing-based IBE by Canetti, Halevi, and Katz [12], a similar analogy can be made from the "all-at-once" pairing-based IBE by Boneh and Boyen [9], as a more efficient way to build a provably secure IBE in the standard model. The full analysis is due to Agrawal *et al.* [1] and is quite involved, but the construction is based on a simple principle.

Here, the recipient identity is encoded into the Regev matrix $\mathbf{A}$ all at once, without decomposing it bit by bit. Specifically, the Regev encryption matrix becomes (for constant $\mathbf{A}_0, \mathbf{A}_1, \mathbf{A}_2 \in \mathbb{Z}_q^{n\times m}$),

$$\mathbf{A}_{id} = \big[\mathbf{A}_0 \big| \mathbf{A}_1 + id \cdot \mathbf{A}_2\big]$$

when the identity $id \in \mathbb{Z}_q$, or even, in all generality,

$$\mathbf{A}_{id} = \big[\mathbf{A}_0 \big| \mathbf{A}_1 + \mathbf{M}_{id} \cdot \mathbf{A}_2\big]$$

when the identity $id \in \mathbb{Z}_q^n$, based on a straightforward deterministic encoding into a regular square matrix $\mathbf{M}_{id} \in \mathbb{Z}_q^{n\times n}$, such that any non-trivial difference $\mathbf{M}_{id_1} - \mathbf{M}_{id_2}$ is itself non-singular.

In the real system, the central authority will have a trapdoor for A_0, and thus be able to find short solutions $\boldsymbol{d}_{id}$ for every requested *id* in the equation (for constant $\boldsymbol{u} \in \mathbb{Z}_q^n$):

$$\mathbf{A}_{id} \cdot \boldsymbol{d}_{id} = \boldsymbol{u} \pmod q$$

In the simulation for the security reduction, one sets things up so that the simulator can extract private keys for all identities id except the challenge id^*. The matrix $\mathbf{A}_0$ is imposed from an external LWE challenge, and thus without a trapdoor. We set $A_1 = \mathbf{A}_0 \cdot \mathbf{R} - \mathbf{M}_{id^*} \cdot \mathbf{A}_2$, for some random $\mathbf{R} \in \{-1, 1\}^{m \times m}$. It follows that for all non-challenge identities, the encryption matrix reads:

$$\mathbf{A}_{id} = \left[\mathbf{A}_0 \middle| \mathbf{A}_0 \cdot \mathbf{R} + (\mathbf{M}_{id} - \mathbf{M}_{id^*}) \cdot \mathbf{A}_2\right]$$

For the challenge identity, the factor $(\mathbf{M}_{id} - \mathbf{M}_{id^*})$ in parentheses vanishes, and what is left is:

$$\mathbf{A}_{id^*} = \left[\mathbf{A}_0 \middle| \mathbf{A}_0 \cdot \mathbf{R}\right]$$

Agrawal *et al.* [1] give an algorithm to find short preimages under matrices $\mathbf{A}_{id}$ of this form,*without* a trapdoor for $\mathbf{A}_0$, provided one knows a trapdoor for $(\mathbf{M}_{id} - \mathbf{M}_{id^*}) \cdot \mathbf{A}_2$, which will simply be that of $\mathbf{A}_2$ provided that the factor in parentheses is regular. Their algorithm exploits the appearance of multiples of $\mathbf{A}_0$ on both sides of the concatenation to engineer a cancellation. Note that the role of the matrix $\mathbf{R}$ is to blind the simulation setup, so that it looks indistinguishable from the real system to an attacker. In the case where $id = id^*$, the term in $\mathbf{A}_2$ vanishes, and so with it any beneficial use of its trapdoor.

For completeness, we describe their system as follows:

System setup. Pick a suitable modulus q, and sample a random Ajtai matrix $\mathbf{A}_0 \in \mathbb{Z}_q^{n \times m}$ with associated trapdoor $\mathbf{B}_0 \in \mathbb{Z}^{m \times m}$. Also sample two random matrices $\mathbf{A}_1, \mathbf{A}_2 \in \mathbb{Z}_q^{n \times m}$, and a random vector $\boldsymbol{u} \in \mathbb{Z}_q^n$. The public parameters and master secret key are:

$$\mathsf{PP} = (q, \mathbf{A}_0, \mathbf{A}_1, \mathbf{A}_2, \boldsymbol{u}) \qquad \mathsf{MK} = \mathbf{B}_0$$

Private key extraction. To extract a private key corresponding to a public identity id, define its matrix encoding $\mathbf{M}_{id}$ and its encryption matrix $\mathbf{A}_{id} = [\mathbf{A}_0 | \mathbf{A}_1 + \mathbf{M}_{id} \cdot \mathbf{A}_2]$. Using the trapdoor $\mathbf{B}_0$, sample a low-norm vector $\boldsymbol{d}_{id}$ solution of $\mathbf{A}_{id} \cdot \boldsymbol{d}_{id} = \boldsymbol{u} \pmod q$. The private key is: $\mathsf{SK}_{\mathsf{id}} = d_{id}$.

Encryption. Proceed as in the Regev system substituting $\mathbf{A}_{id}$ for $\mathbf{A}$.

Decryption. Proceed as in the Regev system, substituting $\boldsymbol{d}_{id}$ for $\boldsymbol{d}$.

4.3 Adaptive or "Full" Security

A drawback of the previous systems is their need to relax the security notion, from a *bona fide* adaptive-identity attack to a less realistic selective-identity one, in order to achieve a reduction in the standard model (*sans* random oracle).

In [11], we propose a scheme and accompanying proof technique that address this limitation. The general idea is to set up the simulator to fail not on one but several possible challenge queries, using an efficient key-space partitioning technique that is quite specific to lattices. The full version of [1] describes the fully secure system and its proof.

5 Delegation and Hierarchies

A classic generalization of the notion of IBE is that of hierarchical IBE, where private-key holders can serve as local authorities to issuing private keys to any identity below them in the hierarchical tree of identities.

5.1 Concatenation-Based Delegation

The first inroad into HIBE from lattice is due to Cash *et al.* [13], who in the same paper leverage their bit-by-bit IBE approach into a hierarchical scheme thanks to a trapdoor delegation mechanism of their design.

The principle is as follows. Let an Ajtai matrix $\mathbf{A}_0$ and its associated "good" trapdoor $\mathbf{T}_0$. Let $\mathbf{A}_1$ be an arbitrary matrix that is dimension-compatible with $\mathbf{A}_0$. Cash *et al.* provide an algorithm that transforms $\mathbf{A}_0$'s trapdoor $\mathbf{T}_0$ into a trapdoor $\mathbf{T}$ for the concatenated matrix $\mathbf{A} = [A_0|A_1]$, and in such a way that the new trapdoor $\mathbf{T}$ has only a slightly higher norm than the originating trapdoor $\mathbf{T}_0$. (While the norm might not increase at all under a naïve delegation process, the degradation of quality is a by-product of a necessary re-randomization step to ensure that the delegated basis cannot be used to reconstruct the delegator basis).

The Cash-Hofheinz-Kiltz-Peikert HIBE. Based on this delegation algorithm, Cash *et al.* [13] extend their bit-by-bit IBE scheme into a hierarchical scheme in a straightforward manner: subordinate identities are constructed by extending an identity prefix with additional bits; the corresponding encryption matrices are likewise constructed by concatenating additional sub-matrices to the right; and the corresponding private keys are obtained by invoking the delegation algorithm for such concatenations.

The (first) Agrawal-Boneh-Boyen HIBE. Based on the same CHKP delegation algorithm, Agrawal *et al.* [1] likewise extend their all-at-once IBE scheme into a hierarchical scheme, in the same straightforward manner.

5.2 Multiplicative In-Place Delegation

A second approach to delegation and HIBE, due to Agrawal *et al.* [2], relies not on concatenation, but on multiplication by invertible low-norm matrices. They propose a delegation mechanism that operates "in place", *i.e.*, without increasing the dimensions of the lattices or the number of elements in the matrices defining them.

Given a good basis $\mathbf{T}_\mathbf{A}$ for an Ajtai lattice $\Lambda^\perp(\mathbf{A})$, they show how to create a (slightly less) good basis $\mathbf{T}_\mathbf{B}$ for another lattice $\Lambda^\perp(\mathbf{B})$, whose defining matrix $\mathbf{B}$ has the same dimension as $\mathbf{A}$ and can be deterministically and publicly computed from $\mathbf{A}$. The delegation mechanism furthermore ensures that given $\mathbf{A}, \mathbf{B}$ and $\mathbf{T}_\mathbf{B}$, it is difficult to recover $\mathbf{T}_\mathbf{A}$ or any other a short basis for $\Lambda^\perp(\mathbf{A})$, thus ensuring the "one-wayness" of the delegation process.

Very informally, the delegated matrix $\mathbf{B} \in \mathbb{Z}_q^{n\times m}$ is defined from the delegator matrix $\mathbf{A} \in \mathbb{Z}_q^{n\times m}$ and a low-norm invertible public delegation matrix $\mathbf{R} \in \mathbb{Z}_q^{m\times m}$, as the product:

$$\mathbf{B} = \mathbf{A} \cdot \mathbf{R}^{-1}$$

Since $\mathbf{T_A}$ is a trapdoor for $\mathbf{A}$, *i.e.*, a short basis for $\Lambda^{\perp}(\mathbf{A})$, it follows that $\mathbf{A} \cdot \mathbf{T_A} = \mathbf{0} \pmod q$. Hence, we also have that $(\mathbf{A} \cdot \mathbf{R}^{-1}) \cdot (\mathbf{R} \cdot \mathbf{T_A}) = \mathbf{0} \pmod q$. Hence, $\mathbf{R} \cdot \mathbf{T_A} \in \mathbb{Z}_q^{m\times m}$ defines a basis for $\Lambda^{\perp}(\mathbf{B})$, and a "good" one since $\mathbf{R}$ has low norm. A final re-randomization step will ensure that the delegation cannot be undone, ensuring its "one-wayness".

The (second) Agrawal-Boneh-Boyen HIBE. Equipped with their delegation tool, Agrawal *et al.* [2] construct an efficient HIBE system, with provable security from the LWE assumption [19], and where the dimension of the keys and ciphertexts does not increase with the depth of the hierarchy. In particular, for shallow hierarchies, the efficiency of their system is directly comparable to the random-oracle non-hierarchical system of [15]. For deep hierarchies, the number of private key and ciphertext elements remains the same, but the bit-size of the modulus needs to increase linearly. This results in an HIBE system whose space complexity is only linear in the depth of the hierarchy.

6 Attributes and Predicates

To conclude this tour, we note a couple of brand new results, that concurrently demonstrated that encryption systems even more expressive than (H)IBE could be constructed from lattices — thereby breaking the "IBE barrier".

Lattice-based "fuzzy IBE". One system, due to Agrawal et al. [4], is a *Fuzzy IBE* system. Fuzzy IBE, a notion originally defined and constructed from bilinear maps in [20], was the first instance of what is now referred under the umbrella of *attribute-based encryption.* In Fuzzy IBE, decryption is conditioned upon an approximate rather than exact match between recipient attributes stated in the ciphertext, and those actually present in the actual recipient's private key.

Lattice-based "fuzzy IBE". The other system, due to Agrawal et al. [5], is an instance of *Predicate-based* encryption system, where decryption is controlled by the (non-)vanishing of the inner product of two vectors of attributes: one from the ciphertext, the other from the private key.

7 Conclusion

While a great many technical and conceptual challenges remain unsolved, if there is a lesson to be drawn from the many recent exciting developments in just a few focused areas of investigation, is that lattice-based cryptography is poised to jump from the sidelines to the mainstream, and find its place into all manners of real-world applications in the coming decades. We certainly look forward to this transformation.

References

1. Agrawal, S., Boneh, D., Boyen, X.: Efficient Lattice (H)IBE in the Standard Model. In: Gilbert, H. (ed.) EUROCRYPT 2010. LNCS, vol. 6110, pp. 553–572. Springer, Heidelberg (2010)
2. Agrawal, S., Boneh, D., Boyen, X.: Lattice Basis Delegation in Fixed Dimension and Shorter-Ciphertext Hierarchical IBE. In: Rabin, T. (ed.) CRYPTO 2010. LNCS, vol. 6223, pp. 98–115. Springer, Heidelberg (2010)
3. Agrawal, S., Boyen, X.: Identity-based encryption from lattices in the standard model (July 2009) (manuscript), `http://www.cs.stanford.edu/~xb/ab09/`
4. Agrawal, S., Boyen, X., Vaikuntanathan, V., Voulgaris, P., Wee, H.: Fuzzy identity based encryption from lattices. Cryptology ePrint Archive, Report 2011/414 (2011), `http://eprint.iacr.org/`
5. Agrawal, S., Freeman, D.M., Vaikuntanathan, V.: Functional Encryption for Inner Product Predicates from Learning with Errors. In: Lee, D.H. (ed.) ASIACRYPT 2011. LNCS, vol. 7073, pp. 21–40. Springer, Heidelberg (2011)
6. Ajtai, M.: Generating hard instances of lattice problems (extended abstract). In: Proceedings of STOC 1996, pp. 99–108. ACM, New York (1996)
7. Ajtai, M., Dwork, C.: A public-key cryptosystem with worst-case/average-case equivalence. In: STOC, pp. 284–293 (1997)
8. Alwen, J., Peikert, C.: Generating shorter bases for hard random lattices. In: STACS, pp. 75–86 (2009)
9. Boneh, D., Boyen, X.: Efficient selective identity-based encryption withoutrandom oracles. J. Cryptology 24(4), 659–693 (2011); Abstract in EUROCRYPT 2004
10. Boneh, D., Franklin, M.: Identity-Based Encryption from the Weil Pairing. In: Kilian, J. (ed.) CRYPTO 2001. LNCS, vol. 2139, pp. 213–229. Springer, Heidelberg (2001)
11. Boyen, X.: Lattice Mixing and Vanishing Trapdoors – a Framework for Fully Secure Short Signatures and More. In: Nguyen, P.Q., Pointcheval, D. (eds.) PKC 2010. LNCS, vol. 6056, pp. 499–517. Springer, Heidelberg (2010)
12. Canetti, R., Halevi, S., Katz, J.: A forward-secure public-key encryption scheme. J. Cryptology 20(3), 265–294 (2007); Abstract in EUROCRYPT 2003
13. Cash, D., Hofheinz, D., Kiltz, E., Peikert, C.: Bonsai Trees or, How to Delegate a Lattice Basis. In: Gilbert, H. (ed.) EUROCRYPT 2010. LNCS, vol. 6110, pp. 523–552. Springer, Heidelberg (2010)
14. Gentry, C.: Fully homomorphic encryption using ideal lattices. In: STOC, pp. 169–178 (2009)
15. Gentry, C., Peikert, C., Vaikuntanathan, V.: Trapdoors for hard lattices and new cryptographic constructions. In: STOC, pp. 197–206. ACM (2008)
16. Halevi, S.: Fully Homomorphic Encryption. Slides from Tutorial Session. In: Rogaway, P. (ed.) CRYPTO 2011. LNCS, vol. 6841, Springer, Heidelberg (2011), `http://www.iacr.org/conferences/crypto2011/slides/Halevi.pdf`
17. Micciancio, D., Regev, O.: Worst-case to average-case reductions based on gaussian measures. In: Proceedings of FOCS 2004, pp. 372–381. IEEE Computer Society, Washington, DC, USA (2004)
18. Peikert, C.: Bonsai trees (or, arboriculture in lattice-based cryptography). Cryptology ePrint Archive, Report 2009/359 (2009), `http://eprint.iacr.org/`
19. Regev, O.: On lattices, learning with errors, random linear codes, and cryptography. In: Proceedings of STOC 2005, pp. 84–93. ACM, New York (2005)
20. Sahai, A., Waters, B.: Fuzzy Identity-Based Encryption. In: Cramer, R. (ed.) EUROCRYPT 2005. LNCS, vol. 3494, pp. 457–473. Springer, Heidelberg (2005)

Breaking Fully-Homomorphic-Encryption Challenges

Phong Q. Nguyen

INRIA, France and Tsinghua University, China
http://www.di.ens.fr/~pnguyen/

Abstract. Following Gentry's breakthrough work [7], there is currently great interest on fully-homomorphic encryption (FHE), which allows to compute arbitrary functions on encrypted data. Though the area has seen much progress recently (such as [10,11,5,2,1,8,6]), it is still unknown if fully-homomorphic encryption will ever become truly practical one day, or if it will remain a theoretical curiosity. In order to find out, several FHE numerical challenges have been proposed by Gentry and Halevi [9], and by Coron *et al.* [5], which provide concrete parameters whose efficiency and security can be studied. We report on recent attempts [3,4] at breaking FHE challenges, and we discuss the difficulties of assessing precisely the security level of FHE challenges, based on the state-of-the-art. It turns out that security estimates were either missing or too optimistic.

References

1. Brakerski, Z., Vaikuntanathan, V.: Efficient fully homomorphic encryption from (standard) LWE. Cryptology ePrint Archive, Report 2011/344 (2011), http://eprint.iacr.org/
2. Brakerski, Z., Vaikuntanathan, V.: Fully Homomorphic Encryption from Ring-LWE and Security for Key Dependent Messages. In: Rogaway, P. (ed.) CRYPTO 2011. LNCS, vol. 6841, pp. 505–524. Springer, Heidelberg (2011)
3. Chen, Y., Nguyen, P.Q.: BKZ 2.0: Better Lattice Security Estimates. In: Lee, D.H. (ed.) ASIACRYPT 2011. LNCS, vol. 7073, pp. 1–20. Springer, Heidelberg (2011)
4. Chen, Y., Nguyen, P.Q.: Faster algorithms for approximate common divisors: Breaking fully-homomorphic-encryption challenges over the integers. Cryptology ePrint Archive, Report 2011/436 (2011), http://eprint.iacr.org/
5. Coron, J.-S., Mandal, A., Naccache, D., Tibouchi, M.: Fully Homomorphic Encryption over the Integers with Shorter Public-Keys. In: Rogaway, P. (ed.) CRYPTO 2011. LNCS, vol. 6841, pp. 487–504. Springer, Heidelberg (2011)
6. Coron, J.-S., Naccache, D., Tibouchi, M.: Optimization of fully homomorphic encryption. Cryptology ePrint Archive, Report 2011/440 (2011), http://eprint.iacr.org/
7. Gentry, C.: Fully homomorphic encryption using ideal lattices. In: Proc. STOC 2009, pp. 169–178. ACM (2009)
8. Gentry, C., Halevi, S.: Fully homomorphic encryption without squashing using depth-3 arithmetic circuits. Cryptology ePrint Archive, Report 2011/279 (2011), http://eprint.iacr.org/

D. Lin, G. Tsudik, and X. Wang (Eds.): CANS 2011, LNCS 7092, pp. 13–14, 2011.

9. Gentry, C., Halevi, S.: Implementing Gentry's Fully-Homomorphic Encryption Scheme. In: Paterson, K.G. (ed.) EUROCRYPT 2011. LNCS, vol. 6632, pp. 129–148. Springer, Heidelberg (2011)
10. Smart, N.P., Vercauteren, F.: Fully Homomorphic Encryption with Relatively Small Key and Ciphertext Sizes. In: Nguyen, P.Q., Pointcheval, D. (eds.) PKC 2010. LNCS, vol. 6056, pp. 420–443. Springer, Heidelberg (2010)
11. van Dijk, M., Gentry, C., Halevi, S., Vaikuntanathan, V.: Fully Homomorphic Encryption over the Integers. In: Gilbert, H. (ed.) EUROCRYPT 2010. LNCS, vol. 6110, pp. 24–43. Springer, Heidelberg (2010)

Cube Cryptanalysis of Hitag2 Stream Cipher

Siwei Sun[1], Lei Hu[1], Yonghong Xie[1], and Xiangyong Zeng[2]

[1] State Key Laboratory of Information Security,
Graduate School of Chinese Academy of Sciences, Beijing 100049, China
[2] Faculty of Mathematics and Computer Science,
Hubei University, Wuhan 430062, China
{swsun,hu,yhxie}@is.ac.cn, xzeng@hubu.edu.cn

Abstract. Hitag2 is a lightweight LFSR-based stream cipher with a 48-bit key and a 48-bit internal state. As a more secure version of the Crypto-1 cipher which has been employed in many Mifare Classic RFID products, Hitag2 is used by many car manufacturers for unlocking car doors remotely. Until now, except the brute force attack, only one cryptanalysis on this cipher was released by Courtois, O'Neil and Quisquater, which broke Hitag2 by an SAT solver within several hours. However, little theoretical analysis and explanation were given in their work. In this paper, we show that there exist many low dimensional cubes of the initialization vectors such that the sums of the outputs of Hitag2 for the corresponding initialization vectors are linear expressions in secret key bits, and hence propose an efficient black- and white-box hybrid cube attack on Hitag2. Our attack experiments show that the cipher can be broken within one minute on a PC. The attack is composed of three phases: a black-box attack of extracting 32 bits of the secret key, a white-box attack to get several other key bits, and a brute force search for the remaining key bits.

Keywords: Hitag2 stream cipher, cube attack, black-box and white-box attack.

1 Introduction

Hitag2 is a lightweight stream cipher with a 48-bit secret key used in many RFID products [17,21,22]. It was designed by Phillips Semiconductors, the same company who designed the Crypto1 stream cipher which is widely employed in Mifare Classic cards.

Hitag2 was first analyzed by Courtois, O'Neil and Quisquater in [10] with algebraic attack. They broke the cipher by writing down algebraic representations, converting them to a logical satisfiability (SAT) problem and then solving the problem by an SAT solver. It turns out that the entire key can be recovered within several hours. However, these researchers did not find any concrete algebraic weakness of the cipher and they were surprised by this result, as quoted as follows:

D. Lin, G. Tsudik, and X. Wang (Eds.): CANS 2011, LNCS 7092, pp. 15–25, 2011.

> "... the boolean function used is quite large and has good 'Algebraic Immunity' of at least 4 ..."; and
>
> "This surprised us, as nothing in the description of Hitag2 allows to believe that it will be weak, ... So there is a real mystery here that remains unsolved: why is this cipher comparatively quite weak? We don't answer this question, just demonstrate the weakness experimentally and compare to the tweaked version."

In this paper, we present a black- and white-box hybrid cube attack on Hitag2. A cube attack is a chosen initialization vector (IV) attack. It tries to find one or high order differentials (or superpolys in terminology of cube attack) of the outputs of the target cipher which are linear in the secret key bits. If an output differential is linear in the key bits of the cipher, then a linear relation in the key bits is found. Sufficiently many linear relations can be used to reduce the effective key size of the target cipher. Previous cube attacks [11,1,3] treat the target cipher as a black box and do not care about the explicit polynomial description of the cipher. In our attack, we recover 32 bits of the key by a black-box cube attack, then we execute a white-box cube attack with explicit polynomial expressions to get some other key bits. This white-box attack results in more exploitable superpolys for the cube attack. Finally, the remanning key bits can be exhaustively searched. Our attack experiments show that the 48-bit secret key of Hitag2 can be recovered within one minute on a PC.

In Sections 2 and 3, we briefly describe the Hitag2 stream cipher and the cube attack. In Section 4, we present our attack on Hitag2. The last section is the conclusion.

2 Hitag2 Stream Cipher

The Hitag2 stream cipher is a LFSR-based stream cipher. The output key stream of Hitag2 is dependent on 3 parameters: a 48-bit secret key Key, a 32-bit initialization vector IV, and a 32-bit serial number $Serial$. The Hitag2 stream cipher starts to output the key stream after a 32-round initialization process.

At the beginning of the initialization process, the 48-bit internal state S is filled with 32 bits of $Serial$ and 16 bits of Key, i.e., the initial internal state is given as follows:

$$S[i] = Serial[i],\ i \in \{0, \ldots, 31\}$$
$$S[32+i] = Key[i],\ i \in \{0, \ldots, 15\}$$

Then the internal state is left shifted and the 47th bit of the state $S[47]$ is updated according to the following rules:

$$S[i-1] = S[i],\ i \in \{1, \ldots, 47\}$$
$$S[47] = NF(S) \oplus IV[j] \oplus Key[15+j]$$

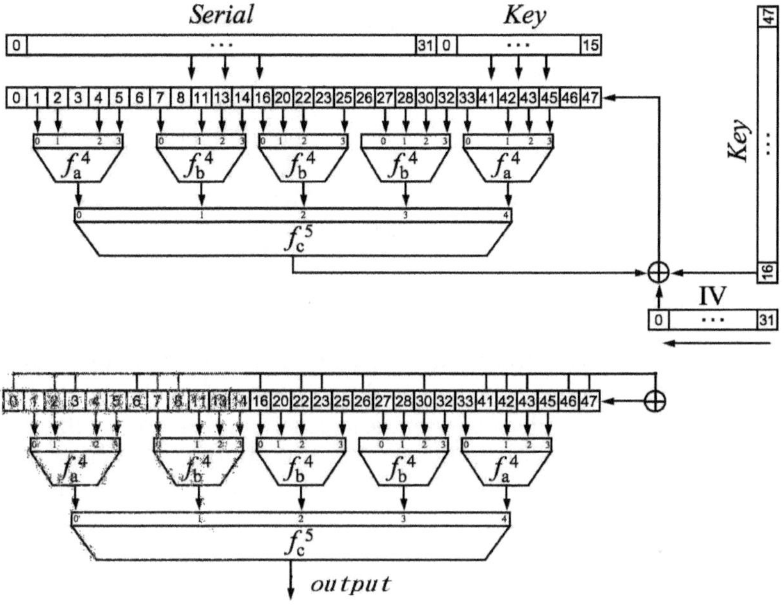

Fig. 1. Hitag2 stream cipher: initialization phase (upper) and working phase (lower)

The above two instructions are repeated 32 times as j goes from 0 to 31 and NF is a nonlinear filter function which is built up by two layers of boolean functions (see Fig.1) where

$$\begin{aligned}
f_a^4(x_3,x_2,x_1,x_0) &= x_0x_1x_2 + x_0x_2 + x_0x_3 + x_1x_2 + x_0 + x_1 + x_3 + 1,\\
f_b^4(x_3,x_2,x_1,x_0) &= x_0x_1x_3 + x_0x_2x_3 + x_1x_2x_3 + x_0x_1 + x_0x_2 + x_1x_2\\
&\quad + x_0 + x_1 + x_3 + 1,\\
f_c^5(x_4,x_3,x_2,x_1,x_0) &= x_0x_1x_2x_4 + x_0x_1x_3x_4 + x_0x_1 + x_0x_2x_3 + x_0x_3x_4\\
&\quad + x_1x_2x_3 + x_1x_2x_4 + x_1x_2 + x_1x_3 + x_1x_4 + x_1\\
&\quad + x_2x_3x_4 + x_2x_4 + x_3x_4 + x_3 + 1.
\end{aligned}$$

Equivalently, in truth table forms, these three functions are

$$\begin{aligned}
f_a^4 = (0010\,1100\,0111\,1001),\ f_b^4 = (0110\,0110\,0111\,0001),\\
f_c^5 = (0111\,1001\,0000\,0111\,0010\,1000\,0111\,1011),
\end{aligned}$$

where the truth table of an n-variable Boolean function f is a binary string $(f(1,\cdots,1), f(1,\cdots,1,0),\cdots,f(0,\cdots,0,1), f(0,\cdots,0))$ of length 2^n.

After the 32-round initialization phase, the cipher enters into the working phase. At each clock, the internal state is updated as a pure LFSR and the cipher outputs a keystream bit using the nonlinear filtering function NF:

$$\begin{aligned} Z &= S[0] \oplus S[2] \oplus S[3] \oplus S[6] \oplus S[7] \oplus S[8] \oplus S[16] \oplus S[22] \\ &\quad \oplus S[23] \oplus S[26] \oplus S[30] \oplus S[41] \oplus S[42] \oplus S[43] \oplus S[46] \oplus S[47], \\ S[i-1] &= S[i],\ i \in \{1, \ldots, 47\}, \\ S[47] &= Z, \\ output &= NF(S) \end{aligned}$$

These instructions are executed repeatedly sufficiently many times until sufficiently many keystream bits are generated.

3 Cube Attack

Cube attack was formally introduced by Ita Dinur and Adi Shamir in Eurocrypt 2009 [11]. According to the comments and arguments of some researchers, cube attack has been studied under other names such as higher order differential attack [14] and algebraic IV differential attack [19,20]. In the following, we describe the method of cube attack for stream ciphers with two inputs, one input is the secret key, Key, and the other is the public initialization vector, IV, which can be manipulated by an attacker.

Let $Key = (k_0, \cdots, k_{n-1})$ and $IV = (v_0, \cdots, v_{m-1})$ be the n-bit key and m-bit initialization vector respectively. Let $y^{(j)} = f^{(j)}(Key, IV)$, $j \geq 0$, be the j-th output bit of the cipher. It is well known that $y^{(j)}$ can be treated as a binary multivariate polynomial, i.e.,

$$y^{(j)} = p^{(j)}(k_0, \ldots, k_{n-1}, v_0, \ldots, v_{m-1}) \in \mathbb{F}_2[k_0, \ldots, k_{n-1}, v_0, \ldots, v_{m-1}].$$

In a chosen IV attack, an attacker is able to choose his own IVs and feed them into the cipher and observe the output bits. The goal of a cube attack is to find many sets V of IVs such that the sum

$$L(k_0, \ldots, k_{n-1}) = \sum_{(v_0, \ldots, v_{m-1}) \in V} p^{(j)}(k_0, \ldots, k_{n-1}, v_0, \ldots, v_{m-1})$$

is linear with respect to the key bits k_i.

Definition 1. *Assume $p(k_0, \ldots, k_{n-1}, v_0, \ldots, v_{m-1})$ is a polynomial and $I \subseteq \{0, \cdots, m-1\}$ is a set of indices to the bits of IV. Let $t_I = \prod_{i \in I} v_i$. Then factoring p by t_I yields*

$$p(k_0, \ldots, k_{n-1}, v_0, \ldots, v_{m-1}) = t_I p_{s(I)} + q(k_0, \ldots, k_{n-1}, v_0, \ldots, v_{m-1}),$$

where q is a linear combination of terms which are not divided by t_I. We call $p_{s(I)}$ the superpoly of I in p.

Definition 2. *A maxterm is a term t_I with a corresponding superpoly $p_{s(I)}$ such that $deg(p_{s(I)}) = 1$, where $p_{s(I)}$ is viewed as a polynomial in the ring $\mathbb{F}_2[k_0, \ldots, k_{n-1}]$, i.e., $p_{s(I)}$ is a nonconstant linear polynomial in the key variables.*

Definition 3. *A size d subset I of $\{0, \ldots, m-1\}$ defines a d-dimensional cube C_I in $\mathbb{F}_2^m$, where the components of the vectors in C_I indexed by I are arbitrarily assigned to 0 or 1 and the other components are set to 0. This cube is a set of 2^d vectors. For a vector $v \in C_I$, we define $p|_v$ to be a derivation of p in which the IV variables indexed by I are set to the values in v. We also define p_I to be the sum $\sum_{v \in C_I} p|_v$ as a polynomial in $\mathbb{F}_2[k_0, \cdots, k_{n-1}]$. If p_I is a nonconstant linear polynomial in the key variables, we call the corresponding I a good cube and its elements cube indices.*

Theorem 1. *([11]) For any polynomail $p(k_0, \ldots, k_{n-1}, v_0, \ldots, v_{m-1})$ and $I \subseteq \{0, \ldots, m-1\}$, $p_I = p_{s(I)}$ holds as polynomials in $\mathbb{F}_2[k_0, \ldots, k_{n-1}, v_0, \ldots, v_{m-1}]$.*

Theorem 1 shows that for every given I, the corresponding superpoly can be computed by the sum $\sum_{v \in C_I} p|_v$. Furthermore, this theorem indicates that the value of the superpoly $p_{s(I)}$ can be computed by an attacker without the explicit polynomial expression of p provided the attacker can control the IV and observe the output of the stream cipher corresponding to p. This is a cube attack with only black-box access to the target cipher, and the paper [11] entitled it as a cube attack on tweakable black box polynomials.

A cube attack can be split into two phases: offline precomputation and online key recovery.

In the precomputation phase, the objective of an attacker is to find r index sets $I_0, I_1, \cdots, I_{r-1}$, which lead to r linear independent superpolys $p_{s(I_0)}$, $p_{s(I_1)}$, $\cdots$, $p_{s(I_{r-1})}$. To accomplish this, the attacker randomly chooses a set I and checks whether the following equality

$$\begin{aligned} &p_I(\bar{k}_0 + \bar{k}_0', \cdots, \bar{k}_{n-1} + \bar{k}_{n-1}') + p_I(\bar{k}_0, \cdots, \bar{k}_{n-1}) \\ &+ p_I(\bar{k}_0', \cdots, \bar{k}_{n-1}') + p_I(0, \cdots, 0) = 0 \end{aligned} \tag{1}$$

holds for several randomly chosen key pairs $(\bar{k}_0, \cdots, \bar{k}_{n-1})$ and $(\bar{k}_0', \cdots, \bar{k}_{n-1}')$. If the above test passes continuously for many times for the particular I, the corresponding superpoly $p_{s(I)}$ is probably a polynomial of degree one in the key bits. Now, the attacker is ready to derive the algebraic structure of the superpoly $p_{s(I)}$ by the following strategy. For every $i \in \{0, \ldots, n-1\}$, compute

$$\lambda_i = p_I(e_i) + p_I(0, \cdots, 0),$$

where e_i is the j-th unit vector, and the expression of the superpoly $p_{s(I)}$ is

$$p_{s(I)} = \lambda_0 k_0 + \lambda_1 k_1 + \cdots + \lambda_{n-1} k_{n-1} + p_I(0, \cdots, 0).$$

Assume the attacker has found r cube sets $I_0, \cdots, I_{r-1}$, now he can perform the second phase of online key recovery by solving the following system of linear equations:

$$\begin{cases} L_0(k_0, \ldots, k_{n-1}) = p_{I_0}(k_0, \cdots, k_{n-1}) \\ L_1(k_0, \ldots, k_{n-1}) = p_{I_1}(k_0, \cdots, k_{n-1}) \\ \cdots \\ L_{r-1}(k_0, \ldots, k_{n-1}) = p_{I_{r-1}}(k_0, \cdots, k_{n-1}). \end{cases}$$

By solving this equation system, the attacker reduces the effective secret key size from n to $n-r$ bits and the remaining key bits may be exhaustively searched.

4 Cube Attack on Hitag2

Previous cube attacks [1,3,11] treated a target cipher as a black-box with secret input Key and public input IV, which we call a black-box cube attack. But for a lightweight cipher such as Hitag2, it may be possible to inspect the explicit polynomial expressions of the output bits, therefore, a white-box cube attack with explicit polynomial expressions is possible. The advantage of this type of attack using explicit polynomial expressions is twofold. Firstly and obviously, we get deeper understanding of the target cipher; secondly, the probabilistic linearity test (1) can be omitted and more exploitable superpolys can be discovered. For a toy example, assume a cipher is explicitly described as a polynomial:

$$\begin{aligned} & p(k_0,k_1,k_2,v_0,v_1,v_2) \\ =\ & v_0v_2k_1k_2+v_0v_2k_1+v_0v_2k_2+v_0v_2+v_0k_0k_1k_2+v_2k_0k_2+v_1k_1k_2+k_0+1 \\ =\ & v_0v_2(k_1k_2+k_1+k_2+1)+v_0k_0k_1k_2+v_2k_0k_2+v_1k_1k_2+k_0+1. \end{aligned}$$

According to the theory of cube attack, we have

$$\begin{aligned} & p(k_0,k_1,k_2,0,v_1,0)+p(k_0,k_1,k_2,0,v_1,1) \\ & +p(k_0,k_1,k_2,1,v_1,0)+p(k_0,k_1,k_2,1,v_1,1)=k_1k_2+k_1+k_2+1, \end{aligned}$$

obviously, if we view p as a black box and choose $I=\{0,2\}$, the corresponding superpoly $p_{s(I)}$ can not pass the linearity test (1) in the precomputation phase. However, this superpoly can be further exploited since

$$p_{s(I)}=k_1k_2+k_1+k_2+1=(k_1+1)(k_2+1),$$

and if $p_{s(I)}=1$ holds in the online key recovery phase, then it must be the case: $k_1=0$ and $k_2=0$. Actually we do have encountered similar cases in our attack experiments on Hitag2. On the other hand, even if the corresponding $p_{s(I)}$ does not possess such special structure, its low degree is also a good property for some other cryptanalysis method such as algebraic attack, and one may find some bits of the key by using equation solvers such as Gröbner base, XL and F4. Yet, if we treat the target cipher as a black-box, this kind of exploitable superpolys can not be discovered easily.

Therefore, it is a natural idea to perform a cube attack by treating the target cipher as a white-box. To automate the polynomial computation involved in this kind of attack, we reimplement the C code of Hitag2 [21] in Magma over the ring

$$\Omega=\mathbb{F}_2[k_0,\ldots,k_{47}]/<k_0^2-k_0,\cdots,k_{47}^2-k_{47}>,$$

and we treat every bit as an element of the ring Ω. The known bits of IV and $Serial$ are represented by 0 or 1, and the unknown secret bits of key are represented by k_i. For convenience, we call this new implementation Ω-code.

For example, if we set $k = (k_0, k_1, k_2, 0, \cdots, 0) \in \Omega^{48}$, $IV = (0, \cdots, 0) \in \Omega^{32}$, and $Serial = (0, \cdots, 0) \in \Omega^{32}$, then it can be verified that the 0th output bit of Hitag2 is

$$k_0 k_1 k_2 + k_0 k_2 + k_1 k_2$$

However, if many bits of the secret key are unknown, our code fails to produce output due to huge memory consumption, and we can not execute a white-box cube attack in this situation.

Therefore, our attack will be divided into three phases. In the first phase, we recover 32 bits of the key, denoted by $\{k_{i_0}, \cdots, k_{i_{31}}\}$, by a black-box cube attack; then a part of the remaining unknown key bits are extracted by a white-box cube attack in the second phase; and finally in the third phase all remaining key bits are recovered by an exhaustive search.

4.1 First Phase: Black-Box Attack

The smaller the dimension of the cube, the more practical the attack is. Therefore, we search good cubes of low dimension d by inspecting different keystream bits $p^{(j)}$ (j is small) in the precomputation phase of the black-box cube attack. If we fail to find useful good cubes of dimension d, we repeat the same process with cube dimension set to $d + 1$. That is, lower dimension is a priority in our attack.

In our experiment, we search good cubes of dimensions from 1 to 6, and repeat the linearity test (1) 50 times for each cube. Every time a new good cube is found, we check whether its corresponding superpoly is linearly independent with the superpolys that have been derived from previous good cubes. If this condition is not satisfied, the corresponding cube is discarded. Good cubes we found and their corresponding maxterms and the targeted output bits are listed in Table 1 (rows are listed according to the lexicographic order of dimension of cube, keystream clock, and index set).

Using this result, we can recover 32 bits of the secret key: k_0, k_1, k_2, k_3, k_4, k_5, k_6, k_7, k_8, k_9, k_{10}, k_{11}, k_{12}, k_{13}, k_{14}, k_{16}, k_{17}, k_{18}, k_{20}, k_{21}, k_{22}, k_{23}, k_{24}, k_{25}, k_{26}, k_{27}, k_{28}, k_{29}, k_{30}, k_{32}, k_{33}, k_{35} by totally 464 times of evaluations of the Hitag2 stream cipher. Along with these recovered key bits, a linear relation in the remaining key bits is also extracted: $k_{34} + k_{36} = v$, where v is a known value which can be determined in the online phase of the cube attack.

4.2 Second Phase: White-Box Attack

Now the remaining unknown key bits can be recovered by exhaustive search directly, however here we propose an alternative method which may be also applicable under other situations.

We initialize the unknown bits of the key with ring elements: k_{15}, k_{19}, k_{31}, k_{34}, k_{36}, k_{37}, k_{38}, k_{39}, k_{40}, k_{41}, k_{42}, k_{43}, k_{44}, k_{45}, k_{46}, $k_{47} \in \Omega$, and the key bits recovered are set to 0 or $1 \in \Omega$ according to the result of the first phase

Table 1. Maxterms and superpolys for Hitag2

superpoly	index set	keystream clock
k_6	4,15	0
k_3	7,10	0
k_0	5,7	1
k_5	3,13	2
k_8	3,14	2
k_{17}	4,6	2
k_{21}	1,15,16	0
$k_2 + k_8$	4,15,19	0
k_{12}	10,12,16	0
$k_0 + k_{13}$	5,6,23	1
k_{27}	4,10,24	2
k_{25}	1,13,8	3
k_{14}	2,7,31	3
$k_1 + k_{10}$	3,31,13	4
k_7	6,8,31	4
k_{22}	3,6,8	6
$k_9 + k_{12} + k_{14} + k_{18}$	3,7,11	6
k_9	2,7,9	6
k_{16}	7,11,14	6
$k_{28} + k_{33}$	1,8,16	7
$k_{23} + k_{30} + k_{33}$	5,10,13	7
$k_6 + k_{11} + k_{26}$	1,15,20,24	0
$k_{21} + k_{23}$	4,5,15,21,	0
$k_2 + k_6 + k_{21} + k_{26}$	4,5,15,24	0
k_{29}	26,21,11,7	0
k_4	5,9,17,25	1
$k_{12} + k_{14} + k_{34} + k_{36}$	3,7,18,31	3
$k_9 + k_{13} + k_{28}$	6,15,26,20,3	0
k_1	8,5,22,28,16	0
k_{35}	5,7,10,20,19	1
k_{24}	7,11,13,15,20	1
$k_{27} + k_{32}$	7,4,10,28,30	2
$k_5 + k_9 + k_{16} + k_{18} + k_{20} + k_{23} + k_{26} + k_{28}$	4,16,3,20,17,26	0

black-box attack. Then we try to find some superpolys with simple expressions by symbolically summing the outputs of the cipher over some other very low dimensional cubes using the Ω-code. The expressions of the superpolys are dependent on the values of *Serial* and *key*, we have done many experiments, all of which show that it is very easy to use the white-box method to find by hand superpolys with very simple expressions for very low dimensional cubes.

The following is a concrete example of our experiments, where we set $Serial = (1,0,0,1,0,1,1,0,1,1,1,0,1,0,1,0,1,1,0,0,0,0,1,0,1,0,0,1,0,0,1,0)$ and $key = (0,1,0,1,\ldots,0,1)$. Then we employ our Ω-code to compute the superpoly $p^{(j)}_{s(\{31\})}$ and we discovered that

$$p_{s(\{31\})}^{(0)} = k_{15}k_{31} + 1 = 0,$$
$$p_{s(\{31\})}^{(2)} = k_{36} = 0.$$

Combined with the linear expression derived in the black-box attack:

$$p_{s(\{3,7,18,31\})}^{(3)} = k_{12} + k_{14} + k_{34} + k_{36} = 0,$$

we can get that $k_{15} = 1$, $k_{31} = 1$, $k_{36} = 0$, and $k_{34} = 0$. In a similar way, we can find that $k_{19} = 1$ by the following expression

$$p_{s(\{28\})}^{(2)} = k_{19} + 1 = 0.$$

Also, by

$$p_{s(\{31\})}^{(3)} = k_{37}k_{46} + k_{37} + 1 = 0,$$
$$p_{s(\{30\})}^{(3)} = k_{37}k_{46} + 1 = 1,$$

we can find that $k_{37} = 1$ and $k_{46} = 0$.

Note that in the phase of white-box attack, a linearity or quadraticity test is not required since we can inspect the symbolic expressions of the superpolys directly.

4.3 Third Phase: Exhaustive Search Attack

The remaining key bits k_{38}, k_{39}, k_{40}, k_{41}, k_{42}, k_{43}, k_{44}, k_{45}, and k_{47} can be exhaustively searched with little effort.

4.4 Experimental Results

We have randomly chosen 100 different secret keys, and tried to recover them on a PC with Intel(R) Core(TM) Quad CPU (2.83GHz, 3.25GB RAM, Windows XP) by the aforementioned attack. Experiments showed that the secret key can be recovered in all cases in no more than one minute. We give a comparison between our attack and previous attacks in [10] in Table 2, where N_{KI}, N_{CI} and N_K denote the numbers of the known IVs, chosen IVs and keystream bits needed in the attack, respectively.

Table 2. Comparison with the attack in [10]

source	technique	N_{KI}	N_{CI}	N_K	attack time
[10]	SAT solver	1	0	50	<11h
[10]	SAT solver	4	0	32	<48h
[10]	SAT solver	0	16	32	<6h
this paper	cube attack	0	< 500	8	<1min

5 Conclusion

In this paper we broke the Hitag2 stream cipher by black- and white-box hybrid cube attack. Our attack combined black-box and white-box methods to find more linear relations in the key bits, and our experiments showed that it is an advantage to know the explicit polynomial description of a target cipher for a cube attack.

There seems to be some obvious weakness and flaws in the design of Hitag2. Firstly, the 48-bit key is too short even for a lightweight cipher. Secondly, the structure of the linear feedback and the nonlinear filtering function is also too simple for cryptanalysis, which is theoretically shown by cubes of low dimensions of initialization vectors, even though the filtering function has good nonlinearity and algebraic immunity against correlation attack and standard algebraic attack. Finally, the small number (namely 32) of initialization rounds, in our opinion, can not sufficiently confuse the bits of the key. In fact, we can recover 32 bits of the secret key in the phase of black-box attack, the same number of the initialization rounds and the same number of the key bits which do not feed into the LFSR state before the initialization process. Due to the restriction of low cost, designing secure lightweight ciphers still seems to be a challenging work.

Acknowledgement. The authors would like to thank the anonymous reviewers for their valuable comments and suggestions. The work of the first three authors was supported by the Natural Science Foundation of China (NSFC) under grants 61070172 and 10990011, and the National Basic Research Program of China (2007CB311201). The work of the fourth author was supported by the NSFC under Grant 60973130.

References

1. Aumasson, J.-P., Dinur, I., Meier, W., Shamir, A.: Cube Testers and Key Recovery Attacks on Reduced-Round MD6 and Trivium. In: Dunkelman, O. (ed.) FSE 2009. LNCS, vol. 5665, pp. 1–22. Springer, Heidelberg (2009)
2. Bogdanov, A.: Attacks on the KeeLoq Block Cipher and Authentication System. In: RFIDSec 2007 (2007)
3. Bedi, S., Pillai, R.: Cube Attacks on Trivium. IACR Cryptology ePrint Archive, 15 (2009)
4. Biham, E., Dunkelman, O., Indesteege, S., Keller, N., Preneel, B.: How to Steal Cars – A Practical Attack on KeeLoq. In: Smart, N.P. (ed.) EUROCRYPT 2008. LNCS, vol. 4965, pp. 1–18. Springer, Heidelberg (2008)
5. Courtois, N.: The Dark Side of Security by Obscurity and Cloning MiFare Classic Rail and Building Passes Anywhere, Anytime. In: SECRYPT 2009: International Conference on Security and Cryptography, Milan, Italy, July 7-10 (2009)
6. Courtois, N.T., Bard, G.V., Wagner, D.: Algebraic and Slide Attacks on KeeLoq. In: Nyberg, K. (ed.) FSE 2008. LNCS, vol. 5086, pp. 97–115. Springer, Heidelberg (2008), `http://eprint.iacr.org/2007/062`

7. Courtois, N., Meier, W.: Algebraic Attacks on Stream Ciphers with Linear Feedback. In: Biham, E. (ed.) EUROCRYPT 2003. LNCS, vol. 2656, pp. 345–359. Springer, Heidelberg (2003)
8. Courtois, N., Nohl, K., O'Neil, S.: Algebraic Attacks on MiFare RFID Chips, http://www.nicolascourtois.com/papers/mifare_rump_ec08.pdf
9. Courtois, N., Nohl, K., O'Neil, S.: Algebraic Attacks on the Crypto-1 Stream Cipher in MiFare Classic and Oyster Cards. Short paper, http://eprint.iacr.org/2008/166
10. Courtois, N.T., O'Neil, S., Quisquater, J.-J.: Practical Algebraic Attacks on the Hitag2 Stream Cipher. In: Samarati, P., Yung, M., Martinelli, F., Ardagna, C.A. (eds.) ISC 2009. LNCS, vol. 5735, pp. 167–176. Springer, Heidelberg (2009)
11. Dinur, I., Shamir, A.: Cube Attacks on Tweakable Black Box Polynomials. In: Joux, A. (ed.) EUROCRYPT 2009. LNCS, vol. 5479, pp. 278–299. Springer, Heidelberg (2009)
12. de Koning Gans, G., Hoepman, J.-H., Garcia, F.D.: A Practical Attack on the MIFARE Classic. In: Grimaud, G., Standaert, F.-X. (eds.) CARDIS 2008. LNCS, vol. 5189, pp. 267–282. Springer, Heidelberg (2008)
13. Garcia, F.D., de Koning Gans, G., Muijrers, R., van Rossum, P., Verdult, R., Schreur, R.W., Jacobs, B.: Dismantling MIFARE Classic. In: Jajodia, S., Lopez, J. (eds.) ESORICS 2008. LNCS, vol. 5283, pp. 97–114. Springer, Heidelberg (2008)
14. Lai, X.: Higher Order Derivatives and Differential Cryptanalysis. Communications and Cryptography: Two Sides of One Tapestry, 227 (1994)
15. Nohl, K.: Cryptanalysis of Crypto-1. Short paper, http://www.cs.virginia.edu/kn5f/Mifare.Cryptanalysis.htm
16. Nohl, K., Evans, D., Starbug, S., Plötz, H.: Reverse-engineering a cryptographic RFID tag. In: USENIX Security 2008 (2008)
17. Philips Semiconductors Corporation: Philips Semiconductors Data Sheet, HT2 Transponder Family, Communication Protocol, Reader, HITAG2(R) Transponder, Product Specification, Version 2.1, http://www.phreaker.ru/showthread.php?p=226
18. Saarinen, M.: Chosen-IV statistical attacks on eStream ciphers. In: SECRYPT 2006, pp. 260–266. INSTICC Press (2006)
19. Vielhaber, M.: Breaking ONE.TRIVIUM by AIDA and Algebraic IV Differential Attack. IACR Cryptology ePrint Archive, 413 (2007)
20. Vielhaber, M.: AIDA Breaks (BIVIUM A and B) in 1 Minute Dual Core CPU Time. IACR Cryptology ePrint Archive, 402 (2009)
21. Wiener, I.: Hitag2 specification, reference implementation and test vectors, http://cryptolib.com/ciphers/hitag2
22. Transponder Table, a list of cars and transponders used in these cars, http://www.keeloq.boom.ru/table.pdf

New Impossible Differential Cryptanalysis of Reduced-Round Camellia*

Leibo Li[1,2], Jiazhe Chen[1,2], and Keting Jia[3]

[1] Key Laboratory of Cryptologic Technology and Information Security, Ministry of Education, Shandong University, Jinan 250100, China
[2] School of Mathematics, Shandong University, Jinan 250100, China
{lileibo,jiazhechen}@mail.sdu.edu.cn
[3] Institute for Advanced Study, Tsinghua University, Beijing 100084, China
ktjia@mail.tsinghua.edu.cn

Abstract. Camellia is one of the widely used block ciphers, which has been selected as an international standard by ISO/IEC. This paper introduces a 7-round impossible differential of Camellia including FL/FL^{-1} layer. Utilizing impossible differential attack, 10-round Camellia-128 is breakable with $2^{118.5}$ chosen plaintexts and $2^{123.5}$ 10 round encryptions. Moreover, 10-round Camellia-192 and 11-round Camellia-256 can also be analyzed, the time complexity are about $2^{130.4}$ and $2^{194.5}$, respectively. Comparing with known attacks on reduced round Camellia including FL/FL^{-1} layer, our results are better than all of them.

Keywords: Camellia, Block Cipher, Impossible Differential, Cryptanalysis.

1 Introduction

Camellia is a 128-bit block cipher with variable key length of 128, 192, 256, which are denoted as Camellia-128, Camellia-192 and Camellia-256, respectively. It was proposed by NTT and Mitsubishi in 2000 [1], and then was selected as an e-government recommended cipher by CRYPTREC in 2002 [5] and NESSIE block cipher portfolio in 2003 [15]. In 2005, it was selected as an international standard by ISO/IEC [8].

Because of the high level of security, Camellia has drawn a great amount of attention from worldwide cryptology researchers. In the past years, a great number of efficient results of reduced round Camellia were published. Such as linear and differential attacks [16], truncated differential attack [9,11,17], higher

* Supported by the National Natural Science Foundation of China (Grant No. 60931160442), and the Tsinghua University Initiative Scientific Research Program (2009THZ01002).

D. Lin, G. Tsudik, and X. Wang (Eds.): CANS 2011, LNCS 7092, pp. 26–39, 2011.

order differential attack [7], collision attack [12,18], square attack [12], square-like attack [6] and impossible differential attack [13,14,17,19].

An important property of the Camellia's structure is that FL/FL^{-1} layers insert every 6 rounds. This design could provide non-regularity across rounds [1] and destroy the differential property. To our knowledge, several attacks were proposed to analyze the reduced round Camellia including FL/FL^{-1} layers. Square attack [12] was used to analyze 9-round Camellia-128 and 10-round Camellia-256, the complexities are about 2^{122} and 2^{210} encryptions, respectively. Higher order differential attack [7] was applied to analyze 11-round Camellia-256 with complexity of about $2^{255.6}$. Recently, Chen *et al.* use the impossible differential attack to analyze 10-round Camellia-192 and 11-round Camellia-256 [4], the time complexities are about $2^{175.3}$ and $2^{206.8}$, respectively (including whitening).

Impossible differential cryptanalysis was first introduced independently by Biham [3] and Knudsen [10]. The idea of this method is using differential that holds with probability zero to discard the wrong keys and keep the right key.

In this paper, we present a 7-round impossible differential of Camellia with FL/FL^{-1} layer. Based on the 7-round impossible differential, we construct an attack on 10-round Camellia-128, the data complexity is about $2^{118.5}$ chosen plaintexts, and the time complexity is about $2^{123.5}$ 10-round encryptions. We are not aware of any other published attacks that can break 10-round Camellia-128 with FL/FL^{-1} layer faster than exhaustively search. Furthermore, we present the new results of attacks on 10-round Camellia-192 and 11-round Camellia-256, the time complexities are about $2^{130.4}$ and $2^{194.5}$, respectively. Table 1 summarizes our results along with previous known results of reduced-round Camellia including FL/FL^{-1} layer, where * represents the attack don't including the whitening layers.

Table 1. Summarizes of Attack on Camellia with FL/FL^{-1} Layer

Cipher	Rounds	Attack Type	Date	Time	Source
Camellia-128	9*	Square	2^{48}	2^{122}	[12]
	10*	Impossible Diff	$2^{118.5}$	$2^{123.5}$	Section 4
Camellia-192	10*	Impossible Diff	2^{121}	2^{144}	[4]
	10	Impossible Diff	2^{121}	2^{175}	[4]
	10	Impossible Diff	$2^{118.7}$	$2^{130.4}$	Section 5.1
Camellia-256	10*	Square	2^{48}	2^{210}	[12]
	11	Impossible Diff	2^{121}	$2^{206.8}$	[4]
	last 11 round	Higher Order Diff	2^{93}	$2^{255.6}$	[7]
	11	Impossible Diff	$2^{118.7}$	$2^{194.5}$	Section 5.2

The rest of this paper is organized as follows. Section 2 provides a brief description of Camellia. A 7-round impossible differential of Camellia is introduced in the section 3. Our proposed impossible differential attack on 10-round

Camellia-128 is presented in the section 4. Section 5 describes the impossible differential attack on 10-round Camellia-192 and 11-round Camellia-256. Finally, we conclude the paper in section 6.

2 Preliminaries

2.1 Notations

The following notations will be used in this paper:

L_{r-1}, L'_{r-1}	: the left 64-bit half of the r-th round input,
R_{r-1}, R'_{r-1}	: the right 64-bit half of the r-th round input,
ΔL_{r-1}	: the difference of L_{r-1} and L'_{r-1},
ΔR_{r-1}	: the difference of R_{r-1} and R'_{r-1},
S_r, S'_r	: the output value of the S-box layer in the r-th round,
ΔS_r	: the difference of S_r and S'_r,
P_r, P'_r	: the output value of P function layer in the r-th round
K_r	: the subkey used in the r-th round,
$A^{(l)}$	: the l-th byte of a 64-bit value ($l = 1, \ldots, 8$),
$x\|y$	: bit string concatenation of x and y,
$\oplus$, $\cap$, $\cup$	: bitwise exclusive OR(XOR), AND, OR.

2.2 A Brief Description of Camellia

Camellia [1] is a Feistel structure block cipher that uses key sizes of 128 for 18-round, 192 and 256 for 24-round. Figure 1 shows the encryption procedure of Camellia-128.

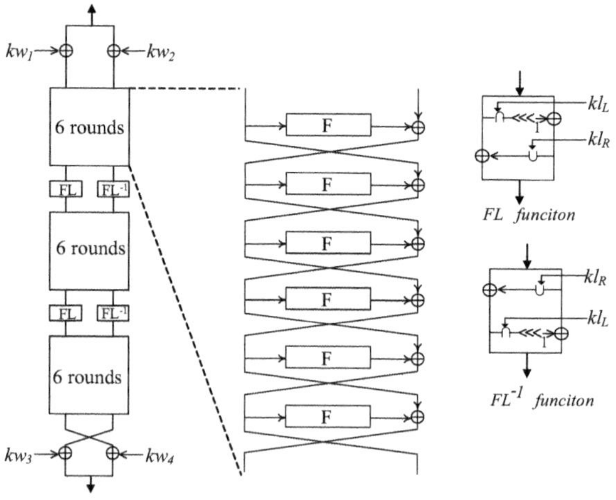

Fig. 1. Encryption procedure of Camellia-128

Firstly, a 128-bit plaintext M is XORed with $(kw_1|kw_2)$ to get two 64-bit data L_0 and R_0. Then, for $r = 1$ to 18, expect for $r = 6$ and 12, the following is carried out:

$$L_r = R_{r-1} \oplus F(L_{r-1}, K_r), \quad R_r = L_{r-1}.$$

For $r = 6$ and 12, do the following:

$$L'_r = R_{r-1} \oplus F(L_{r-1}, K_r), \qquad R'_r = L_{r-1},$$
$$L_r = FL(L'_r, kl_{r/3-1}), \qquad R_r = FL(R'_r, kl_{r/3}).$$

Lastly, the 128-bit ciphertext C is computed as: $C = (R_{18}|L_{18}) \oplus (kw_3|kw_4)$.

The round function F is composed of a key-addition layer, a substitution transformation S and a diffusion layer P. There are four types of 8×8 S-boxes s_1, s_2, s_3 and s_4 in the S transformation layer, and a 64-bit data is substituted as follows:

$$S(x_1|x_2|x_3|x_4|x_5|x_6|x_7|x_8)$$
$$= s_1(x_1)|s_2(x_2)|s_3(x_3)|s_4(x_4)|s_2(x_5)|s_3(x_6)|s_4(x_7)|s_1(x_8).$$

The linear transformation $P : (\{0,1\}^8)^8 \rightarrow (\{0,1\}^8)^8$ maps $(y_1, \cdots, y_8) \rightarrow (z_1, \cdots, z_8)$, this transformation and its inverse P^{-1} are defined as follows:

$$\begin{array}{ll}
z_1 = y_1 \oplus y_3 \oplus y_4 \oplus y_6 \oplus y_7 \oplus y_8 & y_1 = z_2 \oplus z_3 \oplus z_4 \oplus z_6 \oplus z_7 \oplus z_8 \\
z_2 = y_1 \oplus y_2 \oplus y_4 \oplus y_5 \oplus y_7 \oplus y_8 & y_2 = z_1 \oplus z_3 \oplus z_4 \oplus z_5 \oplus z_7 \oplus z_8 \\
z_3 = y_1 \oplus y_2 \oplus y_3 \oplus y_5 \oplus y_6 \oplus y_8 & y_3 = z_1 \oplus z_2 \oplus z_4 \oplus z_5 \oplus z_6 \oplus z_8 \\
z_4 = y_2 \oplus y_3 \oplus y_4 \oplus y_5 \oplus y_7 \oplus y_8 & y_4 = z_1 \oplus z_2 \oplus z_3 \oplus z_5 \oplus z_6 \oplus z_7 \\
z_5 = y_1 \oplus y_2 \oplus y_6 \oplus y_7 \oplus y_8 & y_5 = z_1 \oplus z_2 \oplus z_5 \oplus z_7 \oplus z_8 \\
z_6 = y_2 \oplus y_3 \oplus y_5 \oplus y_7 \oplus y_8 & y_6 = z_2 \oplus z_3 \oplus z_5 \oplus z_6 \oplus z_8 \\
z_7 = y_3 \oplus y_4 \oplus y_5 \oplus y_6 \oplus y_8 & y_7 = z_3 \oplus z_4 \oplus z_5 \oplus z_6 \oplus z_7 \\
z_8 = y_1 \oplus y_4 \oplus y_5 \oplus y_6 \oplus y_7 & y_8 = z_1 \oplus z_4 \oplus z_6 \oplus z_7 \oplus z_8
\end{array}$$

The FL function is defined as $(X_L|X_R, kl_L|kl_R) \mapsto (Y_L\ Y_R)$, where

$$Y_R = ((X_L \cap kl_L) \lll_1) \oplus X_R, \quad Y_L = (Y_R \cup kl_R) \oplus X_L.$$

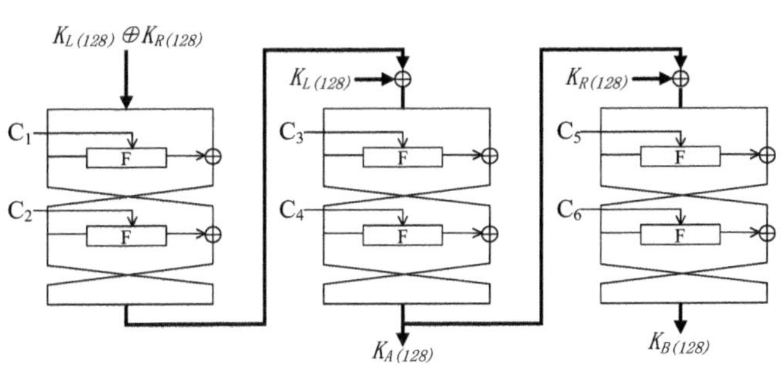

Fig. 2. Key Schedule of Camellia

Similarly to Camellia-128, Camellia-192/256 have 24-round Feistel structure, and the FL/FL^{-1} function layer are inserted in 6, 12 and 18 round. Before the first round and after the last round, there are pre- and post- whitening layers.

Key Schedule. Fig.2 shows the key schedule of Camellia. Two 128-bit variables K_A and K_B are generated form K_L and K_R. For Camellia-128, K_L is the 128-bit K, and K_R is 0. For Camellia-192, K_L is the left 128-bit of K, the concatenation of the right 64-bit of K and its complement used as K_R. For Camellia-256, K_L is the left 128-bit of K, and K_R is the right 128-bit of K. $C_i(i = 1 \cdots 6)$ are 64-bit constants. The round keys of Camellia are rotations of K_L, K_R, K_A, K_B, where K_B only used in Camellia-192/256. For details of Camellia, we refer to [1,2].

3 7-Round Impossible Differential of Camellia

In this section, we introduce a 7-round impossible differential of Camellia including FL/FL^{-1} layer.

Observation 1. *If the output difference of FL^{-1} function is $\Delta X = (0|0|0|0|a|0|0|0)$, then the input difference of FL^{-1} function should satisfies with $\Delta Y^{(2,3,4,6,7,8)} = 0$, and $\Delta Y^{(5)} = a$.*

Proof. According to the definition of FL^{-1} function, the right half of output difference is

$$X_R \oplus X'_R = ((X_L \cap kl_L) \lll_1) \oplus Y_R \oplus ((X'_L \cap kl_L) \lll_1) \oplus Y'_R,$$

where X_R and X_L are the left half and the right half of 64-bit value X, kl is the 64-bit subkey used in FL^{-1} function. On the basis of the condition $\Delta X = (0|0|0|0|a|0|0|0)$, we can conclude that

$$\begin{aligned}
\Delta Y_R &= ((X_L \cap kl_L) \lll_1) \oplus X_R \oplus ((X'_L \cap kl_L) \lll_1) \oplus X'_R \\
&= X_R \oplus X'_R \oplus (((X_L \cap kl_L) \oplus (X'_L \cap kl_L)) \lll_1) \\
&= \Delta X_R = (a|0|0|0), \\
\Delta Y_L &= (Y_R \cup kl_R) \oplus X_L \oplus (Y'_R \cup kl_R) \oplus X'_L \\
&= \Delta X_L \oplus (Y_R \cup kl_R) \oplus (Y'_R \cup kl_R) \\
&= (N|0|0|0),
\end{aligned}$$

where N is a unknown byte.

Observation 2. *Given a 7-round Camellia encryption and a FL/FL^{-1} layer inserted between the sixth and seventh round. If the input difference of the first round is $(0|0|0|0|0|0|0|0, a|0|0|0|0|0|0|0)$, then through the 7 rounds encryption, the output difference $(0|0|0|0|d|0|0|0, 0|0|0|0|0|0|0|0)$ is impossible, where a and d are any non-zero values (see Fig.3).*

Proof. Firstly, we analyze the first three rounds. The input difference

$$(\Delta L_0, \Delta R_0) = (0|0|0|0|0|0|0|0, a|0|0|0|0|0|0|0)$$

becomes

$$(\Delta L_1, \Delta R_1) = (a|0|0|0|0|0|0|0, 0|0|0|0|0|0|0|0)$$

through the first round transformation. After the F function and XOR operation in the second round, it becomes

$$(\Delta L_2, \Delta R_2) = (b|b|b|0|b|0|0|b, b|0|0|0|0|0|0|0),$$

where b is non-zero value. Through the key addition and substitution layer of the third round, the output difference of S-box layer in the third round is $\Delta S_3 = (b_1|b_2|b_3|0|b_5|0|0|b_8)$. After the linear function P and XOR operation, the input difference of the forth round is

$$(\Delta L_3, \Delta R_3) = (f_1 \oplus a|f_2|f_3|f_4|f_5|f_6|f_7|f_8, b|b|b|0|b|0|0|b),$$

where b_i are non-zero values, f_i are unknown values.

Secondly, we consider the inverse direction. If the output difference of the seventh round

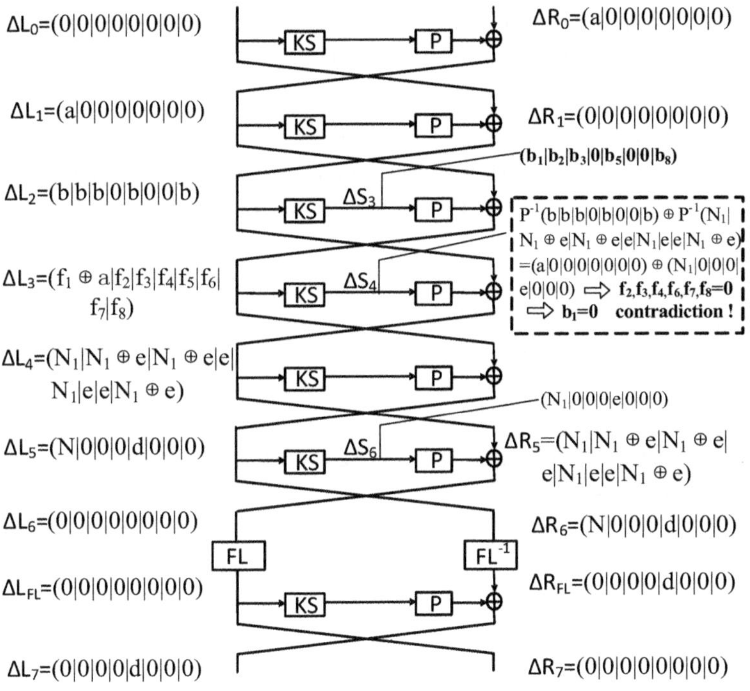

Fig. 3. 7-Round Impossible Differential of Camellia

$$(\Delta L_7, \Delta R_7) = (0|0|0|0|d|0|0|0, 0|0|0|0|0|0|0|0),$$

then the output difference of FL/FL^{-1} layer should be

$$(\Delta L_{FL}, \Delta R_{FL}) = (0|0|0|0|0|0|0|0, 0|0|0|0|d|0|0|0),$$

where d is a non-zero value. According to the observation 1, we can conclude that the input difference of FL^{-1} function should satisfy with $\Delta R_6 = (N|0|0|0|d|0|0|0)$. After key addition and substitution of the sixth round, the output difference of S-box layer in the sixth round is $\Delta S_6 = (N_1|0|0|0|e|0|0|0)$, where N_1 is an unknown value and e is a non-zero value. Through the P function and XOR operation, the input difference of the sixth round should be

$$(\Delta L_5, \Delta R_5) = (N|0|0|0|d|0|0|0, N_1|N_1 \oplus e|N_1 \oplus e|e|N_1|e|e|N_1 \oplus e).$$

Therefore, the left half of input difference in the fifth round is $\Delta L_4 = (N_1|N_1 \oplus e|N_1 \oplus e|e|N_1|e|e|N_1 \oplus e)$.

Finally, we observe the forth round. The output difference of P function layer in the forth round is

$$\Delta P_4 = \Delta R_3 \oplus \Delta L_4 = (b|b|b|0|b|0|0|b) \oplus (N_1|N_1 \oplus e|N_1 \oplus e|e|N_1|e|e|N_1 \oplus e).$$

According to the definition of linear function P^{-1}, the output difference of S-box layer in the forth round is

$$\begin{aligned}\Delta S_4 &= P^{-1}(b|b|b|0|b|0|0|b) \oplus P^{-1}(N_1|N_1 \oplus e|N_1 \oplus e|e|N_1|e|e|N_1 \oplus e)\\ &= (a|0|0|0|0|0|0|0) \oplus (N_1|0|0|0|e|0|0|0)\\ &= (a \oplus N_1|0|0|0|e|0|0|0)\end{aligned}$$

Because S-box are nonlinear and permutation, we can get $f_2 = f_3 = f_4 = f_6 = f_7 = f_8 = 0$. From the expression in the third round, we know $b_1 = f_2 \oplus f_3 \oplus f_4 \oplus f_6 \oplus f_7 \oplus f_8 = 0$, which contradicts with $b_1 \neq 0$. Similarly, we obtain other three impossible differential of 7-round Camellia:

$$(0|0|0|0|0|0|0|0, 0|a|0|0|0|0|0|0) \nrightarrow (0|0|0|0|0|d|0|0, 0|0|0|0|0|0|0|0),$$
$$(0|0|0|0|0|0|0|0, 0|0|a|0|0|0|0|0) \nrightarrow (0|0|0|0|0|0|d|0, 0|0|0|0|0|0|0|0),$$
$$(0|0|0|0|0|0|0|0, 0|0|0|a|0|0|0|0) \nrightarrow (0|0|0|0|0|0|0|d, 0|0|0|0|0|0|0|0).$$

4 Impossible Differential Attack on 10-Round Camellia-128

In this section, we present an impossible differential attack on 10-round Camellia-128, which add three rounds on the bottom of above 7-round impossible differential (Fig. 4). For the sake of reducing the time complexity, we use precomputation technique and consider key redundancy.

Relation between K_8 and K_{10}. The subkey bits to be considered in our attack are $K_8^{(5)}|K_9^{(1,2,3,4,6,7,8)}|K_{10}^{(1,2,3,4,5,6,7,8)}$, which are 128 bits. According to the key schedule of camellia-128, if the master key K_L is denoted by its

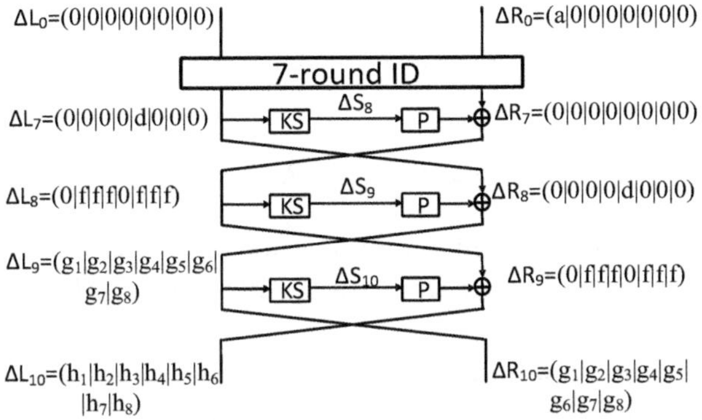

Fig. 4. Impossible Differential Attack on 10-Round Camellia-192

bits as $K_L = \kappa_1|\kappa_2|\ldots|\kappa_{128}$, then target subkeys of round 8 and round 10 are equivalent to

$$K_8^{(5)} = \kappa_{14}|\kappa_{15}|\ldots|\kappa_{21}, \quad K_{10} = \kappa_{125}|\kappa_{126}|\kappa_{127}|\kappa_{128}|\kappa_1|\kappa_2|\ldots|\kappa_{60}.$$

Apparently, there are 8 bits' redundancy in $K_8^{(5)}$ and K_{10}, which is advantageous for us to reduce the complexity of attack.

Precomputation. We need to construct precomputation tables $T_{1,i}$, T_2, to obtain input or output values of S-box layer in our attack.

Table $T_{1,i}$, $i = 1, 2, 3, 4$: This type of table is used to obtain the proper input and output pair of s_i-box with fixed input and output difference. For all of the 2^{16} possible pairs (x, x'), compute $(s_i(x), s_i(x'))$ and their difference $\Delta s_i(x, x')$, store values $(x, s_i(x))$ in a hash table $T_{1,i}$ indexed by 16-bit parameter $(\Delta x|\Delta s_i(x, x'))$.

Table T_2: This table is applied to get the partial input value of S-box layer with fixed input difference and output difference. For all of 2^{64} possible pairs (X, X'), which satisfy $\Delta X = (0|f|f|f|\lambda|f|f|f)$, where λ represents the byte that we don't consider in this table.[1] Compute the values $(S(X)$, $S(X'))$ and their difference $\Delta S(X, X') = (0|f_2|f_3|f_4|\lambda|f_6|f_7|f_8)$. Then compute the value

$$y = S(X)^{(1)} \oplus S(X)^{(2)} \oplus S(X)^{(6)} \oplus S(X)^{(7)} \oplus S(X)^{(8)},$$

where y equals to the value $P(S(X))^{(5)}$. Store the significant 56-bit value of $X = (x_1|x_2|x_3|x_4|x_6|x_7|x_8)$ in a hash table T_2 indexed by 64-bit parameter $(f|f_2|f_3|f_4|f_6|f_7|f_8|y)$. Apparently, there are 2^{64} rows in this table, and on average one pair lies in each of these rows.

---

[1] There are $2^8 - 1$ possible values of f, and for each value of f, there are 2^{56} possible pairs of (X, X'). Totally, there are about 2^{64} possible pairs to be considered.

Attack Procedure

1. Choose 2^n structures of plaintexts, and each structure contains 2^8 plaintexts (L_0, R_0) with
$$L_0 = (x_1|x_2|x_3|x_4|x_5|x_6|x_7|x_8),\ R_0 = (\alpha_1|y_2|y_3|y_4|y_5|y_6|y_7|y_8),$$
where x_i and y_i are fixed values and α_1 takes all the possible values. Thus we can collect 2^{n+15} plaintext pairs with the difference $(\Delta L_0, \Delta R_0) = (0|0|0|0|0|0|0|0, a|0|0|0|0|0|0|0)$.
2. For each structure, obtain the 2^8 ciphertexts and store them in a hash table H indexed by the value $R_{10}^{(2)} \oplus R_{10}^{(3)} \oplus R_{10}^{(4)} \oplus R_{10}^{(6)} \oplus R_{10}^{(7)} \oplus R_{10}^{(8)}$. Then each two texts which lie in the same row of H form a proper pair, which satisfy
$$(\Delta L_{10}, \Delta R_{10}) = (h_1|h_2|h_3|h_4|h_5|h_6|h_7|h_8, g_1|g_2|g_3|g_4|g_5|g_6|g_7|g_8),$$
and $g_2 \oplus g_3 \oplus g_4 \oplus g_6 \oplus g_7 \oplus g_8 = 0$, where g_i, h_i are unknown values. Since this step preform a 8-bit filtration from the all 2^{n+15} pairs, at the end of this step, the expected number of proper pairs is 2^{n+7}.
3. For each of the proper pairs,compute the intermediate value
$P^{-1}(\Delta L_{10}) = P^{-1}(h_1|h_2|h_3|h_4|h_5|h_6|h_7|h_8) = (h_1'|h_2'|h_3'|h_4'|h_5'|h_6'|h_7'|h_8')$.
$P^{-1}(\Delta R_{10}) = P^{-1}(g_1|g_2|g_3|g_4|g_5|g_6|g_7|g_8) = (0|g_2'|g_3'|g_4'|g_5'|g_6'|g_7'|g_8')$.[2]
4. This step consider the ciphertexts of every proper pair.
   (a) For $l = 1, 2, 3, 4, 6, 7, 8$, access the row with index $(g_l|h_l')$ in table $T_{1,i}$, where i denote the type of s-box used in S-box layer (see section 2). For the pair $(x, s_i(x))$ in that row, choose the value $(x \oplus R_{10}^{(l)})$ as a candidate of $K_{10}^{(l)}$, and the value $s_i(x)$ as a candidate of $S_{10}^{(l)}$. According to the structure of table $T_{1,i}$, we expect to obtain about one candidate of $K_{10}^{(l)}$ and $S_{10}^{(l)}$ for each l.
   (b) For all of $2^8 - 1$ possible non-zero value of f, access the row with index $(g_5|h_5' \oplus f)$ in table $T_{1,2}$. Consider the pair $(x, s_2(x))$ in that row, choose the value $(x \oplus R_{10}^{(5)})$ as a candidate of $K_{10}^{(5)}$, and the value $s_2(x)$ as a candidate of $S_{10}^{(5)}$.
   (c) Partially decrypt S_{10} to get the intermediate value $P(S_{10}) \oplus L_{10} = R_9$. Apparently, the candidate of $R_9' = R_9 \oplus (0|f|f|f|0|f|f|f)$. At the end of this step, we could get about 2^8 candidates of K_{10} for each of proper pairs.
5. For each of 2^8 candidates of K_{10} for every proper pair, perform the following substeps.

[2] To guarantee the differential characteristic holds, the difference of intermediate value ΔS_9 should equals to $P^{-1}(\Delta L_9 \oplus \Delta R_8) = (g_1'|g_2' \oplus d|g_3' \oplus d|g_4' \oplus d|g_5' \oplus d|g_6' \oplus d|g_7' \oplus d|g_8')$. Since $\Delta L_8 = \Delta R_9 = (0|f|f|f|0|f|f|f)$, then there must be $g_1' = 0$ and $g_5' = d$.

(a) Access the row with index $(g_5'|f)$ in table $T_{1,2}$ (where $g_5' = d$). For the pair $(x, s_2(x))$ in that row, select the value $(x \oplus K_8^{(5)})$ as a candidate of $R_8^{(5)}$, where the equivalent value of $K_8^{(5)}$ is obtained in step 2.

(b) Access the row with index $(f|g_2'|g_3'|g_4'|g_6'|g_7'|g_8'|R_8^{(5)} \oplus R_{10}^{(5)})$ in table T_2. For the 56-bit value $(x_1|x_2|x_3|x_4|x_6|x_7|x_8)$ in that row, select $(x_i \oplus R_9^{(i)})$ as the candidates of $K_9^{(i)}$ $(i = 1, 2, 3, 4, 6, 7, 8)$. Obviously, this 56-bit value $K_9^{(1,2,3,4,6,7,8)}$ along with 64-bit value $K_{10}^{(1,2,3,4,5,6,7,8)}$ can result in the impossible differential. Remove this 120-bit value from the list of all the 2^{120} possible values.

6. After analyzing all structures, announce the remaining values in the list are the candidates of 120-bit target subkey ($K_L : 1 \sim 60, 125 \sim 128; K_A : 46 \sim 77, 86 \sim 109$). Then search for the remaining 64-bit value of master key K_L and obtain the correct key by trial encryption.

Complexity. In step 7, for each of the 2^{n+7} proper pairs, we can remove on average 2^8 values out of the 2^{120} possible values of target subkey. Thus the remaining values in the list is $\epsilon = 2^{120} \times (1 - \frac{2^8}{2^{120}})^{2^{n+7}}$. If we choose $\epsilon = 2^{56}$, n will be 110.5, then this attack requires $2^{n+8} = 2^{118.5}$ chosen plaintexts.

Table 2 plots the time complexity of different steps of attack procedure. In step 4 and substep 5(a), the value of candidates K_{10}, S_{10} and $R_8^{(5)}$ are obtained by performing 32-bit memory access in table $T_{1,i}$, so the time complexity of each of these operations is less than $\frac{1}{4}$ one round encryption. In substep 5(b), we utilize the 120-bit memory access in table T_2 to get the candidate of K_9, the time complexity of this operation should be less than one round encryption. Therefore, the time complexity of this attack is dominated by step 5(b), and the total time complexity is about $2^{123.5}$ 10 rounds encryptions.

The memory complexity of attack is dominated by excluding the wrong values of target subkey, which needs 2^{120} 120-bit memory blocks to list all 2^{120} possible values. Thus the memory complexity is about 2^{120} 120-bit blocks of memory.

Table 2. Time complexity in attack procedure

Step	Time Complexity	for n=110.5
2	$2^{n+8} + 5 \times 2^{n+8} \times \frac{1}{32} \times \frac{1}{10} \approx 2^{n+8}$E	$2^{118.5}$E
3	$2 \times 2^{n+7} \times \frac{1}{10} \times \frac{16}{32} = 2^{n+3.7}$E	$2^{114.2}$E
4(a)	$7 \times 2^{n+7} = 2^{n+9.8}MA\approx 2^{n+4.5}$E	2^{115}E
4(b)	$2^8 \times 2^{n+7} = 2^{n+15}MA\approx 2^{n+9.7}$E	$2^{120.2}$E
4(c)	$2^8 \times 2^{n+7} \times \frac{24}{32} \times \frac{1}{10} = 2^{n+11.2}$E	$2^{121.9}$E
5(a)	$2^8 \times 2^{n+7} = 2^{n+15}MA\approx 2^{n+9.7}$E	$2^{120.2}$E
5(b)	$2^8 \times 2^{n+7} = 2^{n+15}MA\approx 2^{n+11.7}$E	$2^{122.2}$E
6	$2^{56} \times 2^{64} \times (1 + \frac{4}{10})$E	$2^{120.5}$E

5 Attack on 10-Round Camellia-192 and 11-Round Camellia-256

In this section, we will give the new results of attack on 10-round Camellia-192, and extend the attack to 11-round Camellia-256.

5.1 Attack on 10-Round Camellia-192

According to the key schedule of Camellia-192/256, there are no redundancy in subkeys of round 8, 9 and 10. So we have to consider all of the 128-bit target subkey

$$K_8^{(5)}|K_9^{(1,2,3,4,6,7,8)}|K_{10}^{(1,2,3,4,5,6,7,8)}$$

for attack on 10-round Camellia-192. Consequently, compared with attack on 10-round Camellia-128 in above section, in step 6(b), for each candidate of K_{10} and S_{10}, we should enumerate all of possible values of $K_8^{(5)}$, and get 2^8 candidates of $R_8^{(5)}$ by memory access. Then in step 6(c), we will perform 2^8 memory accesses in table T_2 to get 2^8 candidates of $K_9^{(1,2,3,4,6,7,8)}$. Thus for every proper pair, we can discard about 2^{16} values from the list of 2^{128} possible values.

Complexity. The complexity of step 6(b) is 2^{n+23} 32-bit memory accesses, which is about $2^{n+17.7}$ 10 rounds encryptions. The complexity of step 6(c) is 2^{n+23} 120-bit memory accesses, which is about $2^{n+19.7}$ 10 rounds encryptions. Totally, if $n = 110.7$, the complexity of this attack is about $2^{130.4}$ 10 rounds encryptions. The memory complexity is about 2^{128} 128-bit blocks of memory. We note that the adjunction of whitening layers don't effect the complexity in this attack, because the master key could be retrieved by the guess of equivalent keys [4].

5.2 Attack on 11-Round Camellia-256

We add one round on the bottom of the 10-round attack, and give a 11-round attack on Camellia-256. The attack algorithm is as follows:

1. Choose 2^n structures of plaintexts, and each structure contains 2^8 plaintexts (L_0, R_0) with

$$L_0 = (x_1|x_2|x_3|x_4|x_5|x_6|x_7|x_8),\ R_0 = (\alpha_1|y_2|y_3|y_4|y_5|y_6|y_7|y_8).$$

   Obtain the corresponding ciphertext for each plaintext, then collect 2^{n+15} pairs which satisfy $(\Delta L_0, \Delta R_0) = (0|0|0|0|0|0|0|0, a|0|0|0|0|0|0|0)$.

2. Guess the 64 bit value of K_{11}, partially decrypt every ciphertext pair to get the intermediate value (R_{10}, R'_{10}) and their difference. Keep only the pairs which satify

$$\Delta R_{10}{}^{(2)} \oplus \Delta R_{10}{}^{(3)} \oplus \Delta R_{10}{}^{(4)} \oplus \Delta R_{10}{}^{(6)} \oplus \Delta R_{10}{}^{(7)} \oplus \Delta R_{10}{}^{(8)} = 0.$$

   This step preform a 8-bit filtration form all of 2^{n+15} pairs, then the expected number of remaining pairs is 2^{n+7} at the end of this step.
3. **Application of 10 Round Attack:** for each of remaining pairs, perform the following substeps:
   (a) Compute the intermediate values $P^{-1}(\Delta L_{10}) = (h'_1|h'_2|h'_3|h'_4|h'_5|h'_6|h'_7|h'_8)$ and $P^{-1}(\Delta R_{10}) = (0|g'_2|g'_3|g'_4|g'_5|g'_6|g'_7|g'_8)$.
   (b) For $l = 1, 2, 3, 4, 6, 7, 8$, access the row with index $(\Delta R_{10}{}^{(l)}|h'_l)$ in table $T_{1,i}$, and get the candidates of $K_{10}^{(l)}$ and $S_{10}^{(l)}$.
   (c) For all of $2^8 - 1$ possible non-zero value of f, access the row with index $(\Delta R_{10}{}^{(5)}|h'_5 \oplus f)$ in table $T_{1,2}$, and get the candidate of $K_{10}^{(5)}$ and $S_{10}^{(5)}$. Then we obtain about 2^8 candidates of K_{10} and S_{10} for each of remaining pairs.
   (d) For each candidate of K_{10} and S_{10}, partially decrypt S_{10} to get the intermediate value $P(S_{10}) \oplus L_{10} = R_9$.
   (e) For each of the 2^8 possible values of $K_8^{(5)}$, access the row with index $(g'_5|f)$ in table $T_{1,2}$, and get the candidates of $R_8^{(5)}$.
   (f) Access the row with index $(f|g'_2|g'_3|g'_4|g'_6|g'_7|g'_8|R_8^{(5)} \oplus R_{10}^{(5)})$ in table T_2, and obtain the candidates of $K_9^{(i)}$ $(i = 1, 2, 3, 4, 6, 7, 8)$. Such 56-bit value $K_9^{(1,2,3,4,6,7,8)}$ along with 64-bit value $K_{10}^{(1,2,3,4,5,6,7,8)}$, 8-bit value $K_8^{(5)}$ and 64-bit guessed value K_{11} can result in the impossible differential. Remove the 128-bit value $K_8^{(5)}|K_9^{(1,2,3,4,6,7,8)}|K_{10}^{(1,2,3,4,5,6,7,8)}$ from the list of all the 2^{128} possible values. For every remaining pair, there are 2^{16} 128-bit values being removed in this step.
4. After analyze all of remaining pairs, announce the remaining values in the list along with 64-bit guessed key K_{11} are the candidates 192-bit target subkey $(K_L : 1 \sim 77, 86 \sim 128; K_A : 46 \sim 109, K_B : 1 \sim 6, 127, 128)$. Searching for the remaining 64-bit value of K_A and 8-bit value of K_L, then compute the 128-bit master key K_R and get the correct key by trial encryption. Otherwise, return to step 2 to try the other 64-bit value guess.

Complexity. In step 4, the remaining values in the list is $\epsilon = 2^{192} \times (1 - \frac{2^{16}}{2^{128}})^{2^{n+7}}$, if we choose $\epsilon = 2^{120}$, n will be 110.7, then this attack requires $2^{n+8} = 2^{118.7}$ chosen plaintexts.

The time complexity is dominated by substep 3(e) and 3(f). In substep 3(e), we need 2^{n+87} 32-bit memory accesses, which is about $2^{n+81.5}$ 11 rounds encryptions. Substep 3(f) requires 2^{n+87} 120-bit memory accesses, which is about $2^{n+83.5}$ 11 rounds encryptions. The total complexity of this attack is about $2^{194.5}$

11 rounds encryptions. The memory complexity is about 2^{128} 128-bit blocks of memory.

6 Conclusion

In this paper, we first introduced a 7-round impossible differential of Camellia with FL/FL^{-1} layer, based on this impossible differential, we proposed an impossible differential attack on 10-round Camellia-128, which requires $2^{118.5}$ chosen plaintexts and about $2^{123.5}$ 10-round encryptions. Furthermore, we presented the new result of attack on 10-round Camellia-192 and 11-round Camellia-256. To the best of our knowledge, these results are better than all the previously published attacks on reduced round Camellia with FL/FL^{-1} layer.

References

1. Aoki, K., Ichikawa, T., Kanda, M., Matsui, M., Moriai, S., Nakajima, J., Tokita, T.: *Camellia*: A 128-Bit Block Cipher Suitable for Multiple Platforms - Design and Analysis. In: Stinson, D.R., Tavares, S. (eds.) SAC 2000. LNCS, vol. 2012, pp. 39–56. Springer, Heidelberg (2001)
2. Aoki, K., Ichikawa, T., Kanda, M., Matsui, M., Moriai, S., Nakajima, J., Tokita, T.: Specification of Camellia-a 128-bit Block Cipher. version 2.0 (2001), `http://info.isl.ntt.co.jp/crypt/eng/camellia/specifications.html`
3. Biham, E., Biryukov, A., Shamir, A.: Cryptanalysis of Skipjack Reduced to 31 Rounds Using Impossible Differentials. In: Stern, J. (ed.) EUROCRYPT 1999. LNCS, vol. 1592, pp. 12–23. Springer, Heidelberg (1999)
4. Chen, J., Jia, K., Yu, H., Wang, X.: New Impossible Differential Attacks of Reduced-Round Camellia-192 and Camellia-256. In: Parampalli, U., Hawkes, P. (eds.) ACISP 2011. LNCS, vol. 6812, pp. 16–33. Springer, Heidelberg (2011)
5. CRYPTREC-Cryptography Research and Evaluation Committees, report, Archive (2002), `http://www.ipa.go.jp/security/enc/CRYPTREC/index-e.html`
6. Duo, L., Li, C., Feng, K.: Square Like Attack on Camellia. In: Qing, S., Imai, H., Wang, G. (eds.) ICICS 2007. LNCS, vol. 4861, pp. 269–283. Springer, Heidelberg (2007)
7. Hatano, Y., Sekine, H., Kaneko, T.: igher Order Differential Attack of Camellia (II). In: Nyberg, K., Heys, H.M. (eds.) SAC 2002. LNCS, vol. 2595, pp. 129–146. Springer, Heidelberg (2003)
8. International Standardization of Organization (ISO), International Standard - ISO/IEC 18033-3, Information technology - Security techniques - Encryption algorithms - Part 3: Block ciphers (2005)
9. Kanda, M., Matsumoto, T.: Security of Camellia against Truncated Differential Cryptanalysis. In: Matsui, M. (ed.) FSE 2001. LNCS, vol. 2355, pp. 119–137. Springer, Heidelberg (2002)
10. Knudsen, L.R.: DEAL C a 128-bit Block Cipher. Technical report, Department of Informatics, University of Bergen, Norway (1998)

11. Lee, S., Hong, S., Lee, S., Lim, J., Yoon, S.: Truncated Differential Cryptanalysis of Camellia. In: Kim, K.-c. (ed.) ICISC 2001. LNCS, vol. 2288, pp. 32–38. Springer, Heidelberg (2002)
12. Lei, D., Li, C., Feng, K.: New Observation on Camellia. In: Preneel, B., Tavares, S. (eds.) SAC 2005. LNCS, vol. 3897, pp. 51–64. Springer, Heidelberg (2006)
13. Lu, J., Kim, J.-S., Keller, N., Dunkelman, O.: Improving the Efficiency of Impossible Differential Cryptanalysis of Reduced Camellia and MISTY1. In: Malkin, T. (ed.) CT-RSA 2008. LNCS, vol. 4964, pp. 370–386. Springer, Heidelberg (2008)
14. Mala, H., Shakiba, M., Dakhilalian, M., Bagherikaram, G.: New Results on Impossible Differential Cryptanalysis of Reduced–Round Camellia–128. In: Jacobson Jr., M.J., Rijmen, V., Safavi-Naini, R. (eds.) SAC 2009. LNCS, vol. 5867, pp. 281–294. Springer, Heidelberg (2009)
15. NESSIE-New European Schemes for Signatures, Integrity, and Encryption, final report of European project IST-1999-12324. Archive (1999), `https://www.cosic.esat.kuleuven.be/nessie/Bookv015.pdf`
16. Shirai, T.: Differential, Linear, Boomerang and Rectangle Cryptanalysis of Reduced-Round Camellia. In: Proceedings of 3rd NESSIE Workshop (2002)
17. Sugita, M., Kobara, K., Imai, H.: Security of Reduced Version of the Block Cipher Camellia against Truncated and Impossible Differential Cryptanalysis. In: Boyd, C. (ed.) ASIACRYPT 2001. LNCS, vol. 2248, pp. 193–207. Springer, Heidelberg (2001)
18. Wu, W., Feng, D., Chen, H.: Collision Attack and Pseudorandomness of Reduced-Round Camellia. In: Handschuh, H., Hasan, M.A. (eds.) SAC 2004. LNCS, vol. 3357, pp. 252–266. Springer, Heidelberg (2004)
19. Wu, W., Zhang, W., Feng, D.: Impossible Differential Cryptanalysis of Reduced-Round ARIA and Camellia. Journal of Computer Science and Technology 22(3), 449–456 (2007)

The Initialization Stage Analysis of ZUC v1.5*,**

Chunfang Zhou[1,2], Xiutao Feng[1], and Dongdai Lin[1]

[1] State Key Laboratory of Information Security, Institute of Software, Chinese Academy of Sciences, Beijing, 100190, China
[2] Graduate University of the Chinese Academy of Science, Beijing, 100049, China
{cfzhou,fengxt,ddlin}@is.iscas.ac.cn

Abstract. The ZUC algorithm is a new stream cipher, which is the core of the standardised 3GPP confidentiality and integrity algorithms 128-EEA3 & 128-EIA3. In this paper, we analyze the initialization stage of ZUC v1.5. First of all, we study the differential properties of operations in ZUC v1.5, including the bit-reorganization, exclusive-or and addition modulo 2^n, bit shift and the update of LFSR. And then we give a differential trail covering 24 rounds of the initialization stage of ZUC v1.5 with probability $2^{-23.48}$, which extends the differential given in the design and evaluation report of ZUC v1.5 to four more rounds. Nevertheless, the study shows that the stream cipher ZUC v1.5 can still resist against chosen-IV attacks.

Keywords: ZUC, initialization, chosen-IV attack, differential trail.

1 Introduction

The ZUC algorithm [1] is the core of the standardised 3GPP confidentiality and integrity algorithms 128-EEA3 & 128-EIA3 [2]. The ZUC algorithm has been evaluated by two independent work groups. It can resist against many cryptanalytic attacks such as weak key attacks, guess-and-determine attacks, linear distinguishing attacks, algebraic attacks, etc. Now the version 1.5 of ZUC is in the second public evaluation phase.

The chosen-IV/Key attack [4,5,6], targeting at the initialization state of stream ciphers, is one of the most important attacks of stream ciphers. For a good stream cipher, after the initialization, each bit of the IV/Key should contribute to every bit of the states of the ciphers and any differential of the IV/Key will result in an almost-uniform and unpredictable differential of the internal states. Comparing with the frequency of changing the key, the change of IV is more frequent. And since the IV is known to the public, so the chosen-IV attack is more feasible.

* This work was supported by the National Natural Science Foundation of China (Grant No. 60970152, 60833008 and 60902024), the National 973 Program of China (Grant No. 2011CB302400 and 2007CB807902) and Grand Project of Institute of Software (Grant No. YOCX285056).

** This work had been presented informally at the Second International Workshop on ZUC Algorithm and Related Topics without proceedings.

D. Lin, G. Tsudik, and X. Wang (Eds.): CANS 2011, LNCS 7092, pp. 40–53, 2011.

Bing Sun et al. [9] at the 2010 ZUC workshop and Hongjun Wu [10] at the rump session of ASIACRYPT 2010 independently pointed out that there exists entropy leakage in the initialization state of the version 1.4 of ZUC. Hongjun Wu gave a chosen-IV attack to ZUC v1.4. In order to amend the above flaw, the designers of ZUC modified ZUC v1.4 and presented a new version of ZUC, namely ZUC v1.5. Wu's attack dose not work on ZUC v1.5. A 24-round chosen-Key differential of ZUC v1.4 was given by Ji Li at the first international workshop on ZUC [7]. The author claimed that the chosen-Key differential still worked on ZUC v1.5 and extended it to one more round [8]. She introduced a difference in the key and fixed the differences after the fourth round. Then she gave the 25-round differential. So the differential probability she given was just conditional probability and the probability of the 4-round differential which did not give out by her was not high by our test. A 20-round chosen-IV differential of ZUC v1.5 is given in [3].

In [11] Mouha et al. proposed the concept of S-function ("state function" in short), which is a special T-function and widely used in the design of stream ciphers, block ciphers and hash functions. Based on graph theory, Mouha et al. presented a fully generic and efficient framework to determine the differential properties of S-functions, which can be efficiently calculated using matrix multiplications.

Both modular addition and exclusive-or(XOR) are S-functions. The composition of them is still an S-function, which is used in ZUC. We study the differential properties of the operation combined modular addition and XOR by the methods proposed by Mouha et al. Together with the differential properties of other functions in ZUC, we analyze initialization stage of ZUC v1.5 and extend the chosen-IV differential given in [3] to four more rounds. The differential probability is $2^{-23.48}$. It shows that ZUC v1.5 is still secure under our new analysis.

The outline of the paper is as follows. In Sect. 2, we give a brief introduction to ZUC v1.5 and S-functions. We study the differential properties of operations of ZUC v1.5 and analyze the initialization of ZUC v1.5 in Sect. 3. We conclude in Sect. 4.

2 Preliminaries

2.1 ZUC v1.5

We recall ZUC v1.5 briefly in this section, for details please refer to [1]. ZUC v1.5 is a word-oriented stream cipher, which takes 128-bit key and 128-bit IV as inputs and outputs a 32-bit word key sequence. ZUC v1.5 has three logic layers, see Fig. 1. The top layer is a linear feedback shift register (LFSR) of 16 cells, the middle layer is for bit-reorganization, and the bottom layer is a nonlinear function F.

The LFSR has 16 of 31-bit cells $(s_0, s_1, \cdots, s_{15})$. The characteristic polynomial of the LFSR is a primitive polynomial over the prime field $\mathbb{F}_{2^{31}-1}$.

$$f(x) = x^{16} - (2^{15}x^{15} + 2^{17}x^{13} + 2^{21}x^{10} + 2^{20}x^4 + (2^8 + 1)).$$

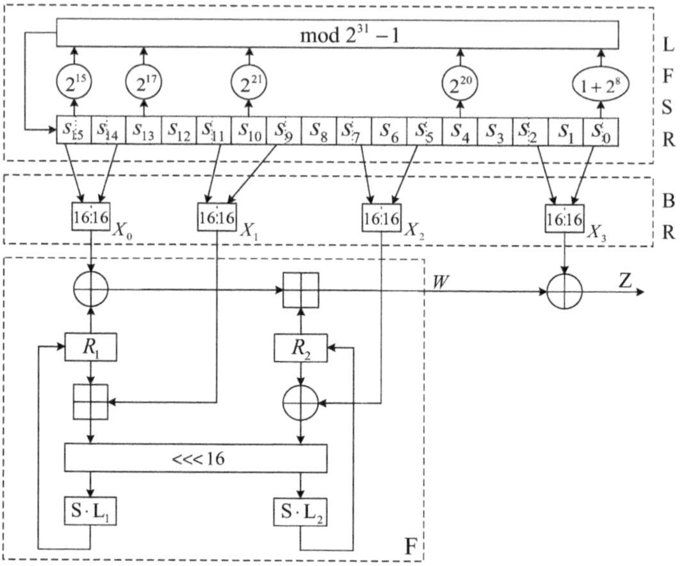

Fig. 1. The Structure of ZUC v1.5

Each cell $s_i (0 \leq i \leq 15)$ is restricted to take values from the set $\{1, 2, \cdots, 2^{31}-1\}$.

The bit-reorganization layer extracts 128 bits from the LFSR and forms 4 of 32-bit words, where the first three words will be used by the nonlinear function F in the bottom layer, and the last word will be involved in producing the keystream. The bit-reorganization forms 4 words X_0, X_1, X_2, X_3 as follows:

$$X_0 = s_{15H} \parallel s_{14L}, X_1 = s_{11L} \parallel s_{9H}, X_2 = s_{7L} \parallel s_{5H}, X_3 = s_{2L} \parallel s_{0H},$$

where s_{iH} means bits $30 \cdots 15$ of s_i, s_{iL} means bits $15 \cdots 0$ of s_i, for $0 \leq i \leq 15$. "$a \parallel b$" denotes the concatenation of two strings a and b.

The nonlinear function F has 2 of 32-bit memory cells R_1 and R_2. Let the inputs to F be X_0, X_1 and X_2, the function F outputs a 32-bit word W. The detailed process of F is as follows:

$$W = (X_0 \oplus R_1) \boxplus R_2, W_1 = R_1 \boxplus X_1, W_2 = R_2 \oplus X_2,$$
$$R_1 = S(L_1(W_{1L} \parallel W_{2H})), R_2 = S(L_2(W_{2L} \parallel W_{1H})),$$

where "$\boxplus$" denotes the addition modulo 2^n. S is a 32×32 S-box. L_1 and L_2 are linear transforms. For details please refer to [1].

Let the 128-bit initial key k and the 128-bit initial vector iv be $k = k_0 \parallel k_1 \parallel \cdots \parallel k_{15}$ and $iv = iv_0 \parallel iv_1 \parallel \cdots \parallel iv_{15}$ respectively, where k_i and iv_i, $0 \leq i \leq 15$, are all bytes. The first stage of ZUC is initialization stage. The algorithm calls the key loading procedure to load k and iv into the LFSR, $s_i = k_i \parallel d_i \parallel iv_i$, where d_i is a 15-bit constant, $0 \leq i \leq 15$. The 32-bit memory cells R_1 and R_2 are set to be all 0. There are 32 rounds of iterations in the initialization stage, see Fig. 2. The update of the LFSR is as below:

$$s_{16} = 2^{15}s_{15} + 2^{17}s_{13} + 2^{21}s_{10} + 2^{20}s_4 + (1+2^8)s_0 + (W \gg 1) \bmod (2^{31}-1).$$

After the initialization stage, the algorithm moves into the working stage. For each iteration, a 32-bit word Z is produced as $Z = W \oplus X_3$.

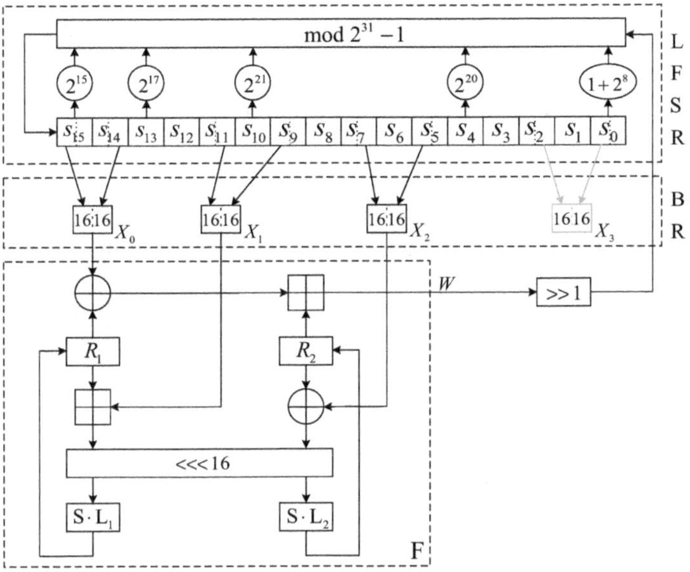

Fig. 2. The Structure of ZUC v1.5 During Initialization Stage

2.2 S-Functions

S-function was introduced by Mouha et al. in 2010 to compute the additive differential probability of XOR [11]. An S-function accepts k n-bit words $a_1, a_2, \cdots, a_k$ and outputs an n-bit word b. A series of states $S[i] (0 \le i \le n)$ are used to calculate the output. The i-th bit of the output and the $(i+1)$-th state can be calculated using only the i-th bits of inputs and the i-th state, see Fig. 3.

$$(b[i], S[i+1]) = f(a_1[i], a_2[i], \cdots, a_k[i], S[i]), \quad 0 \le i < n,$$

where $b[i]$ denotes the i-th bit of b, $0 \le i < n$. Usually, the initial state is set to be zero.

S-functions such as addition, subtraction, multiplication by a constant (all modulo 2^n), XOR and so on, are widely used in cryptographic primitives.

Mouha et al. presented a general framework to analyze S-function efficiently. The frame is used to calculate the probability that given input differences lead to given output differences, such as the additive differential probability of XOR, the XOR differential probability of modular addition, the additive differential probability of the following sequence of operations: addition, bit rotation and XOR [12].

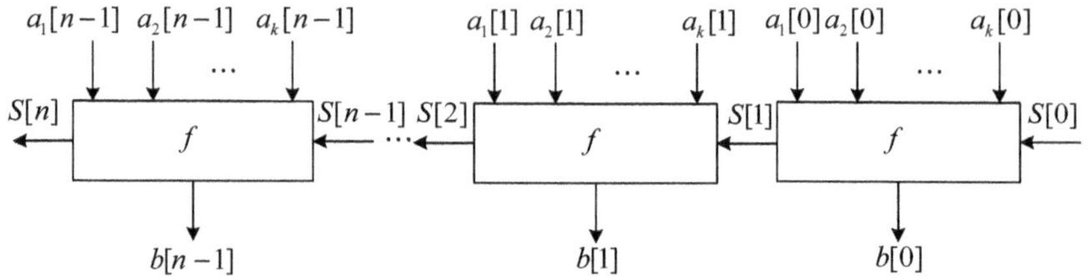

Fig. 3. Representation of an S-function

3 The Chosen-IV Attack of ZUC v1.5

The main idea of chosen-IV attack is to choose some differences in some IV bits and study the propagation of the differences during the initialization of the cipher. We concern on how many rounds are iterated until the differences in the memory cells become random and unpredictable. If the differences in some memory cells are still not random after the initialization stage of the cipher, the cipher is fragile to the chosen-IV attack.

3.1 The Definition of Differences

There are variables with different length in ZUC v1.5, so we define different differences for variables with different length. For 8-bit or 16-bit variables, the differences are defined to be the result of integer subtraction of variables. For 31-bit variables, the differences are defined to be the subtraction modulo $2^{31}-1$. For 32-bit variables, the differences are defined to be the subtraction modulo 2^{32}. We denote the differences of m-bit variables by Δ^m, $m = \{8, 16, 31, 32\}$.

3.2 An Chosen-IV Differential Path of ZUC v1.5

The 128-bit IV is divided into 16 parts. We introduce an 8-bit difference a in iv_3. The difference in iv_3 leads to a difference in s_3, the differences in other LFSR cells, R_1 and R_2 are set to be zero. It has been pointed out in the evaluation report [3] that when the 3-rd byte of the difference of IV's is chosen to be non-zero and the remaining 15 bytes are all 0, the differences in both the LFSR and the two memory cells R_1 and R_2 propagate slowest. We extend the differential in [3] to more rounds.

Given that $\Delta^8 iv_3 = a > 0$, then for given iv_3, we have $\Delta^{31} s_3 = a$. Table 1 gives the differences of the 16 cells of the LFSR and two registers R_1 and R_2 after i rounds of iteration ($i = 0, 1, \cdots, 25$). Differences marked with asterisk " $*$ " indeed have good randomness. It is easy to see that in round 1-3, the difference a dose not propagate and just shifts in the LFSR cells. After the third round, the difference a has shifted to s_0. The difference a propagates to s_{15} in the fourth round. In round 4-9, the differences in the LFSR cells are not affected by the S-box and linear transforms of the F function, so the variables b, c, d, e, f, g

may can be predicted with high probability. After 9 rounds of initialization, the differences propagate into R_2, which makes the differences difficult to be predicted. We have $\Delta^{31}s_0 = g$ after 18 rounds. The differences in the LFSR cells, R_1 and R_2 will be random after 25 rounds of initialization.

3.3 The Differential Properties of Operations in ZUC v1.5

In order to estimate the probability of the differential, we study the differential properties of the operations in ZUC. There are four operations, bit-reorganization, XOR and addition modulo 2^{32}, bit shift and the update of LFSR. We discuss these four operations in turn. In this section, we assume that the inputs of these operations are independent and uniformly distributed.

The Bit-Reorganization. Recall that there are three steps in bit-reorganization. First of all, we get the most significant 16 bits of a 31-bit variable. Assume that the difference of two 31-bit variables a_1 and a_2 is $\Delta^{31}a$. We denote the most significant important 16 bits of $\Delta^{31}a$ to be $(\Delta^{31}a)_H$, and the difference of a_{1H} and a_{2H} to be $\Delta^{16}a_H$. There are four possibilities for $\Delta^{16}a_H$:

$$\Delta^{16}a_H = (\Delta^{31}a)_H - u2^{16} + v, \qquad u, v \in \{0, 1\}.$$

The calculation of the probabilities to take these values are given in Appendix B.

The second step is to get the least significant 16 bits of a 31-bit variable. Assume that the difference of two 31-bit variables b_1 and b_2 is $\Delta^{31}b$. We denote the least significant 16 bits of $\Delta^{31}b$ to be $(\Delta^{31}b)_L$, and the difference of b_{1L} and b_{2L} to be $\Delta^{16}b_L$. There are four possibilities for $\Delta^{16}b_L$:

$$\Delta^{16}b_L = (\Delta^{31}b)_L - u2^{16} + v, \qquad u, v \in \{0, 1\}.$$

The calculation of the probabilities to take these values are similar to the calculation in Appendix B.

The third step is to concatenate two 16-bit words to a 32-bit word. We concern on the relationship between the differences of the 16-bit words and the difference of the 32-bit words. Assume that $\Delta^{32}x = a_{2H}\|b_{2L} - a_{1H}\|b_{1L} \mod 2^{32}$. Then we have:

$$\Delta x^{32} = \begin{cases} \Delta^{16}a_H \parallel \Delta^{16}b_L, & \text{if } \Delta^{16}a_H \geq 0 \text{ and } \Delta^{16}b_L \geq 0, \\ (\Delta^{16}a_H + 2^{16}) \parallel \Delta^{16}b_L, & \text{if } \Delta^{16}a_H < 0 \text{ and } \Delta^{16}b_L \geq 0, \\ (\Delta^{16}a_H + 2^{16} - 1) \parallel (\Delta^{16}b_L + 2^{16}), & \text{if } \Delta^{16}a_H \leq 0 \text{ and } \Delta^{16}b_L < 0, \\ (\Delta^{16}a_H - 1) \parallel (\Delta^{16}b_L + 2^{16}), & \text{if } \Delta^{16}a_H > 0 \text{ and } \Delta^{16}b_L < 0. \end{cases}$$

Assume that the probabilities of differential propagations in the first step and second step are p_1 and p_2, then probability of differential propagation in bit-reorganization is $p_1 * p_2$.

Table 1. The 24-Round Differential Path of ZUC v1.5

Round	s_{15}	s_{14}	s_{13}	s_{12}	s_{11}	s_{10}	s_9	s_8	s_7	s_6	s_5	s_4	s_3	s_2	s_1	s_0	R_1	R_2
0	0	0	0	0	0	0	0	0	0	0	0	0	a	0	0	0	0	0
1	0	0	0	0	0	0	0	0	0	0	0	0	0	a	0	0	0	0
2	0	0	0	0	0	0	0	0	0	0	0	0	0	0	a	0	0	0
3	0	0	0	0	0	0	0	0	0	0	0	0	0	0	0	a	0	0
4	b	0	0	0	0	0	0	0	0	0	0	0	0	0	0	0	0	0
5	c	b	0	0	0	0	0	0	0	0	0	0	0	0	0	0	0	0
6	d	c	b	0	0	0	0	0	0	0	0	0	0	0	0	0	0	0
7	e	d	c	b	0	0	0	0	0	0	0	0	0	0	0	0	0	0
8	f	e	d	c	b	0	0	0	0	0	0	0	0	0	0	0	0	0
9	g	f	e	d	c	b	0	0	0	0	0	0	0	0	0	0	0	*
10	*	g	f	e	d	c	b	0	0	0	0	0	0	0	0	0	*	*
11	*	*	g	f	e	d	c	b	0	0	0	0	0	0	0	0	*	*
12	*	*	*	g	f	e	d	c	b	0	0	0	0	0	0	0	*	*
13	*	*	*	*	g	f	e	d	c	b	0	0	0	0	0	0	*	*
14	*	*	*	*	*	g	f	e	d	c	b	0	0	0	0	0	*	*
15	*	*	*	*	*	*	g	f	e	d	c	b	0	0	0	0	*	*
16	*	*	*	*	*	*	*	g	f	e	d	c	b	0	0	0	*	*
17	*	*	*	*	*	*	*	*	g	f	e	d	c	b	0	0	*	*
18	*	*	*	*	*	*	*	*	*	g	f	e	d	c	b	0	*	*
19	*	*	*	*	*	*	*	*	*	*	g	f	e	d	c	b	*	*
20	*	*	*	*	*	*	*	*	*	*	*	g	f	e	d	c	*	*
21	*	*	*	*	*	*	*	*	*	*	*	*	g	f	e	d	*	*
22	*	*	*	*	*	*	*	*	*	*	*	*	*	g	f	e	*	*
23	*	*	*	*	*	*	*	*	*	*	*	*	*	*	g	f	*	*
24	*	*	*	*	*	*	*	*	*	*	*	*	*	*	*	g	*	*
25	*	*	*	*	*	*	*	*	*	*	*	*	*	*	*	*	*	*

XOR and Addition Modulo 2^{32}. The operation XOR is combined with the addition modulo 2^n. We write them by XA for short. In this section, we set that $n = 32$. Let $w = (x \oplus y) \boxplus z$. Given the input differences $\Delta^{32}x$, $\Delta^{32}y$, $\Delta^{32}z$ and inputs x_1, y_1, z_1, we can calculate the output difference $\Delta^{32}w$ as follows:

$$\begin{aligned}
&x_2 = x_1 \boxplus \Delta^{32}x, \quad y_2 = y_1 \boxplus \Delta^{32}y, \quad z_2 = z_1 \boxplus \Delta^{32}z, \\
&w_2 = (x_2 \oplus y_2) \boxplus z_2, \quad w_1 = (x_1 \oplus y_1) \boxplus z_1, \\
&\Delta^{32}w = w_2 \boxminus w_1,
\end{aligned}$$

where "$\boxminus$" denotes the subtraction modulo 2^{32}. The probability of the differential propagation from input differences α, β, γ to the output difference δ is defined as below:

$$adp^{XA}(\alpha, \beta, \gamma \to \delta) = \frac{\{(x_1, y_1, z_1) | \Delta^{32}x = \alpha, \Delta^{32}y = \beta, \Delta^{32}z = \gamma, \Delta^{32}w = \delta\}}{\{(x_1, y_1, z_1) | \Delta^{32}x = \alpha, \Delta^{32}y = \beta, \Delta^{32}z = \gamma\}}.$$

We consider the special situation that β and γ are equal to zero. This corresponds to the situation that the differences in LFSR cells do not lead to differences in

X_1 and X_2, see Table 1 in Sect. 3.2. So it is enough for us to consider this simple situation. Now the differences $\Delta^{32}w$ can be calculated as follows:

$$x_2 = x_1 \boxplus \Delta^{32}x, \tag{1}$$
$$y_2 = y_1, z_2 = z_1, \tag{2}$$
$$k_2 = x_2 \oplus y_2, k_1 = x_1 \oplus y_1, \tag{3}$$
$$w_2 = k_2 \boxplus z_2, \quad w_1 = k_1 \boxplus z_1, \tag{4}$$
$$\Delta^{32}w = w_2 \boxminus w_1 = k_2 \boxminus k_1 = \Delta^{32}k. \tag{5}$$

Apparently the output difference of XA is equal to the output difference of XOR. The probability of the differential propagation from input differences α to the output difference δ is defined as below:

$$adp^{XA}(\alpha, 0, 0 \to \delta) = \frac{\{(x_1, y_1)|\Delta^{32}x = \alpha, \Delta^{32}y = 0, \Delta^{32}k = \delta\}}{\{(x_1, y_1)|\Delta^{32}x = \alpha, \Delta^{32}y = 0\}}.$$

That is $adp^{XA}(\alpha, 0, 0 \to \delta) = adp^{\oplus}(\alpha, 0 \to \delta)$, where $adp^{\oplus}$ denotes the differential probability of XOR when differences are expressed using addition modulo 2^n, which has been well studied in [11]. Here we study $adp^{\oplus}$ with the restriction that one of the input difference be equal to zero.

Now we consider the pairs (x_1, y_1) satisfying $\Delta^{32}x = \alpha$, $\Delta^{32}y = 0$ and $\Delta^{32}k = \delta$. We denote $c_x[i]$ the i-th carry of the addition in (1) and $c_k[i]$ the i-th borrow of the subtraction in (5), $c_x[i] \in \{0, 1\}$, $c_k[i] \in \{0, -1\}$, $0 \le i \le n$. The formula (1)-(5) can be rewritten on a bit level:

$$x_2[i] = x_1[i] \oplus \Delta^{32}x[i] \oplus c_x[i], \tag{6}$$
$$c_x[i+1] = (x_1[i] + \Delta^{32}x[i] + c_x[i]) \gg 1, \tag{7}$$
$$y_2[i] = y_1[i], \tag{8}$$
$$k_1[i] = x_1[i] \oplus y_1[i], \tag{9}$$
$$k_2[i] = x_2[i] \oplus y_2[i], \tag{10}$$
$$\Delta^{32}k[i] = k_2[i] \oplus k_1[i] \oplus c_k[i], \tag{11}$$
$$c_k[i+1] = (k_2[i] - k_1[i] + c_k[i]) \gg 1. \tag{12}$$

Define the state $S[i] = (c_x[i], c_k[i])$. Then (6)-(12) correspond to the S-function

$$(\Delta^{32}k[i], S[i+1]) = f(x_1[i], y_1[i], \Delta^{32}x[i], S[i]), \quad 0 \le i < n.$$

The computation can be represented by graph, as described in [11]. For $0 \le i \le n$, we can represent every state $S[i]$ as a vertex in a graph. This graph consists of several subgraphs, containing only vertices $S[i]$ and $S[i+1]$ for some i. Set $\Delta^{32}x[i]$ to be equal to $\alpha[i]$. We loop over the all possible values of $(x_1[i], y_1[i], S[i])$, and for each combination, $\Delta^{32}k[i]$ and $S[i+1]$ are uniquely determined. We draw an edge between $S[i]$ and $S[i+1]$ if and only if $\Delta^{32}k[i]$ equals to $\delta[i]$. Every path from $S[0](0,0)$ to any of the four vertices of $S[32]$ corresponds to a pair (x_1, y_1) satisfying the differential propagation. So if we can count the number of paths, then we can calculate the probability of differential.

There are four possible subgraphs totally, see Fig. 4, corresponding to the four possible values of $\alpha[i]$ and $\delta[i]$. The pair $(\alpha[i], \delta[i])$ can be written as a 2-bit string, denoted by $w[i]$. Each $w[i]$ corresponds to one of the four possible subgraphs. Since the subgraphs are bipartite graphs, we can construct their biadjacency matrices $A_{w[i]} = [x_{k,j}]$. The element $x_{k,j}$ is the number of edges connecting vertices $j = S[i]$ and $k = S[i+1]$. All of the four biadjacency matrices are given in Appendix A. Then the number of paths between two vertices can be calculated by means of a matrix multiplication, that is,

$$adp^{XA}(\alpha, 0, 0 \to \delta) = 4^{-n} L A_{w[n-1]} \cdots A_{w[1]} A_{w[0]} C,$$

where $L = [1111]$ is a 1×4 matrix and $C = [1000]^T$ is a 4×1 matrix.

For a given α, now we consider how to choose δ such that $adp^{XA}(\alpha, 0, 0 \to \delta)$ reaches maximum. We call a state $S[i]$ to be uniform, if each vertex of $S[i]$ have only one path coming from $S[i-1]$.

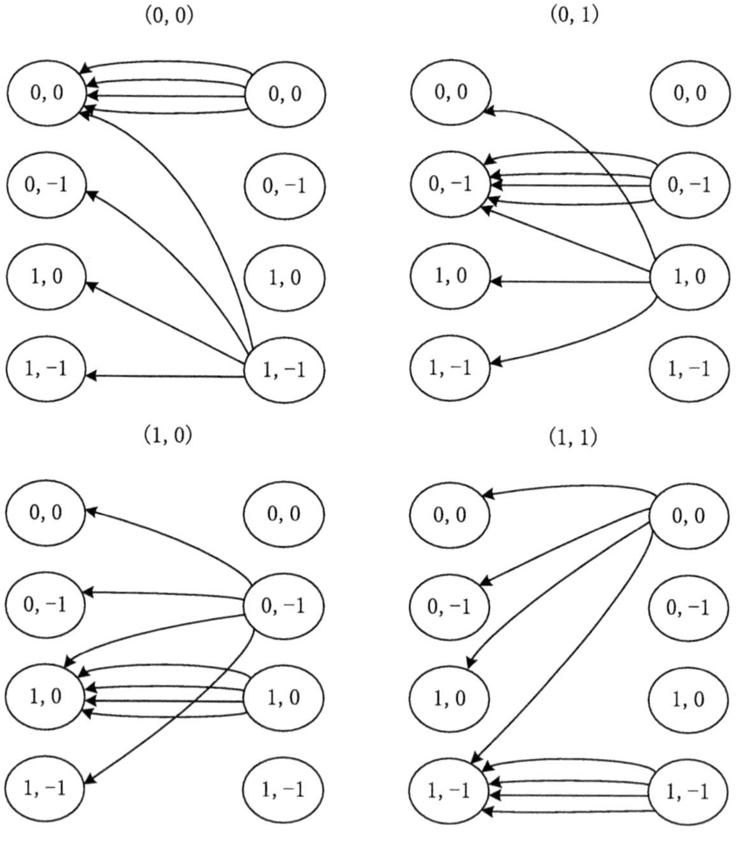

Fig. 4. Four Possible Subgraphs

Lemma 1. *For any n-bit integer $\alpha \neq 0$. Assume that $\alpha[i]$ is the least significant nonzero bit of α, that is $\alpha[0] = \alpha[1] = \cdots = \alpha[i-1] = 0$ and $\alpha[i] = 1$, $0 \leq i < n$. If $S[i+1]$ is uniform and $adp^{XA}(\alpha, 0, 0 \to \delta) \neq 0$, then $\alpha[j] = \delta[j]$, $0 \leq j \leq i$.*

Proof. If $\alpha[0] = 1$, that is $i = 0$, there must be $\delta[0] = 1$. Otherwise, no path starts from vertex $S[0](0,0)$ and $adp^{XA}(\alpha, 0, 0 \to \delta) = 0$. So $S[1]$ is uniform. If $\alpha[0] = 0$, there must be $\delta[0] = 0$. Now all edges starting from vertex $S[0](0,0)$ end to vertex $S[1](0,0)$. So if $\alpha[1] = 0$, there must have $\delta[1] = 0$ in order to draw edges between $S[1]$ and $S[2]$. If $\alpha[0] = \alpha[1] = \cdots = \alpha[i-1] = 0$, then we have $\delta[0] = \delta[1] = \cdots = \delta[i-1] = 0$. Now the paths start from vertex $S[0](0,0)$ and end to vertex $S[i](0,0)$. In order to extend the paths to $S[i+1]$, there must be $\delta[i] = 1$. Then $S[i+1]$ is uniform. ■

Lemma 2. *For any n-bit integer $\alpha \neq 0$. If $S[i+1]$ is uniform and adp^{XA} $(\alpha, 0, 0 \to \delta)$ reaches maximum, then $\alpha[j] = \delta[j]$ for all $i < j < n$ or $\alpha[j] \neq \delta[j]$ for all $i < j < n$.*

Proof. Note that each vertex of $S[i+1]$ has only one edge coming from $S[i]$, so $\delta[i+1] = 0$ has the same contribution to the differential probability as $\delta[i+1] = 1$. We hope that the paths reach $S[n]$ as more as possible. So we should choose the value of $\delta[j]$ such that the vertex of $S[j]$ connecting with most edges from $S[j-1]$ has edges connecting with $S[j+1]$ as more as possible. If $\delta[i+1] \neq \alpha[i+1]$, then for $i+1 < j < n$, there should be $\alpha[j] \neq \delta[j]$. If $\delta[i+1] = \alpha[i+1]$, then for $i+1 < j < n$, we have $\alpha[j] = \delta[j]$. ■

By Lemmas 1 and 2, we can get the following theorem:

Theorem 1. *Given an input difference $\alpha \neq 0$. Assume that $\alpha[i]$ is the least significant nonzero bit of α, that is, $\alpha[0] = \alpha[1] = \cdots = \alpha[i-1] = 0$ and $\alpha[i] = 1$, $0 \leq i < n$. The differential probability $adp^{XA}(\alpha, 0, 0 \to \delta)$ reaches maximum when $\delta = \alpha$ or $\delta = \alpha \oplus (2^n - 2^{i+1})$.*

Bit Shift. We get the most significant 31 bits of a 32-bit variable by shifting it one bit to the right. Assume that the difference of two 32-bit variables W_1 and W_2 is $\Delta^{32}W$. We denote the most significant important 31 bits of $\Delta^{32}W$ to be $(\Delta^{32}W)_{H31}$, and the difference of $u_1 = W_1 \gg 1$ and $u_2 = W_2 \gg 1$ to be $\Delta^{31}u$. There are four possibilities for $\Delta^{31}u$:

$$\Delta^{31}u = \{(\Delta^{32}w)_{H31}, (\Delta^{32}w)_{H31} \pm 1, (\Delta^{32}w)_{H31} - 2^{31} + 1\}.$$

The calculation of the probabilities to take these values are similar to the calculation in Appendix B.

The Update of LFSR. The difference propagation caused by update of LFSR is easy to deal with because the differences of LFSR cells are linear, that is,

$$\begin{aligned}\Delta^{31}s_{16} = 2^{15}\Delta^{31}s_{15} + 2^{17}\Delta^{31}s_{13} + 2^{21}\Delta^{31}s_{10} + 2^{20}\Delta^{31}s_4 \\ +(1+2^8)\Delta^{31}s_0 + \Delta^{31}u \bmod (2^{31}-1).\end{aligned} \tag{13}$$

3.4 The Probability of the Differential

We choose the difference a with hamming weight 1. Then we can get that $b = (a \lll 8) + a$ with probability 1 by (13). We predict the differences c, d, e, f, g according to the properties of operations given in Sect. 3.3 and then estimate the differential probability by two sets of 2^{28} pairs (k, iv). In the first set, all pairs (k, iv) are randomly chosen. And in the second set, all ks in pairs (k, iv) are identical, and ivs are distinct and randomly chosen. The test results show that the probabilities of the differential trail in two test sets are almost identical and are listed in Table 2.

Table 2. The Differential and Differential Probability when $a = 1$

	a	b	c	d	e	f	g
Difference(hex)	01	00000101	00808000	40808100	03030081	0343c585	68070585
Probability($\log_2$)	0	0	-0.011	-6.340	-10.134	-16.967	-23.476

We try to extend our differential trial to more rounds. When we begin with the 1-st round, or the 2-nd round, or the 3-rd round, and backtrack in the reverse order, the steps for a good difference propagation could be added by 1-3 rounds. If the differential before key loading is

$$(0, 0, 0, 0, 0, 0, 0, 0, 0, a, 0, 0, 0, \delta, 0, 0; 0, 0)$$

and $2^{20}a + (1 + 2^8)\delta \mod (2^{31} - 1) = 0$. Then after 3 rounds, the differential will be

$$(0, 0, 0, 0, 0, 0, 0, 0, 0, 0, 0, 0, a, 0, 0, 0; 0, 0).$$

Unfortunately, we can not find such δ.

4 Conclusion

In this paper we analyze the initialization stage of ZUC v1.5. First of all, we study the differential properties of operations in ZUC v1.5, including the bit-reorganization, XOR and addition modulo 2^n, bit shift and the update of LFSR. And then we give a differential trail of ZUC v1.5 which covers 24 rounds of the initialization stage with probability $2^{-23.48}$. We fail to extend this differential trail to more rounds.

Acknowledgement. The authors are grateful to the anonymous reviewers for their constructive comments.

A Matrices for adp^{XA}

$$\mathbf{A_{00}} = \begin{pmatrix} 4\,0\,0\,1 \\ 0\,0\,0\,1 \\ 0\,0\,0\,1 \\ 0\,0\,0\,1 \end{pmatrix}, \mathbf{A_{01}} = \begin{pmatrix} 0\,0\,1\,0 \\ 0\,4\,1\,0 \\ 0\,0\,1\,0 \\ 0\,0\,1\,0 \end{pmatrix}, \mathbf{A_{10}} = \begin{pmatrix} 0\,1\,0\,0 \\ 0\,1\,0\,0 \\ 0\,1\,4\,0 \\ 0\,1\,0\,0 \end{pmatrix}, \mathbf{A_{11}} = \begin{pmatrix} 1\,0\,0\,0 \\ 1\,0\,0\,0 \\ 1\,0\,0\,0 \\ 1\,0\,0\,4 \end{pmatrix}.$$

B The Calculation of Differential Probability in Bit-Reorganization

In this section, we assume that a_1 and a_2 are n-bit integers and take values from the set $\{1, 2, \cdots, 2^n - 1\}$. We denote the most significant h bits of a_i and the least significant l bits of a_i by a_{ih} and a_{il} respectively, that is $a_i = a_{ih} \| a_{il}$, where $0 < h < n$, $l = n - h$, $i = 1, 2$. The difference of a_1 and a_2 is defined to be the subtraction modulo $2^n - 1$ of a_1 and a_2, denoted by $\Delta^n a$. The difference of a_{1h} and a_{2h} is defined to be the integer subtraction of a_{1h} and a_{2h}, denoted by $\Delta^h a_h$. Assume that we have known $\Delta^n a$, we want to predict $\Delta^h a_h$.

By the property of modulo addition, the difference of a_1 and a_2 can be written as follows:

$$\begin{aligned} \Delta^n a &= (a_{1h} - a_{2h} + u2^h - v) \| (a_{1l} - a_{2l} + v2^l - u) \\ &= (\Delta^h a_h + u2^h - v) \| (a_{1l} - a_{2l} + v2^l - u), \qquad u, v \in \{0, 1\}. \end{aligned}$$

The boolean variable u takes value according to the integer subtraction of a_1 and a_2. If $a_1 < a_2$, we should borrow from the position n, then we have $u = 1$. The boolean variable v takes value according to the integer subtraction of the least significant l bits of a_1 and a_2. If $a_{1l} < a_{2l}$, we should borrow from the position l and then we have $v = 1$. More explicitly, we have

$$(u, v) = \begin{cases} (0, 0), & \text{if } a_{1h} \geq a_{2h} \text{ and } a_{1l} \geq a_{2l}, \\ (1, 0), & \text{if } a_{1h} < a_{2h} \text{ and } a_{1l} > a_{2l}, \\ (0, 1), & \text{if } a_{1h} > a_{2h} \text{ and } a_{1l} < a_{2l}, \\ (1, 1), & \text{otherwise.} \end{cases} \tag{14}$$

Obviously, $\Delta^h a_h$ and $\Delta^n a$ have the following relationship:

$$\Delta^h a_h = (\Delta^n a)_h - u2^h + v, \qquad u, v \in \{0, 1\},$$

where $(\Delta^n a)_h$ denotes the most important h bits of $\Delta^n a$ and the least important l bits of $\Delta^n a$ is denoted by $(\Delta^n a)_l$.

So there are four possible values for $\Delta^h a_h$. We can get the probability that (u, v) take certain value by counting the pairs of (a_1, a_2) which satisfy the corresponding condition in (14) and the following equation:

$$a_1 - a_2 \mod (2^n - 1) = \Delta^n a. \tag{15}$$

First of all, the number of pairs of (a_1, a_2) satisfying (15) is $2^n - 1$.

The pairs of (a_1, a_2) satisfying (15) and the first condition in (14) can be calculated as follows:

$$\begin{aligned}
&\sharp\{(a_1,a_2)|a_1 - a_2 \mod (2^n-1) = \Delta^n a, a_{1h} \geq a_{2h}, a_{1l} \geq a_{2l}\} \\
&= \sharp\{(a_1,a_2)|a_1 - a_2 = \Delta^n a, a_{1h} \geq a_{2h}, a_{1l} \geq a_{2l}\} \\
&= \sharp\{(a_1,a_2)|a_{1h} - a_{2h} = (\Delta^n a)_h, a_{1l} - a_{2l} = (\Delta^n a)_l, a_{1h} \geq a_{2h}, a_{1l} \geq a_{2l}\} \\
&= (2^h - (\Delta^n a)_h) \times (2^l - (\Delta^n a)_l) - 1.
\end{aligned}$$

Similarly, the pairs satisfying (15) and the second condition in (14) can be calculated as follows:

$$\begin{aligned}
&\sharp\{(a_1,a_2)|a_1 - a_2 \mod (2^n-1) = \Delta^n a, a_{1h} < a_{2h}, a_{1l} > a_{2l}\} \\
&= \sharp\{(a_1,a_2)|a_1 - a_2 + 2^n - 1 = \Delta^n a, a_{1h} < a_{2h}, a_{1l} > a_{2l}\} \\
&= \sharp\{(a_1,a_2)|a_{1h} - a_{2h} + 2^h = (\Delta^n a)_h, a_{1l} - a_{2l} - 1 = (\Delta^n a)_l\} \\
&= (\Delta^n a)_h \times (2^l - (\Delta^n a)_l - 1).
\end{aligned}$$

The pairs satisfying (15) and the third condition in (14) can be calculated as follows:

$$\begin{aligned}
&\sharp\{(a_1,a_2)|a_1 - a_2 \mod (2^n-1) = \Delta^n a, a_{1h} > a_{2h}, a_{1l} < a_{2l}\} \\
&= \sharp\{(a_1,a_2)|a_1 - a_2 = \Delta^n a, a_{1h} > a_{2h}, a_{1l} < a_{2l}\} \\
&= \sharp\{(a_1,a_2)|a_{1h} - a_{2h} - 1 = (\Delta^n a)_h, a_{1l} - a_{2l} + 2^l = (\Delta^n a)_l\} \\
&= (2^h - (\Delta^n a)_h - 1) \times (\Delta^n a)_l.
\end{aligned}$$

The pairs satisfying (15) and the fourth condition in (14) can be calculated by subtracting the number of pairs satisfying the first three conditions from $2^n - 1$.

References

1. ETSI/SAGE: Specification of the 3GPP Confidentiality and Integrity Algorithms 128-EEA3 & 128-EIA3, Document 2: ZUC Specification, Version 1.5 (January 4, 2011), `http://gsmworld.com/documents/EEA3_EIA3_ZUC_v1_5.pdf`
2. ETSI/SAGE: Specification of the 3GPP Confidentiality and Integrity Algorithms 128-EEA3 & 128-EIA3, Document 1: 128-EEA3 and 128-EIA3 Specification, Version 1.5 (January 4, 2011), `http://gsmworld.com/documents/EEA3_EIA3_specification_v1_5.pdf`
3. ETSI/SAGE: Specification of the 3GPP Confidentiality and Integrity Algorithms 128-EEA3 & 128-EIA3, Document 4: Design and Evaluation Report, Version 1.3 (January 18, 2011), `http://gsmworld.com/documents/EEA3_EIA3_Design_Evaluation_v1_3.pdf`
4. Englund, H., Johansson, T., Sönmez Turan, M.: A Framework for Chosen IV Statistical Analysis of Stream Ciphers. In: Srinathan, K., Rangan, C.P., Yung, M. (eds.) INDOCRYPT 2007. LNCS, vol. 4859, pp. 268–281. Springer, Heidelberg (2007)
5. Fischer, S., Khazaei, S., Meier, W.: Chosen IV Statistical Analysis for Key Recovery Attacks on Stream Ciphers. In: Vaudenay, S. (ed.) AFRICACRYPT 2008. LNCS, vol. 5023, pp. 236–245. Springer, Heidelberg (2008)

6. Biham, E., Shamir, A.: Differential Cryptanalysis of DES-like Cryptosystems. In: Menezes, A., Vanstone, S.A. (eds.) CRYPTO 1990. LNCS, vol. 537, pp. 2–21. Springer, Heidelberg (1991)
7. Li, J.: Improved Differential Paths on ZUC. Appear in the First International Workshop on ZUC Algorithm (December 2010)
8. Li, J.: Differential analysis of ZUC. Appear in the Second International Workshop on ZUC Algorithm and Related Topics (June 2011)
9. Sun, B., Tang, X., Li, C.: Preliminary Cryptanalysis Results of ZUC. Appear in the First International Workshop on ZUC Algorithm (December 2010)
10. Wu, H.: Cryptanalysis of the Stream Cipher ZUC in the 3GPP Confidentiality & Integrity Algorithms 128-EEA3 & 128-EIA3. Appear at the sump session in ASIACRYPT (2010)
11. Mouha, N., Velichkov, V., De Cannière, C., Preneel, B.: The Differential Analysis of S-Functions. In: Biryukov, A., Gong, G., Stinson, D.R. (eds.) SAC 2010. LNCS, vol. 6544, pp. 36–56. Springer, Heidelberg (2011)
12. Velichkov, V., Mouha, N., De Cannière, C., Preneel, B.: The Additive Differential Probability of ARX. In: Joux, A. (ed.) FSE 2011. LNCS, vol. 6733, pp. 342–358. Springer, Heidelberg (2011)

Algebraic Cryptanalysis of the Round-Reduced and Side Channel Analysis of the Full PRINTCipher-48

Stanislav Bulygin[1,2] and Johannes Buchmann[1,2]

[1] Center for Advanced Security Research Darmstadt - CASED
Mornewegstraße 32, 64293 Darmstadt, Germany
{johannes.buchmann,Stanislav.Bulygin}@cased.de
[2] Technische Universität Darmstadt, Department of Computer Science
Hochschulstraße 10, 64289 Darmstadt, Germany
buchmann@cdc.informatik.tu-darmstadt.de

Abstract. In this paper we analyze the recently proposed lightweight block cipher PRINTCipher. Applying algebraic methods and SAT-solving we are able to break 8 rounds of PRINTCipher-48 and 9 rounds under some additional assumptions with only 2 known plaintexts faster than brute force. We show that it is possible to break the full 48-round cipher by assuming a moderate leakage of internal state bits or even just Hamming weights of some three-bit states. Such a simulation side-channel attack has practical complexity.

Keywords: Algebraic cryptanalysis, SAT-solving, PRINTCipher, MiniSAT, CryptoMiniSAT.

1 Introduction

The target of this paper is the lightweight block cipher PRINTCipher proposed in 2010 at CHES [1]. PRINTCipher proposes a security solution for low cost devices such as RFID tags. In particular, PRINTCipher aims to facilitate secure usage of integrated circuit (IC) printing for RFID tags. An interesting feature of such "printing" is that a tag can obtain key dependent circuitry already at the manufacturing phase. As a result, the authors of [1] propose to use the same key for all rounds, but implementing key-dependent S-Boxes that may be "printed" when a tag is manufactured. In cryptographic sense, PRINTCipher pushes even further the limits of lightweight block cipher design set by such ciphers as PRESENT [2] and KATAN/KTANTAN-family [3]. In fact the authors state that e.g. PRINTCipher-48, which is claimed to provide 80-bit security, may be implemented with almost three times less gate equivalents (GEs) than the 80-bit version of PRESENT.

Naturally, the question of security for such an extremely lightweight design arises. The first cryptanalytic results known is the differential cryptanalysis of 22 rounds of the cipher using the entire code-book [4]. A much more powerful

D. Lin, G. Tsudik, and X. Wang (Eds.): CANS 2011, LNCS 7092, pp. 54–75, 2011.

result is presented in [5]. In this paper the authors show that there exist a non-negligible portion (2^{52} for PRINTCipher-48 with 80 bit keys and 2^{102} for PRINTCipher-96 with 160 bit keys) of weak keys. For these keys the cipher is easily broken with only several plaintext pairs, which makes the attack practical for these weak keys. Note, however, that the invariant coset attack of [5] may be easily overcome by adjusting the round counter appropriately, see Section 2.3 of [5]. Therefore, analyzing PRINTCipher still is of interest and serves further and deeper understanding of lightweight design principles, which PRINTCipher is using extensively.

Note that although the IC printing technology itself is "in its infancy", side channel security is still of theoretical interest. The preprint [6] addresses, in particular, security of PRINTCipher against the fault injection. There the authors show that by introducing faults into just one nibble position at some last rounds and having 12–24 effective faulty samples it is possible to dramatically reduce the key space to be searched.

In this paper we propose algebraic cryptanalysis of round-reduced PRINTCipher-48 using SAT-solving. Algebraic cryptanalysis combined with SAT-solving has become a popular tool in analyzing stream (e.g. [7]) and block ciphers (e.g. [8,9,10]). We show that we are able to break 8 rounds of the cipher having only 2 known plaintext pairs faster than the brute force. We are able to break 9 rounds assuming knowledge of 50 key bits out of 80. The success of this attack (i.e. it is faster than the brute force) is around 3/4. Note that although the number of rounds we can break is rather low, the data complexity of these attacks is minimal: exactly two known plaintexts. As indicated in [11], it is important to consider low data complexity attacks on block ciphers due to possible applications of block ciphers other than encryption. Note that it may very well be possible that a linear or differential attack on 8 rounds would be actually faster than the algebraic attack from this paper. Still it would require many more plaintext/ciphertext pairs, which could be hard to get in practice.

Moreover we provide side-channel analysis of PRINTCipher-48. This part comes in line with the research on algebraic methods in side channel analysis, cf. e.g. [12,13]. We show that assuming knowledge of internal state bits at round 4, we are able to break the full 48 rounds of PRINTCipher-48 in practical time. For example, assuming knowing 6 output bits of 2 S-Boxes (3 bits each) at round 4 for 12 known plaintext/ciphertext pairs, we estimate total key recovery in less than 2 hours on average. Relaxing the knowledge of internal state bits by knowing only Hamming weights of inputs and outputs to the 2 first S-Boxes at round 4, we estimate recovery of the 80-bit key in less than 3 days on average using 12 pairs of plaintext/ciphertext.

The paper is structured as follows. Section 2 gives a brief description of PRINTCipher-48. Then in Section 3 we elaborate on algebraic representations of the S-Boxes. Section 4 is a brief overview of the SAT-solving technique and conversion techniques used in this paper. Section 5 provides analysis of tools and conversion techniques that are further used in Section 6 to attack the cipher. Section 6 has several subsections. Section 6.1 gives results of algebraic cryptanalysis of the round-reduced PRINTCipher-48. Section 6.2 observes an interesting property of the cipher, which is then applied to side channel analysis in Section 6.3. We conclude in Section 7 and outline there some open problems. At the end we have three appendices. Appendix A contains details on the S-Box equations. Appendix B provides experimental evidence for the claims made in Section 6.2. Finally Appendix C discusses potential applications of the idea in Section 6.2 to the attack on the full PRINTCipher-48 with 48 rounds.

The main results of the paper are contained in Sections 6.1 and 6.3.

2 PRINTCipher

PRINTCipher is a substitution-permutation network. The cipher is largely inspired by the lightweight block cipher PRESENT [2]. The main differences with PRESENT is absence of the key schedule (all round keys are the same and are equal to the master key) and key-dependent S-Boxes. PRINTCipher comes in two variations: PRINTCipher-48 encrypts 48 bits blocks with an 80 bit key and has 48 rounds, PRINTCipher-96 encrypts 96 bit blocks with a 160 bit key and has 96 rounds. Here we present a short overview of the cipher, referring the reader to [1] for a more detailed description and analysis. In this paper we concentrate on the smaller version PRINTCipher-48.

The encryption process of PRINTCipher-48 is organized as in Algorithm 1.

Algorithm 1. Encryption with PRINTCipher-48
Require:
- 48-bit plaintext p
- 80-bit key $k = (sk_1, sk_2)$, where sk_1 is 48 bits and sk_2 is 32 bits

Ensure: 48-bit ciphertext c

Begin
$state := p$
for $i = 1, \dots, 48$ **do**
  $state := state \oplus sk_1$
  $state := Perm(state)$
  $state := state \oplus RC_i$
  $state := SBOX(state, sk_2)$
end for
$c := state$
return c
End

Some comments to Algorithm 1 follow. The linear diffusion layer $Perm$ implements a bit permutation similar to PRESENT. It allows full dependency on plaintext and key bits already at round 4. RC_i for $i = 1, \dots, 48$ is a 6-bit round counter that is placed in the last two 3-bit nibbles. The S-Box layer $SBOX$ is a layer of 16 3-bit S-Boxes numbered $0, \dots, 15$, where each S-Box is chosen according to the value of two corresponding bits of the subkey sk_2. Therewith there are 4 possible S-Boxes at each position called V_0, V_1, V_2, V_3 in [1]. One may also consider such an S-Box as a composition of a key dependent bit permutation that acts on groups of 3 bits and then followed by the layer of fixed S-Boxes, each one being a 3-bit S-Box with the truth table as in Table 1. This S-Box, called V_0 in [1], is preceded by a key-dependent permutation defined by Table 2. In Table 2 the three input bits are permuted according to the two consecutive key bits from the subkey sk_2 called a_0 and a_1. Figure 1 provides an illustration for one encryption round of PRINTCipher-48.

Table 1. Truth table for the S-Box V_0

x	0	1	2	3	4	5	6	7
$S[x]$	0	1	3	6	7	4	5	2

Table 2. Key depended permutation

a_1	a_0	Permutation
0	0	(0,1,2)
0	1	(0,2,1)
1	0	(1,0,2)
1	1	(2,1,0)

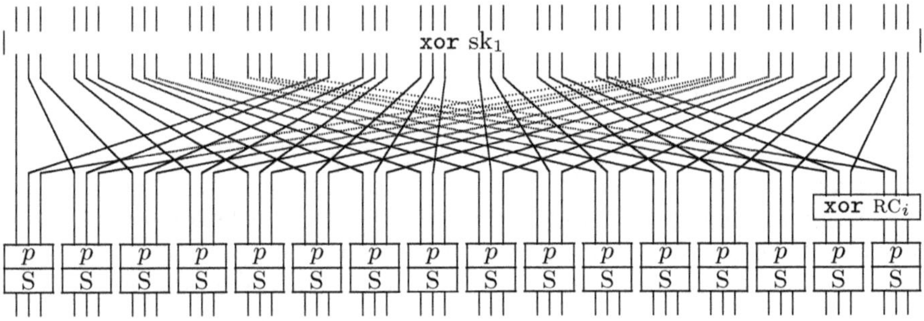

Fig. 1. Round function of PRINTCipher-48, cf. Figure 1 of [1].

As has been already mentioned in Section 1, in [5] the authors propose an attack on PRINTCipher that works for a large set of weak keys. In Section 2.3 of [5] it is indicated that it is possible to overcome the attack by spreading the round counter to e.g. the last three three-bit entries instead of two. We apply this change and use the modified cipher when working with the full 48-round cipher in Sections 6.2 and 6.3 and also Appendices B and C.

In our attacks we extensively use another notion defined for bit-permutations in SP-networks. This is the notion of a *low diffusion trail*. The effect of a low diffusion trail was observed already for PRESENT and was used as a tool for statistical saturation attacks [14]. An observation on low diffusion trails is also made by the authors of [1] in Section 4.4. One fixes certain S-Boxes and computes how many bits stay within the fixed S-Boxes after the application of the bit permutation layer. Low diffusion trails correspond to those S-Boxes with the highest count. We number S-Boxes from 0 to 15 from left to right. For example, looking at Figure 1 one sees that if we fix S-Boxes 0,1 and 5, then 5 out of 9 output bits at round i are input bits to these three S-Boxes at round $i+1$. In fact positions $\{0,1,5\}$ provide a low diffusion trail for 3 S-Boxes. We cite now Table 1 from [1] which provides examples of low diffusion trails and append it with some other trails that we used for the number of positions 2, 9 and 10, see Table 3. Note that many other low diffusion trails exist for every given number of fixed S-Boxes.

Table 3. Low diffusion trails for PRINTCipher-48

# of S-Boxes in trail	Example trail	# bits in the trail	Ratio
2	$\{0,1\}$	2	2/6
3	$\{0,1,5\}$	5	5/9
4	$\{0,1,5,15\}$	7	7/12
5	$\{4,10,12,14,15\}$	9	9/15
6	$\{0,1,2,5,6,7\}$	12	12/18
7	$\{3,8,9,10,11,13,15\}$	14	14/21
8	$\{0,1,4,5,10,12,14,15\}$	18	18/24
9	$\{0,1,2,3,5,10,11,14,15\}$	20	20/27
10	$\{0,1,3,4,5,10,11,12,14,15\}$	24	24/30

3 Algebraic Description

In this section we give an algebraic description of the PRINTCipher-48. As usual in algebraic cryptanalysis, the most interesting part is to describe non-linear transformations, the S-Boxes. In the case of PRINTCipher we have key dependent S-Boxes, so we have to treat this situation.

First of all, let us describe the system of equations for r rounds of PRINTCipher-48. We are working in the Boolean ring with variables $X_i := (X_{i,j}), 0 \le i \le r, 0 \le j \le 47$ *input* variables; $Y_i := (Y_{i,j}), 1 \le i \le r, 0 \le j \le 47$ *output* variables; $K_1 := (K_{1,j}), 0 \le j \le 47$ *key-xor* variables; $K_2 := (K_{2,j}), 0 \le j \le 31$ *key-permutation* variables. Now schematically the system of equations $Sys(p,c)$ is as follows

$$\begin{cases} X_0 = p, \\ X_i = Y_{i-1} \oplus K_1, 1 \le i \le r, \\ Y_i = SBOX(Perm(X_i) \oplus RC_i, K_2), 1 \le i \le r, \\ Y_r = c. \end{cases}$$

Variables X_0 and Y_r do not actually appear, since they are replaced by the values of the corresponding plaintext/ciphertext pair (p,c). Our task is to find solutions for the variables K_1 and K_2 which then give us the key k that encrypts p to c. The variables K_i represent the subkey sk_i for $i = 1, 2$. Some remarks on notation. It is clear that in the second line we simply do bitwise XOR of the variables. The third line assumes that the variables are grouped by three and two corresponding key-permutation variables. So $SBOX$ is the concatenation of 16 non-linear maps. The map $Perm$ is a bit permutation, see Algorithm 1 and Figure 1. For example for the first S-Box we have equations:

$$Y_{i,j} = SBOX_j(X_{i,0}, X_{i,16}, X_{i,32}, K_{2,0}, K_{2,1}), 0 \leq j \leq 2,$$

since RC_i is always zero at those positions. Another important observation is that since the key length of PRINTCipher-48 is larger than its block length, we actually need two plaintext/ciphertext pairs to uniquely determine the key. So actually we need two systems $Sys(p_1, c_1)$ and $Sys(p_2, c_2)$, where the $X-$ and $Y-$variables are different the K_1-, K_2-variables are the same.

Our next goal is to look closely at the $SBOX$ map. This map is a concatenation of 16 smaller maps $sbox(sk_2)$, which depend in the key sk_2. Note that each $sbox$ can be considered as a composition of the key-dependent permutation $SP : \mathbb{F}_2^3 \times \mathbb{F}_2^2 \to \mathbb{F}_2^3$ and the fixed S-Box V_0, see Section 2. We need to find equations that describe the map SP. Introducing local notation, let $X = (X_0, X_1, X_2)$ be the input variables $A = (A_0, A_1)$ be the key-permutation variables and $W = (W_0, W_1, W_2)$ be the output variables. Then SP can be described as $W = SP(X, A) = (f_{ij}(A)) \cdot X^T$, where for each assigned value for $A = (A_0, A_1)$, the matrix $F := (f_{ij}(A))$ is a permutation matrix according to the table in Section 2. By using interpolation techniques, we find that f_{ij}'s are polynomials in A-variables of degree at most 2. The matrix is as follows:

$$F = \begin{pmatrix} A_0 + 1 & A_0A_1 + A_0 & A_0A_1 \\ A_0A_1 + A_0 & A_0 + A_1 + 1 & A_0A_1 + A_1 \\ A_0A_1 & A_0A_1 + A_1 & A_1 + 1 \end{pmatrix}.$$

Therefore, we obtain cubic equations for the map SP: quadratic in key-permutation variables and linear in the input variables. Appendix A.1 contains the equations describing this map.

Now the S-Box V_0 is a "classical" S-Box and standard techniques may be used to describe it. Let $Y = (Y_0, Y_1, Y_2)$ be the output variables of the S-Box, so that $Y = V_0(W)$. Since V_0 is a $\mathbb{F}_2^3 \to \mathbb{F}_2^3$ S-Box, we may obtain explicit quadratic equations for the output variables in terms of the input variables:

$$\begin{aligned} Y_0 &= W_1W_2 + W_0, \\ Y_1 &= W_0W_2 + W_0 + W_1, \\ Y_2 &= W_0W_1 + W_0 + W_1 + W_2. \end{aligned}$$

In total there are 14 linearly independent quadratic equations that describe V_0.

Of course we may write equations for the $sbox$ in terms of X-, A-, and Y-variables only, eliminating the W-variables. Therewith, we obtain quartic equations for $sbox$: quadratic in the key-permutation and the input variables. We list these equations in Appendix A.2.

4 SAT Techniques for Algebraic System Solving

Using SAT-solvers in cryptanalysis is a quite recent trend. The idea is to translate an algebraic representation of a cryptographic primitive, such as block or stream cipher, into a satisfiability problem from propositional logic. Therewith, the problem of solving an algebraic system of equations in a Boolean ring is replaced with finding a satisfiability assignment of variables in a logical formula (or proving that such an assignment does not exist). Whereas equations from an algebraic representation are in the algebraic normal form (ANF), the resulting satisfiability problem is usually in the conjunctive normal form (CNF). In the block cipher cryptanalysis there are examples of using SAT-solving in analyzing KeeLoq [8], DES [9], PRESENT and KTANTAN [15] whereas for stream ciphers an example target is Grain [7]. Although the problem of finding a satisfiability assignment of variables, as well as solving non-linear systems, is NP-hard in general, highly tuned SAT-solvers can solve particular instances occurring in practice amazingly fast. The two most commonly used SAT-solvers in the cryptographic context are MiniSAT2 [16] and recently proposed CryptoMiniSAT2 [17]. The latter uses ideas of MiniSAT adding a lot of new heuristics and the possibility to handle long XOR chains more efficiently.

There exist several methods of transforming an ANF of a polynomial into the CNF form. In our experiments we rely on the two methods: the one due to Courtois, Bard, Jefferson [18] and another one that is based on the truth tables [19,7].

We would like to make a remark about estimating complexity of SAT-solving. Whereas memory consumption of SAT-solvers is very modest (at least compared to algebraic solvers that compute Gröbner bases), estimating time complexity is really an issue. The problem is that due to highly heuristic and randomized nature of SAT solvers (at least those based on the DPLL algorithm, such as MiniSAT and CryptoMiniSAT) the execution time varies very significantly for each trial run and also for runs of similar problems. Sometimes one observes a difference of factor 1000. Since variance for time measurements is so high, it is hard to make estimates even after many runs. Still, in this paper we apply the method of averaging running times of 100 trials for each experiment in question. This is similar to approaches of other authors in the field, e.g. [10,15]. Our experiments with some 1000-trial instances suggest that 100 is quite accurate in terms of the average time. A more advanced way of estimating time complexity would be statistical hypothesis testing. We set using such an estimation tool as a future work.

4.1 Conversion Techniques

Conversion Due to Bard, Courtois, Jefferson

ANF to CNF conversion of Bard, Courtois, Jefferson first represents separate monomials occurring in an algebraic system as a conjunction of clauses, which in turn are disjunctions of variables or their negations. Then one adds new variables corresponding to these monomials, so that all equations become linear in these new variables. The next task is to represent the newly obtained linear polynomials as conjunctions of clauses. The problem here is that representing a linear relation (a XOR-chain) needs exponentially many clauses in the number of variables. Therefore, one "cuts" a linear relation in several ones by introducing new variables. Representing one cut XOR-chain then needs relatively few clauses.

As an example, consider an equation

$$xyz + xy + yz + xz = 0.$$

We introduce new variables $a := xyz, b := xy, c := yz, d := xz$. Then a term $a = xyz$ is described by a logical formula

$$(x \vee \bar{a}) \wedge (y \vee \bar{a}) \wedge (z \vee \bar{a}) \wedge (a \vee \bar{x} \vee \bar{y} \vee \bar{z}).$$

Similarly one can write the terms b, c and d. Resulting linear relation $a+b+c+d = 0$ may be written as

$$(\bar{a} \vee b \vee c \vee d) \wedge (a \vee \bar{b} \vee c \vee d) \wedge (a \vee b \vee \bar{c} \vee d) \wedge (a \vee b \vee c \vee \bar{d}) \wedge$$
$$(\bar{a} \vee \bar{b} \vee \bar{c} \vee d) \wedge (\bar{a} \vee \bar{b} \vee c \vee \bar{d}) \wedge (\bar{a} \vee b \vee \bar{c} \vee \bar{d}) \wedge (a \vee \bar{b} \vee \bar{c} \vee \bar{d}).$$

As we see, the latter formula contains $4 + 4 = 8$ clauses. Similarly, describing a XOR-chain with n variables (n is odd) needs

$$\binom{n}{1} + \binom{n}{3} + \cdots + \binom{n}{n} = 2^{n-1}$$

clauses. Therefore, the concept of cutting makes sense. We could cut the number of variables XORed in our formula above as $e+a+b = 0, f+c+e = 0, g+d+f = 0$ introducing new variables e, f, and g. So one needs 4 clauses for each of the three equations. The cutting number [18] defines, how many variables are in one XOR-chain. In the example given, it is 3. Usually cutting numbers 4–7 are used.

In our experiments we call this conversion technique `BCJ`. More details on this technique may be found in [18]. The convertor is implemented in SAGE computer algebra system [20] by Martin Albrecht and Mate Soos [21].

Conversion Based on the Truth Table

Whereas in the `BCJ` method the number of variables in a logical formula is usually quite large compared to the initial number of variables in the ANF of a polynomial, the following method preserves this number. The method, which we call `TruthTable` in this paper, is based on writing a CNF for the given Boolean function by examining the evaluation (truth) table of the function. We work with a specific implementation of the method described in [19] and implemented by Michael Brickenstein [22]. The core of the method dates back to Karnaugh [23] and is also used in cryptographic context in [7]. Note that the problem of long XOR chains also appears here. So one might want to apply the cutting technique first and then the truth table method for resulting shorter polynomials.

As an example, the polynomial $f = xyz + xy + yz + xz$ from above with the method of [19] has the CNF

$$(x \vee \bar{y} \vee \bar{z}) \wedge (\bar{x} \vee y \vee \bar{z}) \wedge (\bar{x} \vee \bar{y} \vee z).$$

So contains only 3 clauses and 3 variables.

5 Optimal Tools and Strategies for the Attacks

Algebraic cryptanalysis is a lot about trying out certain heuristics. This is especially the case when applying SAT-solving due to highly heuristic nature of the method. In this section we summarize strategies and tools that we used and appear optimal in the case of PRINTCipher-48. This summary comes as a result of experiments with different ideas and tools. Due to space limitations, we do not provide these experiments here, only the results. A thorough discussion appears in the full version of the paper [24].

First of all, the question arises, which algebraic representation for the cipher to choose, since, S-Boxes have different representations. Next, we have to see which ANF-to-CNF conversion technique is better for our application.

The question of bit guessing is an important issue that has to be analyzed. It is known that in order to be able to solve a large system of equations coming from a block cipher, it is a good practice to first guess some key bits and then try to solve easier systems. In particular, we solve one easier system if we assume that we are given the correct key bits. In the cryptanalytic context this means that we need to make sure that if we are given (or guessed) g out of k key bits, the resulting system solving is faster than the brute force of the remaining key space of size 2^{k-g}, see Section 6.1 for more details. Since PRINTCipher does not have a key schedule, i.e. the same key is used in all rounds of the cipher, it is particularly plausible to apply this approach. Moreover, by guessing enough key bits, we are able to bring our timings to some feasible figures. This in turn will give us an opportunity to make estimations of the attack complexity. So, guessing key bits is important. But now we have to answer certain questions in order to understand where and how to guess bits optimally.

- Since the bits in a round are grouped in groups of three bits before applying the S-Box transformations, does it make a difference to guess the key bits in such groups of three or random positions are as good?
- Does it make a difference which groups/positions do we choose for guessing?
- Is it better to first guess the key sk_2 and then some parts of the key sk_1 or one should guess parts of these keys?
- Related to the question above, latter part: should guessing in the keys sk_1 and sk_2 be synchronized (that is, for example, locations of chosen groups of three for sk_1 match guessing the corresponding key-permutation bits of sk_2) or the guesses may be done independently?
- Last, but not least is the question which solver is better for our purposes: MiniSAT2 or CryptoMiniSAT2.

Below we summarize answers to the question we stated above based on our experiments.

- We observed that a false key bits guess results in a slow-down of approximately factor 2 on average compared to correct guesses.
- We always guess (or assume given) key bits (both sk_1 and sk_2) at positions grouped in groups of three, according to S-Box positions.
- Grouped positions correspond to low diffusion trails, see end of Section 2.
- We use explicit equations for the S-Boxes of the form $W = SP(X, A), Y = V_0(W)$.
- Performance of CryptoMiniSAT2 is somewhat better than the one of MiniSAT2.
- `TruthTable` conversion is superior to `BCJ` (different cutting numbers were used for comparison).

Guessing in groups of three bits seems quite natural: we fix an S-Box which makes analysis easier. On the other hand, a non-trivial task is to optimally choose these groups of three bits. Using low diffusion trails turned out to be particularly efficient. The fact that in such a trail many bits stay within the set of chosen S-Boxes, reflects favorably on reducing overall algebraic complexity of our attacks. In fact, the fewer S-Boxes are in a trail, the larger is the efficiency gain compared to the brute force attack. For 9 and 10 S-Boxes in a trail we obtained a speed-up around factor 10 with low diffusion trails compared to random selections of three-bit groups, see Section 6.1. For many instances we were not able to obtain any figures in a reasonable time when using 2 randomly chosen S-Boxes in a trail, whereas using a low diffusion trail made it possible, see Section 6.2. This shows that one object may be beneficial for different cryptanalytic methods, e.g. low diffusion trails are used in statistical saturation attacks and are a handy instrument for algebraic attacks. See also Remark 1.

Computing environment: We do our computations on a server with 4 AMD processors each having 4 cores working at 2.3 GHz. The computations are run using one core. The server has 128 GB of RAM and is running under Debian Linux version 4.3.5-4 Linux kernel version 2.6.32-5-amd64. For our computations we use SAGE version 4.3.3, MiniSAT version 2.0 beta and CryptoMiniSAT version 2.9.0.

Experimental setting: For estimating one timing figure we compute the average time for 100 trial runs. The time in tables is given in seconds, unless is indicated otherwise.

6 Algebraic Analysis of PRINTCipher-48

6.1 Attack on Round-Reduced PRINTCipher-48

In this subsection, basing on our choices from Section 5, we attack the round-reduced PRINTCipher-48. First of all, we have to define what do we mean by an attack here and when such an attack succeeds. In all our attacks a certain number of key bits will be fixed of guessed. We want to compare our method with the brute force attack. Similarly to previous works in the field of algebraic cryptanalysis of block ciphers, e.g. [10,15], we employ the following way of comparing the methods. First, let us assume that g bits out of 80 are given to an attacker, i.e. he/she knows correct values of those bits. Let $T_{correct}$ be an average time an attacker needs to solve a True-instance (corresponds to a correct key guess) using choices from Section 5. Now let $T_{enc,r}$ be the time one needs to check with a trial encryption one key from the remaining 2^{80-g} in a reduced cipher with r rounds. We assume that for such an encryption one needs 2 CPU cycles per round: one for the linear layer and one for the S-Box layer. To translate this figure to seconds, we observe that our experiments are done on a machine with the CPU frequency 2.3 GHz. So we may assume that 1 CPU cycle is done in $\frac{1}{2.3} \cdot 10^{-9} \approx 0.4 \cdot 10^{-9}$ seconds. We have then $T_{enc,r} = 2 \cdot 0.4 \cdot 10^{-9} \cdot r$. Note that this is a rather optimistic estimate.[1] The time complexity for a brute force attacker is therefore

$$T_{bf,r} = 2^{80-g} T_{enc,r}/2 = 2^{80-g} \cdot 0.4r \cdot 10^{-9}$$

seconds. We divide over 2, since on average an attacker has to search only through a half of the remaining key space.[2] For a successful algebraic attack we require that

$$T_{correct,r} < 2^{80-g} \cdot 0.4r \cdot 10^{-9}. \tag{1}$$

Now assume that bits of the key are not given. So we have to guess some g key bits. As we have noted in Section 5, the time needed to solve a False-instance, i.e. the one with an incorrect key guess, is higher than the one for a correct guess. Therefore, in this case our time complexity is determined by the average time needed to solve a False-instance: $T_{wrong,r}$. Assuming that all False-instances are approximately the same in difficulty, we have that the attack is successful if

$$T_{wrong,r} < 2T_{bf,r} = 2^{80-g} \cdot 0.8r \cdot 10^{-9}. \tag{2}$$

Since both methods run into a correct solution on average two times faster than the full search, we put back the factor 2, taken away in (1). For both algebraic and the brute force attack, the memory complexity is quite low, although being higher for the former one. Still, memory is not a bottleneck for SAT-solving, so we ignore it in our analysis. Our first result:

Result 1. *We are able to attack 8 rounds of PRINTCipher-48 with 2 known plaintext/ciphertext pairs faster than the brute force.*

We guess at $g = 45$ bits of the key and the low diffusion trail with 9 S-Boxes, see Table 3. We guess subkey bits that correspond to chosen S-Boxes at positions $\{0, 1, 2, 3, 5, 10, 11, 14, 15\}$. Two random plaintexts are taken for each of the 100 trials. Table 4 summarizes the results.

[1] We do not take into account other possibilities to speed up the brute force attacks, such as using FPGAs or GPUs.

[2] In fact for a portion of 2^{-32} keys one needs to do the second encryption, since 2^{32} out of 2^{80} keys are expected to encrypt a given plaintext to a given ciphertext. Due to negligible probability of this event, we do not take this into account.

Table 4. Attack on 8 rounds, 2 known plaintexts

Guess	Time for CryptoMiniSAT	Time for MiniSAT
Correct	24	30
Wrong	61	60

In order to see that we are really faster than the brute force, in Table 5 we provide the figures of LHS and RHS of (2) for $g = 45$. Here timings for (Crypto)MiniSAT2 correspond to False-instances. As we noted, it is necessary to assess time complexity of the attack. Both solvers are doing the job.

We could not entirely break the 9 rounds, but still we were very close and could achieve partial success. When doing experiments for 9 rounds we took a look at several low diffusion trails with 10 S-Boxes. It turns out that some of them are better than the others. Table 6 summarizes experimental results for 9 rounds made with CryptoMiniSAT2. So we see that the trail $\{0, 1, 3, 4, 5, 10, 11, 12, 14, 15\}$ yields the best results. Still we are not able to get below the threshold for both False- and True- instances.

Table 5. Comparing to brute force

	Time for CMS	Time for MS	Time for brute force
$g = 45$	61	60	239

Table 6. Attack on 9 rounds, 2 known plaintexts, different trails

Trail	Guess	Time	Time for brute force
$\{0, 1, 2, 3, 4, 5, 6, 7, 11, 12\}$	Correct	19.5	4.2
$\{0, 1, 2, 3, 4, 5, 6, 7, 11, 12\}$	Wrong	43.9	8.4
$\{0, 1, 3, 4, 5, 10, 11, 12, 14, 15\}$	Correct	5.8	4.2
$\{0, 1, 3, 4, 5, 10, 11, 12, 14, 15\}$	Wrong	16.3	8.4
$\{2, 3, 4, 6, 7, 8, 9, 11, 12, 13\}$	Correct	14.3	4.2
$\{2, 3, 4, 6, 7, 8, 9, 11, 12, 13\}$	Wrong	25.2	8.4
$\{3, 4, 8, 9, 10, 11, 12, 13, 14, 15\}$	Correct	34.4	4.2
$\{3, 4, 8, 9, 10, 11, 12, 13, 14, 15\}$	Wrong	56.3	8.4

Nevertheless, for True-instances we may ask a question, what is the portion of measurements that fall below the brute force time. Table 7 shows these figures for all four trails. We see that with probability around 3/4 we are able to solve a True-instance faster than the brute force. Therewith,

Result 2. *For 9 rounds, having 2 known plaintext/ciphertext pairs and knowing 50 key bits at three-bit positions* $\{0, 1, 3, 4, 5, 10, 11, 12, 14, 15\}$, *we may find remaining 30 bits faster than brute force with probability around 75%.*

Table 7. Attack on 9 rounds, 2 known plaintexts, percentage of runs faster than the brute force

Trail	%
$\{0,1,2,3,4,5,6,7,11,12\}$	22
$\{0,1,3,4,5,10,11,12,14,15\}$	78
$\{2,3,4,6,7,8,9,11,12,13\}$	56
$\{3,4,8,9,10,11,12,13,14,15\}$	30

Remark 1. In fact when using low diffusion trails, as we have seen, only looking at how many bits stay within a trail is not enough. Actually, one should consider how many bits one gets "for free" when having key bits corresponding to the trail nibble positions, plaintext and ciphertext bits. Here one can observe that, obviously, 3 known input/output bits to an S-Box yield 3 output/input bits to the previous or next rounds. Less obviously it is possible to show that knowing 2 input/output bits yields knowledge of 1 output/input bit on average. Having these observations, one could improve the guessing strategy for example as follows. Since knowing 2 bits yields 1 on average and knowing 3 yields 3, then if one knows 2 input bits to an S-Box it may be worthwhile to guess at the remaining bit. Therewith, a 2-fold slow-down due to guessing will be payed off by a 4-fold speed-up due to knowing 2 additional internal state bits. This should result in a speed up of a factor of 2. All in all, one may investigate how known internal state bits propagate to yield more known bits, maybe with some probability. We do not elaborate on this issue in this paper, although this is something to consider and that may have impact for other methods, like statistical saturation attacks.

6.2 Additional Bits at Round Four

In the previous subsection we presented our approach of attacking the PRINTCipher-48 based on SAT-solving. We could attack 8–9 rounds with it. In this subsection we show that it is possible to attack the full 48 rounds of the cipher by just having moderate amount of additional information. We are able to do so, after discovering quite an interesting property of the cipher. Namely, by just knowing values of output bits at round 4 of several known plaintext/ciphertext pairs and some key bits, it is possible to recover all remaining key bits fast. In this subsection we proceed as follows. We first provide results on attacking the full 48 rounds of the cipher with guessing bits at rounds 4 and 5. Then we try to explain, what makes round 4 so special by providing additional data and observations.

In this section we use CryptoMiniSAT2 for our experiments. Note that for the analysis of the full 48 rounds we adjusted the round counter of the cipher, so that the attack of [5] that works for many weak keys, does not work anymore, cf. Section 2.3 of [5]. In fact, this adjustment does not play any role for our attacks. It will become clear why, after we look deeply into the situation of using the 4th round for guessing.

Table 8 summarizes the results. As before, we guess key bits and internal state bits at rounds 4 and 5 using low diffusion trails. We use parameter x, meaning how many three-bit groups are fixed in a trail, so that we guess at $5x$ key bits and $3xp$ intermediate state bits, where p is the number of known plaintext/ciphertext pairs. As in the previous section we do averaging over 100 trials.

Table 8. Recovering key bits by having some output bits at rounds 4 or 5

r	4	4	4	4	4	4	4	4	5	5	5	5
p	10	10	12	12	2	2	2	2	2	2	2	2
x	2	2	2	2	7	7	6	6	7	7	6	6
Correct guess	True	Wrong	True	Wrong	True	Wrong	True	Wrong	True	Wrong	True	Wrong
Time	116	62	10	8	1.7	0.4	295	75	10	11	934	1426

One of the conclusions that may be drawn from the table is that assuming 6 output bits of round 4 known for 10/12 pairs and knowing only 10 key bits, the remaining 70 key bits can be recovered very fast. The same is true for 21 bits at round 4 for only 2 pairs and knowing 35 key bits: remaining 45 key bits are recovered in a matter of a second. We are still able to solve quite fast by having 18 bits at round 4 for 2 pairs. We could get some results also by inserting values at round 5, but the performance appears to be inferior to what we can get from round 4.

Knowing internal key bits is a strong assumption. Still, it may be feasible if one thinks about side channel security. See Section 6.3 for this. Some thoughts on using the idea of this section to a potential attack on the full cipher are given in Appendix C. Note also that Table 8 suggests that False-instances are solved faster for $r = 4$, which is different from results of the previous sections. This is due to the fact that wrongly assigned bits propagate from round 1 upwards and from round 4 downwards, having a meeting point in the middle. Quite often (in fact in around half of the cases) this leads to an immediate contradiction at this meeting point, thus a computation is extremely fast.

Now we want to shed some light on why guessing/fixing at round 4 is so effective; why cannot we do the same for, e.g., rounds 3 or 5 with comparable efficiency. The main reason appears to be the following. Let us consider the variety $V(r, p, C)$ of keys that after r rounds encrypt p known plaintexts to some ciphertexts that have given values at positions from C ($C \subset \{0, \dots, 47\}$). It turns out that for $r = 3$ this variety is rather large due to incomplete diffusion (recall that the "full" diffusion for PRINTCipher-48 is attained after 4 rounds). On the other hand, from round 4 and on, this variety drastically looses in size. This means that having enough pairs p we cut off many keys starting already at round 4. The second issue is how easy is it to sample elements from $V(r, p, C)$. For $r = 3$ this is extremely easy (fast), for $r = 4$ is somewhat slower, but still quite fast. Starting from round 5, it gets harder, at least in some cases of interest. Therefore, roughly speaking, round 4 yields an optimal trade-off between the number of elements in the variety $V(r, p, C)$ and hardness of computing its elements.

In principle, one may run the following attack for the case one knows internal state bits at round r at positions C. Choose $r < 48$. Let $t(r, p, C)$ be the (average) time needed to compute one element from $V(r, p, C)$. Then one may simply run an exhaustive search on $V(r, p, C)$ and this will have time complexity approximately

$$(t(r, p, C) + T_{enc,48}) \cdot |V(r, p, C)|,$$

where $T_{enc,48} \ll t(r, p, C)$ is the time needed for one trial encryption to check the correctness of the key guess for the full 48 rounds using p known plaintext/ciphertext pairs. Therefore, finding a trade-off as explained above is important. The elements of $V(r, p, C)$ may be sampled again using SAT-solving techniques in assumption that one gets independent random solutions from $V(r, p, C)$ each time. Considering highly heuristic and randomized nature of SAT-solvers this assumption does not appear unreasonable. Results from Table 8 are obtained by just assigning values at round 4 and running CryptoMiniSAT2 for the obtained systems having 48 rounds. Still we expect the solver to take advantage of the above trade-off: inserted values at round 4 give an opportunity to narrow down the search a SAT-solver takes. We provide some experimental evidence in Appendix B that support the above reasoning.

6.3 Side Channel Analysis of the Full PRINTCipher-48

Using results of Section 6.2, we can show that it is possible to break the full PRINTCipher-48 with practical complexity by using side channel analysis. We first take a strong assumption of knowing internal state bits at round 4, namely some output bits of round 4. Then we consider more usual scenario in side channel analysis, when only Hamming weights of inputs and outputs of certain S-Boxes at round 4 are known. We show that due to properties of S-Boxes in PRINTCipher, we are still able to do our attack with only moderate increase in complexity.

Assuming knowledge of some information about internal states of a cipher we end up in the side channel scenario. That is, we assume that we can recover certain internal bits, which in principle can be done by examining an implementation of the cipher's encryption function. Note that actual recovering of bits may be a very challenging task. In this section we only consider a simulation side channel attack, not an actual one, i.e. we simply assume some internal bits to be given.

From the point of view of side channel analysis, it seems reasonable to concentrate on the case, when as few nibbles as possible are spotted for input/output values. In this respect, using $x = 2$ with the nibble positions $\{0, 1\}$ seems appropriate. Note that in this case we are guessing at $5x = 10$ bits of the key: 6 at the subkey sk_1 and 4 at the subkey sk_2. In Table 8 we presented timings for recovering remaining 70 key bits. We also considered the situation, when our key guess was wrong and how much time does it take to realize this. As has been noted, solving False-instances, i.e. when key guess and guess of internal state bits is wrong is actually somewhat faster than solving a True-instance. Since wrong guesses constitute the majority in the search through 10 key bits, we take the time of solving False-instances, T_{wrong}, as a measurement. We multiply the values from Table 8 by 2^9 (i.e. the half of 2^{10} on average) to get an estimated average time for the full key recovery for the entire cipher with 48 rounds. The results are in Table 9. Note that neither here nor further in the subsection did we do actual computations to recover the unknown key.

Table 9. Side channel attack assuming knowledge of internal state bits at round 4: 6 bits for each of the p pairs are known

p	T_{wrong}, sec.	Total Time, hours
10	62	8.8
12	8	1.2

Result 3. *For the full cipher, by having values of* 6 *output bits at round 4 that correspond to the nibble positions* $\{0, 1\}$, *we are able to recover the key on average in approximately 9 hours by using 10 known pairs of plaintext/ciphertext and in less than 2 hours with 12 pairs.*

Note that we need knowledge of $6 \cdot 10 = 60$ and $6 \cdot 12 = 72$ internal bits respectively.

A weaker assumption we take next is the knowledge of Hamming weights instead of exact knowledge of values. Namely, we assume knowledge of Hamming weights of inputs and outputs to the S-Boxes at positions 0 and 1. Small size of the S-Boxes actually does not make our task much more difficult than before. Examining the S-Boxes $V_0, \ldots, V_3$ of PRINTCipher it may be seen that input and output weights uniquely determine the values of inputs and outputs, except for the case when both weights are equal to 2. In this case, for each S-Box, there are exactly two pairs of input/output that yield weight 2. Therefore, such an ambiguity happens with probability 1/4 for each of the S-Boxes $V_0, \ldots, V_3$. By guessing key bits at three-bit positions $\{0, 1\}$, we fix the two S-Boxes at these positions. If for an S-Box the input and output weight is equal to 2, we have to choose one of the two potential values of inputs/outputs. In the case of other weights, we uniquely determine the values, since S-Boxes are fixed now. Therefore, on average, we expect 1/4 of the guesses for the three-bit positions $\{0, 1\}$ at round 4 to be wrong for a correct key guess. Therefore, for one guess of the 10 key bits, we will have to run our attack with p pairs on average around

$$2^{2 \cdot p/4 - 1} = 2^{p/2 - 1}$$

times, since $p/2$ out of $2p$ outputs will have wrong values on average and we expect to attain correct values after $2^{p/2-1}$ times. For estimating the total attack complexity, we use the formula

$$T_{sc-attack} = T_{wrong} \cdot 2^{9+p/2-1} = T_{wrong} \cdot 2^{8+p/2}.$$

This means multiplying the values in Table 9 by $2^{p/2-1}$. Table 10 shows the results.

Table 10. Side channel attack assuming knowledge of Hamming weights of input/output to S-Boxes 0 and 1 at round 4

p	T_{wrong}, sec.	$T_{sc-attack}$, days
10	62	12
12	8	3

Result 4. *For the full cipher we are able to find the key by only knowing Hamming weights for input/output to S-Boxes at positions 0 and 1 at round 4 on average in less than two weeks for* $p = 10$ *and in around 3 days for* $p = 12$.

Note that although the number of bits or Hamming weights we require to be leaked is not so small, the place to get these leaks is localized to only two three-bit positions at round 4, rather than smeared all over the cipher's encryption function. Therewith, it may be possible to better exploit these leaks in practice.

7 Conclusion and Future Work

In this paper we considered algebraic attacks on PRINTCipher-48 with SAT-solving as a tool. We showed that it is possible to attack 8–9 rounds of the cipher using only 2 known plaintexts. Next, we observed an interesting fact on high diffusion / low algebraic complexity at round 4, which enabled us to give a simulation side channel attack on the full cipher with 48 rounds with practical complexity.

As future work we may identify the following points.

- Obtaining symbolically relations among internal states for several plaintext/ciphertext pairs in line with [25].
- Apply the above point to chosen plaintext scenario and also to ideas described in Appendix C to try to attack the full cipher.
- Provide more accurate and statistically sound estimations for time complexity. This work is not specific to PRINTCipher and is of great importance for the entire field of applying SAT-solvers to cryptanalysis.
- More research should be put in studying the effect of using low diffusion trails as indicated in Remark 1. Therewith more advanced guessing techniques could emerge.
- Try to adapt the performance of a SAT solver, e.g. CryptoMiniSAT, to breaking this specific cipher; improving guessing and conflict finding of a solver.

Acknowledgements. The first author is supported by the German Science Foundation (DFG) grant BU 630/22-1. Special thanks go to Denise Demirel for writing a considerable portion of the equation generator for PRINTCipher. The author is grateful to Martin Albrecht for numerous discussions and also for sharing his source codes for relevant matters. Thanks to Mohamed Saied Emam Mohamed for reading this manuscript. The author is also thankful to Mate Soos for assisting with using CryptoMiniSAT.

References

1. Knudsen, L., Leander, G., Poschmann, A., Robshaw, M.J.B.: PRINTCipher: A Block Cipher for IC-Printing. In: Mangard, S., Standaert, F.-X. (eds.) CHES 2010. LNCS, vol. 6225, pp. 16–32. Springer, Heidelberg (2010)
2. Bogdanov, A.A., Knudsen, L.R., Leander, G., Paar, C., Poschmann, A., Robshaw, M.J.B., Seurin, Y., Vikkelsoe, C.: PRESENT: An Ultra-Lightweight Block Cipher. In: Paillier, P., Verbauwhede, I. (eds.) CHES 2007. LNCS, vol. 4727, pp. 450–466. Springer, Heidelberg (2007)
3. De Cannière, C., Dunkelman, O., Knežević, M.: KATAN and KTANTAN — A Family of Small and Efficient Hardware-Oriented Block Ciphers. In: Clavier, C., Gaj, K. (eds.) CHES 2009. LNCS, vol. 5747, pp. 272–288. Springer, Heidelberg (2009)
4. Abdelraheem, M.A., Leander, G., Zenner, E.: Differential Cryptanalysis of Round-Reduced PRINTcipher: Computing Roots of Permutations. In: Joux, A. (ed.) FSE 2011. LNCS, vol. 6733, pp. 1–17. Springer, Heidelberg (2011)
5. Leander, G., Abdelraheem, M.A., AlKhzaimi, H., Zenner, E.: A Cryptanalysis of PRINTCipher: The Invariant Coset Attack. In: Rogaway, P. (ed.) CRYPTO 2011. LNCS, vol. 6841, pp. 206–221. Springer, Heidelberg (2011)
6. Zhao, X., Wang, T., Guo, S.: Fault Propagate Pattern Based DFA on SPN Structure Block Ciphers using Bitwise Permutation, with Application to PRESENT and PRINTCipher. ePrint, `http://eprint.iacr.org/2011/086.pdf`
7. Soos, M.: Grain of Salt - An Automated Way to Test Stream Ciphers through SAT Solvers, `http://www.msoos.org/grain-of-salt`
8. Courtois, N.T., Bard, G.V., Wagner, D.: Algebraic and Slide Attacks on Keeloq. In: Nyberg, K. (ed.) FSE 2008. LNCS, vol. 5086, pp. 97–115. Springer, Heidelberg (2008)
9. Courtois, N.T., Bard, G.V.: Algebraic Cryptanalysis of the Data Encryption Standard. In: Galbraith, S.D. (ed.) Cryptography and Coding 2007. LNCS, vol. 4887, pp. 152–169. Springer, Heidelberg (2007)
10. Bard, G.V., Courtois, N.T., Nakahara Jr, J., Sepehrdad, P., Zhang, B.: Algebraic, AIDA/Cube and Side Channel Analysis of KATAN Family of Block Ciphers. In: Gong, G., Gupta, K.C. (eds.) INDOCRYPT 2010. LNCS, vol. 6498, pp. 176–196. Springer, Heidelberg (2010)
11. Bouillaguet, C., Derbez, P., Dunkelman, O., Keller, N., Fouque, P.-A.: Low Data Complexity Attacks on AES. ePrint, `http://eprint.iacr.org/2010/633.pdf`
12. Renauld, M., Standaert, F.-X.: Algebraic Side-Channel Attacks. In: Bao, F., Yung, M., Lin, D., Jing, J. (eds.) Inscrypt 2009. LNCS, vol. 6151, pp. 393–410. Springer, Heidelberg (2010)
13. Renauld, M., Standaert, F.-X.: Combining Algebraic and Side-Channel Cryptanalysis against Block Ciphers. In: Proceedings of the 30th Symposium on Information Theory in the Benelux (2009)
14. Collard, B., Standaert, F.-X.: A Statistical Saturation Attack against the Block Cipher PRESENT. In: Fischlin, M. (ed.) CT-RSA 2009. LNCS, vol. 5473, pp. 195–211. Springer, Heidelberg (2009)
15. Albrecht, M.: Algorithmic Algebraic Techniques and their Application to Block Cipher Cryptanalysis. Ph.D. thesis. Royal Holloway, University of London, `http://www.sagemath.org/files/thesis/albrecht-thesis-2010.pdf`
16. Een, N., Sorensson, N.: An Extensible SAT-Solver. In: Giunchiglia, E., Tacchella, A. (eds.) SAT 2003. LNCS, vol. 2919, pp. 502–518. Springer, Heidelberg (2004)

17. Soos, M.: CryptoMiniSat – a SAT solver for cryptographic problems, `http://planete.inrialpes.fr/~soos/CryptoMiniSat2/index.php`
18. Bard, G.V.: Algebraic Cryptanalysis. Springer, Heidelberg (2009)
19. Brickenstein, M.: Boolean Gröbner bases – Theory, Algorithms and Applications, Logos Berlin (2010)
20. William Stein, S., et al.: SAGE Mathematics Software. The Sage Development Team (2008), `http://www.sagemath.org`
21. Albrecht, M., Soos, M.: Boolean Polynomial SAT-Solver, `http://bitbucket.org/malb/algebraic_attacks/src/tip/anf2cnf.py`
22. Brickenstein, M.: PolyBoRi's CNF converter, `https://bitbucket.org/malb/algebraic_attacks/src/013dd1b793e8/polybori-cnf-converter.py`
23. Karnaugh, M.: The map method for synthesis of combinational logic circuits. Transactions of American Institute of Electrical Engineers part I 72(9), 593–599 (1953)
24. Bulygin, S.: Algebraic cryptanalysis of the round-reduced and side channel analysis of the full PRINTCipher-48 (2011), `http://eprint.iacr.org/2011/287`
25. Albrecht, M., Cid, C., Dullien, T., Faugère, J.-C., Perret, L.: Algebraic Precomputations in Differential and Integral Cryptanalysis. In: Lai, X., Yung, M., Lin, D. (eds.) Inscrypt 2010. LNCS, vol. 6584, pp. 387–403. Springer, Heidelberg (2011)
26. Gomes, C.P., Sabharwal, A., Selman, B.: Model Counting. In: Handbook of Satisfiability, pp. 633–654. IOS Press (2009)

A Equations for Different Algebraic Representation of a PRINTCipher S-Box

A.1 Explicit Cubic Equations Describing the Key Dependent Permutation SP

$$
\begin{aligned}
W_0 &= A_0A_1X_1 + A_0A_1X_2 + A_0X_0 + A_0X_1 + X_0,\\
W_1 &= A_0A_1X_0 + A_0A_1X_2 + A_0X_0 + A_1X_1 + A_0X_1 + A_1X_2 + X_1,\\
W_2 &= A_0A_1X_0 + A_0A_1X_1 + A_1X_1 + A_1X_2 + X_2.
\end{aligned}
$$

A.2 Explicit Quartic Equations Describing the Key Dependent S-Box

$$
\begin{aligned}
&Y_0 = A_0A_1X_0X_1 + A_0A_1X_0X_2 + A_0A_1X_1 + A_0A_1X_2 + A_0X_0X_2 + A_0X_1X_2+\\
&+A_0X_0 + A_0X_1 + X_1X_2 + X_0,\\
&Y_1 = A_0A_1X_0X_1 + A_0A_1X_1X_2 + A_0A_1X_0 + A_0A_1X_1 + A_0X_0X_2 + A_0X_1X_2+\\
&+A_1X_0X_1 + A_1X_0X_2 + A_1X_1 + A_1X_2 + X_0X_2 + X_0 + X_1,\\
&Y_2 = A_0A_1X_0X_2 + A_0A_1X_1X_2 + A_1X_0X_1 + A_1X_0X_2 + X_0X_1 + X_0 + X_1 + X_2.
\end{aligned}
$$

B Experimental Evidence for Section 6.2

We support the reasoning in Section 6.2 with some experimental data. First we give some experimental values for $t(r, p, C)$ for different r, p and C. Table 11

Table 11. Time $t(r, p, C)$

r	p	C	$t(r, p, C)$
3	8	$\{0, 1\}$	0.06
3	10	$\{0, 1\}$	0.09
3	12	$\{0, 1\}$	0.12
3	16	$\{0, 1\}$	0.15
4	8	$\{0, 1\}$	0.31
4	10	$\{0, 1\}$	1.84
4	12	$\{0, 1\}$	3.38
4	16	$\{0, 1\}$	1.22
5	8	$\{0, 1\}$	4813
3	2	$\{0, 1, 2, 5, 6, 7\}$	< 0.02
4	2	$\{0, 1, 2, 5, 6, 7\}$	0.03
5	2	$\{0, 1, 2, 5, 6, 7\}$	0.25
3	2	$\{3, 8, 9, 10, 11, 13, 15\}$	< 0.02
4	2	$\{3, 8, 9, 10, 11, 13, 15\}$	0.04
5	2	$\{3, 8, 9, 10, 11, 13, 15\}$	1.42

summarizes the results. Abusing notation, we use three-bit positions and not the bit positions. Positions, as usual, correspond to low diffusion trails. The table confirms the claims made before, concerning values of $t(r, p, C)$. We observe quite fast timings for $r = 5, x = 6 - 7$, though.

A more complicated task is to estimate the number of elements in $V(r, p, C)$. One of the techniques ([15,26]) is estimating size of a variety by adding random linear relations on variables. The hope is that a random linear relation cuts off approximately a half of the variety in question. Therefore, if adding s random linear relations yields a solution (a satisfying assignment of variables) and $s + 1$ does not, then with certain probability we may claim that the variety size is 2^s. We applied this approach to estimate the values of $V(r, p, C)$. We added random XOR-chains that contain only key variables suggested by $\{0, \ldots, 48\} \setminus C$ (or $\{0, \ldots, 15\} \setminus C$ in the nibble notation), since these determine the rest of the variables. Also, since adding long XOR chains obstructs a SAT-solver from getting a solution in reasonable time, we have to work with chains of certain length. We chose length 7 for our experiments. The results are summarized in Table 12.[3] The numbers are obtained after averaging over 100 trials. As the table suggests, we get a serious reduction in the variety size when going from $r = 3$ to $r = 4$. We must note, however, that the method used above is not very accurate. By trying out sampling solutions from $V(r, p, C)$ as described in Section 6.2, we noticed that figures in Table 12 are underestimations of the actual values. This is especially true for the case $r = 3$. This may be explained by observing that since we do not yet have full diffusion after 3 rounds, the systems we consider are less random than those for $r \geq 4$. Therefore, the variety $V(3, p, C)$ is not that "homogeneous" and XOR-relations may stop the count too early. Still, we believe that these experimental results provide rough intuition and the basis for our claims made in Section 6.2.

Table 12. Variety size $|V(r, p, C)|$

r	p	C	$\log_2 \lvert V(r,p,C) \rvert$
3	10	$\{0, 1\}$	25
4	10	$\{0, 1\}$	11
5	10	$\{0, 1\}$	N/A
3	2	$\{3, 8, 9, 10, 11, 13, 15\}$	16
4	2	$\{3, 8, 9, 10, 11, 13, 15\}$	8
5	2	$\{3, 8, 9, 10, 11, 13, 15\}$	4

C Towards Cryptanalysis of the Full PRINTCipher-48

Here we investigate the potential of the method described in Section 6.2 to break the full cipher. Since the complexity of the method is not exponential in the number of rounds, but rather in the number of pairs used (i.e. number of internal state

[3] "N/A" means we were not able to get any results in a reasonable time.

bits to be guessed), this approach appears to be worth investigating. Whereas in Sections 6.2 and 6.3 we assumed information about intermediate values been given to us, for the cryptanalysis we have to guess this information. Note that if x three-bit positions are targeted for guessing and p known plaintext/ciphertext pairs are used, we need to guess $5x$ key bits and $3px$ intermediate bits; this sums up to $(3p+5)x$ bits. We estimate the attack complexity by the formula

$$T_{attack} = T_{wrong} \cdot 2^{(3p+5)x},$$

similarly to Section 6.3. Table 13 contains comparisons of the attack for different x and p and the brute force attack. Namely, we compute the ratio of expected time for our attack and the brute force. The table suggests that in the best case we are around 2^{17} slower than the brute force.

Table 13. Comparing our attack with brute force

p	x	$(3p+5)x$	$3px$	$\log_2(T_{attack}/T_{bf,48})$
10	2	70	60	20
12	2	82	72	29
2	7	77	42	21
2	6	66	36	17

The following observation gives a hope that the method may still be applicable. Note that in Section 6 we always assumed known plaintext/ciphertext pairs. Therewith intermediate state bits at round 4 may be assumed to be independent. So we really have to guess at $3px$ bits. It could be possible to apply chosen plaintext scenario with the goal to be able to compute certain intermediate bits via others that have been guessed at already. Ideally, we would like to have a possibility to compute intermediate bits at round 4 for $p-1$ pairs from those bits for just one plaintext/ciphertext. Clearly, plaintexts for the remaining $p-1$ pairs have to be chosen accordingly to allow such a relation. Other possible scenarios could be applied. It may be possible to compute intermediate bits for some $p-s$ pairs having those from s pairs. We may even be satisfied with computing values with some reasonably high probability. It is not clear, though, whether T_{wrong} will stay the same or increase under these additional dependencies. Alternatively, we could increase the number of known or chosen pairs to somehow spot those pairs that yield prescribed values at round 4.

Anyway, results of Table 13 suggest that if we were able e.g. to reduce the number of guessed internal bits from 60 to 39 in the case of $x=2, p=10$ or from 36 to 18 for $x=6, p=2$ we would have been better than the brute force. We set the problem of finding appropriate chosen plaintexts as an open problem.

EPCBC - A Block Cipher Suitable for Electronic Product Code Encryption

Huihui Yap[1,2], Khoongming Khoo[1,2], Axel Poschmann[2,⋆], and Matt Henricksen[3]

[1] DSO National Laboratories, 20 Science Park Drive, Singapore 118230
[2] Division of Mathematical Sciences, School of Physical and Mathematical Sciences Nanyang Technological University, Singapore
[3] Institute for Infocomm Research, A*STAR, Singapore
{yhuihui,kkhoongm}@dso.org.sg, aposchmann@ntu.edu.sg, mhenricksen@i2r.a-star.edu.sg

Abstract. In this paper, we present EPCBC, a lightweight cipher that has 96-bit key size and 48-bit/96-bit block size. This is suitable for Electronic Product Code (EPC) encryption, which uses low-cost passive RFID-tags and exactly 96 bits as a unique identifier on the item level. EPCBC is based on a generalized PRESENT with block size 48 and 96 bits for the main cipher structure and customized key schedule design which provides strong protection against related-key differential attacks, a recent class of powerful attacks on AES. Related-key attacks are especially relevant when a block cipher is used as a hash function. In the course of proving the security of EPCBC, we could leverage on the extensive security analyses of PRESENT, but we also obtain new results on the differential and linear cryptanalysis bounds for the generalized PRESENT when the block size is less than 64 bits, and much tighter bounds otherwise. Further, we analyze the resistance of EPCBC against integral cryptanalysis, statistical saturation attack, slide attack, algebraic attack and the latest higher-order differential cryptanalysis from FSE 2011 [11]. Our proposed cipher would be the most efficient at EPC encryption, since for other ciphers such as AES and PRESENT, it is necessary to encrypt 128-bit blocks (which results in a 33% overhead being incurred). The efficiency of our proposal therefore leads to huge market implications. Another contribution is an optimized implementation of PRESENT that is smaller and faster than previously published results.

Keywords: Electronic Product Code, EPC, PRESENT block cipher, RFID encryption, lightweight cryptography.

1 Introduction

The usage of tiny computing devices is gaining popularity in the consumer market, and they are becoming an integral part of a ubiquitous/pervasive

⋆ The research was supported in part by the Singapore National Research Foundation under Research Grant NRF-CRP2-2007-03, and by NTU SUG.

D. Lin, G. Tsudik, and X. Wang (Eds.): CANS 2011, LNCS 7092, pp. 76–97, 2011.

communications infrastructure. Thus it is crucial to employ well-designed lightweight cryptography for security purposes. A popular lightweight block cipher that fulfils this requirement is the 64-bit block cipher PRESENT with 80-bit/128-bit key [10]. PRESENT takes only $1000 - 1570$ GE to implement [10,44], is secure against differential/linear cryptanalysis (DC/LC) and is resistant against a slew of other block cipher attacks. As a consequence, PRESENT is currently under standardization within the upcoming ISO 29192 Standard on Lightweight Cryptography. In [9], the authors also generalized the PRESENT design concept to n-bit block size, in the context of designing hash functions. There was no description of a generic key schedule and no theoretical results on the security properties of the generalized PRESENT in [9]. In [33], Leander also defined small scale variants of PRESENT for the purpose of investigating the relationship between the running time of certain attacks and the number of rounds. He called these toy ciphers SMALLPRESENT-[n] whose block sizes are $4n$-bit. Their key scheduling algorithms produce $4n$-bit round keys from an 80-bit master key. In this paper, for ease of reference, we shall name the generalized PRESENT block cipher with block size n bits as "PR-n".

PRESENT only has 64-bit block size and this may not be suitable for applications which require lightweight encryption on a larger block size. One example is the upcoming Electronic Product Code (EPC), which is thought to be a replacement for bar codes using low-cost passive RFID-tags, and in its smallest form uses 96 bits as a unique identifier for any physical item [22]. A smaller block size of 64 bits (e.g. PRESENT) requires two consecutive encryptions. On the other hand, the use of a larger block size of 128 bits (e.g. AES) results in a truncation to 96 bits which wastes internal state and effort. Our intention is thus to design a lightweight and efficient 96-bit block cipher for EPC encryption which has huge market implications, and at the same time improves previous analysis of PRESENT for increased confidence in security.

We propose two variants of EPCBC: EPCBC(48,96) which has 48-bit block size and 96-bit key, and, EPCBC(96,96) which has 96-bit block size and 96-bit key. EPCBC(48,96) uses the PR-48 design for the main cipher structure and for the key schedule, it uses an 8-round variant-Feistel structure with 4-round PR-48 as the nonlinear function. EPCBC(96,96) uses the PR-96 design both for the main cipher structure and the key schedule.

The security of EPCBC(96,96) against DC and LC relies on that of PR-96 cipher structure. The DC and LC bounds can easily be inferred from that of PRESENT [10], because the results of PRESENT applies to PR-n for any $n \geq 64$. Our contribution for the analysis of EPCBC(96,96) is that we improve on the bounds of [10]. This allows us to deduce DC and LC bounds of EPCBC(96,96) which are tighter than the bounds obtained by applying the results of [10].

However, in proving the security of EPCBC(48,96) against DC and LC, the DC and LC bounds cannot be inferred from that of PRESENT [10] because the block size $n = 48$ is less than 64. Therefore, we prove new DC/LC bounds for PR-n when $n < 64$. Using these new bounds, we are able to prove the resistance of EPCBC(48,96) against DC and LC.

A recent class of powerful attacks against block ciphers are related-key differential attacks [5,6,4,21] which can break well established standards such as AES-128 and KASUMI. Although the practicality of these attacks is arguable (due to the difficulty in obtaining related keys), resistance against related-key differential attack is especially relevant when these block ciphers are used as hash functions in Davies-Meyer mode (e.g. see [9] and Section 2 of this paper). This is an important issue, since many designer of RFID security protocols assume a lightweight hash function to be available on the tag [1,26,35]. Our customized key schedule design ensures many active S-boxes in the key schedule when there is a non-zero key differential. Consequently and in contrast to PRESENT, we are able to prove resistance against related-key differential attacks for both versions of EPCBC, which enables a secure usage of EPCBC in Davies-Meyer mode as a lightweight hash functions.

Further, we show that EPCBC is resistant against currently best known integral cryptanalysis, statistical saturation attack, slide attack, algebraic attack and the latest higher-order differential cryptanalysis from FSE 2011 [11].

On top of this, EPCBC performs well with respect to lightweight applications. In fact, EPCBC(48,96) has a slightly smaller area footprint than PRESENT-80, while at the same time offering a slightly higher speed, resulting in a 20% higher figure of merit (FOM). Our power estimates of 2.21 μW for EPCBC(48,96) and 3.63 μW for EPCBC(96,96) (at 1.8V and 100 KHz) indicate how well EPCBC is suited for ultra-constrained applications, such as passive RFID tags. As another contribution, we present an optimized hardware implementation of PRESENT-80 that is both smaller and faster than previously published results.

The remainder of this paper is organized as follows: in Section 2 we briefly recall the Electronic Product Code before we propose two variants of EPCBC in Section 3. Then we improve existing and prove new bounds for generalized PR-n in Section 4, which we will use for the security analysis of EPCBC in Section 5. Hardware implementation results are presented in Section 6 and finally the paper is concluded in Section 7.

2 The Electronic Product Code - EPC

The Electronic Product Code (EPC) is an industry standard by EPCglobal, designed to "facilitate business processes and applications that need to manipulate visibility data (i.e. data about observations of physical objects)" [22]. In other words, it is a unique identifier for any physical object. The standard also focuses on EPC class 1 Gen 2 RFID tag [23] as the carrier for the EPC and proposes a 96-bit unique identifier to be stored for low-cost applications. EPCglobal has specified seven application-dependent identification keys for EPC, such as Serialized Global Trade Item Number (SGTIN), Serialized Shipping Container Code (SSCC), or the Global Document Type Identifier (GDTI). These numbers consist at least of a company prefix and a unique serial number commisioned by the company. Some application also comprise additional mandatory fields such as the document type in the case of GDTI. As a consequence, the length of the serial number varies between 36-bits and 62-bits (note that there are further restrictions on the choice of the serial number such as no leading zeros).

An EPC class 1 Gen 2 RFID tag consists of four memory banks: one to store the kill and access passwords of 32-bits each, one for the EPC information, one for the manufacturer information about the tag itself, and one for user data. The tag manufacturer information (TID) may contain a manufacturer ID, a code for the tag model, and a unique serial number, which can be used independently of the EPC.

A globally unique identifier for any physical item certainly promises many benefits, such as optimized supply chains. On the other hand it could be a nightmare for privacy, especially since RFID tags could be read out without notion of its owner, and readers can be easily hidden and positioned at highly frequented locations. In this way unique movement patterns of single tags or a selection of tags can be created, which can be used to reveal the identity of their owner. For this reason a wide variety of protocols with the aim of securing privacy have been proposed. Out of these protocols, there are several which use lightweight cryptographic mechanisms with hash functions [46,42,47,26,20]. There are also some protocols which use pseudo-random number generators such as the privacy protocol proposed by Juels in [29]. In particular, Juels proposes a "minimalist" system in which every tag contains and rotates a short list of pseudonyms, emitting a different pseudonym on each reader query.

A block cipher is a versatile building-block and, besides encrypting data, can also be used as a one-way function, a pseudo random number generator or a compression function. In a straightforward way one would encrypt the whole 96-bit EPC, but one can also envision different scenarios where a user wants to either hide the serial number or the company's identity for the sake of privacy. The former may be the case that a book is borrowed from a public library: it is fine for others to know that you borrow a book, but might not be when the book is on cancer treatment. The latter may be the case that a pharmaceutical product is bought: knowing the pharmaceutical company can reveal very sensitive information, while a serial number might reveal no information about the product.

A wide variety of lightweight block ciphers optimized for ultra-constrained devices, such as passive RFID tags has been proposed over the last couple of years. Most notably are DESXL [34], KATAN [12], or PRESENT [10] among others. However, none of them has either key or block length that suits the uncommon 96 bits of an EPC and with the proposal of EPCBC in the next section we close this gap.

3 A New Block Cipher Suitable for EPC Encryption: EPCBC

We propose two variants of EPCBC: EPCBC(48,96) which has 48-bit block size and 96-bit key, and, EPCBC(96,96) which has 96-bit block size and 96-bit key. (The testvectors are provided in Appendix A.2.)

3.1 EPCBC(48,96) - EPCBC with 48-Bit Block Size and 96-Bit Key Size

The main cipher of EPCBC(48,96) iterates the PR-48 structure for 32 rounds, i.e. it uses the PRESENT description of Section 4.1 with $n = 48$ and $r = 32$. Due to $n = 48$, EPCBC(48,96) will be suited for applications with the key being changed frequently. The choice of $r = 32$ ensures security of the cipher against differential and linear cryptanalysis as explained in Section 5.1.

However, the key schedule of EPCBC(48,96) is different from that of PRESENT. The EPCBC(48,96) key schedule takes in the 96-bit secret key as a 96-bit keystate. The left half of the keystate is taken as the first subkey. Then a variant-Feistel structure is applied to the keystate for 8 rounds. Each nonlinear function F of the Feistel variant consists of 4 rounds of the PR-48 main cipher structure where a subkey is output after each round, giving $8 \times 4 + 1 = 33$ subkeys in total. The following is a description of the EPCBC(48,96) key schedule where $1Round$ denotes one round of PR-48 without subkey addition. Subkey[i] represents the subkey for round i of the main cipher.

```
(LKeystate,RKeystate) = 96-bit key
Subkey[0] ← LKeystate
for i = 0 to 7 do
    temp ← LKeystate ⊕ RKeystate
    for j = 0 to 3 do
        RKeystate ← 1Round(RKeystate) ⊕ (4i + j)
        Subkey[4i + j + 1] ← RKeystate
    end for
    LKeystate ← RKeystate
    RKeystate ← temp
end for
```

3.2 EPCBC(96,96) - EPCBC with 96-Bit Block Size and 96-Bit Key Size

The main cipher of EPCBC(96,96) iterates the PR-96 structure for 32 rounds, i.e. it uses the PRESENT description of Section 4.1 with $n = 96$ and $r = 32$. The choice of $r = 32$ ensures security of the cipher against differential and linear cryptanalysis as explained in Section 5.1. The key schedule of EPCBC(96,96) is also different from the PRESENT key schedule. It takes in the 96-bit secret key as a 96-bit keystate, outputs this keystate as the first subkey and applies the PR-96 main cipher structure[1] to it for 32 rounds where a subkey is output after each round. Let $1Round$ denote one round of PR-96 without subkey addition. Subkey[i] represents the subkey for round i of the main cipher.

---

[1] Note that the hash function WHIRLPOOL[3] and the block cipher SMS4[19] also use an identical function for both their main cipher and key schedule.

```
Keystate = 96-bit key
Subkey[0] ← Keystate
for i = 0 to 31 do
    Keystate ← 1Round(Keystate) ⊕ i
    Subkey[i + 1] ← Keystate
end for
```

4 Improved Differential and Linear Cryptanalyis of PR-n

As we are using the structure of PRESENT for the main cipher of EPCBC, we did a security analysis of n-bit block size PRESENT (PR-n) against differential and linear cryptanalysis, and, we obtained new results. We first give a brief description of PR-n.

4.1 Brief Description of PR-n

The detailed description of PRESENT can be found in [10]. The r-round encryption of the plaintext STATE can be written at a top-level as follows:

```
for i = 1 to r do
    STATE ← STATE ⊕ eLayer(KEY, i)
    STATE ← sBoxLayer(STATE)
    STATE ← pLayer(STATE)
    KEY ← genLayer(KEY, i)
end for
STATE ← STATE ⊕ eLayer(KEY, r + 1)
```

The eLayer describes how a subkey is combined with a cipher STATE, sBoxLayer (S-box layer) and pLayer (permutation layer) describe how the STATE evolves, and genLayer is used to describe the generation of the next subkey.

According to [9], the building blocks sBoxLayer and pLayer can be generalized for n-bit block size PRESENT:

1. **sBoxLayer:** This denotes use of the PRESENT 4-bit to 4-bit S-box S and is applied $\frac{n}{4}$ times in parallel.
2. **pLayer:** This is an extension of the PRESENT bit-permutation[10] and moves bit i of STATE to bit position $P(i)$, where

$$P(i) = \begin{cases} i \cdot \frac{n}{4} \pmod{n-1}, & \text{if } i \in \{0, 1, \cdots, n-2\} \\ n-1, & \text{if } i = n-1. \end{cases}$$

The sBoxLayer. Denote the Fourier coefficient of a S-box S by

$$S_b^W(a) = \sum_{x \in \mathbb{F}_2^4} (-1)^{\langle b, S(x) \rangle + \langle a, x \rangle}.$$

Then the properties of the 4-bit PRESENT S-boxes can be described as follows:

S1. For any fixed non-zero input difference $\Delta_I \in \mathbb{F}_2^4$ and any fixed non-zero output difference $\Delta_O \in \mathbb{F}_2^4$,

$$\#\{x \in \mathbb{F}_2^4 \mid S(x) + S(x + \Delta_I) = \Delta_O\} \leq 4.$$

S2. For any fixed non-zero input difference $\Delta_I \in \mathbb{F}_2^4$ and any fixed non-zero output difference $\Delta_O \in \mathbb{F}_2^4$ such that $\mathrm{wt}(\Delta_I) = \mathrm{wt}(\Delta_O) = 1$,

$$\{x \in \mathbb{F}_2^4 \mid S(x) + S(x + \Delta_I) = \Delta_O\} = \emptyset.$$

S3. For all non-zero $a \in \mathbb{F}_2^4$ and all non-zero $b \in \mathbb{F}_2^4$, it holds that $|S_b^W(a)| \leq 8$.
S4. For all non-zero $a \in \mathbb{F}_2^4$ and all non-zero $b \in \mathbb{F}_2^4$ such that $\mathrm{wt}(a) = \mathrm{wt}(b) = 1$, it holds that $S_b^W(a) = \pm 4$.

As in the case of PRESENT [10], for PR-n where $16|n$, we divide the $\frac{n}{4}$ S-boxes into groups of four as follows:

- Number the S-boxes from 0 to $(\frac{n}{4} - 1)$ in a right-to-left manner.
- Then Group j comprises S-boxes $4(j-1)$, $4(j-1)+1$, $4(j-1)+2$ and $4(j-1)+3$ for $j = 1, 2, \cdots, \frac{n}{4}$.

The pLayer. For $n \geq 64$, the (generalized) pLayer observe the following properties:

P1. The input bits to an S-box come from 4 distinct S-boxes of the same group.
P2. The input bits to a group of four S-boxes come from 16 different S-boxes.
P3. The four output bits from a particular S-box enter four distinct S-boxes, each of which belongs to a distinct group of S-boxes in the subsequent round.
P4. The output bits of S-boxes in distinct groups go to distinct S-boxes.

For PR-48, observe that while P1 and P4 are still true, P2 and P3 do not hold. Instead of P3, PR-48 (and in fact PR-n for all n, $16|n$) obeys the following:

P3′. The four output bits from a particular S-box enter four distinct S-boxes.

In the remaining of this paper, we only consider PR-n where $16|n$.

4.2 Improved Differential and Linear Cryptanalysis

Differential and linear cryptanalysis are among the most powerful techniques available to the cryptanalyst. In order to evaluate the resistance of a block cipher to differential and linear cryptanalysis, we provide a lower bound to the number of active S-boxes involved in a differential/linear characteristic. We first prove new bounds for the differential and linear resistance of PR-n which are of particular interest with regards to $n < 64$.

Theorem 1. *Any 4-round differential characteristic of PR-n has a minimum of 6 active S-boxes.*

Proof. Suppose there are at most 5 active S-boxes for four consecutive rounds. The numbers of active S-boxes in the four consecutive rounds takes up one of the following patterns: 1-1-1-1, 2-1-1-1, 1-2-1-1, 1-1-2-1, or 1-1-1-2. But the patterns 1-1-1 and and 1-2-1 are impossible by virtue of S2 and P3′. The result now follows. □

Theorem 2. *Let ϵ_4 be the maximal bias of a linear approximation of 4 rounds of PR-n. Then $\epsilon_4 \leq 2^{-7}$.*

Proof. The proof is similar to that of [10, Theorem 2]. However note that when $n < 64$, the patterns (denoting the numbers of active S-boxes over four consecutive rounds) 1-2-1-1 and 1-1-2-1 are now allowed, in addition to the existing patterns of [10, Theorem 2]. But due to P3′, we must have at least one active S-box with single-bit approximation over four rounds for all possible patterns. It follows that $\epsilon_4 \leq 2^4 \times (2^{-3}) \times (2^{-2})^4 \leq 2^{-7}$, as desired. □

Remark 1. The new bounds in Theorems 1 and 2 are needed when analyzing the security of EPCBC(48,96) against DC and LC because the block size $n = 48$ is less than 64.

In the remaining of this section, we state the improved and generalized results on the differential and linear probability bounds of [10]. As many technicalities and rigorous arguments are involved, we have included the formal proofs of the theorems in Appendix A.1.

Theorem 3. *For $n \geq 64$, the r-round differential characteristic of PR-n has a minimum of $2r$ active S-boxes for $r \geq 5$.*

Remark 2. Note that if we have used the differential bound in [10, Theorem 1], we would only be able to deduce 10 (differential) active S-boxes every 5 rounds. For example, if there are 14 rounds, [10, Theorem 1] would give 20 active S-boxes from 10 out of 14 rounds, the security margin from the remaining 4 rounds is not captured. In contrast, Theorem 3 would give us 28 active S-boxes from 14 rounds.

Theorem 4. *Let ϵ_r be the maximal bias of a linear approximation of r rounds of PR-n where $n \geq 64$. Then $\epsilon_r \leq 2^{-2r+1}$ for $r = 4, 5, 6, 7, 8$ and 9.*

Remark 3. Note that [10, Theorem 2] only proves the LC bound for 4 rounds. If we consider linear cryptanalysis over e.g. 11 rounds, then applying [10, Theorem 2] would give us:

$$\epsilon_{11} \leq 2 \times \epsilon_4 \times \epsilon_4 \leq 2 \times 2^{-7} \times 2^{-7} = 2^{-13},$$

which only uses 8 out of 11 rounds, and the security margin from the remaining 3 rounds is not captured.

If we apply Theorem 4, we obtain the tighter (better) bound:

$$\epsilon_{11} \leq 2 \times \epsilon_5 \times \epsilon_6 \leq 2 \times 2^{-9} \times 2^{-11} = 2^{-19}.$$

Theorem 4 enables us to use all 11 rounds in deriving the linear probability bound for PR-n, $n \geq 64$.

Remark 4. Note also that the linear probability bound for 8 rounds, which is given by $\epsilon_8 \leq 2^{-15}$, is better than applying the bound for 4 rounds twice, which is given by $2 \times (\epsilon_4)^2 \leq 2^{-13}$.

5 Security Analysis of EPCBC

We now present the results of a security analysis of EPCBC.[2]

5.1 Differential, Linear and Related-Key Differential Cryptanalysis

Analysis of EPCBC(48,96). We assume minus-4 round attacks, so a distinguisher of EPCBC(48,96) is on 28 rounds. For protection against differential cryptanalysis, there are seven 4-round blocks in 28 rounds. By Theorem 1, the differential characteristic probability is at most:

$$\Delta_{28} \leq [(2^{-2})^6]^7 = 2^{-84} < 2^{-48} = 2^{-blocksize}.$$

For protection against linear cryptanalysis, there are seven 4-round blocks in 28 rounds. By Theorem 2, the linear bias is at most:

$$\epsilon_{28} = (2^6) \times (\epsilon_4)^7 \leq (2^6) \times (2^{-7})^7 = 2^{-43} < 2^{-24} = 2^{-blocksize/2}.$$

For protection against related-key differential cryptanalysis, we consider the related key differential characteristic probability, given by $p_{c|k} \times p_k$. $p_{c|k}$ is the differential characteristic probability of the main cipher conditioned on subkey differential and p_k is the differential characteristic probability of the key schedule.

We first bound p_k which is the differential probability of the key schedule. A minus-4 round attack would involve 7 variant-Feistel rounds of the key schedule. By Theorem 1, every 4-round nonlinear function F of the variant-Feistel structure in the key schedule has differential probability at most $(2^{-2})^6 = 2^{-12}$. It is easy to deduce that there are at least two active nonlinear functions F for every three rounds of the variant-Feistel structure. We consider all three possible cases: $(\Delta LKeystate, 0)$, $(0, \Delta RKeystate)$ and $(\Delta LKeystate, \Delta RKeystate) \neq (0, 0)$. Based on the structure of the variant-Feistel, it can then be shown directly that at least two of the three nonlinear functions have non-zero input differentials. Thus over 6 rounds of the key schedule, we have $p_k \leq [(2^{-12})^2]^2 = 2^{-48}$.

By the key schedule design, we see that the differential of four consecutive subkeys are either all zero or all non-zero. In the case when the subkeys have non-zero differential, for a particular round of the main cipher, if the input differential is not cancelled by the key difference, then we have at least an active S-box in the round. Otherwise, that round has no active S-box and zero output difference. Then the subsequent nonzero subkey difference when xored with this zero differential will cause an active S-box in the subsequent round. Hence, there

[2] We leave the analysis of linear hull effect on EPCBC as future work.

are at least two active S-boxes every four rounds of the main cipher when the subkeys have non-zero differential. Hence over 28 rounds of the main cipher, there are at least 6 active S-boxes. This is because there are always at least three blocks of four consecutive rounds of the main cipher with non-zero key differentials. Therefore $p_{c|k} \leq (2^{-2})^6 = 2^{-12}$ and the related key differential probability satisfies $p_{c|k} \times p_k \leq 2^{-12} \times 2^{-48} = 2^{-60}$. Thus the attack complexity of related-key differential attack is at least 2^{60}. Moreover, the adversary needs to obtain ciphertexts corresponding to 2^{48} related keys, which might be infeasible in practice.

Analysis of EPCBC(96,96). We assume minus-4 round attacks, so a distinguisher of EPCBC(96,96) is on 28 rounds. For protection against differential cryptanalysis, we apply Theorem 3 on 28 rounds. The differential characteristic probability is at most:

$$\Delta_{28} \leq (2^{-2})^{28\times 2} = 2^{-112} < 2^{-96} = 2^{-blocksize}.$$

For protection against linear cryptanalysis, there are three 9-round blocks in 27 out of 28 rounds. By Theorem 4, the linear bias is at most:

$$\epsilon_{27} \leq (2^2) \times \epsilon_9^3 \leq (2^2) \times (2^{-17})^3 = 2^{-49} < 2^{-48} = 2^{-blocksize/2}.$$

For protection against related-key differential attack, we need to bound both p_k and $p_{c|k}$ as in the proof for EPCBC(48,96). By Theorem 3, $p_k \leq 2^{-96}$. With some simple argument as before, we can prove that there are at least one active S-box every two rounds of the main cipher because the subkeys have non-zero differential. Thus $p_{c|k} \leq (2^{-2})^{14} = 2^{-28}$ and the related-key differential probability satisfies $p_{c|k} \times p_k \leq 2^{-28} \times 2^{-96} = 2^{-124} < 2^{-96}$. Thus related-key differential attack is infeasible.

5.2 Other Attacks on EPCBC

Integral Attacks. The integral attack [31] is a chosen plaintext attack originally applied to byte-based ciphers such as SQUARE and Rijndael. Lucks [37] ported it as the 'saturation attack' to Twofish, which is fundamentally a byte-based algorithm incorporating bit-oriented rotations. On this basis, the authors of PRESENT, which contains a bit-based permutation, discarded this attack almost out of hand. In 2008, Z'aba et al. [48] developed the bit-based integral attack, which they applied to very reduced-round versions of Noekeon, Serpent and PRESENT.

The attack categorizes each bit or byte across a structure of texts, as to whether or not it is balanced (the sum of its value in each text equals zero). At some point in the evolution of the text through the encryption, the balance property is lost. The attacker can guess parts of the subsequent round key, and partially decrypt this point. If the balance is not restored, the partial round key guess is incorrect.

The details of the attack are driven by the structure of the linear permutation rather than of the S-boxes. The attack works best for ciphers in which the block size is less than the size of the secret key. It applies to seven of PRESENT's 31 rounds, partly due to a weakness in its key schedule that allows 61 bits of round keys 5 and 6 to be deduced by guessing the 64 bits of round key 7.

The integral attacks on EPCBC(48, 96) and EPCBC(96, 96) use very similar differentials to those in the PRESENT attack, since the S-box is identical and permutation scaled to a different block size. In particular, the S-box has a probability-one differential $0x1 \rightarrow w||0x1$ for $w \in \{0x1, 0x3, 0x4, 0x6\}$.

In both cases, the attacker uses a structure of sixteen chosen plaintexts. For EPCBC(48, 96), each text in the structure is of the form $(c_0, c_1, c_2||j)$, where c_0 and c_1 are 16-bit constants, c_2 is a 12-bit constant, and j varies from 0 through to 15. This permits a 3.5 round differential similar to that of PRESENT except that the balance of bits relating to S-boxes 2, 5 and 8 are lost in the third rather than fourth round. The attacker launches an attack on EPCBC(48, 96) reduced to four rounds, recovering 32 bits of the fourth round key with 2^{12} partial decryptions, and brute forcing the remaining 16 bits of the key for a total complexity of $O(2^{16.13})$. The attack can be extended to five rounds, in which the attacker guesses 16 bits of the fifth round key for every four bits of the fourth round key. Due to the structure of the key schedule, the attacker does not need to perform an additional brute force on the fourth round key, so the complexity of the attack on five rounds is $O(2^{27.9})$. As per the PRESENT attack, a seven round attack can be mounted by adding one round at the beginning of the cipher, and brute forcing the seventh round key for a total complexity of $O(2^{91.9})$ (due to the structure of the key schedule, the seventh round key does not allow the attacker shortcuts in deducing parts of the sixth round key).

For EPCBC(96, 96), the attacker uses a structure of 16 chosen plaintexts of form $(c_0, c_1, c_2, c_3, c_4, c_5||j)$ where j varies from 0 to 16. This permits the same 3.5 round differential as PRESENT. The attack on 4 rounds has complexity $O(2^{13.58})$. The structure of the key schedule means that in an attack on five rounds, the attacker can work out the relevant parts of the fourth round key from his guesses on the fifth. The attack can be extended to six rounds with complexity $O(2^{33.6})$. It is possible to extend the attack to seven rounds, guessing on a set of 64 key bits. The remaining 32 bits of the key can be brute forced. The complexity of the attack is $O(2^{88})$.

Because the point at which the balance property is lost in the structure cannot be easily extended, the number of rounds for which the integral attack is applicable cannot be readily increased. So the full version of EPCBC is immune to integral attacks.

Statistical Saturation Attacks. Statistical saturation attack (SSA) was first proposed by Collard and Standaert in [16] to cryptanalyze PRESENT. The attack targets mainly the diffusion properties of the permutation layer. By first fixing some plaintext bits, the attacker extracts information about the key by observing non-uniform distributions in the ciphertexts. This leads to an *estimated* attack against PRESENT up to 24 rounds, using approximately 2^{60}

chosen plaintexts. Collard and Standaert later proposed the use of multiple trails in [15] which provided experimental evidence that SSA can attack up to 15 rounds of PRESENT with $2^{35.6}$ plaintext-ciphertext pairs. However as noted by the authors, this attack is only of theoretical interest and their results do not threaten the security of PRESENT in practice. Since EPCBC and PRESENT use the same main cipher structure (in particular, the same permutation layer), statistical saturation attack does not work on the full version of EPCBC.

Higher Order Differential Attack. Higher order differential attack was introduced by Knudsen in [28]. This attack works especially well on block ciphers with components of low algebraic degree such as the KN-Cipher (see [28]), whereby the ciphers can be represented as Boolean polynomials of low degree in terms of the plaintext. The attack requires $O(2^{d+1})$ chosen plaintext when the cipher has degree d. Hence we are interested in estimating the degree of a composed function $G \circ F$ say. Trivially, $\deg(G \circ F) \leq \deg(F) \times \deg(G)$. A first improvement of the trivial bound was provided by Canteaut and Videau [13] and recently further improved by Boura et al[11]. Recall that the PRESENT S-box has algebraic degree 3. Hence by applying Theorem 2 of [11], from the composition of r S-box layers, the algebraic degree of r rounds of EPCBC is expected to be

$$\min(3^r, blocksize - \frac{blocksize - \deg(R^{r-1})}{3}, blocksize - 1),$$

where R denotes one round of EPCBC. This implies that EPCBC(48, 96) reaches the maximum degree of 47 after 5 rounds while EPCBC(96, 96) reaches the maximum degree of 95 after 6 rounds. Thus it is unlikely that higher order differential attack will work on EPCBC which has 32 rounds.

Slide Attacks. Slide attacks exploit block ciphers that can be broken down into multiple rounds of an identical F function [7,8]. Usually, a block cipher iterates a round function, with a different subkey being xored to each round. Therefore this boils down to the subkeys being cyclic with period t, in which case F consists of t rounds of the cipher. The adversary needs to find a slid pair (P_0, C_0) and (P_1, C_1) such that $P_0 = F(P_1)$ and $C_0 = F(C_1)$. Then with these chosen plaintexts, he only needs to attack t rounds instead of the entire cipher. In EPCBC, the key schedule is essentially a block cipher by itself where the input is the secret key and the output of each round is a subkey. So it is highly unlikely that the subkeys will be periodic to allow a slide attack. We have also added different round constants to each round to ensure that the subkeys do not repeat.

Algebraic Attacks. In 2002, it was claimed in [17] that the XSL method is able to break AES by expressing the cipher as a sparse system of quadratic equations and solving it. However in 2005, it was proven in [14] that the XSL attack does not work. Instead, practical results on algebraic cryptanalysis of block ciphers have been obtained by applying the Buchberger and F_4 algorithms within MAGMA [39]. Therefore, the authors of PRESENT applied the F_4 algorithm on MAGMA to solve a mini version of PRESENT [10]. They found that even when considering a system consisting of seven S-boxes, i.e. a block size of 28 bits, they were unable to obtain a solution in a reasonable time to even a two-round version of the reduced cipher. Therefore they conclude that algebraic attacks are unlikely to pose a threat to PRESENT, which can be written as a system of 11067 quadratic equations in 4216 variables, arising from 527 S-boxes. The EPCBC cipher also uses the PRESENT S-box which can be described by 21 quadratic equations in 8 input/output-bit variables over $GF(2)$ [10]. There are $12 \times 32 + 12 \times 33 = 780$ S-boxes in EPCBC(48,96). Thus it can be expressed as a system of $780 \times 21 = 16380$ quadratic equations in $780 \times 8 = 6240$ variables. In a similar way, EPCBC(96,96) has 1560 S-boxes and it can be expressed as a system of 32760 quadratic equations in 12480 variables. Hence both versions of EPCBC result in a more complex system of quadratic equations than that of PRESENT. Therefore we do not expect algebraic attacks to be a threat to EPCBC too.

6 Implementation of EPCBC

To demonstrate the efficiency of our proposal we have implemented EPCBC(48,96) and EPCBC(96,96) in VHDL and used *Synopsys DesignVision 2007.12* to synthesize them using the *Virtual Silicon* (VST) standard cell library *UMCL18G212T3*, which is based on the *UMC L180 0.18μm 1P6M* logic process and has a typical voltage of 1.8 Volt [45].

Figure 1 depicts serialized hardware architectures for EPCBC(48,96) (top) and EPCBC(96,96) (bottom). Components that contain mainly sequential logic are presented in rectangles while purely combinational components are presented in ovals. Naturally the architecture of EPCBC is very similar to a serialized PRESENT architecture, as published previously e.g. in [44,43]. A significant difference is in the key schedule of EPCBC(48,96), as it does not perform any operation on the left halve of the key (i.e. LKey) in every round. This allows to store LKey in the simplest flip-flops available (4.67 GE per bit) contrary to the State and RKey, which have to be stored in flip-flops with two inputs (6 GE).

Another optimization is that every round requires only $n/4$ clock cycles, n being the block size, as compared to $n/4 + 1$ clock cycles e.g. [44,43]. This can be achieved by by simply wiring the S-box output directly as input to the Permutation layer, thus combining the execution of the S-box look-up of the last chunk of a round with the Permutation layer into one clock cycle. Note that this optimization can also be applied to PRESENT. Thus, a second contribution of this paper are optimized serialized PRESENT-80 and PRESENT-128 implementations that requires only 516 and 528 clock cycles (compared to previously 547 and 559, respectively [44]). Note that in order to apply this speed-up trick a second S-box has to be implemented (22.3 GE[3]), while the MUX (11 GE) for the S-box input can be saved. In principal this would result in a 6 GE *larger*

[3] We hereby acknowledge the support of Dag Arne Osvik to derive a more compact S-box.

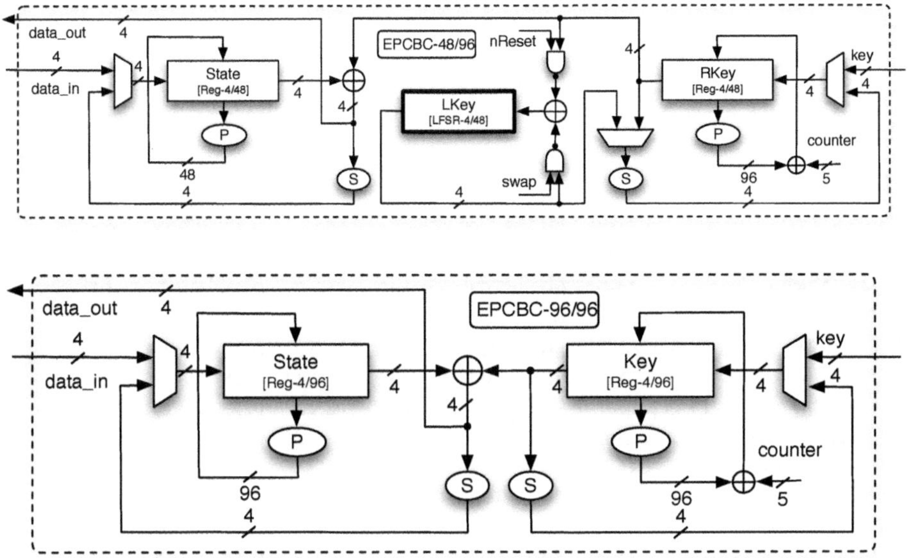

Fig. 1. Serial hardware architectures

area requirement compared to the design strategy of [44]. However, since we optimized the control logic significantly, we were able to decrease the area requirements to 1,030 GE and 1,343 GE, respectively. It is crucial to note that the storage of the internal state already takes up 864 GE (1,152 GE), which leaves only 211 GE (239 GE) to improve. Our results relate to an improvement of more than 20% for those parts of the implementations that can be improved by better design decisions. To the best of our knowledge these are the smallest and fastest PRESENT implementations in hardware.

We used *Synopsys PowerCompiler* version *A-2007.12-SP1* to estimate the power consumption of our implementations. The power estimates for the smallest wire-load model (10K GE) at a supply voltage of 1.8 Volt and a frequency of 100 KHz are between 2.21 μW for EPCBC(48,96) and 3.63 μW for EPCBC (96,96), which indicates how well EPCBC is suited for ultra-constrained applications, such as passive RFID tags. However, the accuracy level of simulated power figures greatly depends on the simulation tools and parameters used. Furthermore, the power consumption also strongly depends on the target library used. Thus to have a fair comparison, we do not include any power figures in Table 1. Instead we use a figure of merit (FOM), proposed by [2] –which reflects the time-area-power trade-off inherent in any hardware implementation– as a somewhat *fair* metric for comparison of the energy efficiency of different block cipher implementations. Table 1 lists also latency, area requirements, and

Table 1. Hardware implementation results of some lightweight block ciphers and the AES

Algorithm	Ref.	key size	block size	cycles/ block	T'put (@100 KHz)	Tech. [μm]	Area [GE]	FOM block [$\frac{bits \times 10^9}{clk \cdot GE^2}$]	FOM 96-bits
KLEIN-64	[25]	64	64	207	30.9	0.18	1,220	208	156
PRINTcipher-48*	[32]	80	48	768	6.25	0.18	402	387	387
KATAN48	[12]	80	48	255	18.8	0.13	927	219	219
KLEIN-80	[25]	80	64	271	23.62	0.18	1,478	108	81
PRESENT-80	[44]	80	64	547	11.7	0.18	1,075	101	76
PRESENT-80	This paper	80	64	516	12.4	0.18	1,030	117	88
EPCBC-48	This paper	96	48	396	12.12	0.18	1,008	119	119
KLEIN-96	[25]	96	64	335	19.1	0.18	1,528	82	61
EPCBC-96	This paper	96	96	792	12.12	0.18	1,333	68	68
PRESENT-128	This paper	128	64	528	12.12	0.18	1,339	68	51
PRESENT-128	[43]	128	64	559	11.45	0.18	1,391	59	44
AES	[40]	128	128	226	56.64	0.13	2,400	98	74
PRINTcipher-96*	[32]	160	96	3,072	3.13	0.18	726	59	59
DESXL	[34]	184	64	144	44.4	0.18	2,168	95	71

* Hard-wired Keys

throughput of *unbroken* block ciphers that have a smaller area footprint than AES.[4] It is sorted after increasing key length and block size.

It can be seen that EPCBC(48,96) has a slightly smaller area footprint then PRESENT-80, while having a slightly lower speed resulting in a somewhat similar FOM. However, this is only the case if we consider messages that match the block length of the cipher. In that sense it is a best case scenario for every algorithm. If we focus on the EPC scenario with a given block length of 96-bit, the picture changes. The efficiency of PRESENT, AES and any other algorithm, for which the block length is not 96 bits (or that divides 96), drops significantly. As one can see in the last column, EPCBC's efficiency stays the same (as it does for the other algorithms with similar block lengths). If the block cipher is going to be used as a compression function, e.g. in Davies-Meyer or Hirose Mode, the same drop in efficiency for 96-bit messages can be observed.

No implementation figures for MESH have been published so far, but for MESH-96 at least 1,345 GE are required (in our technology) to store the 288 bits internal state (GE), which is already more than what is required for EPCBC(96,96). As MESH-96 operates on 16-bit words, a 16-bit datapath seems a natural choice, but, given the rather complex round function of MESH-96, it is not clear, if this would be optimal with regards to compact area.

[4] Please note that this excludes 3-WAY [18], which can be broken with one related key and about 2^{22} chosen plaintexts [30], and SEA with a 96-bit key and 96-bit block size which requires far more area than the AES (3,758 GE [38]).

7 Conclusion

In this paper, we designed the EPCBC block ciphers which use a 96-bit key and are provable secure against related-key differential/boomerang attacks. When evaluating the security of EPCBC we could leverage on the extensive analyses published for PRESENT, providing a good "trust" starting point for our design, contrary to other exotic, i.e. not easy to analyze, lightweight block cipher designs. Nevertheless during the security evaluation of EPCBC(96,96), we improved the bounds for the differential/linear resistance of PR-n, $n \geq 64$, and proved new results on the DC/LC resistance of PR-n, $n < 64$, when evaluating the security of EPCBC(48,96). For the envisioned scenario of EPC applications, we showed that the chosen block sizes of 48 and 96 bits allow EPCBC to outperform other lightweight or standardized algorithms, such as KLEIN, PRESENT and AES, regardless if used as a block cipher or as a compression function. We also presented two optimized serialized PRESENT architectures that are both smaller and faster than previous results.

It is noteworthy to stress that EPCBC's key schedule (as opposed to PRESENT) is optimized against related key differential attacks, which allows a secure usage of EPCBC in such scenarios. Furthermore, EPCBC has a larger key length than PRESENT-80, which indicates a higher security level.

References

1. Avoine, G., Oechslin, P.: A Scalable and Provably Secure Hash-Based RFID Protocol. In: PerCom Workshops, pp. 110–114. IEEE Computer Society Press (2005)
2. Badel, S., Dağtekin, N., Nakahara Jr., J., Ouafi, K., Reffé, N., Sepehrdad, P., Sušil, P., Vaudenay, S.: ARMADILLO: A Multi-purpose Cryptographic Primitive Dedicated to Hardware. In: Mangard, S., Standaert, F.-X. (eds.) CHES 2010. LNCS, vol. 6225, pp. 398–412. Springer, Heidelberg (2010)
3. Barreto, P., Rijmen, V.: The Whirlpool Hashing Function, `http://www.larc.usp.br/~pbarreto/WhirlpoolPage.html`
4. Biham, E., Dunkelman, O., Keller, N.: A Related-Key Rectangle Attack on the Full KASUMI. In: Roy, B. (ed.) ASIACRYPT 2005. LNCS, vol. 3788, pp. 443–461. Springer, Heidelberg (2005)
5. Biryukov, A., Khovratovich, D., Nikolić, I.: Distinguisher and Related-Key Attack on the Full AES-256. In: Halevi, S. (ed.) CRYPTO 2009. LNCS, vol. 5677, pp. 231–249. Springer, Heidelberg (2009)
6. Biryukov, A., Khovratovich, D.: Related-Key Cryptanalysis of the Full AES-192 and AES-256. In: Matsui, M. (ed.) ASIACRYPT 2009. LNCS, vol. 5912, pp. 1–18. Springer, Heidelberg (2009)
7. Biryukov, A., Wagner, D.: Slide Attacks. In: Knudsen, L.R. (ed.) FSE 1999. LNCS, vol. 1636, pp. 245–259. Springer, Heidelberg (1999)
8. Biryukov, A., Wagner, D.: Advanced Slide Attacks. In: Preneel, B. (ed.) EUROCRYPT 2000. LNCS, vol. 1807, pp. 589–606. Springer, Heidelberg (2000)
9. Bogdanov, A., Leander, G., Paar, C., Poschmann, A., Robshaw, M.J.B., Seurin, Y.: Hash Functions and RFID Tags: Mind the Gap. In: Oswald, E., Rohatgi, P. (eds.) CHES 2008. LNCS, vol. 5154, pp. 283–299. Springer, Heidelberg (2008)

10. Bogdanov, A.A., Knudsen, L.R., Leander, G., Paar, C., Poschmann, A., Robshaw, M.J.B., Seurin, Y., Vikkelsoe, C.: PRESENT: An Ultra-Lightweight Block Cipher. In: Paillier, P., Verbauwhede, I. (eds.) CHES 2007. LNCS, vol. 4727, pp. 450–466. Springer, Heidelberg (2007)
11. Boura, C., Canteaut, A., De Cannière, C.: Higher-Order Differential Properties of KECCAK and *Luffa*. In: Joux, A. (ed.) FSE 2011. LNCS, vol. 6733, pp. 252–269. Springer, Heidelberg (2011)
12. De Cannière, C., Dunkelman, O., Knežević, M.: KATAN and KTANTAN — A Family of Small and Efficient Hardware-Oriented Block Ciphers. In: Clavier, C., Gaj, K. (eds.) CHES 2009. LNCS, vol. 5747, pp. 272–288. Springer, Heidelberg (2009)
13. Canteaut, A., Videau, M.: Degree of Composition of Highly Nonlinear Functions and Applications to Higher Order Differential Cryptanalysis. In: Knudsen, L.R. (ed.) EUROCRYPT 2002. LNCS, vol. 2332, pp. 518–533. Springer, Heidelberg (2002)
14. Cid, C., Leurent, G.: An Analysis of the XSL Algorithm. In: Roy, B. (ed.) ASIACRYPT 2005. LNCS, vol. 3788, pp. 333–352. Springer, Heidelberg (2005)
15. Collard, B., Standaert, F.-X.: Multi-trail Statistical Saturation Attacks. In: Zhou, J., Yung, M. (eds.) ACNS 2010. LNCS, vol. 6123, pp. 123–138. Springer, Heidelberg (2010)
16. Collard, B., Standaert, F.-X.: A Statistical Saturation Attack against the Block Cipher PRESENT. In: Fischlin, M. (ed.) CT-RSA 2009. LNCS, vol. 5473, pp. 195–210. Springer, Heidelberg (2009)
17. Courtois, N.T., Pieprzyk, J.: Cryptanalysis of Block Ciphers with Overdefined Systems of Equations. In: Zheng, Y. (ed.) ASIACRYPT 2002. LNCS, vol. 2501, pp. 267–287. Springer, Heidelberg (2002)
18. Daemen, J., Govaerts, R., Vandewalle, J.: A New Approach to Block Cipher Design. In: Anderson, R. (ed.) FSE 1993. LNCS, vol. 809, pp. 18–32. Springer, Heidelberg (1994)
19. Diffe, W., Ledin, G.: SMS4 Encryption Algorithm for Wireless Networks. Cryptology ePrint Archive: Report, 329 (2008)
20. Dimitriou, T.: A lightweight RFID protocol to protect against traceability and cloning attacks. In: Proc. IEEE Intern. Conf. on Security and Privacy in Communication Networks, SECURECOMM 2005. IEEE Press (2005)
21. Dunkelman, O., Keller, N., Shamir, A.: A Practical-Time Related-Key Attack on the KASUMI Cryptosystem Used in GSM and 3G Telephony. In: Rabin, T. (ed.) CRYPTO 2010. LNCS, vol. 6223, pp. 393–410. Springer, Heidelberg (2010)
22. EPCglobal. EPC Tag Data Standard Version 1.5. EPCglobal Specification (August 2010), `www.gs1.org/gsmp/kc/epcglobal/tds/`
23. EPCglobal. EPC Radio-Frequency Identity Protocols Class-1 Generation-2 UHF RFID Protocol for Communications at 860 MHz - 960 MHz Version 1.2.0. EPCglobal Specification (May 2008), `www.gs1.org/gsmp/kc/epcglobal/uhfc1g2`
24. Gong, Z., Nikova, S., Law, Y.-W.: KLEIN, a new family of lightweight block ciphers, `http://doc.utwente.nl/73129/`
25. Gong, Z., Nikova, S., Law, Y.-W.: KLEIN, a new family of lightweight block ciphers. In: Proceedings of The 7th Workshop on RFID Security and Privacy 2011. LNCS. Springer, Heidelberg (2011), `http://rfid-cusp.org/rfidsec/`
26. Henrici, D., Müller, P.: Hash-based enhancement of location privacy for radio-frequency identification devices using varying identifers. In: Proc. IEEE Intern. Conf. on Pervasive Computing and Communications, pp. 149–153 (2004)

27. Hong, D., Sung, J., Hong, S.H., Lim, J.-I., Lee, S.-J., Koo, B.-S., Lee, C.-H., Chang, D., Lee, J., Jeong, K., Kim, H., Kim, J.-S., Chee, S.: HIGHT: A New Block Cipher Suitable for Low-Resource Device. In: Goubin, L., Matsui, M. (eds.) CHES 2006. LNCS, vol. 4249, pp. 46–59. Springer, Heidelberg (2006)
28. Jakobsen, T., Knudsen, L.R.: Attacks on Block Ciphers of Low Algebraic Degree. Journal of Cryptology 14, 197–210 (2001)
29. Juels, A.: Minimalist Cryptography for Low-Cost RFID Tags. In: Blundo, C., Cimato, S. (eds.) SCN 2004. LNCS, vol. 3352, pp. 149–164. Springer, Heidelberg (2005)
30. Kelsey, J., Schneier, B., Wagner, D.: Related-Key Cryptanalysis of 3-WAY, Biham-DES, CAST, DES-X, NewDES, RC2, and TEA. In: Han, Y., Quing, S. (eds.) ICICS 1997. LNCS, vol. 1334, pp. 233–246. Springer, Heidelberg (1997)
31. Knudsen, L., Wagnger, D., Daemen, J., Rijmen, V.: Integral Cryptanalysis. In: Daemen, J., Rijmen, V. (eds.) FSE 2002. LNCS, vol. 2365, pp. 112–127. Springer, Heidelberg (2002)
32. Knudsen, L.R., Leander, G., Robshaw, M.J.B.: PRINTcipher: A Block Cipher for IC-Printing. In: Mangard, S., Standaert, F.-X. (eds.) CHES 2010. LNCS, vol. 6225, pp. 16–32. Springer, Heidelberg (2010)
33. Leander, G.: Small Scale Variants of the Block Cipher PRESENT. Cryptology ePrint Archive, Report 2010/143 (2010), `http://eprint.iacr.org/2010/143.pdf`
34. Leander, G., Paar, C., Poschmann, A., Schramm, K.: New Lightweight DES Variants. In: Biryukov, A. (ed.) FSE 2007. LNCS, vol. 4593, pp. 196–210. Springer, Heidelberg (2007)
35. Lee, S.M., Hwang, Y.J., Lee, D.-H., Lim, J.-I.: Efficient Authentication for Low-Cost RFID Systems. In: Gervasi, O., Gavrilova, M.L., Kumar, V., Laganá, A., Lee, H.P., Mun, Y., Taniar, D., Tan, C.J.K. (eds.) ICCSA 2005. LNCS, vol. 3480, pp. 619–627. Springer, Heidelberg (2005)
36. Lim, C.H., Korkishko, T.: mCrypton - a Lightweight Block Cipher for Security of Low-Cost RFID Tags and Sensors. In: Song, J.-S., Kwon, T., Yung, M. (eds.) WISA 2005. LNCS, vol. 3786, pp. 243–258. Springer, Heidelberg (2006)
37. Lucks, S.: The Saturation Attack - A Bait for Twofish. In: Matsui, M. (ed.) FSE 2001. LNCS, vol. 2355, pp. 1–15. Springer, Heidelberg (2002)
38. Mace, F., Standaert, F.-X., Quisquater, J.-J.: ASIC Implementations of the Block Cipher SEA for Constrained Applications. In: RFID Security - RFIDsec 2007, Workshop Record, Malaga, Spain, pp. 103–114 (2007)
39. MAGMA v2.12. Computational Algebra Group, School of Mathematics and Statistics, University of Sydney (2005), `http://magma.maths.usyd.edu.au`
40. Moradi, A., Poschmann, A., Ling, S., Paar, C., Wang, H.: Pushing the Limits: A Very Compact and a Threshold Implementation of AES. In: Paterson, K.G. (ed.) EUROCRYPT 2011. LNCS, vol. 6632, pp. 69–88. Springer, Heidelberg (2011)
41. Nakahara, J., Rijmen, V., Preneel, B., Vandewalle, J.: The MESH Block Ciphers. In: Chae, K.-J., Yung, M. (eds.) WISA 2003. LNCS, vol. 2908, pp. 458–473. Springer, Heidelberg (2004)
42. Ohkubo, M., Suzuki, K., Kinoshita, S.: Cryptographic approach to "privacy-friendly" tags. In: Proc. RFID Privacy Workshop (2003)
43. Poschmann, A.: Lightweight Cryptography - Cryptographic Engineering for a Pervasive World. IT Security. Europäischer Universitätsverlag, vol. 8 (2009) Ph.D. Thesis, Ruhr University Bochum

44. Rolfes, C., Poschmann, A., Leander, G., Paar, C.: Ultra-Lightweight Implementations for Smart Devices – Security for 1000 Gate Equivalents. In: Grimaud, G., Standaert, F.-X. (eds.) CARDIS 2008. LNCS, vol. 5189, pp. 89–103. Springer, Heidelberg (2008)
45. Virtual Silicon Inc. 0.18 μm VIP Standard Cell Library Tape Out Ready, Part Number: UMCL18G212T3, Process: UMC Logic 0.18 μm Generic II Technology: 0.18μm (July 2004)
46. Sarma, S., Weiss, S.A., Engels, D.W.: RFID Systems and Security and Privacy Implications. In: Kaliski Jr., B.S., Koç, Ç.K., Paar, C. (eds.) CHES 2002. LNCS, vol. 2523, pp. 454–469. Springer, Heidelberg (2003)
47. Weis, S.A., Sarma, S.E., Rivest, R.L., Engels, D.W.: Security and Privacy Aspects of Low-Cost Radio Frequency Identification Systems. In: Hutter, D., Müller, G., Stephan, W., Ullmann, M. (eds.) Security in Pervasive Computing. LNCS, vol. 2802, pp. 201–212. Springer, Heidelberg (2004)
48. Z'aba, M.R., Raddum, H., Henricksen, M., Dawson, E.: Bit-Pattern Based Integral Attack. In: Nyberg, K. (ed.) FSE 2008. LNCS, vol. 5086, pp. 363–381. Springer, Heidelberg (2008)

A Appendix

A.1 Improved Differential and Linear Cryptanalysis of PR-*n*

We first clarify the notations used. We use an unordered sequence to specify the number of differential/linear (depending on the context) active S-boxes in each round. For example, $\{1, 1, 2, 3\}$ is used to denote that there are two rounds with one active S-box each, one round with two active S-boxes and one round with three active S-boxes. On the other hand, the ordered sequence 1-1-2-3 is used to denote that there is one active S-box in the first and second rounds, third round has two active S-boxes while the fourth round has three active S-boxes.

Differential Cryptanalysis

Lemma 1. *Consider PR-n where $n \geq 64$. Let D_j be the number of active S-boxes in round j. Suppose $D_i = 1$ and $D_{i+1} \geq 2$. Denote the total number of active S-boxes from round i to round $(i + k - 1)$ (both rounds inclusive) by n_k, for $k \geq 1$. Then $n_1 = 1$, $n_2 \geq 3$, $n_3 \geq 7$, $n_4 \geq 11$, $n_5 \geq 13$, $n_6 \geq 14$, and $n_k \geq 2k + 1$ for $k = 7, 8, 9$.*

Proof. $n_1 = 1$ and $n_2 \geq 3$ are trivial. According to P3, the output bits from the active S-box in round i goes to different S-boxes of distinct groups. In particular, the active S-boxes of round $i + 1$ have a single bit difference in their inputs. Together with P4, $D_{i+2} \geq 4$. This implies that $n_3 \geq 7$. Again, because of P4, we see that the active S-boxes in round $i + 2$ have a single bit difference in their inputs. Further, in round $i + 2$, due to P3, there are at least two groups of S-boxes which contain at least one active S-box. This means that $D_{i+3} \geq 4$. So, $n_4 \geq 11$. Because of P3, there must exist two active S-boxes in round $i + 3$ such that these two S-boxes belong to different groups. By P4, $D_{i+4} \geq 2$ and this yields $n_5 \geq 13$. Since $D_{i+5} \geq 1$ trivially holds, $n_6 \geq 14$.

$n_6 \geq 14$ implies that $n_7 \geq 15$. Suppose $n_8 = 16$. Note that the sequence of n_i is strictly increasing. Hence $n_7 = 15$ and $n_6 = 14$. It follows that $D_{i+6} = D_{i+7} = 1$. From the earlier argument, we see that the worst case of $n_6 = 14$ is obtained when $D_{i+5} = 1$. However by S2 and P3, there cannot be exactly one active S-box for three consecutive rounds. Hence we must have $D_{i+7} \geq 2$, i.e. $n_8 \geq 17$.

Finally, in a similar but slightly more complicated fashion, suppose for a contradiction that $n_9 = 18$. Then $n_8 = 17$ and we have the following cases:

Case 1: $n_6 = 14, n_7 = 15$. Then $D_{i+8} \geq 4$ by applying the previous argument. So $n_8 \geq 21$.
Case 2: $n_6 = 14, n_7 = 16$. Then $D_{i+5} = 1$, $D_{i+6} = 2$ and $D_{i+7} = 1$, but this violates S2 and is thus impossible.
Case 3: $n_6 = 15, n_7 = 16$. Then there are three consecutive rounds with exactly one active S-box each, which is again not possible.
Therefore $n_k \geq 2k+1$ for $k = 7, 8, 9$. □

Lemma 2. *Consider PR-n where $n \geq 64$. Let D_j be the number of active S-boxes in round j. Suppose $D_i = 1$ and $D_{i-1} \geq 2$. Denote the total number of active S-boxes from round i to round $(i-k+1)$ (both rounds inclusive) by n_k, for $k \leq i$. Then $n_1 = 1$, $n_2 \geq 3$, $n_3 \geq 7$, $n_4 \geq 11$, $n_5 \geq 13$, $n_6 \geq 14$ and $n_k \geq 2k+1$ for $k = 7, 8, 9$.*

Proof. The proof is similar to that of Lemma 1 and by using P1 and P2 instead. □

Lemma 3. *For $n \geq 64$, the r-round differential characteristic of PR-n has a minimum of $2r$ active S-boxes for $r = 5, 6, 7, 8, 9$.*

Proof. With no loss of generality, we consider the first round to the r-th round. Let D_i be the number of active S-boxes in round i. It is proven in [10] that the case for $r = 5$ holds. Consider 6-round differential characteristic. Since there are at least 10 active S-boxes in the first five rounds, if there are at least two active S-boxes in the sixth round, then we are done. Otherwise, suppose that $D_6 = 1$. If $D_5 \geq 2$, then applying Lemma 2 from round 6 to round 1, there is a minimum of 14 active S-boxes. Otherwise $D_5 = 1$. Then we must have a single bit difference to the output of the active S-box in round 5, else it will contradict P1. Because of S2, $D_4 \geq 2$. Now we can apply Lemma 2 from round 5 to round 1, yielding $\sum_{i=1}^{6} D_i \geq 13 + 1 = 14$. For $r = 7, 8, 9$, we apply similar argument as before. □

As a direct consequence of Lemma 3, we have the following result.

Theorem 3. *For $n \geq 64$, the r-round differential characteristic of PR-n has a minimum of $2r$ active S-boxes for $r \geq 5$.*

Linear Cryptanalysis

Lemma 4. *Consider linear approximations of PR-n where $n \geq 64$. The active S-boxes over consecutive rounds cannot form the following patterns:*
1-2-1, 1-3-1, 1-3-2, 1-4-1, 1-4-2, 1-4-3, 1-i, i-1, *for* $i \geq 5$.

Proof. For $i \geq 5$, the patterns 1-i and i-1 clearly cannot happen since the S-boxes are four-bit. If the pattern 1-2-1 were to happen, then the active S-boxes in the middle round are activated by the same S-box and must therefore belong to two different groups. However, they cannot activate only one S-box in the following round. Hence the pattern 1-2-1 is impossible. In general, we see that the active S-boxes in the middle round must activate at least an equal number of S-boxes in the following round. □

Definition 1. *Let a and b be the input and output mask to a S-box respectively. If $wt(a) = wt(b) = 1$, then the S-box is said to have single-bit approximation.*

In particular, with reference to S4, we see that the bias of any single-bit approximation is less than 2^{-3}. Further, we define n_s to be the number of active S-boxes over r rounds with single-bit approximations.

Lemma 5. *Consider r-round linear approximations of PR-n where $n \geq 64$.*

1. *For $r = 3$ and the pattern 1-j-j, where $1 \leq j \leq 4$, $n_s = j$.*
2. *For $r = 3$ and the pattern 1-2-3, $n_s = 1$.*
3. *For $r = 3$ and the pattern 1-3-4, $n_s = 2$.*
4. *For $r = 9$ and $\{1,1,1,1,1,1,1,1,i\}$ where $i \geq 2$, $n_s \geq 6$.*
5. *For $r = 9$ and $\{1,1,1,1,1,1,1,i,j\}$ where $i \geq 2$ and $j \geq 2$, $n_s \geq 3$.*
6. *For $r = 9$ and $\{1,1,1,1,1,1,i,j,k\}$ where $i \geq 2$, $j \geq 2$ and $k \geq 2$, $n_s \geq 2$.*

Proof. (1) to (3) follows easily from P3 and P4, while (4) to (6) can be easily deduced from Lemma 4. □

Lemma 6. *Consider nine-round linear approximations of PR-n where $n \geq 64$. Suppose there are exactly k rounds with one active S-box each and for each of the remaining rounds, there are exactly 2 active S-boxes, where $3 \leq k \leq 8$. Then $n_s \geq k - 2$.*

Proof. The result for $k = 8$ follows directly from Lemma 5.4. For $k = 7$, by Lemma 4, we either have two active S-boxes in the first and the last rounds, or, we have 1-2-2 occurring in the pattern. For the former, $n_s = 5$. For the latter, by virtue of Lemma 5, each time 1-2-2 occurs in the pattern, there are two active S-boxes with single-bit approximations. Hence, $n_s \geq 5$. For $k = 6$, again, because of Lemma 4, either the pattern contains 1-2-2, or is of the form 2-2-2-1-· · ·-1 or 2-2-1-1-· · ·-1-2. Together with Lemma 5, it can thus be easily checked that $n_s \geq 4$. We can apply the same argument to the cases when $k = 5, 4$ and 3. □

Theorem 4. *Let ϵ_r be the maximal bias of a linear approximation of r rounds of PR-n where $n \geq 64$. Then $\epsilon_r \leq 2^{-2r+1}$ for $r = 4, 5, 6, 7, 8$ and 9.*

Proof. Since the argument is similar for $r = 4, 5, 6, 7, 8$ and 9, we shall just consider the (most complicated) case of $r = 9$.

Let ϵ_9^j denote the bias of a linear approximation over nine rounds involving j active S-boxes. We consider the following cases.

Case 1: Each round of a nine-round linear approximation has exactly one active S-box. Then there are at least 7 active S-boxes with single-bit approximations. By virtue of Matsui's piling-up lemma, we have

$$\epsilon_9^9 \leq 2^8 \times (2^{-3})^7 \times (2^{-2})^2 = 2^{-17}.$$

Case 2: There are exactly ten active S-boxes over nine rounds. Then by Lemma 6, we have that

$$\epsilon_9^{10} \leq 2^9 \times (2^{-3})^6 \times (2^{-2})^4 = 2^{-17}.$$

Case 3: There are exactly eleven active S-boxes over nine rounds. There are two possiblities: $\{1,1,1,1,1,1,1,2,2\}$ and $\{1,1,1,1,1,1,1,1,3\}$. It can be deduced from Lemma 6 and Lemma 5.4 respectively that $n_s \geq 5$. Hence, we have

$$\epsilon_9^{11} \leq 2^{10} \times (2^{-3})^5 \times (2^{-2})^6 = 2^{-17}.$$

Case 4: There are exactly twelve active S-boxes over nine rounds. We consider: (a) $\{1,1,1,1,1,1,2,2,2\}$ (b) $\{1,1,1,1,1,1,1,2,3\}$ (c) $\{1,1,1,1,1,1,1,1,4\}$ For (a) and (c), we apply Lemma 6 and Lemma 5.4 respectively to deduce that $n_s \geq 4$. For (b), note that by Lemma 4, 1-3-2, 1-2-1 and 1-3-1 cannot occur in the pattern. So either 1-2-3 occurs in the pattern, or the pattern is of the form 2-3-1-$\cdots$-1, 3-2-1-$\cdots$-1, 2-1-$\cdots$-1-3 or 3-1-$\cdots$-1-2. Together with Lemma 5, $n_s \geq 4$. Therefore,

$$\epsilon_9^{12} \leq 2^{11} \times (2^{-3})^4 \times (2^{-2})^8 = 2^{-17}.$$

Case 5: There are exactly thirteen active S-boxes over nine rounds. Note that the patterns 1-5 and 5-1 are impossible by Lemma 4. Thus it suffices to consider the patterns $\{1,1,1,1,1,2,2,2,2\}$, $\{1,1,1,1,1,1,1,2,4\}$, $\{1,1,1,1,1,1,1,3,3\}$ and $\{1,1,1,1,1,1,2,2,3\}$. Similar to Case 4, it can be shown that $n_s \geq 3$.
Case 6: There are exactly fourteen active S-boxes over nine rounds. By first noting that the patterns 1-6 and 6-1 are impossible by Lemma 4 and following a similar argument to Cases 4 and 5, it can be shown that $n_s \geq 2$.
Case 7: There are exactly fifteen active S-boxes over nine rounds. In a similar fashion to Cases 4, 5 and 6, we have $n_s \geq 1$.
Hence for Case 5($j = 13, n_s \geq 3$), Case 6($j = 14, n_s \geq 2$) and Case 7($j = 15, n_s \geq 1$), we have

$$\epsilon_9^j \leq 2^{j-1} \times (2^{-3})^{n_s} \times (2^{-2})^{j-n_s} = 2^{-17}.$$

Case 8: There are more than 15 active S-boxes over nine rounds. Then

$$\epsilon_9^j \leq 2^{j-1} \times (2^{-2})^j = 2^{-j-1} \leq 2^{-17}, \text{ for } j > 15.$$ □

A.2 Testvectors

	EPCBC(48,96)	EPCBC(96,96)
plaintext	0123456789AB	0123456789ABCDEF01234567
key	0123456789ABCDEF01234567	0123456789ABCDEF01234567
ciphertext	0B46B67143DC	408C65649781E6A5C9757244

On Permutation Layer of Type 1, Source-Heavy, and Target-Heavy Generalized Feistel Structures

Shingo Yanagihara and Tetsu Iwata

Dept. of Computational Science and Engineering, Nagoya University, Japan
s_yanagi@echo.nuee.nagoya-u.ac.jp, iwata@cse.nagoya-u.ac.jp

Abstract. The Generalized Feistel Structure (GFS) generally uses the sub-block-wise cyclic shift in the permutation layer, the layer between the two F function layers. For Type 2 GFS, at FSE 2010, Suzaki and Minematsu showed that a better diffusion property can be obtained if one uses some other sub-block-wise permutation. In this paper, we consider Type 1, Source-Heavy (SH), and Target-Heavy (TH) GFSs, and study if their diffusion properties can be improved by changing the sub-block-wise cyclic shift. For Type 1 GFS, we show that it achieves better diffusion for many cases, while this is not the case for SH and TH GFSs, i.e., the diffusion property of SH and TH GFSs does not change even if we change the sub-block-wise cyclic shift. We also experimentally derive optimum permutations in terms of diffusion, and evaluate the security of the resulting schemes against saturation, impossible differential, differential, and linear attacks.

Keywords: Blockcipher, generalized Feistel structure (GFS), permutation layer, computer experiment.

1 Introduction

Background. Building a secure and efficient blockcipher has been studied extensively, and the Generalized Feistel Structure, which we write GFS, is one of the widely used structures adopted in many practical constructions. In the classical Feistel structure used, e.g., in DES [14] or Camellia [3], the plaintext is divided into two halves, while in GFS, the plaintext is divided into d sub-blocks for $d > 2$. These structures have advantage over the SP network used in AES [13] in that the encryption and decryption algorithms are similar and thus allow small implementations on hardware. There are several types of GFS known in literature [42,29]. For example, Type 1, Type 2, Type 3, Source-Heavy (SH), Target-Heavy (TH), Alternating, and Nyberg's GFS are known. We list examples of the constructions that are based on these types of GFS. Type 1 GFS is used in CAST-256 [1], as well as the blockcipher used in Lesamnta [16]. Type 2 GFS is used in RC6 [33], HIGHT [18], and CLEFIA [37]. SH GFS is used in RC2 [32], SPEED [41], and the blockcipher used in SHA-1 and SHA-2 [15], and TH GFS is used in MARS [19]. BEAR/LION [2] is an example that uses the Alternating GFS.

D. Lin, G. Tsudik, and X. Wang (Eds.): CANS 2011, LNCS 7092, pp. 98–117, 2011.

The security of these structures has been extensively evaluated. The pseudorandomness of Type 1, Type 2, Type 3, and Alternating GFSs is analyzed in [42,27,17]. The pseudorandomness of unbalanced Feistel structure is proved in [23,28]. [29] shows the differential attack and linear attack against Nyberg's GFS, and [34] shows the differential attack and linear attack against unbalanced Feistel structure. The distinguishing attacks against SH and TH GFSs are presented in [30] and in [20,31], respectively. The security of Type 1 GFS against the impossible differential attack is analyzed in [21]. In [6], the security of SH GFS against differential and linear attacks is studied. The lower bounds on the number of active S-boxes for Type 1 and Type 2 GFSs with SP-functions and single-round diffusion are proved in [40] and [35], respectively. The lower bounds for Type 1 and Type 2 GFSs with SP-functions and multiple-round diffusions are proved in [36]. The security of Type 1, Type 2, and TH GFS against impossible boomerang attack is analyzed in [11]. [7,9,8] analyze the security of Type 1 and Type 2 GFSs with double SP-functions and single-round diffusion. [22,26] study the provable security of various types of GFSs against the differential attack.

Generally, GFS uses a sub-block-wise cyclic shift in the permutation layer, a layer between the two F function layers. At FSE 2010, Suzaki and Minematsu considered to use a permutation which is not a cyclic shift, and demonstrated that the security of Type 2 GFS is actually improved [39]. More precisely, they introduced the notion of the maximum diffusion round, which we write $\mathrm{DR}_{\max}(\pi)$, to evaluate the diffusion property of Type 2 GFS that uses a sub-block-wise permutation π in the permutation layer. Intuitively, $\mathrm{DR}_{\max}(\pi)$ indicates how many rounds are needed to achieve the full diffusion, i.e., each output sub-block depends on all input sub-blocks. They provided a construction of a permutation that has a good diffusion property. They also experimentally derived the optimum permutation in terms of diffusion for $d \leq 16$. For the resulting schemes, they analyzed their pseudorandomness as well as the security against the saturation attack [12], impossible differential attack [4], differential attack [5], and linear attack [24].

Our Contributions. In this paper, we closely look at Type 1, SH, and TH GFSs to see if changing the permutation in their permutation layer from the sub-block-wise cyclic shift improves their diffusion property or security against various attacks. After encrypting several rounds, we expect that each output sub-block depends on all input sub-blocks, and the blockcipher achieves the full diffusion. However, some permutation makes the cipher weak. For example, if we use the identity mapping in the permutation layer, then the full diffusion cannot be achieved. For Type 1, SH, and TH GFSs, we first identify a necessary condition on the permutation so that the blockcipher achieves the full diffusion. For Type 1 GFS, we next introduce two parameters, which we write r_{01} and r_{10}. These are the numbers of rounds so that the 0th sub-block reaches the 1st sub-block, and vice-versa. Basing on these parameters, we obtain a lower bound on $\mathrm{DR}_{\max}(\pi)$. Our main result on Type 1 GFS is the explicit construction of the optimum π in terms of diffusion for odd d, that is, our permutation tightly meets the lower bound on $\mathrm{DR}_{\max}(\pi)$. For SH and TH GFSs, we show that changing the permutation layer does not change the diffusion property as long as they achieve the full diffusion. In other words, basing on our definition of equivalence, as long as $\mathrm{DR}_{\max}(\pi)$ is finite, we show that the resulting schemes are equivalent to the scheme that uses the cyclic shift.

We then experimentally search over all permutations. For Type 1 GFS, we search for $3 \le d \le 16$, and for SH and TH GFSs, we search for $3 \le d \le 8$, and list all permutations (but omitting the equivalent permutations) that are better than the sub-block-wise cyclic shift in terms of diffusion. We also evaluate the security of the resulting schemes against saturation, impossible differential, differential, and linear attacks. As a result, we find that for Type 1 GFS and for $3 \le d \le 16$ and $d \ne 3, 4, 6$, the diffusion property can be improved if one changes the permutation from the cyclic shift. Furthermore, the security against saturation, differential, and linear attacks improves in many cases, and does not get worse. On the other hand, for $d = 3, 4, 6$, the diffusion property does not change. For all cases, the security against the impossible differential attack does not change. For SH and TH GFSs, the cyclic shift is optimum in terms of diffusion, and changing the permutation does not change the security against saturation, impossible differential, differential, and linear attacks.

2 Preliminaries

For two bit strings X and Y of the same length, $X \oplus Y$ is their xor (exclusive-or). For an integer $n \ge 1$, $\{0,1\}^n$ is the set of all bit strings of n bits.

2.1 Generalized Feistel Structure (GFS)

In Fig. 1, we illustrate the overall structure of GFS. It takes an N-bit plaintext x and a secret key as inputs and outputs an N-bit ciphertext y. It is parameterized by an integer d, which we call the number of sub-blocks, where each sub-block is n bits, and we thus have $N = dn$. GFS has an iterated structure and it consists of several round functions. The number of rounds is denoted by R. The round function itself consists of the $\mathcal{F}$-layer and the Π-layer. The $\mathcal{F}$-layer has a key dependent F function and the xor operation, and its structure depends on the types of GFS. The Π-layer, which we also call the permutation layer, is a permutation π over the sub-blocks, i.e., $\pi : (\{0,1\}^n)^d \to (\{0,1\}^n)^d$ is a permutation over the d sub-blocks. We assume that the final round, the R-th round, consists of only the $\mathcal{F}$-layer. The decryption is done by using $\mathcal{F}^{-1}$-layer and Π^{-1}-layer in an obvious way.

The sub-blocks are numbered sequentially as $0, 1, \ldots, d-1$ from left to right, and instead of treating π as a permutation over $(\{0,1\}^n)^d$, we regard it as a permutation over $\{0, 1, \ldots, d-1\}$ and write $\pi(i)$ for the index of the sub-block after applying π to the i-th sub-block. For example, in Fig. 1, we have $\pi(0) = 3$, $\pi(1) = 0$, $\pi(2) = 1$, and $\pi(3) = 2$, and they are collectively written as $\pi = (3, 0, 1, 2)$. For an integer $r \ge 1$, $\pi^r(i)$ is the index of the sub-block after applying the permutation π on the i-th sub-block for r times. Similarly, $\pi^{-r}(i)$ is the one

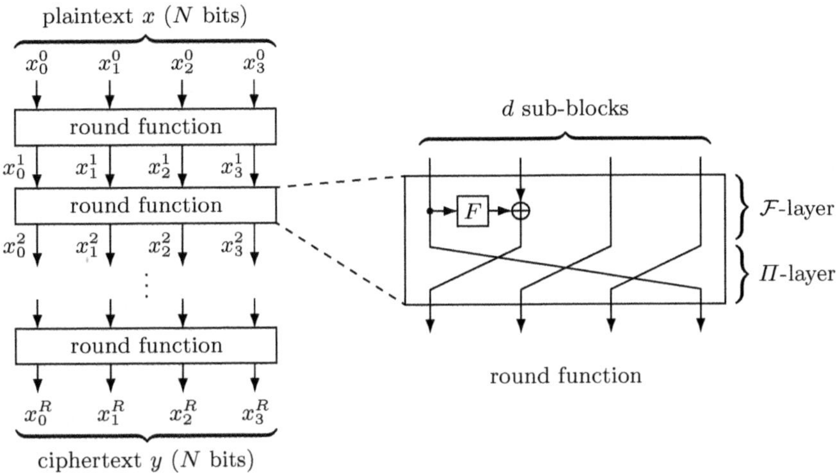

Fig. 1. Overview of GFS ($d = 4$)

after applying π^{-1} on the i-th sub-block for r times. We use the convention that $\pi^0(i) = i$. Let π_s be defined as $\pi_s = (d-1, 0, 1, 2, \ldots, d-2)$, i.e., π_s corresponds to the sub-block-wise left cyclic shift.

Let x^r be the intermediate result of the encryption after r rounds, where x^0 is the plaintext and x^R is the ciphertext. Let $x^r = (x_0^r, \ldots, x_{d-1}^r) \in (\{0,1\}^n)^d$ be the partition of x^r into n-bit strings, i.e., x_i^r is the i-th sub-block after r round. Similarly, let y^r be the intermediary result of the decryption after r rounds, where y^0 is the ciphertext and y^R is the plaintext. We write $y^r = (y_0^r, \ldots, y_{d-1}^r)$ for its n-bit partition.

Next, we review three types of GFS.

Type 1 GFS. Let $E^{\mathrm{T1}}(\pi)$ be Type 1 GFS that uses π in the Π-layer. For $E^{\mathrm{T1}}(\pi)$, x_i^r is defined as

$$x_i^r = \begin{cases} F(x_0^{r-1}) \oplus x_1^{r-1} & \text{if } \pi^{-1}(i) = 1, \\ x_{\pi^{-1}(i)}^{r-1} & \text{otherwise,} \end{cases}$$

where $F : \{0,1\}^n \to \{0,1\}^n$. Figure 2 (left) shows $E^{\mathrm{T1}}(\pi_s)$ with $d = 4$.

Source-Heavy GFS. Let $E^{\mathrm{SH}}(\pi)$ be Source-Heavy (SH) GFS that uses π. For $E^{\mathrm{SH}}(\pi)$, x_i^r is defined as

$$x_i^r = \begin{cases} F(x_0^{r-1}, \ldots, x_{d-2}^{r-1}) \oplus x_{d-1}^{r-1} & \text{if } \pi^{-1}(i) = d-1, \\ x_{\pi^{-1}(i)}^{r-1} & \text{otherwise,} \end{cases}$$

where $F : \{0,1\}^{(d-1)n} \to \{0,1\}^n$. Figure 2 (middle) shows $E^{\mathrm{SH}}(\pi_s)$ with $d = 4$.

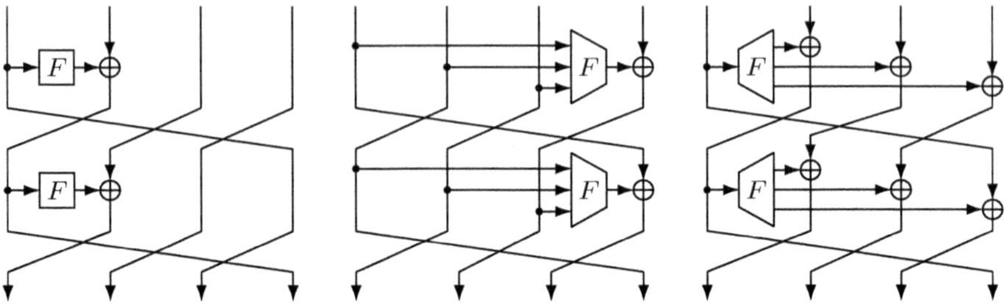

Fig. 2. Type 1 GFS (left), SH GFS (middle), and TH GFS (right), where $d = 4$ and $\pi = (3, 0, 1, 2)$ $(= \pi_s)$

Target-Heavy GFS. Let $E^{\mathrm{TH}}(\pi)$ be Target-Heavy (TH) GFS that uses π. For $E^{\mathrm{TH}}(\pi)$, x_i^r is defined as

$$x_i^r = \begin{cases} x_{\pi^{-1}(i)}^{r-1} & \text{if } \pi^{-1}(i) = 0, \\ F_{\pi^{-1}(i)}(x_0^{r-1}) \oplus x_{\pi^{-1}(i)}^{r-1} & \text{otherwise,} \end{cases}$$

where $F : \{0,1\}^n \to \{0,1\}^{(d-1)n}$ is a function such that $F = (F_1, \ldots, F_{d-1})$ for $F_j : \{0,1\}^n \to \{0,1\}^n$. Figure 2 (right) shows $E^{\mathrm{TH}}(\pi_s)$ with $d = 4$.

Other types of GFS include Type 2 GFS, Type 3 GFS, Alternating GFS, and Nyberg's GFS.

2.2 Diffusion of GFS

In this subsection, following [39], we introduce the notion of $\mathrm{DR}_{\max}(\pi)$ to evaluate the diffusion property of GFS.

Data Dependent Variables. We define variables $X^r \in \{0,1\}^d$ and $Y^r \in \{0,1\}^d$, which we call *data dependent variables.* For the encryption, when we consider the diffusion of the i-th sub-block of the input data (plaintext), we let $X^0 = (X_0^0, X_1^0, \ldots, X_{d-1}^0)$, where $X_i^0 = 1$ and $X_{i'}^0 = 0$ for $\forall i' \neq i$. Intuitively, if the j-th sub-block after encrypting r rounds depends on the i-th input sub-block, then we let $X_j^r = 1$, and $X_j^r = 0$ otherwise (the precise definition is given below). We write $X^r = (X_0^r, X_1^r, \ldots, X_{d-1}^r)$.

Similarly, the data dependent variable Y^r is defined for the decryption. When we consider the diffusion of the i-th sub-block of the ciphertext, we let $Y^0 = (Y_0^0, Y_1^0, \ldots, Y_{d-1}^0)$, where $Y_i^0 = 1$ and $Y_{i'}^0 = 0$ for $\forall i' \neq i$. If the j-th sub-block after decrypting r rounds depends on the i-th ciphertext sub-block, we let $Y_j^r = 1$, and $Y_j^r = 0$ otherwise. We write $Y^r = (Y_0^r, Y_1^r, \ldots, Y_{d-1}^r)$.

Let $|X^r|$ be the Hamming weight of X^r, that is, $|X^r|$ is the number of bit "1" in X^r. We say that the i-th sub-block is *active* if $X_i^r = 1$. If there exists $r \geq 0$ such that all sub-blocks are active, i.e., $|X^r| = d$, then we say that the input X^0 *achieves FD (Full Diffusion).* If all X^0 such that $|X^0| = 1$ achieve FD, then we say that the blockcipher achieves FD.

Definition of X^r and Y^r of Type 1, SH, and TH GFSs. For $E^{\mathrm{T1}}(\pi)$, given $X^0 = (X_0^0, X_1^0, \ldots, X_{d-1}^0) \in \{0,1\}^d$ and $Y^0 = (Y_0^0, Y_1^0, \ldots, Y_{d-1}^0) \in \{0,1\}^d$, X^r and Y^r for $r \geq 1$ are successively defined as follows.

$$X_i^r = \begin{cases} X_0^{r-1} \vee X_1^{r-1} & \text{if } \pi^{-1}(i) = 1 \\ X_{\pi^{-1}(i)}^{r-1} & \text{otherwise} \end{cases} \qquad Y_i^r = \begin{cases} Y_0^{r-1} \vee Y_1^{r-1} & \text{if } \pi(i) = 1 \\ Y_{\pi(i)}^{r-1} & \text{otherwise} \end{cases}$$

We note that $a \vee b$ is the or operation of a and b.

Next, for $E^{\mathrm{SH}}(\pi)$, X^r and Y^r for $r \geq 1$ are defined as follows.

$$X_i^r = \begin{cases} X_0^{r-1} \vee X_1^{r-1} \vee \cdots \vee X_{d-1}^{r-1} & \text{if } \pi^{-1}(i) = d-1 \\ X_{\pi^{-1}(i)}^{r-1} & \text{otherwise} \end{cases}$$

$$Y_i^r = \begin{cases} Y_0^{r-1} \vee Y_1^{r-1} \vee \cdots \vee Y_{d-1}^{r-1} & \text{if } \pi(i) = d-1 \\ Y_{\pi(i)}^{r-1} & \text{otherwise} \end{cases}$$

Finally, for $E^{\mathrm{TH}}(\pi)$, X^r and Y^r for $r \geq 1$ are similarly defined as follows.

$$X_i^r = \begin{cases} X_0^{r-1} & \text{if } \pi^{-1}(i) = 0 \\ X_0^{r-1} \vee X_{\pi^{-1}(i)}^{r-1} & \text{otherwise} \end{cases} \qquad Y_i^r = \begin{cases} Y_0^{r-1} & \text{if } \pi(i) = 0 \\ Y_0^{r-1} \vee Y_{\pi(i)}^{r-1} & \text{otherwise} \end{cases}$$

Definition of $\mathrm{DR}_{\max}(\pi)$. Next, we define $\mathrm{DR}_{\max}(\pi)$, which is used to characterize the diffusion property of GFS using π in the Π-layer. Roughly speaking, this value is defined as the minimum number of round so that every sub-block depends on all the input sub-blocks.

More precisely, it is defined as $\mathrm{DR}_{\max}(\pi) \stackrel{\mathrm{def}}{=} \max\{\mathrm{DR}_{\max}^E(\pi), \mathrm{DR}_{\max}^D(\pi)\}$. To define $\mathrm{DR}_{\max}^E(\pi)$, we first define $\mathrm{DR}_i^E(\pi)$, which is the minimum number of round such that the i-th input sub-block diffuses to all the sub-blocks in the encryption direction, i.e., it is defined as

$$\mathrm{DR}_i^E(\pi) \stackrel{\mathrm{def}}{=} \min\{r \mid \forall i' \neq i, X_{i'}^0 = 0, X_i^0 = 1, |X^r| = d\}.$$

Then $\mathrm{DR}_{\max}^E(\pi)$ is defined as the maximum of $\mathrm{DR}_i^E(\pi)$ over all $0 \leq i \leq d-1$ as follows.

$$\mathrm{DR}_{\max}^E(\pi) \stackrel{\mathrm{def}}{=} \max\{\mathrm{DR}_i^E(\pi) \mid 0 \leq i \leq d-1\}$$

Next, $\mathrm{DR}_{\max}^D(\pi)$ is similarly defined for the decryption direction. First, $\mathrm{DR}_i^D(\pi)$ is defined as

$$\mathrm{DR}_i^D(\pi) \stackrel{\mathrm{def}}{=} \min\{r \mid \forall i' \neq i, Y_{i'}^0 = 0, Y_i^0 = 1, |Y^r| = d\},$$

and $\mathrm{DR}_{\max}^D(\pi)$ is defined as

$$\mathrm{DR}_{\max}^D(\pi) \stackrel{\mathrm{def}}{=} \max\{\mathrm{DR}_i^D(\pi) \mid 0 \leq i \leq d-1\}.$$

We note that smaller $\mathrm{DR}_{\max}(\pi)$ implies the better diffusion, and $\mathrm{DR}_{\max}(\pi) = \infty$ implies that there exists an input sub-block X_i^0 such that, after any rounds, some output sub-block is independent of the i-th input sub-block. We also remark that the above definitions are given for the sub-block-wise dependency, and that "linear dependency" is sufficient. That is, the definition of the full diffusion does not guarantee that every bit in every sub-block depends on all the input bits.

There are $d!$ permutations over $\{0, 1, \ldots, d-1\}$ in total, and we say that the permutation π is *optimum* in terms of diffusion if $\mathrm{DR}_{\max}(\pi)$ is the minimum among all the $d!$ permutations. We note that the optimum π may not be unique.

3 Equivalence of GFSs

In this section, we define the equivalence of GFSs.

Case for Type 1 GFS. Let $E^{\mathrm{T1}}(\pi)$ and $E^{\mathrm{T1}}(\pi')$ be Type 1 GFSs using π and π', respectively, in the Π-layer. We say that $E^{\mathrm{T1}}(\pi)$ and $E^{\mathrm{T1}}(\pi')$ are *equivalent* if there exists $\pi^* = (a_0, \ldots, a_{d-1})$ such that $\pi' = \pi^* \circ \pi \circ (\pi^*)^{-1}$ and

$$a_0 = 0, a_1 = 1, \text{ and } \{a_2, \ldots, a_{d-1}\} = \{2, 3, \ldots, d-1\},$$

where $g \circ f(x)$ is $g(f(x))$. That is, $E^{\mathrm{T1}}(\pi)$ and $E^{\mathrm{T1}}(\pi')$ are equivalent if π' is obtained from π by permuting the last $d-2$ elements under π^*. Notice that we have $E^{\mathrm{T1}}(\pi') = \pi^* \circ E^{\mathrm{T1}}(\pi) \circ (\pi^*)^{-1}$, i.e., $(\pi^*)^{-1}$ can be moved to the input of $E^{\mathrm{T1}}(\pi)$, and π^* can be moved to the output. For example, let $d = 6$ and consider $E^{\mathrm{T1}}(\pi)$ and $E^{\mathrm{T1}}(\pi')$, where $\pi = (5, 0, 1, 2, 3, 4)$ and $\pi' = (4, 0, 1, 5, 3, 2)$. Then we see that these two Type 1 GFSs are equivalent since $\pi' = \pi^* \circ \pi \circ (\pi^*)^{-1}$ for $\pi^* = (0, 1, 2, 5, 3, 4)$.

Case for SH GFS. Let $E^{\mathrm{SH}}(\pi)$ and $E^{\mathrm{SH}}(\pi')$ be SH GFSs using π and π'. We say that they are *equivalent* if there exists $\pi^* = (a_0, \ldots, a_{d-1})$ such that $\pi' = \pi^* \circ \pi \circ (\pi^*)^{-1}$ and

$$\{a_0, \ldots, a_{d-2}\} = \{0, 1, \ldots, d-2\} \text{ and } a_{d-1} = d-1.$$

In other words, $E^{\mathrm{SH}}(\pi)$ and $E^{\mathrm{SH}}(\pi')$ are equivalent if π' is obtained from π by permuting the first $d-1$ elements under π^*.

Case for TH GFS. Let $E^{\mathrm{TH}}(\pi)$ and $E^{\mathrm{TH}}(\pi')$ be TH GFSs using π and π'. We say that they are *equivalent* if there exists $\pi^* = (a_0, \ldots, a_{d-1})$ such that $\pi' = \pi^* \circ \pi \circ (\pi^*)^{-1}$ and

$$a_0 = 0 \text{ and } \{a_1, \ldots, a_{d-1}\} = \{1, 2, \ldots, d-1\},$$

i.e., if π' is obtained from π by permuting the last $d-1$ elements under π^*.

4 Analysis on $\mathrm{DR}_{\max}(\pi)$

In the section, we present our theoretical analyses on $\mathrm{DR}_{\max}(\pi)$ for Type 1, SH and TH GFSs. For Type 1 GFS, in Lemma 1 and Lemma 2, we first identify a necessary condition of π so that $\mathrm{DR}_{\max}(\pi)$ is finite. We then introduce two new parameters, r_{01} and r_{10}, associated to π. Basing on these parameters, in Lemma 3, we obtain a lower bound on $\mathrm{DR}_{\max}(\pi)$. In Lemma 4, we present a sufficient condition of π for odd d to achieve the lowest value of $\mathrm{DR}_{\max}(\pi)$. In Lemma 5, we show a construction of π for given r_{01} and r_{10}. Our main result is in Theorem 1, which shows an explicit construction of optimum π in terms of diffusion for odd d.

For SH and TH GFSs, we first identify in Lemma 6 and in Lemma 8 a necessary condition of π so that $\mathrm{DR}_{\max}(\pi)$ is finite. We then show in Lemma 7 and in Lemma 9 that, if $\mathrm{DR}_{\max}(\pi)$ is finite, then the resulting scheme is equivalent to $E^{\mathrm{SH}}(\pi_s)$ or $E^{\mathrm{TH}}(\pi_s)$. These results show that, in SH and TH GFSs, the diffusion does not improve even if we change the permutation in the Π-layer.

4.1 Type 1 GFS

First, we present the following lemma.

Lemma 1. *If there does not exist $r \geq 0$ such that $\pi^r(i) = 0$ for some $0 \leq i \leq d-1$, then $\mathrm{DR}_{\max}(\pi) = \infty$.*

A proof is obvious as the 0th sub-block does not depend on the i-th sub-block, and hence the full diffusion cannot be achieved. For example, consider Type 1 GFS with $\pi = (3, 0, 4, 1, 2)$. For $i = 2$, one can easily verify that $\pi^r(i)$ is either 2 or 4, and therefore we have $\mathrm{DR}_{\max}(\pi) = \infty$ for this π. From this lemma, for all $0 \leq i \leq d-1$, there must exist $r \geq 0$ such that $\pi^r(i) = 0$ so that $\mathrm{DR}_{\max}(\pi)$ is finite. In other words, Lemma 1 gives a necessary condition of $\mathrm{DR}_{\max}(\pi)$ being finite.

The next lemma gives the equivalent condition of the necessary condition of Lemma 1.

Lemma 2. *For all $0 \leq i \leq d-1$, there exists $r \geq 0$ such that $\pi^r(i) = 0$, if and only if for all $0 \leq i \leq d-1$, we have $\{\pi^1(i), \pi^2(i), \ldots, \pi^{d-1}(i)\} = \{0, 1, \ldots, d-1\} \setminus \{i\}$ and $\pi^d(i) = i$.*

Proof. We first prove the "only if" direction. Consider a permutation π that, for all $0 \leq i \leq d-1$, there exists $r \geq 0$ that satisfies $\pi^r(i) = 0$. Let $I[j]$ be the index of the sub-block after applying the permutation π on the 0th sub-block for j times, i.e., $I[j] = \pi^j(0)$. Note that $I[0] = 0$. For $l \geq 1$, let $\mathcal{L}_l = \{I[0], I[1], \ldots, I[l-1]\}$ be the list of $I[j]$. In the proof, we first identify a property satisfied by the permutation, namely, we show that $\{I[0], \ldots, I[d-1]\} = \{0, \ldots, d-1\}$ and $I[d] = I[0]$ hold. We then show the lemma based on the property by a mathematical induction on i.

If $\pi(I[0]) = 0$, this contradicts against the fact that for all $i' \notin \mathcal{L}_1$, there exists $r \geq 0$ such that $\pi^r(i') = 0$. Therefore, we have $\pi(I[0]) \neq 0$, and hence $\pi(I[0]) = I[1] \notin \mathcal{L}_1$. Next, if $\pi(I[1]) = 0$, this contradicts against the fact that for all $i' \notin \mathcal{L}_2$, there exists $r \geq 0$ such that $\pi^r(i') = 0$. Therefore, we have $\pi(I[1]) \neq 0$. Besides, since π is a permutation, we have $\pi(I[1]) \neq I[1]$. These two facts imply $\pi(I[1]) = I[2] \notin \mathcal{L}_2$. Similarly, let $1 \leq k \leq d-1$ and suppose that $\pi(I[k-1]) = 0$. Then we see that this contradicts against the fact that for all $i' \notin \mathcal{L}_k$, there exists $r \geq 0$ such that $\pi^r(i') = 0$, and hence we have $\pi(I[k-1]) \neq 0$. Considering that π is a permutation, we have $\pi(I[k-1]) \neq I[1], I[2], \ldots, I[k-1]$. Therefore, $\pi(I[k-1]) = I[k] \notin \mathcal{L}_k$.

At this point, we have $\mathcal{L}_d = \{I[0], \ldots, I[d-1]\} = \{0, \ldots, d-1\}$. From the fact that π is a permutation, we necessary have $\pi(I[d-1]) = I[0]$, and hence we obtain that $I[d] = I[0]$.

Next, we consider the mathematical induction part. As we have shown that $\{I[0], \ldots, I[d-1]\} = \{0, \ldots, d-1\}$ holds, instead of proving the lemma for $i = 0, 1, \ldots, d-1$, we show the proof for $i = I[0], I[1], \ldots, I[d-1]$.

First, we prove the base case of $i = I[0]$. From the definition of $I[j]$, we have

$$\begin{aligned}\{\pi^1(I[0]), \pi^2(I[0]), \ldots, \pi^{d-1}(I[0])\} &= \{\pi^1(0), \pi^2(0), \ldots, \pi^{d-1}(0)\} \\ &= \{I[1], I[2], \ldots, I[d-1]\} \\ &= \{I[0], I[1], I[2], \ldots, I[d-1]\} \setminus \{I[0]\} \\ &= \{0, 1, \ldots, d-1\} \setminus \{0\},\end{aligned}$$

where the last equality follows from the property that we have proved above, and we have $\pi^d(I[0]) = I[d] = I[0]$, which also follows from the above mentioned property. This concludes the proof for the base case.

We next consider the induction step. Fix $0 \leq j \leq d-2$ and suppose that the lemma holds for $i = I[j]$. We show the lemma for $i = I[j+1]$. Because $I[j+1] = \pi^1(I[j])$, we have

$$\begin{aligned}&\{\pi^1(I[j+1]), \pi^2(I[j+1]), \ldots, \pi^{d-1}(I[j+1])\} \\ &= \{\pi^2(I[j]), \pi^3(I[j]), \ldots, \pi^d(I[j])\} \\ &= (\{\pi^1(I[j]), \pi^2(I[j]), \pi^3(I[j]), \ldots, \pi^{d-1}(I[j])\} \cup \{\pi^d(I[j])\}) \setminus \{\pi^1(I[j])\} \\ &= ((\{0, 1, \ldots, d-1\} \setminus \{I[j]\}) \cup \{I[j]\}) \setminus \{\pi^1(I[j])\} \\ &= \{0, 1, \ldots, d-1\} \setminus \{I[j+1]\},\end{aligned}$$

where the third equality follows from the induction hypothesis, and we also have $\pi^d(I[j+1]) = \pi^{d+1}(I[j]) = \pi(\pi^d(I[j])) = \pi(I[j]) = I[j+1]$, where we used the induction hypothesis at the third equality. This completes the proof for the induction step, and we obtain the "only if" direction of the lemma.

We next consider the "if" direction. First, we obviously have $\pi^0(0) = 0$ and hence we have $r \geq 0$ such that $\pi^r(i) = 0$ for $i = 0$. Next, for $1 \leq i \leq d-1$, we have $r \geq 0$ such that $\pi^r(i) = 0$ since $\{\pi^1(i), \pi^2(i), \ldots, \pi^{d-1}(i)\} = \{0, 1, \ldots, d-1\} \setminus \{i\}$ and hence $0 \in \{0, 1, \ldots, d-1\} \setminus \{i\}$. □

In the rest of the analysis of Type 1 GFS, we only consider a permutation π that satisfies the condition in Lemma 2. We next introduce r_{01} and r_{10} which are associated to π and will be used to characterize the diffusion property of Type 1 GFS that uses π.

Definition 1. *For any permutation π over $\{0, 1, \ldots, d-1\}$, let r_{01} be the smallest $r \geq 1$ such that $\pi^r(0) = 1$. Similarly, let r_{10} be the smallest $r \geq 1$ such that $\pi^r(1) = 0$.*

It is easy to see that if π satisfies the condition in Lemma 2, we have $r_{01}+r_{10} = d$. We remark that, for $\pi_s = (d-1, 0, 1, \ldots, d-2)$, we have $r_{01} = d-1$ and $r_{10} = 1$.

In the next lemma, we characterize the lower bound of $\mathrm{DR}_{\max}(\pi)$ by using r_{01} and r_{10}.

Lemma 3. *For any permutation π over $\{0, 1, \ldots, d-1\}$ that satisfies the condition in Lemma 2, we have*

$$\mathrm{DR}_{\max}(\pi) \geq \max\{r_{01}, r_{10}\} \times (d-2) + d.$$

Proof. We see that if the 0th sub-block is active and the 1st sub-block is not, then the number of active sub-blocks increases. We also see that this is the only situation that the number of active sub-blocks increases. In other words, we have $|X^{r+1}| = |X^r| + 1$ if and only if $X_0^r = 1$ and $X_1^r = 0$. The condition "$X_0^r = 1$ and $X_1^r = 0$" is referred to as the *increasing condition.*

We first consider $\mathrm{DR}^E_{\max}(\pi)$, and we see that $\mathrm{DR}^E_{\max}(\pi) \geq R_1 + R_2$ holds, where R_1 is the sufficient number of rounds so that the 0th sub-block being active, and R_2 is the necessary number of rounds so that the 0th sub-block (i.e., $X^r = (1, 0, \ldots, 0)$) achieves FD, since the number of active sub-blocks increases only after the 0th sub-block being active.

We first consider R_1. Consider the input X^0 such that $X^0_{\pi^{-(d-1)}(0)} = 1$ and $\forall i \neq \pi^{-(d-1)}(0)$, $X_i^0 = 0$. Then we see that R_1 is $d-1$, which is the maximum over all inputs. In particular, $\{X^0, \ldots, X^{d-1}\}$ becomes the set of all d-bit vectors of Hamming weight 1, and hence it is enough to evaluate R_2 for this X^0.

When $r = d-1$, we have $X_0^r = 1$, $X_1^r = 0$, and $X_{i'}^r = 0$ for $i' \neq 0, 1$, and the increasing condition is satisfied. Therefore, when $r = d$, we have $X^r_{\pi(0)} = 1$, $X^r_{\pi(1)} = 1$, and $X^r_{i'} = 0$ for $i' \neq \pi(0), \pi(1)$. We also obtain $|X^r| = 2$, i.e., the number of active sub-blocks increases.

We proceed to encrypt X^0, and the number of active sub-blocks is 2 until $r = d + r_{10} - 1$. When $r = d + r_{10} - 1$, we have $X_0^r = 1$, $X^r_{\pi^{r_{10}}(0)} = 1$, and $X^r_{i'} = 0$ for $i' \neq 0, \pi^{r_{10}}(0)$. If the diffusion works ideally, we may assume that $\pi^{r_{10}}(0) \neq 1$, and hence the increasing condition is satisfied. The implies that when $r = d + r_{10}$, we have $X^r_{\pi(0)} = 1$, $X^r_{\pi(1)} = 1$, $X_{\pi^{r_{10}+1}(0)} = 1$, and $X^r_{i'} = 0$ for $i' \neq \pi(0), \pi(1), \pi^{r_{10}+1}(0)$, and we obtain that $|X^r| = 3$.

Similarly, when $r = d + kr_{10} - 1$, where $k \geq 1$, we have $X_0^r = 1$. If $X_1^r = 0$, then the increasing condition is satisfied. Now we consider the sufficient condition for $X_1^r = 0$ by focusing on the 0th and 1st sub-blocks. If $X_0^{r-r'} = 1$ for some $r' = ld + mr_{10} + r_{01}$, where $l \geq 0$ and $m \geq 0$, we have $X_1^r = 1$ from the diffusion of $X_0^{r-r'}$.

If $r' \neq ld + mr_{10} + r_{01}$, we see that X_1^r depends on $X_2^{r-r'}, X_3^{r-r'}, \ldots, X_{d-1}^{r-r'}$, but it does not depend on $X_0^{r-r'}$ nor $X_1^{r-r'}$. Recall that $X_0^{d-1} = 1$ and $X_i^{d-1} = 0$ for $2 \leq i \leq d-1$. Therefore, the sufficient condition for $X_1^r = 0$ is summarized as follows: There does not exist (l, m) such that

$$l \geq 0, m \geq 0, \text{ and } kr_{10} = ld + mr_{10} + r_{01}. \tag{1}$$

We proceed to encrypt X^0, and we see that if there does not exist (l, m) satisfying (1) for every $1 \leq k \leq d-2$, then we have $|X^r| = d$ when $r = d + (d-2)r_{10}$. Therefore, we have $\mathrm{DR}_{\max}^E(\pi) \geq d + (d-2)r_{10}$.

Similarly, the lower bound on $\mathrm{DR}_{\max}^D(\pi)$ can be obtained by using r_{01} instead of r_{10} and we have $\mathrm{DR}_{\max}^D(\pi) \geq d + (d-2)r_{01}$.

Finally, we obtain the result as $\mathrm{DR}_{\max}(\pi) = \max\{\mathrm{DR}_{\max}^E(\pi), \mathrm{DR}_{\max}^D(\pi)\} \geq \max\{r_{10}, r_{01}\} \times (d-2) + d$. □

In the next lemma, for odd d, we characterize a permutation π that achieves the lower bound in Lemma 3 for the smallest value of $\max\{r_{01}, r_{10}\} \times (d-1) + d$. In other words, Lemma 4 shows a sufficient condition to obtain an optimum permutation in terms of diffusion.

Lemma 4. *Let d be an odd integer and let π be a permutation over $\{0, 1, \ldots, d-1\}$ such that $(r_{01}, r_{10}) = ((d+1)/2, (d-1)/2)$ or $((d-1)/2, (d+1)/2)$. Then we have*

$$\mathrm{DR}_{\max}(\pi) = \frac{d(d+1)}{2} - 1.$$

Proof. If there does not exist (l, m) satisfying (1) for every $1 \leq k \leq d-2$, then we have $\mathrm{DR}_{\max}(\pi) = \max\{r_{01}, r_{10}\} \times (d-2) + d$.

We first consider $\mathrm{DR}_{\max}^E(\pi)$. For any $1 \leq k' \leq d-2$, if there does not exist an integer $0 \leq l' \leq k'-1$ such that $k'r_{10} = l'd + r_{01}$, then we see that there does not exist (l, m) satisfying (1). By substituting $r_{01} = (d \pm 1)/2$ and $r_{10} = (d \mp 1)/2$, we have $2l' = (k'-1) \mp (k'+1)/d$. Since $0 < (k'+1)/d < 1$, there is no integer l' that satisfies the equality. Therefore, there does not exist (l, m) satisfying (1) for every $1 \leq k \leq d-2$, and we have $\mathrm{DR}_{\max}^E(\pi) = r_{10} \times (d-2) + d$.

Next, we consider $\mathrm{DR}_{\max}^D(\pi)$, and it can be proved that $\mathrm{DR}_{\max}^D(\pi) = r_{01} \times (d-2) + d$ similarly to the analysis of $\mathrm{DR}_{\max}^E(\pi)$.

Finally, we have $\mathrm{DR}_{\max}(\pi) = \max\{r_{01}, r_{10}\} \times (d-2) + d = d(d+1)/2 - 1$. □

Next, we show that for any d (even or odd) and for any integers a and b such that $a + b = d$, one can obtain a permutation such that $r_{01} = a$ and $r_{10} = b$.

Lemma 5. *Let a and b be integers such that $a, b > 0$ and $a + b = d$. If $a < b$, then we have $r_{01} = a$ and $r_{10} = b$ for*

$$\pi = (2, 3, \ldots, 2a-1, 1, 2a, 2a+1, \ldots, d-1, 0).$$

If $a \geq b$, then we have $r_{01} = a$ and $r_{10} = b$ for

$$\pi = (2, 3, \ldots, 2b, 0, 2b+1, 2b+2, \ldots, d-1, 1).$$

Furthermore, both permutations satisfy the condition in Lemma 2.

Proof. We prove this lemma for the case $a < b$, as another case can be proved similarly. We write $i \xrightarrow{\pi} j$ if $\pi(i) = j$. Then the 0th sub-block is permuted as

$$0 \xrightarrow{\pi} 2 \xrightarrow{\pi} 4 \xrightarrow{\pi} \cdots \xrightarrow{\pi} 2a-2 \xrightarrow{\pi}$$
$$1 \xrightarrow{\pi} 3 \xrightarrow{\pi} 5 \xrightarrow{\pi} \cdots \xrightarrow{\pi} 2a-1 \xrightarrow{\pi}$$
$$2a \xrightarrow{\pi} 2a+1 \xrightarrow{\pi} \cdots \xrightarrow{\pi} d-1 \xrightarrow{\pi} 0.$$

We see that $\pi^a(0) = 1$ and $\pi^b(1) = 0$. Furthermore, for all $0 \le i \le d-1$, there exists $r \ge 0$ such that $\pi^r(i) = 0$. □

Finally, we obtain the following theorem from Lemma 4 and Lemma 5, which shows the construction of the optimum permutation for any odd d.

Theorem 1. *Let d be an odd integer. Then $E^{\mathrm{T1}}(\pi)$ with*

$$\pi = (2, 3, \ldots, d-2, 1, d-1, 0) \quad or \quad \pi = (2, 3, \ldots, d-2, d-1, 0, 1)$$

satisfies $\mathrm{DR}_{\max}(\pi) = d(d+1)/2 - 1$.

4.2 Source-Heavy GFS

Let $E^{\mathrm{SH}}(\pi)$ be SH GFS that uses π. In the next lemma, we present a necessary condition of π so that $\mathrm{DR}_{\max}(\pi)$ being finite.

Lemma 6. *In $E^{\mathrm{SH}}(\pi)$, if $\mathrm{DR}_{\max}(\pi) \neq \infty$, then for all $0 \le i \le d-1$, we have $\{\pi^1(i), \pi^2(i), \ldots, \pi^{d-1}(i)\} = \{0, 1, \ldots, d-1\} \setminus \{i\}$ and $\pi^d(i) = i$.*

A proof is similar to that of Lemma 2. Specifically, the proof is obtained by changing the definition of $I[j] = \pi^j(0)$ in Lemma 2 to $I[j] = \pi^j(d-1)$.

In the following lemma, we show that if π satisfies the condition of Lemma 6, then $E^{\mathrm{SH}}(\pi)$ is equivalent to $E^{\mathrm{SH}}(\pi_s)$.

Lemma 7. *If we have $\{\pi^1(d-1), \pi^2(d-1), \ldots, \pi^{d-1}(d-1)\} = \{0, 1, \ldots, d-2\}$ and $\pi^d(d-1) = d-1$, then $E^{\mathrm{SH}}(\pi)$ is equivalent to $E^{\mathrm{SH}}(\pi_s)$.*

Proof. In order to show that $E^{\mathrm{SH}}(\pi)$ is equivalent to $E^{\mathrm{SH}}(\pi_s)$, it is enough to show that there exists π^* such that $\pi = \pi^* \circ \pi_s \circ (\pi^*)^{-1}$. Let π^* be

$$\pi^* = (\pi^{d-1}(d-1), \pi^{d-2}(d-1), \ldots, \pi(d-1), d-1).$$

In what follows, we show that this π^* satisfies $\pi(k) = \pi^* \circ \pi_s \circ (\pi^*)^{-1}(k)$ for all $0 \le k \le d-1$. First, we see that $\pi^*(d-1-k) = \pi^k(d-1)$, which is equivalent to $(\pi^*)^{-1}(\pi^k(d-1)) = d-1-k$. From the assumption on π, any $0 \le k \le d-1$ can be written as $k = \pi^x(d-1)$ for some $0 \le x \le d-1$. Basing on these observations, we have

$$\pi^* \circ \pi_s \circ (\pi^*)^{-1}(k) = \pi^* \circ \pi_s \circ (\pi^*)^{-1}(\pi^x(d-1)) = \pi^* \circ \pi_s(d-1-x).$$

If $x \neq d-1$, then $\pi^* \circ \pi_s(d-1-x) = \pi^*(d-1-(x+1)) = \pi^{x+1}(d-1) = \pi(\pi^x(d-1)) = \pi(k)$. If $x = d-1$, then $\pi^* \circ \pi_s(d-1-x) = \pi^*(d-1) = d-1 = \pi^d(d-1) = \pi(\pi^{d-1}(d-1)) = \pi(k)$. Therefore, we have $\pi = \pi^* \circ \pi_s \circ (\pi^*)^{-1}$. □

These two results imply that if $\mathrm{DR}_{\max}(\pi) \neq \infty$, then $E^{\mathrm{SH}}(\pi)$ is equivalent to $E^{\mathrm{SH}}(\pi_s)$.

4.3 Target-Heavy GFS

We present the following result, which can be proved similarly to Lemma 2 and Lemma 6 by using $I[j] = \pi^j(0)$.

Lemma 8. *In* $E^{\text{TH}}(\pi)$*, if* $\text{DR}_{\max}(\pi) \neq \infty$*, then for all* $0 \leq i \leq d-1$*, we have* $\{\pi^1(i), \pi^2(i), \ldots, \pi^{d-1}(i)\} = \{0, 1, \ldots, d-1\} \setminus \{i\}$ *and* $\pi^d(i) = i$*.*

Finally, we obtain the following result. A proof is similar to Lemma 7.

Lemma 9. *If we have* $\{\pi^1(0), \pi^2(0), \ldots, \pi^{d-1}(0)\} = \{1, \ldots, d-1\}$ *and* $\pi^d(0) = 0$*, then* $E^{\text{TH}}(\pi)$ *is equivalent to* $E^{\text{TH}}(\pi_s)$*.*

Therefore, if $\text{DR}_{\max}(\pi) \neq \infty$, then $E^{\text{TH}}(\pi)$ is equivalent to $E^{\text{TH}}(\pi_s)$.

5 Experimental Results

In this section, we present our experimental results on computing $\text{DR}_{\max}(\pi)$ for $E^{\text{T1}}(\pi)$, $E^{\text{SH}}(\pi)$, and $E^{\text{TH}}(\pi)$. For $E^{\text{T1}}(\pi)$, we computed $\text{DR}_{\max}(\pi)$ for $3 \leq d \leq 16$ and for all permutations π over $\{0, 1, \ldots, d-1\}$ up to the equivalent classes. For $E^{\text{SH}}(\pi)$ and $E^{\text{TH}}(\pi)$, we computed $\text{DR}_{\max}(\pi)$ for $3 \leq d \leq 8$. We also evaluated the security against the saturation attack [12], impossible differential attack [4], differential attack [5], and linear attack [24].

$\text{DR}_{\max}(\pi)$. The results on $\text{DR}_{\max}(\pi)$ are presented in the "FD" column in Table 1–4. In the tables, we list all π such that $\text{DR}_{\max}(\pi) \leq \text{DR}_{\max}(\pi_s)$, i.e., those permutations that are better than π_s in terms of diffusion, but only the lexicographically first permutation in the equivalent class is presented. We note that a permutation with $*$ indicates that it is equivalent to π_s. To obtain the results, we first divide all permutations into equivalent classes, and list the lexicographically first permutations. We then derived the values by the actual computation of $\text{DR}_{\max}(\pi)$. For $E^{\text{T1}}(\pi)$, we present in Fig. 3 (in Appendix A) a graph showing the values of $\text{DR}_{\max}(\pi_s)$ and $\text{DR}_{\max}(\pi)$ for optimum π in terms of diffusion.

Saturation Attack. The results on the saturation attack are presented in the "SC" column in Table 1–4. The figures indicate the number of round of the longest saturation path. We briefly recall the concept based on [38]. Let $X = \{X_i \mid X_i \in \{0,1\}^n, 0 \leq i < 2^n\}$ be a set of 2^n strings. The set X is categorized into one of the following four states:

- Constant (C): for all $0 \leq i < j < 2^n$, $X_i = X_j$
- All (A): for all $0 \leq i < j < 2^n$, $X_i \neq X_j$
- Balance (B): $\bigoplus_{0 \leq i < 2^n} X_i = 0$
- Unknown (U): Otherwise

Based on the above notation, let $\alpha \in \{C, A\}^d$ be the set of 2^n plaintexts such that it has one A and the rest is C. Let β be the result of encrypting α for r rounds. If $\beta \neq U^d$, it constitutes a saturation path $\alpha \xrightarrow{r} \beta$ of r rounds.

We obtained the results based on Table 5 and by making the following assumptions.

Table 1. Results on $E^{\mathrm{T1}}(\pi)$ for $3 \leq d \leq 12$

d	π	FD	SC	IDC	DAF	LAF
3	(1,2,0)	5	9	11	7	7
	(2,0,1)*	5	9	11	7	7
4	(1,2,3,0)	10	16	19	12	12
	(2,0,3,1)*	10	16	19	12	12
5	(1,2,3,4,0)	17	25	29	17	17
	(2,0,3,4,1)*	17	25	29	17	17
	(2,3,1,4,0)	14	21	29	16	16
	(2,3,4,0,1)	14	21	29	16	16
6	(1,2,3,4,5,0)	26	36	41	22	22
	(2,0,3,4,5,1)*	26	36	41	22	22
7	(1,2,3,4,5,6,0)	37	49	55	28	28
	(2,0,3,4,5,6,1)*	37	49	55	28	28
	(2,3,1,4,5,6,0)	32	43	55	28	27
	(2,3,4,0,5,6,1)	32	43	55	28	27
	(2,3,4,5,1,6,0)	27	37	55	27	27
	(2,3,4,5,6,0,1)	27	37	55	27	27
8	(1,2,3,4,5,6,7,0)	50	64	71	36	36
	(2,0,3,4,5,6,7,1)*	50	64	71	36	36
	(2,3,4,5,1,6,7,0)	38	50	71	34	34
	(2,3,4,5,6,0,7,1)	38	50	71	34	34
9	(1,2,3,4,5,6,7,8,0)	65	81	89	42	42
	(2,0,3,4,5,6,7,8,1)*	65	81	89	42	42
	(2,3,1,4,5,6,7,8,0)	58	73	89	40	40
	(2,3,4,0,5,6,7,8,1)	58	73	89	40	40
	(2,3,4,5,6,7,1,8,0)	44	57	89	40	40
	(2,3,4,5,6,7,8,0,1)	44	57	89	40	40
10	(1,2,3,4,5,6,7,8,9,0)	82	100	109	50	50
	(2,0,3,4,5,6,7,8,9,1)*	82	100	109	50	50
	(2,3,4,5,1,6,7,8,9,0)	66	82	109	47	47
	(2,3,4,5,6,0,7,8,9,1)	66	82	109	47	47
11	(1,2,3,4,5,6,7,8,9,10,0)	101	121	131	61	61
	(2,0,3,4,5,6,7,8,9,10,1)*	101	121	131	61	61
	(2,3,1,4,5,6,7,8,9,10,0)	92	111	131	57	57
	(2,3,4,0,5,6,7,8,9,10,1)	92	111	131	57	57
	(2,3,4,5,1,6,7,8,9,10,0)	83	101	131	54	54
	(2,3,4,5,6,0,7,8,9,10,1)	83	101	131	54	54
	(2,3,4,5,6,7,1,8,9,10,0)	74	91	131	54	54
	(2,3,4,5,6,7,8,0,9,10,1)	74	91	131	54	54
	(2,3,4,5,6,7,8,9,1,10,0)	65	81	131	54	54
	(2,3,4,5,6,7,8,9,10,0,1)	65	81	131	54	54
12	(1,2,3,4,5,6,7,8,9,10,11,0)	122	144	155	68	68
	(2,0,3,4,5,6,7,8,9,10,11,1)*	122	144	155	68	68
	(2,3,4,5,6,7,8,9,1,10,11,0)	82	100	155	60	60
	(2,3,4,5,6,7,8,9,10,0,11,1)	82	100	155	60	60

Table 2. Results on $E^{\mathrm{T1}}(\pi)$ for $13 \leq d \leq 16$

d	π	FD	SC	IDC	DAF	LAF
13	(1,2,3,4,5,6,7,8,9,10,11,12,0)	145	169	181	77	77
	(2,0,3,4,5,6,7,8,9,10,11,12,1)*	145	169	181	77	77
	(2,3,1,4,5,6,7,8,9,10,11,12,0)	134	157	181	76	76
	(2,3,4,0,5,6,7,8,9,10,11,12,1)	134	157	181	76	76
	(2,3,4,5,1,6,7,8,9,10,11,12,0)	123	145	181	73	73
	(2,3,4,5,6,0,7,8,9,10,11,12,1)	123	145	181	73	73
	(2,3,4,5,6,7,1,8,9,10,11,12,0)	112	133	181	70	70
	(2,3,4,5,6,7,8,0,9,10,11,12,1)	112	133	181	70	70
	(2,3,4,5,6,7,8,9,1,10,11,12,0)	101	121	181	70	70
	(2,3,4,5,6,7,8,9,10,0,11,12,1)	101	121	181	70	70
	(2,3,4,5,6,7,8,9,10,11,1,12,0)	90	109	181	71	71
	(2,3,4,5,6,7,8,9,10,11,12,0,1)	90	109	181	71	71
14	(1,2,3,4,5,6,7,8,9,10,11,12,13,0)	170	196	209	84	84
	(2,0,3,4,5,6,7,8,9,10,11,12,13,1)*	170	196	209	84	84
	(2,3,4,5,1,6,7,8,9,10,11,12,13,0)	146	170	209	83	83
	(2,3,4,5,6,0,7,8,9,10,11,12,13,1)	146	170	209	83	83
	(2,3,4,5,6,7,8,9,1,10,11,12,13,0)	122	144	209	79	79
	(2,3,4,5,6,7,8,9,10,0,11,12,13,1)	122	144	209	79	79
15	(1,2,3,4,5,6,7,8,9,10,11,12,13,14,0)	197	225	239	99	99
	(2,0,3,4,5,6,7,8,9,10,11,12,13,14,1)*	197	225	239	99	99
	(2,3,1,4,5,6,7,8,9,10,11,12,13,14,0)	184	211	239	93	93
	(2,3,4,0,5,6,7,8,9,10,11,12,13,14,1)	184	211	239	93	93
	(2,3,4,5,6,7,1,8,9,10,11,12,13,14,0)	158	183	239	90	90
	(2,3,4,5,6,7,8,0,9,10,11,12,13,14,1)	158	183	239	90	90
	(2,3,4,5,6,7,8,9,10,11,12,13,1,14,0)	119	141	239	88	88
	(2,3,4,5,6,7,8,9,10,11,12,13,14,0,1)	119	141	239	88	88
16	(1,2,3,4,5,6,7,8,9,10,11,12,13,14,15,0)	226	225	271	108	108
	(2,0,3,4,5,6,7,8,9,10,11,12,13,14,15,1)*	226	225	271	108	108
	(2,3,4,5,1,6,7,8,9,10,11,12,13,14,15,0)	198	211	271	102	102
	(2,3,4,5,6,0,7,8,9,10,11,12,13,14,15,1)	198	211	271	102	102
	(2,3,4,5,6,7,8,9,1,10,11,12,13,14,15,0)	170	183	271	101	101
	(2,3,4,5,6,7,8,9,10,0,11,12,13,14,15,1)	170	183	271	101	101
	(2,3,4,5,6,7,8,9,10,11,12,13,1,14,15,0)	142	141	271	98	98
	(2,3,4,5,6,7,8,9,10,11,12,13,14,0,15,1)	142	141	271	98	98

Table 3. Results on $E^{\mathrm{SH}}(\pi)$

d	π	FD	SC	IDC	DAF	LAF
3	(1,2,0)*	3	5	5	6	8
4	(1,2,3,0)*	4	6	6	7	14
5	(1,2,3,4,0)*	5	7	7	8	18
6	(1,2,3,4,5,0)*	6	8	8	9	27
7	(1,2,3,4,5,6,0)*	7	9	9	10	32
8	(1,2,3,4,5,6,7,0)*	8	10	10	11	44

Table 4. Results on $E^{\mathrm{TH}}(\pi)$

d	π	FD	SC	IDC	DAF	LAF
3	(1,2,0)*	3	5	5	8	6
4	(1,2,3,0)*	4	6	7	14	7
5	(1,2,3,4,0)*	5	7	9	18	8
6	(1,2,3,4,5,0)*	6	8	11	27	9
7	(1,2,3,4,5,6,0)*	7	9	13	32	10
8	(1,2,3,4,5,6,7,0)*	8	10	15	44	11

Table 5. Output of xor and F function for the saturation path. In Type 1 GFS, $F_j(x)$ corresponds to $F(x)$.

input x	$x \oplus C$	$x \oplus A$	$x \oplus B$	$x \oplus U$	$F_j(x)$
C	C	A	B	U	C
A	A	B	B	U	A
B	B	B	B	U	U
U	U	U	U	U	U

Table 6. Output of xor and F function for the impossible differential characteristic. In Type 1 GFS, $F_j(x)$ corresponds to $F(x)$.

input x	$x \oplus Z$	$x \oplus G$	$x \oplus D$	$x \oplus D \oplus G$	$x \oplus R$	$F_j(x)$
Z	Z	G	D	$D \oplus G$	R	Z
G	G	Z	$D \oplus G$	D	R	D
D	D	$D \oplus G$	R	R	R	D
$D \oplus G$	$D \oplus G$	D	R	R	R	R
R	R	R	R	R	R	R

- For $E^{\mathrm{T1}}(\pi)$, the function $F : \{0,1\}^n \to \{0,1\}^n$ is a permutation over $\{0,1\}^n$.
- For $E^{\mathrm{SH}}(\pi)$, the function $F : \{0,1\}^{(d-1)n} \to \{0,1\}^n$ can be written as $F(x_0^r, x_1^r, \ldots, x_{d-2}^r) = F_0(x_0^r) \oplus F_1(x_1^r) \oplus \cdots \oplus F_{d-2}(x_{d-2}^r)$, where F_j is a permutation over $\{0,1\}^n$.
- For $E^{\mathrm{TH}}(\pi)$, the function $F : \{0,1\}^n \to \{0,1\}^{(d-1)n}$ can be written as $F(x_0^r) = (F_1(x_0^r), F_2(x_0^r), \ldots, F_{d-1}(x_0^r))$, where F_j is a permutation over $\{0,1\}^n$.

We note that our results do not include the case that A or B is defined for $\{0,1\}^{kn}$ for $k > 1$.

Impossible Differential Attack. The results on the impossible differential attack are presented in the "IDC" column in Table 1–4. The figures indicate the number of round of the longest impossible differential characteristic, which is the differential characteristic of probability zero.

In order to obtain these results, we make the same assumptions as in the analysis of the saturation attack. In searching the impossible differential characteristic, every sub-block is categorized into one of the following five states: Z (zero difference), G (non-zero fixed difference), D (non-zero unfixed difference), $D \oplus G$ (xor of D and G), and R (unfixed difference). Results are obtained based on Table 6 and by following [21].

Differential Attack. The results on the differential attack are presented in the "DAF" column in Table 1–4. The figures indicate the number of round such that the minimum number of differentially active F functions is equal to or larger than the threshold. We adopted $d+1$ as the threshold, and if N_D is the number of differentially active F functions, then we list the number of round such that $N_D \geq d+1$ holds[1]. The results are obtained by following [36].

Linear Attack. The results on the linear attack are presented in the "LAF" column in Table 1–4. The figures indicate the number of round such that the minimum number of linearly active F functions is equal to or larger than the threshold, which was chosen to be $d+1$ as is the case for the differential attack. Therefore, if N_L is the number of linearly active F functions, then the number of round such that $N_L \geq d+1$ is listed[2]. We obtained the results by using the duality between differential and linear attacks [25,10], and by following [36].

Observations. We summarize the observations made from the tables.

- When $3 \leq d \leq 16$ and $d \neq 3,4,6$, the diffusion of Type 1 GFS can be improved if one changes the permutation from π_s. Furthermore, the security against saturation, differential, and linear attacks improves in many cases, and does not get worse.
- In Type 1 GFS, when $d = 3,4,6$, π_s is the optimum permutation in terms of diffusion, and hence changing the permutation does not improve the diffusion property.
- In Type 1 GFS, changing the permutation does not change the security against the impossible differential attack, and hence this attack is required to be handled by some other means.
- In SH GFS and TH GFS, π_s is the optimum permutation in terms of diffusion as shown in the previous section. Furthermore, changing the permutation does not change the security against other attacks.

6 Conclusions

In this paper, we studied the effect of changing the permutation layer used in Type 1, SH, and TH GFSs. For Type 1 GFS, we introduced r_{01} and r_{10} that are useful in characterizing the diffusion property. We also presented the explicit construction of the optimum permutation in terms of diffusion for odd d. For SH and TH GFSs, we showed that changing the permutation layer does not change the diffusion property as long as they achieve the full diffusion. Then, we presented our experimental results. For Type 1 GFS, we searched for $3 \leq d \leq 16$, and for SH and TH GFSs, we searched for $3 \leq d \leq 8$. We listed all permutations π such that $\mathrm{DR}_{\max}(\pi) \leq \mathrm{DR}_{\max}(\pi_s)$ (but omitting the equivalent permutations). We also evaluated the security of the resulting schemes against saturation, impossible differential, differential, and linear attacks.

[1] The threshold is chosen only for the purpose of a comparison. In particular, DAF does not suggest a sufficient number of rounds so that the resulting cipher is secure against the differential attack.

[2] As is the case for the differential attack, LAF does not suggest a sufficient number of rounds so that the resulting cipher is secure against the linear attack.

Acknowledgments. The authors would like to thank the anonymous reviewers for useful and insightful comments. A part of this work was supported by MEXT KAKENHI, Grant-in-Aid for Young Scientists (A), 22680001.

References

1. Adams, C., Gilchrist, J.: The CAST-256 Encryption Algorithm. Network Working Group RFC 2612 (June 1999), `http://www.ietf.org/rfc/rfc2612.txt`
2. Anderson, R.J., Biham, E.: Two Practical and Provably Secure Block Ciphers: BEAR and LION. In: Gollmann, D. (ed.) FSE 1996. LNCS, vol. 1039, pp. 113–120. Springer, Heidelberg (1996)
3. Aoki, K., Ichikawa, T., Kanda, M., Matsui, M., Moriai, S., Nakajima, J., Tokita, T.: The 128-Bit Block Cipher Camellia. IEICE Trans. Fundamentals E85-A(1), 11–24 (2002)
4. Biham, E., Biryukov, A., Shamir, A.: Cryptanalysis of Skipjack Reduced to 31 Rounds Using Impossible Differentials. In: Stern, J. (ed.) EUROCRYPT 1999. LNCS, vol. 1592, pp. 12–23. Springer, Heidelberg (1999)
5. Biham, E., Shamir, A.: Differential Cryptanalysis of DES-like Cryptosystems. In: Menezes, A., Vanstone, S.A. (eds.) CRYPTO 1990. LNCS, vol. 537, pp. 2–21. Springer, Heidelberg (1991)
6. Bogdanov, A.: On Unbalanced Feistel Networks with Contracting MDS Diffusion. Des. Codes Cryptography 59(1-3), 35–58 (2011)
7. Bogdanov, A., Shibutani, K.: Analysis of 3-Line Generalized Feistel Networks with Double SD-functions. Inf. Process. Lett. 111(13), 656–660 (2011)
8. Bogdanov, A., Shibutani, K.: Double SP-Functions: Enhanced Generalized Feistel Networks. In: Parampalli, U., Hawkes, P. (eds.) ACISP 2011. LNCS, vol. 6812, pp. 106–119. Springer, Heidelberg (2011)
9. Bogdanov, A., Shibutani, K.: Generalized Feistel Networks Revisited. In: WCC 2011 (2011)
10. Chabaud, F., Vaudenay, S.: Links Between Differential and Linear Cryptanalysis. In: De Santis, A. (ed.) EUROCRYPT 1994. LNCS, vol. 950, pp. 356–365. Springer, Heidelberg (1995)
11. Choy, J., Yap, H.: Impossible Boomerang Attack for Block Cipher Structures. In: Takagi, T., Mambo, M. (eds.) IWSEC 2009. LNCS, vol. 5824, pp. 22–37. Springer, Heidelberg (2009)
12. Daemen, J., Knudsen, L.R., Rijmen, V.: The Block Cipher SQUARE. In: Biham, E. (ed.) FSE 1997. LNCS, vol. 1267, pp. 149–165. Springer, Heidelberg (1997)
13. Daemen, J., Rijmen, V.: The Design of Rijndael: AES - The Advanced Encryption Standard. Springer, Heidelberg (2002)
14. FIPS: Data Encryption Standard. National Institute of Standards and Technology (1999)
15. FIPS: Secure Hash Standard. National Institute of Standards and Technology (2002)
16. Hirose, S., Kuwakado, H., Yoshida, H.: SHA-3 Proposal: Lesamnta (2008), `http://www.hitachi.com/rd/yrl/crypto/lesamnta/index.html`
17. Hoang, V.T., Rogaway, P.: On Generalized Feistel Networks. In: Rabin, T. (ed.) CRYPTO 2010. LNCS, vol. 6223, pp. 613–630. Springer, Heidelberg (2010)

18. Hong, D., Sung, J., Hong, S.H., Lim, J.-I., Lee, S.-J., Koo, B.-S., Lee, C.-H., Chang, D., Lee, J., Jeong, K., Kim, H., Kim, J.-S., Chee, S.: HIGHT: A New Block Cipher Suitable for Low-Resource Device. In: Goubin, L., Matsui, M. (eds.) CHES 2006. LNCS, vol. 4249, pp. 46–59. Springer, Heidelberg (2006)
19. IBM Corporation: MARS–A Candidate Cipher for AES (September 1999), `http://domino.research.ibm.com/comm/research_projects.nsf/pages/security.mars.html`
20. Jutla, C.S.: Generalized Birthday Attacks on Unbalanced Feistel Networks. In: Krawczyk, H. (ed.) CRYPTO 1998. LNCS, vol. 1462, pp. 186–199. Springer, Heidelberg (1998)
21. Kim, J., Hong, S., Sung, J., Lee, C., Lee, S.: Impossible Differential Cryptanalysis for Block Cipher Structures. In: Johansson, T., Maitra, S. (eds.) INDOCRYPT 2003. LNCS, vol. 2904, pp. 82–96. Springer, Heidelberg (2003)
22. Kim, J., Lee, C., Sung, J., Hong, S., Lee, S., Lim, J.: Seven New Block Cipher Structures with Provable Security against Differential Cryptanalysis. IEICE Trans. Fundamentals E91-A(10), 3047–3058 (2008)
23. Lucks, S.: Faster Luby-Rackoff Ciphers. In: Gollmann, D. (ed.) FSE 1996. LNCS, vol. 1039, pp. 189–203. Springer, Heidelberg (1996)
24. Matsui, M.: Linear Cryptanalysis Method for DES Cipher. In: Helleseth, T. (ed.) EUROCRYPT 1993. LNCS, vol. 765, pp. 386–397. Springer, Heidelberg (1994)
25. Matsui, M.: On Correlation between the Order of S-Boxes and the Strength of DES. In: De Santis, A. (ed.) EUROCRYPT 1994. LNCS, vol. 950, pp. 366–375. Springer, Heidelberg (1995)
26. Minematsu, K., Suzaki, T., Shigeri, M.: On Maximum Differential Probability of Generalized Feistel. In: Parampalli, U., Hawkes, P. (eds.) ACISP 2011. LNCS, vol. 6812, pp. 89–105. Springer, Heidelberg (2011)
27. Moriai, S., Vaudenay, S.: On the Pseudorandomness of Top-Level Schemes of Block Ciphers. In: Okamoto, T. (ed.) ASIACRYPT 2000. LNCS, vol. 1976, pp. 289–302. Springer, Heidelberg (2000)
28. Naor, M., Reingold, O.: On the Construction of Pseudorandom Permutations: Luby-Rackoff Revisited. J. Cryptology 12(1), 29–66 (1999)
29. Nyberg, K.: Generalized Feistel Networks. In: Kim, K.-c., Matsumoto, T. (eds.) ASIACRYPT 1996. LNCS, vol. 1163, pp. 91–104. Springer, Heidelberg (1996)
30. Patarin, J., Nachef, V., Berbain, C.: Generic Attacks on Unbalanced Feistel Schemes with Contracting Functions. In: Lai, X., Chen, K. (eds.) ASIACRYPT 2006. LNCS, vol. 4284, pp. 396–411. Springer, Heidelberg (2006)
31. Patarin, J., Nachef, V., Berbain, C.: Generic Attacks on Unbalanced Feistel Schemes with Expanding Functions. In: Kurosawa, K. (ed.) ASIACRYPT 2007. LNCS, vol. 4833, pp. 325–341. Springer, Heidelberg (2007)
32. Rivest, R.L.: A Description of the RC2(r) Encryption Algorithm. Network Working Group RFC 2268 (March 1998), `http://www.ietf.org/rfc/rfc2268.txt`
33. Rivest, R.L., Robshaw, M.J.B., Sidney, R., Yin, Y.L.: The RC6 block cipher. Specification 1.1 (August 1998), `http://people.csail.mit.edu/rivest/Rc6.pdf`
34. Schneier, B., Kelsey, J.: Unbalanced Feistel Networks and Block Cipher Design. In: Gollmann, D. (ed.) FSE 1996. LNCS, vol. 1039, pp. 121–144. Springer, Heidelberg (1996)
35. Shibutani, K.: On the Diffusion of Generalized Feistel Structures Regarding Differential and Linear Cryptanalysis. In: Biryukov, A., Gong, G., Stinson, D.R. (eds.) SAC 2010. LNCS, vol. 6544, pp. 211–228. Springer, Heidelberg (2011)
36. Shirai, T., Araki, K.: On Generalized Feistel Structures Using a Diffusion Switching Mechanism. IEICE Trans. Fundamentals E91-A(8), 2120–2129 (2008)

37. Shirai, T., Shibutani, K., Akishita, T., Moriai, S., Iwata, T.: The 128-Bit Blockcipher CLEFIA (Extended Abstract). In: Biryukov, A. (ed.) FSE 2007. LNCS, vol. 4593, pp. 181–195. Springer, Heidelberg (2007)
38. Sony Corporation: The 128-bit Blockcipher CLEFIA, Security and Performance Evaluations (2007) revision 1.0, http://www.sony.net/Products/cryptography/clefia/
39. Suzaki, T., Minematsu, K.: Improving the Generalized Feistel. In: Hong, S., Iwata, T. (eds.) FSE 2010. LNCS, vol. 6147, pp. 19–39. Springer, Heidelberg (2010)
40. Wu, W., Zhang, W., Lin, D.: Security on Generalized Feistel Scheme with SP Round Function. I. J. Network Security 3(3), 215–224 (2006)
41. Zheng, Y.: The SPEED Cipher. In: Hirschfeld, R. (ed.) FC 1997. LNCS, vol. 1318, pp. 71–90. Springer, Heidelberg (1997)
42. Zheng, Y., Matsumoto, T., Imai, H.: On the Construction of Block Ciphers Provably Secure and Not Relying on Any Unproved Hypotheses. In: Brassard, G. (ed.) CRYPTO 1989. LNCS, vol. 435, pp. 461–480. Springer, Heidelberg (1990)

A Graph of $\mathrm{DR}_{\max}(\pi)$ of Type 1 GFS

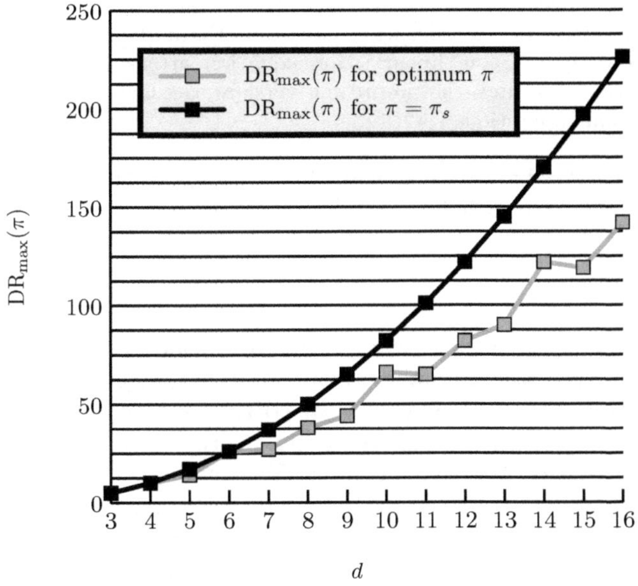

Fig. 3. $\mathrm{DR}_{\max}(\pi)$ of Type 1 GFS

Security Analysis of an Improved MFE Public Key Cryptosystem

Xuyun Nie[1,3,4], Zhaohu Xu[1], Li Lu[1], and Yongjian Liao[1,2,3]

[1] School of Computer Science and Engineering,
University of Electronic Science and Technology of China, Chengdu 611731, China
[2] Shanghai Key Laboratory of Integrate Administration Technologies
for Information Security
[3] Network and Data Security Key Laboratory of Sichuan Province
[4] State Key Laboratory of Information Security,
Graduate University of Chinese Academy of Sciences, Beijing 100049, China
{xynie,zhhxu,luli2009,liaoyj}@uestc.edu.cn

Abstract. MFE is a multivariate public key encryption scheme. In 2007, MFE was broken by Ding et al using high order linearization equation attack. In 2009, Huang et al gave an improvement of MFE. They claimed that the improved MFE is secure against high order linearization equation attack. However, through theoretical analysis, we find that there are many first order linearization equations(FOLEs) satisfied by this improved version. Using linearization equation attack can break this version. We also find the improved version satisfied Second Order Linearization Equantions (SOLEs).

Keywords: Algebraic attack, MFE, Linearization equation, Multivariate public key cryptography.

1 Introduction

Multivariate public key cryptosystem (MPKC) is one of the promising alternatives to public keys based number theory problems such as RSA in resisting quantum algorithms attack. The security of MPKC relies on the difficulty of solving systems of nonlinear polynomial equations with many variables (MQ), and the latter is a NP-hard problem in general. The public key of MPKC is mostly a set of quadratic polynomials. These polynomials are derived from composition of maps. Compared with RSA public key cryptosystems, the computation in MPKC can be very fast because it is operated on a small finite field.

The Medium Field Equation (MFE) multivariate public key cryptosystem [WYH06] is proposed by Lih-Chung Wang et al in CT-RSA 2006 conference. MFE can be viewed as a triangular MPKC due to the structure of its central map. Compared with TTM[Moh99], MFE hidden its triangular structure by rational maps instead of high order polynomials. In 2007, Ding et al found that MFE satisfied some Second Order Linearization Equations (SOLEs)[DHN07]. For given public key and a valid ciphertext, corresponding plaintext can be recovered within 2^{23} $\mathbb{F}_{2^{16}}$-operations after 2^{52} $\mathbb{F}_{2^{16}}$-operations pre-computations.

D. Lin, G. Tsudik, and X. Wang (Eds.): CANS 2011, LNCS 7092, pp. 118–125, 2011.

In 2009, Jiasen Huang et al. gave an improvement [HWO09] of MFE by redesigning the central map of MFE. They modified the matrix equations used in MFE and claimed that the improved scheme can resist second order linearization equation (SOLE) attack. However, our analysis shows that the scheme still satisfy second order linearization equation. Furthermore, we found there are many First order linearization equations satisfied by the improvement. Using FOLEs attack, For given public key and a valid ciphertext, we can recover its corresponding plaintext.

The paper is organized as follows. We introduce MFE and its improvement in Section 2 and present the FOLE attack and SOLE attack in Section 3. Finally, in Section 4, we conclude the paper.

2 MFE and Its Improvement

We use the same notation as in [WYH06]. Let $\mathbb{K}$ be a finite field of "medium" size and $\mathbb{L}$ be its degree r extension field and be a large field. Let $q = |\mathbb{K}|$, $l = |\mathbb{L}|$. In MFE and its improvement, we always identify $\mathbb{L}$ with $\mathbb{K}^r$ by a $\mathbb{K}$-linear isomorphism $\pi : \mathbb{L} \to \mathbb{K}^r$. Namely we take a basis of $\mathbb{L}$ over $\mathbb{K}$, $\{\theta_1, \cdots, \theta_r\}$, and define$\pi$ by $\pi(a_1\theta_1 + \cdots + a_r\theta_r) = (a_1, \cdots, a_r)$ for any $a_1, \cdots, a_r \in \mathbb{K}$. It is natural to extend π to two $\mathbb{K}$-linear isomorphisms $\pi_1 : \mathbb{L}^{12} \to \mathbb{K}^{12r}$and $\pi_2 : \mathbb{L}^{15} \to \mathbb{K}^{15r}$.

2.1 MFE Cryptosystem

In MFE, its encryption map $F : K^{12r} \to K^{15r}$ is a composition of three maps ϕ_1, ϕ_2, ϕ_3. Let

$$(x_1, \cdots, x_{12r}) = \phi_1(m_1, \cdots, m_{12r}),$$
$$(y_1, \cdots, y_{15r}) = \phi_2(x_1, \cdots, x_{12r}),$$
$$(z_1, \cdots, z_{15r}) = \phi_3(y_1, \cdots, y_{15r}).$$

where ϕ_1 and ϕ_3 are invertible affine maps, ϕ_2 is its central map, which is equal to$\pi_1 \circ \bar{\phi_2} \circ \pi_2^{-1}$.

ϕ_1 and ϕ_3 are taken as the private key, while the expression of the map $(z_1, \cdots, z_{15r}) = F(m_1, ..., m_{12r})$ is the public key. The map $\bar{\phi_2} : \mathbb{L}^{12} \to \mathbb{L}^{15}$ is defined as follows.

$$\begin{cases} Y_1 = X_1 + X_5X_8 + X_6X_7 + Q_1; & \\ Y_2 = X_2 + X_9X_{12} + X_{10}X_{11} + Q_2; & \\ Y_3 = X_3 + X_1X_4 + X_2X_3 + Q_3; & \\ Y_4 = X_1X_5 + X_2X_7; & Y_5 = X_1X_6 + X_2X_8; \\ Y_6 = X_3X_5 + X_4X_7; & Y_7 = X_3X_6 + X_4X_8; \\ Y_8 = X_1X_9 + X_2X_{11}; & Y_9 = X_1X_{10} + X_2X_{12}; \\ Y_{10} = X_3X_9 + X_4X_{11}; & Y_{11} = X_3X_{10} + X_4X_{12}; \\ Y_{12} = X_5X_9 + X_7X_{11}; & Y_{13} = X_5X_{10} + X_7X_{12}; \\ Y_{14} = X_6X_9 + X_8X_{11}; & Y_{15} = X_6X_{10} + X_8X_{12}. \end{cases} \quad (1)$$

Here Q_1, Q_2, and Q_3 form a triple (Q_1, Q_2, Q_3) which is a triangular map from $\mathbb{K}^{3r}$ to itself, more detail please see [WYH06]. The encryption of MFE is the evaluation of public-key polynomials, namely given a plaintext $(m_1, \cdots, m_{12r})$, its ciphertext is

$$(z_1, \cdots, z_{15r}) = (F_1(m_1, \cdots, m_{12r}), \cdots, F_{15r}(m_1, \cdots, m_{12r})).$$

Given a valid ciphertext $(z_1, \cdots, z_{15r})$, the decryption of the scheme is to calculating in turn $\phi_1^{-1} \circ \pi_1 \circ \bar{\phi_2}^{-1} \circ \pi_2^{-1} \circ \phi_3^{-1}(m_1, \cdots, m_{8r})$. The key point is how to invert $\bar{\phi}_2$. Write $X_1, \cdots, X_8, Y_4, \cdots, Y_{15}$ as four 2×2 matrices.

Write $X_1, \cdots, X_{12}, Y_4, \cdots, Y_{15}$ as six 2×2 matrices:

$$\begin{aligned} M_1 &= \begin{pmatrix} X_1 & X_2 \\ X_3 & X_4 \end{pmatrix}, M_2 = \begin{pmatrix} X_5 & X_6 \\ X_7 & X_8 \end{pmatrix}, M_3 = \begin{pmatrix} X_9 & X_{10} \\ X_{11} & X_{12} \end{pmatrix}, \\ Z_3 &= M_1 M_2 = \begin{pmatrix} Y_4 & Y_5 \\ Y_6 & Y_7 \end{pmatrix}, Z_2 = M_1 M_3 = \begin{pmatrix} Y_8 & Y_9 \\ Y_{10} & Y_{11} \end{pmatrix}, \\ Z_1 &= M_2^T M_3 = \begin{pmatrix} Y_{12} & Y_{13} \\ Y_{14} & Y_{15} \end{pmatrix}. \end{aligned} \tag{2}$$

Then

$$\begin{cases} \det(M_1) \cdot \det(M_2) = \det(Z_3), \\ \det(M_1) \cdot \det(M_3) = \det(Z_2), \\ \det(M_2) \cdot \det(M_3) = \det(Z_1). \end{cases}$$

When M_1, M_2, and M_3 are all invertible, we can get values of $\det(M_1)$, $\det(M_2)$, and $\det(M_3)$ from $\det(Z_1)$, $\det(Z_2)$, and $\det(Z_3)$, for instance, $\det(M_1) = \left(\det(Z_2) \cdot \det(Z_3)/\det(Z_1)\right)^{1/2}$.

With values of $\det(M_1)$, $\det(M_2)$, and $\det(M_3)$, we can use the triangular form of the central map to get $X_1, X_2, \ldots, X_{12}$ in turn. then we can recover the ciphertext. More detail of decryption are presented in [WYH06].

2.2 Improvement of MFE

The origin MFE cryptosystems were broken by Ding et al. through SOLEs attack. Denote by M^* the associated matrix of a square matrix; for $M = \begin{pmatrix} a & b \\ c & d \end{pmatrix}$, its associated matrix is $M^* = \begin{pmatrix} d & -b \\ -c & a \end{pmatrix}$. From

$$Z_3 = M_1 M_2, \qquad Z_2 = M_1 M_3. \tag{3}$$

we can derive

$$M_3 M_3^* M_1^* M_1 M_2 = M_3 (M_1 M_3)^* (M_1 M_2) = M_3 Z_2^* Z_3,$$

$$M_3 M_3^* M_1^* M_1 M_2 = (M_3 M_3^*)(M_1 M_1^*) M_2 = \det(M_3) \det(M_1) M_2 = \det(Z_2) M_2,$$

and hence,

$$M_3 Z_2^* Z_3 = \det(Z_2) M_2, \tag{4}$$

Expanding it, many second order linearization equation appealed. After finding all SOLEs, the attack can find the corresponding plaintext for a given valid cipheretext.

To avoid the SOLE, Jiasen Huang et al. proposed a modification of MFE. They modified only the matrix equations as follows after some analysis.

M_1, M_2 and M_3 are defined as same as the origin MFE, while Z_1, Z_2 and Z_3 are defined as follows:

$$Z_3 = M_1 M_2^* = \begin{pmatrix} Y_4 & Y_5 \\ Y_6 & Y_7 \end{pmatrix}, Z_2 = M_1^* M_3 = \begin{pmatrix} Y_8 & Y_9 \\ Y_{10} & Y_{11} \end{pmatrix}, Z_1 = M_2^T M_3^* = \begin{pmatrix} Y_{12} & Y_{13} \\ Y_{14} & Y_{15} \end{pmatrix}. \tag{5}$$

where M_i^* $(1 \leq i \leq 3)$ are the associated matrices of M_i^*.

These matrices are also satisfied

$$\begin{cases} \det(M_1) \cdot \det(M_2) = \det(Z_3), \\ \det(M_1) \cdot \det(M_3) = \det(Z_2), \\ \det(M_2) \cdot \det(M_3) = \det(Z_1). \end{cases}$$

so the decryption process is very similar to the origin MFE. See [HWO09] for more detail.

3 Linearization Equation Attack

The authors of [HWO09] claimed their modified MFE can resist SOLEs attack. Through analysis, we found that there are many FOLEs and SOLEs satisfied by the above improved MFE. Using ciphertext-only attack, we can get the corresponding plaintext.

3.1 First Order Linearnation Equation

The FOLE is of the form

$$\sum_{i,j} a_{ij} m_i z_j + \sum_i b_i m_i + \sum_j c_j z_j + d = 0$$

where m_i are plaintext variables and z_j are ciphertext variables. Clearly, if we have a valid cipheretext and substitute it into FOLE, we should get a linear equation in plaintext variables.

Note that, for any square matrices M_1 and M_2, we have

$$(M_1^*)^* = M_1, (M_1 M_2)^* = M_2^* M_1^*,$$

$$(M_1^*)^T = (M_1^T)^*$$

Finding FOLEs. From

$$Z_3 = M_1 M_2^*, Z_2 = M_1^* M_3$$

we can derive

$$M_3^* Z_3 = M_3^* M_1 M_2^* = (M_1^* M_3)^* M_2^* = Z_2^* M_2^*$$

and hence,

$$Z_2^* M_2^* = M_3^* Z_3 \tag{6}$$

Expanding it, we have

$$\begin{pmatrix} Y_{11} & -Y_9 \\ -Y_{10} & Y_8 \end{pmatrix} \begin{pmatrix} X_8 & -X_6 \\ -X_7 & X_5 \end{pmatrix} = \begin{pmatrix} X_{12} & -X_{10} \\ -X_{11} & X_9 \end{pmatrix} \begin{pmatrix} Y_4 & Y_5 \\ Y_6 & Y_7 \end{pmatrix}$$

then

$$\begin{pmatrix} X_8 Y_{11} + X_7 Y_9 & -X_6 Y_{11} - X_9 Y_5 \\ -X_8 Y_{10} - X_7 Y_8 & X_6 Y_{10} + X_5 Y_8 \end{pmatrix} = \begin{pmatrix} X_{12} Y_4 - X_{10} Y_6 & X_{12} Y_5 - X_{10} Y_7 \\ -X_{11} Y_4 + X_9 Y_6 & -X_{11} Y_5 + X_9 Y_7 \end{pmatrix}$$

that is,

$$\begin{cases} X_8 Y_{11} + X_7 Y_9 = X_{12} Y_4 - X_{10} Y_6; \\ -X_6 Y_{11} - X_9 Y_5 = X_{12} Y_5 - X_{10} Y_7; \\ -X_8 Y_{10} - X_7 Y_8 = -X_{11} Y_4 + X_9 Y_6; \\ X_6 Y_{10} + X_5 Y_8 = -X_{11} Y_5 + X_9 Y_7. \end{cases} \tag{7}$$

Substituting $(X_1, \cdots, X_{12}) = \pi_1 \circ \phi_1(u_1, \cdots, u_{12r})$ and $(Y_1, \cdots, Y_{15}) = \pi_2^{-1} \circ \phi_3^{-1}(z_1, \cdots, z_{15r})$ into (7), we get $4r$ equations of the form

$$\sum_{i,j} a_{ij} m_i z_j + \sum_i b_i m_i + \sum_j c_j z_j + d = 0 \tag{8}$$

where the coefficients $a_{ij}, b_i, c_j, d \in \mathbb{K}$, and the summations are respectively over $1 \leq i \leq 12r$ and $1 \leq j \leq 15r$. These equations are exactly **first order linearization equations (FOLEs)**. Furthermore, we can show these $4r$ equations are linearly independent.

Similarly, we can derive other $8r$ SOLEs. Note that

$$Z_1 M_1 = M_2^T M_3^* M_1 = M_2^T Z_2^*$$

$$Z_1^* M_1^T = (M_2^T M_3^*)^* M_1^T = M_3 (M_2^T)^* M_1^T = M_3 (M_2^*)^T M_1^T = M_3 Z_3^T$$

That is,

$$\begin{aligned} Z_1 M_1 &= M_2^T Z_2^* \\ Z_1^* M_1^T &= M_3 Z_3^T \end{aligned} \tag{9}$$

Expanding them and substituting $(X_1, \cdots, X_{12}) = \pi_1 \circ \phi_1(u_1, \cdots, u_{12r})$ and $(Y_1, \cdots, Y_{15}) = \pi_2^{-1} \circ \phi_3^{-1}(z_1, \cdots, z_{15r})$ into them, we get another linearly independent $8r$ FOLEs.

To continue our attack, we must find all FOLEs. The equation (8) is equivalent to a system of equations on the coefficients a_i, b_i, c_j, and d. To find all equations is equivalent to find a basis of V, a linear space spanned by all vectors $(a_{1,1}, \cdots, a_{12r,15r}, b_1, \cdots, b_{12r}, c_1, \cdots, c_{15r}, d)$. The number of unknown coefficients in these equations is equal to

$$12r \times 15r + 12r + 15r + 1 = 180r^2 + 27r + 1.$$

We made computer simulations to find all linearization equations. In one of our experiments, we choose $\mathbb{K} = GF(2^{16})$, $r = 4$. In this case, the number of unknown coefficients is equal to 2989.

To find a basis of V, we can randomly select slightly more than 2989, say 3000, plaintexts $(m_1, \cdots, m_{48})$ and substitute them in (8) to get a system of 3000 linear equations and then solve it. Let $\{(a_{ij}^{(k)}, b_i^{(k)}, c_j^{(k)}, d^{(k)}), 1 \leq k \leq D\}$ be the coefficient vectors corresponding to a basis of V, where i, and j stand for $1 \leq i \leq 48$, $1 \leq j \leq 60$, respectively. Hence, we derive D linearly independent equations in m_i and z_j. Let $E_k (1 \leq k \leq D)$ denote the equations:

$$\sum_{i=1,j=1}^{12r,15r} a_{ij}^{(k)} m_i z_j + \sum_{i=1}^{12r} b_i^{(k)} m_i + \sum_{j=1}^{15r} c_j^{(k)} z_j + d^{(k)} = 0 \qquad (10)$$

The computation complexity is

$$(3000)^3 \leq 2^{35}.$$

Our experiments show that it take about 22 minutes on the execution of this step and $D = 48$.

Note that, this step is independent of the value of the ciphertext and can be done once for a given public key.

Ciphertext-Only Attack. Now we have derived all FOLEs. Our goal is to find corresponding plaintext $(m'_1, \cdots, m'_{12r})$ for a given valid ciphertext $(z'_1, \cdots, z'_{15r})$.

Substitute $(z'_1, \cdots, z'_{15r})$ into basis equations E_k, we can get k equations in following form:

$$\begin{cases} \sum_{i,j} a_{ij}^{(k)} m_i z'_j + \sum_i b_i^{(k)} m_i + \sum_j c_j^{(k)} z'_j + d^{(k)} = 0 \\ 1 \leq k \leq D \end{cases} \qquad (11)$$

Suppose the dimension of the basis of the system (11) solution space is s'. Then, we can represent s' variables of $m_1, \cdots, m_{12r}$ by linear combinations of other $12r - s'$. Denote $w_1, \cdots w_{12r-s'}$ are remainder variables. Our experiments show $s' = 32$, when $r = 4$.

Now substitute the expressions obtained above into $F_j(m_1, \cdots, m_{12r})$,we can get $15r$ new quadratic functions $\tilde{F}_j(w_1, \cdots w_{12r-s'})$, $j = 1, \cdots, 12r$. Then, our attack turn to solve the following system:

$$\begin{cases} \tilde{F}_i(w_1, \cdots w_{12r-s'}) = z'_i \\ 1 \leq i \leq 15r \end{cases} \qquad (12)$$

There are $4r$ unknowns and $15r$ equations in system (12). We can solve this system by F_4 and recover the corresponding plaintext.

Our experiments show, it takes about 6 second to solve the system (12) and recover the corresponding plaintext.

All of our experiments were performed on a normal computer, with Genuine Intel(R) CPU T2300@1.66GHz, 504MB RAM by magma.

3.2 Second Order Linearization Equation

Moreover, we derive many SOLEs satisfied by the modified MFE.

From

$$Z_3 = M_1 M_2^*, Z_2 = M_1^* M_3$$

we can derive

$$Z_3 M_2 Z_2 = M_1 M_2^* M_2 M_1^* M_3 = det(Z_3) M_3$$

That is

$$\begin{pmatrix} Y_4 & Y_5 \\ Y_6 & Y_7 \end{pmatrix} \begin{pmatrix} X_5 & X_6 \\ X_7 & X_8 \end{pmatrix} \begin{pmatrix} Y_8 & Y_9 \\ Y_{10} & Y_{11} \end{pmatrix} = (Y_4 Y_7 - Y_5 Y_6) \begin{pmatrix} X_9 & X_{10} \\ X_{11} & X_{12} \end{pmatrix}$$

Expanding it,we get equations of the form

$$\sum a_{ijk} X_i Y_j X_k = 0 \tag{13}$$

which hold for any $(X_1, \cdots, X_{12}, Y_4, \cdots, Y_{15})$. After substituting $(X_1, \cdots, X_{12}) = \pi_1 \circ \phi_1(u_1, \cdots, u_{12r})$ and $(Y_1, \cdots, Y_{15}) = \pi_2^{-1} \circ \phi_3^{-1}(z_1, \cdots, z_{15r})$ into it. We get $4r$ SOLEs.

Similar to the process above, we can derive

$$\begin{aligned}
M_1^* M_3 M_3^* M_1 M_2^* &= Z_2 M_3^* Z_3 = \det(Z_3) M_3 \\
M_2^T M_3^* M_3 (M_2^T)^* M_1^T &= Z_1 M_3 Z_3^T = \det(Z_1) M_1^T \\
M_3 (M_2^T)^* M_2^T M_3^* M_1 &= Z_1^* M_2^T Z_2^* = \det(Z_1) M_1 \\
M_3^* M_1 M_1^* M_3 (M_2^T)^* &= Z_2^* M_1^* Z_1^* = \det(Z_2)(M_2^T)^* \\
(M_2^T)^* M_1^T (M_1^T)^* M_2^T M_3^* &= Z_3^T (M_1^T)^* Z_1 = \det(Z_3) M_3^*
\end{aligned}$$

Hence, we get another twenty equations of form (13). That is to say, we can get at least 24r SOLEs. So, the modified MFE can not resist SOLEs attack.

4 Conclusion

We use FOLE attack method to break an improved MFE proposed by Jiasen Huang et al in this paper. For a given ciphertext, our method can find its corresponding plaintext. Moreover, we tried different way to make the improved design work, but it seems that it does not work anyway. Currently, there are three other improvements on MFE [W07, WW09, WZY07]. They also have some defect and have been broken in another paper [CNH10].

Acknowledgements. This work is supported by the Fundamental Research Funds for the Central Universities under Grant ZYGX2010J069, the National Natural Science Foundation of China (No. 60903155) and the Opening Project of Shanghai Key Laboratory of Integrate Administration Technologies for Information Security (No. AGK2010007).

References

[CNH10] Cao, W., Nie, X., Hu, L., Tang, X., Ding, J.: Cryptanalysis of Two Quartic Encryption Schemes and One Improved MFE Scheme. In: Sendrier, N. (ed.) PQCrypto 2010. LNCS, vol. 6061, pp. 41–60. Springer, Heidelberg (2010)

[DHN07] Ding, J., Hu, L., Nie, X., et al.: High Order Linearization Equation (HOLE) Attack on Multivariate Public Key Cryptosystems. In: Okamoto, T., Wang, X. (eds.) PKC 2007. LNCS, vol. 4450, pp. 233–248. Springer, Heidelberg (2007)

[GJ79] Garey, M., Johnson, D.: Computers and intractability, A Guide to the theory of NP-compuleteness. W.H. Freeman (1979)

[HWO09] Huang, J., Wei, B., Ou, H.: An Improved MFE Scheme Resistant against SOLE Attacks. In: PrimeAsia 2009 Asia Pacific Conference on Postgraduate Research in Microelectronics Electronics, pp. 157–160. IEEE (2009)

[Moh99] Moh, T.: A fast public key system with signature and master key functions. Lecture Notes at EE department of Stanford University, (May 1999), `http://www.usdsi.com/ttm.html`

[W07] Wang, Z.: An Improved Medium-Field Equation (MFE) Multivariate Public Key Encryption Scheme. IIH-MISP (2007), `http://bit.kuas.edu.tw/iihmsp07/acceptedlistgeneralsession.html`

[WW09] Wang, X., Wang, X.: A More Secure MFE Multivariate Public Key Encryption Scheme. International Journal of Computer Science and Applications, Technomathematics Research Foundation 6(3), 1–9 (2009), `http://www.tmrfindia.org/ijcsa/v6i31.pdf`

[WYH06] Wang, L., Yang, B., Hu, Y., et al.: A Medium-Field Multivariate Public key Encryption Scheme. In: Pointcheval, D. (ed.) CT-RSA 2006. LNCS, vol. 3860, pp. 132–149. Springer, Heidelberg (2006)

[WZY07] Wang, Z.-w., Zheng, S.-h., Yang, Y.-x., et al.: Improved Medium-Field Multivariate Public Key Encryption. Journal of University of Electonic Science an Technology of China 36(6), 1152–1154 (2007) (in Chinese)

A New Lattice-Based Public-Key Cryptosystem Mixed with a Knapsack*

Yanbin Pan[1], Yingpu Deng[1], Yupeng Jiang[1], and Ziran Tu[2]

[1] Key Laboratory of Mathematics Mechanization
Academy of Mathematics and Systems Science, Chinese Academy of Sciences
{panyanbin,dengyp,jiangyupeng}@amss.ac.cn
[2] Henan University of Science and Technology
naturetu@gmail.com

Abstract. In SAC'98, Cai and Cusick proposed an efficient lattice-based public-key cryptosystem mixed with a knapsack. However, a ciphertext-only attack given by Pan and Deng shows that it is not secure. In this paper, we present a new efficient lattice-based public-key cryptosystem mixed with a knapsack, which can resist Pan and Deng's attack well. What's more, it has reasonable key size, quick encryption and decryption. However, we have to point out that the new cryptosystem has no security proof.

Keywords: Lattice, Public-Key Cryptosystem, Knapsack.

1 Introduction

In 1996, Ajtai [1] presented a family of one-way hash functions based on the worst-case hardness of several lattice problems. After his seminal work, cryptographic constructions based on lattices have drawn considerable attention.

Lattice-based cryptosystems are usually considered as post-quantum cryptosystems. They have resisted the cryptanalysis by quantum algorithms by now whereas RSA [25] and ECC [14,18] can not since Shor's [26] quantum algorithms can factor integers and compute discrete logarithms efficiently.

The first lattice-based public-key cryptosystem was proposed by Ajtai and Dwork [3] in 1997, whose security is based on the worst-case hardness assumptions. Several other lattice-based cryptosystems, such as [7,12,5,23,24,2,8,22,16,28], have been proposed after their work. Roughly, we can classify these lattice-based public-key cryptosystems into two classes by whether they have security proofs or not.

The lattice-based public-key cryptosystems in the class without security proofs, including GGH [7], NTRU [12] and the Cai-Cusick cryptosystems [5], are usually very efficient.

* This work was supported in part by the NNSF of China (No. 11071285 and No. 60821002) and in part by 973 Project (No. 2011CB302401).

D. Lin, G. Tsudik, and X. Wang (Eds.): CANS 2011, LNCS 7092, pp. 126–137, 2011.

GGH [7] was proposed by Goldreich, Goldwasser and Halevi in Crypto'97. It has efficient encryption and decryption and a natural signature scheme. Its security is related to the hardness of approximating the CVP in a lattice. However, it can't provide sufficient security without being impractical due to a major flaw found by Nguyen [20].

NTRU [12] is one of the most practical schemes known to date. It was proposed by Hoffstein, Pipher, Silverman and now is an IEEE 1363.1 Standard [13]. It features reasonably short, easily created keys, high speed, and low memory requirements. The security of NTRU is related to the hardness of some lattice problems by the results of Coppersmith and Shamir [6]. Most of the ciphertext-only attacks [17,10,9] against NTRU depend on the special cyclic structure of its underlying lattice.

By the heuristic attack of Nguyen and Stern [19], in order to be secure, the implementations of the Ajtai-Dwork cryptosystem would require very large keys, making it impractical in a real-life environment. To increase the efficiency of the Ajtai-Dwork cryptosystem, Cai and Cusick [5] proposed a new efficient cryptosystem called the Cai-Cusick cryptosystem in 1998. By mixing the Ajtai-Dwork cryptosystem with a knapsack, the Cai-Cusick cryptosystem has much less data expansion. However, compared with the vectors in the public keys, the perturbation vector in the ciphertext is too short to provide enough security. By the observation, Pan and Deng [21] proposed an efficient ciphertext-only attack to show that it's not secure.

On the other hand, the lattice-based public-key cryptosystems in the other class have security proofs. Their average-case security can be based on the worst-case hardness of some lattice problems. Most of the early cryptosystems with security proofs have less efficiency. They usually need long keys and ciphertexts. The late cryptosystems with security proofs do much better. By using structured lattices, one can even obtain cryptosystems with quasi-optimal asymptotic performances, for example, [16,28].

No doubt, a security proof raises our confidence in the security of a cryptosystem. However, none of the lattice problems, which most of the existed provably secure cryptosystems are based on, has been proved to be NP-hard. Most of them are assumed to be hard. Especially, when it comes to cryptosystems intended to be post-quantum secure, the hardness of these lattice problems under quantum computers should be further studied. In addition, from the point of view of complexity theory, the hardness of a problem does make sense when the security parameter n is big enough, so further work should be done to measure the security of these provably secure cryptosystems for fixed n. As a result, it is still well worth constructing an efficient lattice-based public-key cryptosystem, which may have no security proof.

In this paper, we propose a new lattice-based public key cryptosystem mixed with a knapsack. Like the Cai-Cusick cryptosystem, our cryptosystem also involves a knapsack. However, we use a totally different way to generate the keys. All the computations can be completed over $\mathbb{Z}$, but not $\mathbb{R}$ as in the Cai-Cusick cryptosystem. What's more, a small modulus p is used to bound the entries of the keys, so the size of the public keys can not be too large. All these makes the cryptosystem not only more efficient but also resist the attack by Pan and Deng [21].

Similar to GGH, our cryptosystem uses a matrix $H \in \mathbb{Z}^{m\times m}$ as its public key to encrypt a message $t \in \{0,1\}^m$. A similar direct lattice-based attack can also be used to recover a encrypted message. However, we need solve a CVP in a $2m$-dimensional lattice instead of an m-dimensional lattice which occurs in attacking GGH. This may allow us to use small dimensional matrix as public key to provide sufficient security. Moreover, all the entries of H is bounded by a small positive integer p, so the public key size is usually smaller than in GGH.

The key size in our cryptosystem is bigger than in NTRU, one of the most practical lattice-based public-key cryptosystem known to date. However, there may be no obvious attack to obtain the private key of our cryptosystem, whereas the private key of NTRU can be obtained by finding the short vector of the so-called NTRU-lattice. Furthermore, the underlying lattice in our cryptosystem has no special cyclic structure like in NTRU. This makes our system resist some similar attacks against NTRU which are based on its cyclic structure.

We have to point out that the new cryptosystem has no security proof. However, the heuristic analysis shows that it may resist the direct lattice attack well. We just give an original public-key cryptosystem. To resist the adaptive chosen-ciphertext attacks, the cryptosystem may need some additional padding. Another drawback is that the new cryptosystem does not have a natural signature scheme.

The remainder of the paper is organized as follows. In Section 2, we give some preliminaries needed. In Section 3, we describe our lattice-based public key cryptosystem. In Section 4, we present some details in the practical implementation. In Section 5, we give the security analysis and some experimental evidence. Finally, we give a short conclusion in Section 6.

2 Preliminaries

2.1 Knapsack Problem

Given positive integers $N_1, N_2, \cdots, N_n$ and s, the knapsack or subset sum problem is to find, if there exists, variables $a_1, a_2, \cdots, a_n$, with $a_i \in \{0,1\}$, such that

$$\sum_{i=1}^{n} a_i N_i = s.$$

The problem is known to be NP-complete. However, if $[N_1, N_2, \cdots, N_n]$ is a superincreasing sequence, i.e. $N_i > \sum_{j=1}^{i-1} N_j$ for $i = 2, 3, \ldots, n$, there is an efficient greedy algorithm to recover $a_1, a_2, \cdots, a_n$ from s. It is easy to see that

$$a_n = \begin{cases} 1, & \text{if } s \geq N_n; \\ 0, & \text{otherwise.} \end{cases}$$

After having a_n, we then substitute s by $s - a_n N_n$ and find a_{n-1} similarly. Obviously the process can be continued until all a_i's are found.

2.2 Lattice

An integer lattice $\mathcal{L}$ is a discrete additive subgroup of $\mathbb{Z}^n$. It is well-known that there must be d linearly independent vectors $b_1, b_2 \cdots, b_d \in \mathbb{Z}^n$ in an integer lattice $\mathcal{L}$, such that

$$\mathcal{L} = \{\sum_{i=1}^{d} a_i b_i | a_i \in \mathbb{Z}\}.$$

We can call that $\mathcal{L}$ is spanned by $b_1, b_2 \cdots, b_d$ and $B = [b_1, b_2 \cdots, b_d]$ is a basis of $\mathcal{L}$.

A lattice is full rank if $d = n$. For a full rank lattice $\mathcal{L}$, the determinant $\det(\mathcal{L})$ is equal to the absolute value of determinant of the basis B. If A is a matrix with d linearly independent columns, we denote by $\mathcal{L}(A)$ the lattice spanned by $A_1, A_2, \cdots, A_d$, where A_i the i-th column of A.

There are two main lattice problems. One is the shortest vector problem (SVP) which refers the question to find the shortest non-zero vector in a given lattice. SVP is known to be NP-hard under random reduction. The celebrated LLL algorithm [15] runs in polynomial time and approximates the shortest vector within a factor of $2^{n/2}$. The other is the closest vector problem (CVP) which is to find a lattice vector minimizing the distance to a given vector. CVP is known to be NP-complete. Babai [4] also gave an polynomial-time algorithm that approximates the closest vector by a factor of $(3/\sqrt{2})^n$, which is usually called Nearest Plane Algorithm.

Denote by $\|v\|$ the Euclidean l_2-norm of a vector v and by $\lambda_1(\mathcal{L})$ the length of the shortest non-zero vector in the lattice $\mathcal{L}$. By the Gaussian Heuristic, $\lambda_1(\mathcal{L}) \approx \sqrt{\frac{n}{2\pi e}}\det(\mathcal{L})^{\frac{1}{n}}$ for an n-dimensional random lattice $\mathcal{L}$. Similarly, most closest vector problems for $\mathcal{L}$ have a solution whose size is approximately $\sqrt{\frac{n}{2\pi e}}\det(\mathcal{L})^{\frac{1}{n}}$. If we want to find a short vector v in $\mathcal{L}$ (*resp. a vector v in $\mathcal{L}$ closest to the target vector t*), then experiences tell us the smaller $\frac{\|v\|}{\sqrt{\frac{n}{2\pi e}}\det(\mathcal{L})^{\frac{1}{n}}}$ (*resp.* $\frac{\|v-t\|}{\sqrt{\frac{n}{2\pi e}}\det(\mathcal{L})^{\frac{1}{n}}}$) is, the more easily we will find v in practice.

3 Description of Our Cryptosystem

3.1 The Basic Cryptosystem

Parameter: m
Key Generation:
Let $n = 2m$.

Step 1. Choose a superincreasing sequence $N_1, N_2, \cdots, N_n$ where $N_1 = 1$.
Step 2. Randomly choose a permutation τ on n letters with $\tau^{-1}(1) \leq m$.

Step 3. For $i = m+1, m+2, \cdots, n$, represent $N_{\tau(i)}$ as

$$N_{\tau(i)} = \sum_{j=1}^{m} b_{i-m,j} N_{\tau(j)}$$

where $b_{i-m,j} \in \mathbb{Z}$ and we expect their absolute values to be as small as possible.
Define $A \in \mathbb{Z}^{m \times n}$ as belows:

$$A = \begin{pmatrix} 1 & 0 & \cdots & 0 & b_{1,1} & b_{2,1} & \cdots & b_{m,1} \\ 0 & 1 & \cdots & 0 & b_{1,2} & b_{2,2} & \cdots & b_{m,2} \\ \vdots & \vdots & \ddots & \vdots & \vdots & \vdots & \ddots & \vdots \\ 0 & 0 & \cdots & 1 & b_{1,m} & b_{2,m} & \cdots & b_{m,m} \end{pmatrix}.$$

So, $(N_{\tau(1)}, N_{\tau(2)}, \cdots, N_{\tau(m)})A = (N_{\tau(1)}, \cdots, N_{\tau(m)}, N_{\tau(m+1)}, \cdots, N_{\tau(n)})$.
Step 4. Let

$$\begin{aligned} l_{i,1} &= \textstyle\sum_{\substack{j=1,\cdots,n \\ A_{i,j}<0}} A_{i,j}, \ i = 1, 2, \cdots, m \\ l_{i,2} &= \textstyle\sum_{\substack{j=1,\cdots,n \\ A_{i,j}>0}} A_{i,j}, \ i = 1, 2, \cdots, m \\ q &= \max_{i=1,\cdots,m} \{l_{i,2} - l_{i,1}\} \\ p &= \text{ the smallest prime greater than } q. \end{aligned}$$

Step 5. Randomly choose a permutation σ on n letters, such that the matrix $S = [A_{\sigma(1)}, A_{\sigma(2)}, \cdots, A_{\sigma(m)}]$ is invertible in $\mathbb{Z}_p^{m \times m}$.
Step 6. Let $H = S^{-1}[A_{\sigma(m+1)}, A_{\sigma(m+2)}, \cdots, A_{\sigma(n)}] \bmod p$.

Public Key: H, p.
Private Key: $N_1, N_2, \cdots, N_n$, S, τ, $\sigma^{-1}\tau^{-1}$, $l_{i,1}$, $l_{i,2}$ $(i = 1, 2, \cdots, m)$.

Encryption: For any message $t \in \{0,1\}^m$, first, we uniformly randomly choose a vector r from $\{0,1\}^m$, then compute the ciphertext:

$$c = Ht + r \bmod p.$$

Decryption: Let $v = \begin{pmatrix} r \\ t \end{pmatrix}$, and $v' = \begin{pmatrix} v_{\sigma^{-1}(1)} \\ \vdots \\ v_{\sigma^{-1}(n)} \end{pmatrix}$. We first compute

$$\begin{aligned} c' &= Sc && \bmod p \\ &= Sr + SHt && \bmod p \\ &= S[I|H]v && \bmod p \\ &= [S|SH]v && \bmod p \\ &= Av' && \bmod p. \end{aligned}$$

By the fact that every entry of v' is in $\{0,1\}$, we know the i-th entry of Av' is in the interval from $l_{i,1}$ to $l_{i,2}$. So we can choose i-th the entries of c' in the interval from $l_{i,1}$ to $l_{i,2}$ since $p > q$ and get $c' = Av'$. Then we compute

$$(N_{\tau(1)}, N_{\tau(2)}, \cdots, N_{\tau(m)})Av' = \sum_{i=1}^{n} v_{\sigma^{-1}(i)} N_{\tau(i)} = \sum_{i=1}^{n} v_{\sigma^{-1}\tau^{-1}(i)} N_i.$$

and easily get $v_{\sigma^{-1}\tau^{-1}(i)}$ by the greedy algorithm. Finally the message t can be found by the known permutation $\sigma^{-1}\tau^{-1}$.

4 Implementations of Our Cryptosystem

4.1 Choosing the Superincreasing Sequence and τ

To generate a superincreasing sequence, we can first give a bound $d \in \mathbb{Z}^+$, then select $N_1 = 1$ and generate N_i $(2 \leq i \leq n)$ inductively as follows: after having $N_1, N_2, \cdots, N_k$, we uniformly randomly choose an integer $e \in (0, d)$ and let $N_{k+1} = \sum_{j=0}^{k} N_j + e$. Experiments show that the bigger d is, the bigger the final p is.

We don't uniformly randomly choose a permutation τ on n letters directly, but use the following way:

1. For $1 \leq i \leq m$, we uniformly randomly choose a permutation ρ on $\{1, 2, \cdots, m\}$, and let $\tau(i) = 2\rho(i) - 1$.
2. For $m + 1 \leq i \leq n$, we independently uniformly randomly choose another permutation ρ' on $\{1, 2, \cdots, m\}$, and let $\tau(i) = 2\rho'(i - m)$.

The reason to set $N_1 = 1$ and $\tau^{-1}(1) \leq m$ is mainly to ensure that $N_{\tau(i)}(m + 1 \leq i \leq n)$ can be represented as the integer linear combination of $N_{\tau(1)}, N_{\tau(2)}, \cdots, N_{\tau(m)}$, since $1 \in \{N_{\tau(1)}, N_{\tau(2)}, \cdots, N_{\tau(m)}\}$. Notice that even we don't choose $N_1 = 1$, the probability that $N_{\tau(i)}(m + 1 \leq i \leq n)$ can be represented as an integer linear combination of $N_{\tau(1)}, N_{\tau(2)}, \cdots, N_{\tau(m)}$ is also very large.

As we will see in the next subsection, if we choose such τ, we can reduce some operations for sorting $N_{\tau(i)}$'s, and reduce the size of p which affects the security of the system.

4.2 Finding Integer Linear Combination with Small Coefficients

First, we give an algorithm to represent an integer y as an integer linear combination of $T_1, T_2, \cdots, T_k \in \mathbb{Z}$ with small coefficients by the lattice reduction algorithm.

Algorithm. *LS*
Input: $y, T_1, T_2, \cdots, T_k$.
Output: $b_1, b_2, \cdots, b_k \in \mathbb{Z}$ which are small and $y = \sum_{i=1}^{k} b_i T_i$

1. Choose $b'_1, b'_2, \cdots, b'_k \in \mathbb{Z}$, such that $y = \sum_{i=1}^{k} b'_i T_i$.
2. Let $\mathcal{L}$ be the lattice $\{(x_1, x_2, \cdots, x_k)^T \in \mathbb{Z}^k | \sum_{i=1}^{k} x_i T_i = 0\}$, and use Babai's Nearest Plane Algorithm to find $(x'_1, x'_2, \cdots, x'_k)^T \in \mathcal{L}$ close to $(b'_1, b'_2, \cdots, b'_k)$.
3. Let $b_i := b'_i - x'_i$ for $1 \leq i \leq k$.
4. Output $b_1, b_2, \cdots, b_k$.

We don't use the algorithm LS to find $b_{i-m,j}(1 \leq j \leq m)$ directly, because it costs too much time when m is large. To make the algorithm more efficient, we involve a greedy strategy.

Let $T_i = N_{\tau(i)}$, $i = 1, 2, \cdots, m$. We first sort $T_1, T_2, \cdots, T_m$ in ascending order. Assume $T_{\varphi(1)} < T_{\varphi(2)} < \cdots < T_{\varphi(m)}$, where φ is some permutation on m letters. Notice that if we choose τ as in Subsection 4.1, we can easily get the permutation φ since $T_{\varphi(i)} = N_{2i-1}$. Then we use the algorithm below to find $b_{i-m,j}(1 \leq j \leq m)$ for $N_{\tau(i)}(m+1 \leq i \leq n)$ with parameters δ and k, where $\delta, k \in \mathbb{Z}^+$ and $k \leq m$.

Algorithm. Finding Integer Linear Combination with Small Coefficients
Input: δ, k, $T_{\varphi(1)}, T_{\varphi(2)}, \cdots, T_{\varphi(m)}$, φ and $N_{\tau(i)}$ where $i > m$
Output: Small $b_{i-m,1}, \cdots, b_{i-m,m} \in \mathbb{Z}$ such that $N_{\tau(i)} = \sum_{j=1}^{m} b_{i-m,j}T_j$

1. for j from 1 to m do
   uniformly choose an integer $a \in [-\delta, \delta]$
   $N_{\tau(i)} := N_{\tau(i)} - aT_{\varphi(j)}$
   $b_{i-m,\varphi(j)} := a$
2. end for
3. for $j = m$ to $k+1$ do
   compute $N_{\tau(i)} = qT_{\varphi(j)} + r$, where $q, r \in \mathbb{Z}$ and $|r| \leq \frac{T_{\varphi(j)}}{2}$
   $N_{\tau(i)} := r$
   $b_{i-m,\varphi(j)} := b_{i-m,\varphi(j)} + q$
4. end for
5. Compute $(x_1, x_2, \cdots, x_k)^T := LS(N_{\tau(i)}, T_{\varphi(1)}, T_{\varphi(2)}, \cdots, T_{\varphi(k)})$, and let $b_{i-m,\varphi(i)} := b_{i-m,\varphi(i)} + x_i$ for $1 \leq i \leq k$.
6. Output $b_{i-m,1}, \cdots, b_{i-m,m}$

In the algorithm, we first do some initial randomization (adding or subtracting small multiples of $T_{\varphi(j)}$) to make the coefficients of the linear combination look more random. Then, a greedy strategy is involved. By $T_{\varphi(1)} < T_{\varphi(2)} < \cdots < T_{\varphi(m)}$ and the choice of τ, the quotient q will be usually very small. Finally, we use LS in a k-dimensional lattice to get the last k coefficients. Notice that $T_{\varphi(1)} = 1$, so LS always return a solution. If we did not use LS, or equivalently $k = 0$, the last coefficient, namely the remainder when some integer is divided by $T_{\varphi(2)}$ since $T_{\varphi(1)} = 1$, would be usually large.

4.3 Some Experimental Results

We implemented the cryptosystem on an Intel(R) Core(TM) 2 Duo E8400 2.99 GHz PC using Shoup's NTL library version 5.4.1 [27]. In all our experiments, we let the bound d in Subsection 4.1 be 40, $\delta = 2$, and we used the function LatticeSolve in NTL directly instead of implementing the Algorithm LS. For $m = 100, 200, 300, 400, 500$, we let $k = 10, 20, 30$ respectively. For each m and k, 200 instances were tested. The results are stated as below.

How Large p can Be?. Since p decides the size of the key directly, it is necessary to study how large it can be. For each m and k, we give the minimum, the maximum, the mean and the standard deviation(ST) of those p's in our 200 instances in Table 1.

It is reasonable to assume that $p \approx 1.7m$. We would like to point out that p is not necessary a prime. We can just choose $q + \alpha$ as p where α is some small positive integer. This can decrease the key size. However, choosing p as a prime can increase the probability that S is invertible.

The Probability that S is Invertible. In our experiments, we uniformly chose a permutation σ on n letters, and S was always invertible. Hence, it is also reasonable to believe that S is invertible with very high probability.

Table 1. The Statistical Information on p

m	100			200			300			400			500		
k	10	20	30	10	20	30	10	20	30	10	20	30	10	20	30
min	157	149	149	317	311	307	479	479	457	617	631	631	773	757	761
max	239	211	227	449	449	439	677	673	653	907	877	857	1069	1087	1087
mean	183.11	180.36	181.29	359.2	355.31	354.64	540.01	527.68	529.92	710.31	704.14	702.84	889.08	872.96	870.81
ST	14.33	13.71	15.24	23.43	22.87	21.63	32.00	33.21	33.52	42.23	41.25	43.59	48.13	52.79	54.06

The Key Size and Speed. Since $p \approx 1.7m$, we give some basic information about the new system:

Table 2. Some Basic Information

Parameter	m
Public Key Size	$O(m^2 \log m)$
Private Key Size	$O(m^2 \log m)$
Message Size	$O(m)$
Ciphertext Size	$O(m \log m)$
Encryption Speed	$O(m^2 \log m)$
Decryption Speed	$O(m^2 \log^2 m)$

5 Security Analysis

5.1 Knapsack Structure

Almost all knapsack-based public-key cryptosystems have been broken. However, we did not try to hide the trapdoor in our cryptosystem by transforming a superincreasing knapsack into a general one. A different way is used. The knapsack structure is hidden behind the linear combinations so that no obvious knapsack structure does appear in our cryptosystem. It seems hard to attack our system with the successful attacks against knapsack-based cryptosystems. What's more, even A has been obtained, there may be no obvious attack to recover the superincreasing sequence since τ is unknown.

5.2 Message Security

An attacker can recover the message by trying all possible $t \in \{0,1\}^m$ to check if all the entries of $c - Ht \bmod p$ are in $\{0,1\}$. This can be done in $O(2^m)$.

The direct lattice attack to recover the message is to find a vector in the lattice spanned by

$$B = \begin{pmatrix} \alpha I & 0 \\ \hline H & pI \end{pmatrix}$$

close to the target vector $\begin{pmatrix} 0 \\ \hline c \end{pmatrix}$, because there exists a vector $u \in \mathbb{Z}^m$, such that

$$\begin{pmatrix} \alpha I & 0 \\ \hline H & pI \end{pmatrix} \begin{pmatrix} t \\ \hline u \end{pmatrix} - \begin{pmatrix} 0 \\ \hline c \end{pmatrix} = \begin{pmatrix} \alpha t \\ \hline -r \end{pmatrix},$$

where t is the corresponding message and r is the random vector selected in encryption and $\| \begin{pmatrix} \alpha t \\ \hline -r \end{pmatrix} \|$ is small.

By the Gaussian Heuristic, the size of the solution of the closest vector problems is approximately $\sqrt{\frac{n}{2\pi e}} \det(\mathcal{L}(B))^{\frac{1}{n}} = \sqrt{\frac{\alpha p m}{\pi e}}$. For any message t and random vector r, to minimize

$$c(t,r) = \frac{\sqrt{\alpha^2 \|t\|^2 + \|r\|^2}}{\sqrt{\frac{\alpha p m}{\pi e}}} = \sqrt{\frac{(\alpha^2\|t\|^2 + \|r\|^2)\pi e}{\alpha p m}},$$

we get $\alpha = \|r\| / \|t\|$ and

$$c(t,r) = \sqrt{\frac{2\pi e \|t\| \|r\|}{pm}}.$$

$c(t,r)$ gives a measure of the vulnerability of an individual message to a lattice attack. An encrypted message is most vulnerable when $c(t,r)$ is small, and becomes less as $c(t,r)$ gets closer to 1.

Notice that it can be expected that $\|t\|\|r\| \approx \frac{m}{2}$, $c(t,r)$ is approximately $\sqrt{\frac{\pi e}{p}}$. The mean and the standard deviation of the values $c_{msg} = \sqrt{\frac{\pi e}{p}}$ in our experiments are given in Table 3.

Table 3. The Statistical Information on c_{msg}

m	100			200			300			400			500		
k	10	20	30	10	20	30	10	20	30	10	20	30	10	20	30
mean	0.216	0.218	0.218	0.154	0.155	0.155	0.126	0.127	0.127	0.110	0.110	0.110	0.098	0.099	0.099
ST	0.008	0.008	0.008	0.005	0.005	0.005	0.004	0.004	0.004	0.003	0.003	0.003	0.003	0.003	0.003

An attack was performed for $m = 150$. We generated the keys and got $p = 307$. We randomly generated a message t and a random vector r. Using the embedding method, we tried to find short vector in $\mathcal{L}(\begin{pmatrix} I & 0 & 0 \\ \hline H & pI & c \\ \hline 0 & 0 & 1 \end{pmatrix})$ with BKZ_QP1 in NTL.

We let block size= 20 and prune=0, but failed to get the information about t and r after 39962.8 seconds. The shortest vector we found was the vector which had only one non-zero entry 307.

5.3 Key Security

It seems there is no obvious attack to obtain the whole private key in our cryptosystem.

However, using the direct lattice attack to obtain A from H, we need find m short vectors in the lattice spanned by $\left(\begin{array}{c|c} I & 0 \\ \hline H^T & pI \end{array}\right)$, since every column of $A' = [A_{\sigma(1)}, A_{\sigma(2)}, \cdots, A_{\sigma(n)}]^T$ is in the lattice. It is obviously very different to find all m correct short vectors. However, we still discuss the attack.

We denote by l the mean of $\|A_i'\|$'s for $1 \leq i \leq m$. By the Gaussian Heuristic, the size of the solution of the shortest vector problems is approximately $\sqrt{\frac{n}{2\pi e}}\det(\mathcal{L}(B))^{\frac{1}{n}} = \sqrt{\frac{pm}{\pi e}}$. So we get the value of $c_{key} = \frac{l}{\sqrt{\frac{pm}{\pi e}}}$. The smaller c_{key} is, the more easily A_i''s may be found. As it gets closer to 1, to find A_i''s may be more difficult. We give the values of l's, the mean and the standard deviation of c_{key}'s in our experiments in Table 4.

Table 4. The Statistical Information on c_{key}

m	100			200			300			400			500		
k	10	20	30	10	20	30	10	20	30	10	20	30	10	20	30
l	17.69	17.72	17.79	24.92	24.94	25.00	30.47	30.52	30.55	35.16	35.20	35.22	39.30	39.32	39.34
mean	0.383	0.386	0.387	0.272	0.274	0.275	0.222	0.224	0.224	0.193	0.194	0.194	0.172	0.174	0.174
ST	0.014	0.015	0.016	0.008	0.008	0.008	0.006	0.007	0.007	0.006	0.005	0.006	0.005	0.005	0.005

5.4 Remarks

Notice that c_{msg} and c_{key} become smaller slowly as m increase. From this we can not conclude that it is easy to attack the cryptosystem with big m, since bigger dimension of the lattice leads less efficiency of the attack. In Table 5, we compare the constant c_{msg} in our cryptosystem with that in GGH and NTRU, since c_{msg} is usually smaller than c_{key}.

What's more, compared with GGH, to recover the message using direct lattice reduction, we need solve a CVP for a $2m$-dimensional lattice instead of m-dimensional in GGH. This may allow us to use small dimensional matrix as public key to provide sufficient security.

Table 5. Constants in Some Lattice-Based Public-Key Cryptosystems

m	100			200			300			400			500		
k	10	20	30	10	20	30	10	20	30	10	20	30	10	20	30
Ours	0.216	0.218	0.218	0.154	0.155	0.155	0.126	0.127	0.127	0.110	0.110	0.110	0.098	0.099	0.099
GGH[20]	-			0.103			0.105			0.104			-		
NTRU[11]	-			0.175			0.145			0.125			0.112		

Compared with NTRU, there may no obvious attack to obtain the private key in our cryptosystem whereas the private key of NTRU can be obtained by finding the short vector in NTRU-lattice. Moreover, the underlying lattice of our cryptosystem has no special cyclic structure like NTRU. This makes our system resist some similar attacks against NTRU which are based on its cyclic structure.

6 Conclusion

We present a new lattice-based public-key cryptosystem mixed with a knapsack, and give some algorithms to implement the cryptosystem in details. By our implementation, we show it has reasonable key size, quick encryption and quick decryption. The security analysis shows that it may resist the direct lattice attack well. Moreover, more improvement on finding integer linear combination with smaller coefficients will yield smaller p, which may improve the cryptosystem's efficiency and security.

Acknowledgments. We thank the anonymous referees for their suggestions and discussions about the paper and further work.

References

1. Ajtai, M.: Gennerating hard instances of lattice problems. In: The 28th STOC, pp. 99–108. ACM, New York (1996)
2. Ajtai, M.: Representing hard lattices with $O(n \log n)$ bits. In: The 37th STOC, pp. 94–103. ACM, New York (2005)
3. Ajtai, M., Dwork, C.: A public-key cryptosystem with worst-case/average-case equivalence. In: The 29th STOC, pp. 284–293. ACM, New York (1997)
4. Babai, L.: On Lovász lattice reduction and the nearest lattice point problem. Combinatorica 6, 1–13 (1986)
5. Cai, J.-Y., Cusick, T.W.: A Lattice-Based Public-Key Cryptosystem. In: Tavares, S., Meijer, H. (eds.) SAC 1998. LNCS, vol. 1556, pp. 219–233. Springer, Heidelberg (1999)
6. Coppersmith, D., Shamir, A.: Lattice Attacks on NTRU. In: Fumy, W. (ed.) EUROCRYPT 1997. LNCS, vol. 1233, pp. 52–61. Springer, Heidelberg (1997)
7. Goldreich, O., Goldwasser, S., Halevi, S.: Public-Key Cryptosystems from Lattice Reduction Problems. In: Kaliski Jr., B.S. (ed.) CRYPTO 1997. LNCS, vol. 1294, pp. 112–131. Springer, Heidelberg (1997)
8. Gentry, C., Peikert, C., Vaikuntanathan, V.: Trapdoors for hard lattices and new cryptographic constructions. In: The 40th STOC, pp. 197–206. ACM, New York (2008)
9. Howgrave-Graham, N.: A Hybrid Lattice-Reduction and Meet-in-the-Middle Attack against NTRU. In: Menezes, A. (ed.) CRYPTO 2007. LNCS, vol. 4622, pp. 150–169. Springer, Heidelberg (2007)
10. Howgrave-Graham, N., Silverman, J.H., Whyte, W.: A Meet-In-The-Meddle Attack on an NTRU Private Key. Technical report, `http://www.ntru.com/cryptolab/technotes.htm#004`

11. Howgrave-Graham, N., Silverman, J.H., Whyte, W.: Choosing Parameter Sets for NTRUEncrypt with NAEP and SVE-3. In: Menezes, A. (ed.) CT-RSA 2005. LNCS, vol. 3376, pp. 118–135. Springer, Heidelberg (2005)
12. Hoffstein, J., Pipher, J., Silverman, J.H.: NTRU: A Ring-Based Public Key Cryptosystem. In: Buhler, J.P. (ed.) ANTS 1998. LNCS, vol. 1423, pp. 267–288. Springer, Heidelberg (1998)
13. IEEE. P1363.1 Public-Key Cryptographic Techniques Based on Hard Problems over Lattices (June 2003), `http://grouper.ieee.org/groups/1363/lattPK/index.html`
14. Koblitz, N.: Elliptic curve cryptosystems. Mathematics of Computation 48(177), 203–209 (1987)
15. Lenstra, A.K., Lenstra, H.W., Lovász, L.: Factoring polynomials with rational coeffcients. Math. Ann. 261, 515–534 (1982)
16. Lyubashevsky, V., Peikert, C., Regev, O.: On Ideal Lattices and Learning with Errors over Rings. In: Gilbert, H. (ed.) EUROCRYPT 2010. LNCS, vol. 6110, pp. 1–23. Springer, Heidelberg (2010)
17. May, A., Silverman, J.H.: Dimension Reduction Methods for Convolution Modular Lattices. In: Silverman, J.H. (ed.) CaLC 2001. LNCS, vol. 2146, pp. 110–125. Springer, Heidelberg (2001)
18. Miller, V.S.: Use of Elliptic Curves in Cryptography. In: Williams, H.C. (ed.) CRYPTO 1985. LNCS, vol. 218, pp. 417–426. Springer, Heidelberg (1986)
19. Nguyên, P.Q., Stern, J.: Cryptanalysis of the Ajtai-Dwork Cryptosystem. In: Krawczyk, H. (ed.) CRYPTO 1998. LNCS, vol. 1462, pp. 223–242. Springer, Heidelberg (1998)
20. Nguyên, P.Q.: Cryptanalysis of the Goldreich-Goldwasser-Halevi Cryptosystem from Crypto'97. In: Wiener, M. (ed.) CRYPTO 1999. LNCS, vol. 1666, pp. 288–304. Springer, Heidelberg (1999)
21. Pan, Y., Deng, Y.: A Ciphertext-Only Attack Against the Cai-Cusick Lattice-Based Public-Key Cryptosystem. IEEE Transactions on Information Theory 57, 1780–1785 (2011)
22. Peikert, C.: Public-Key Cryptosystems from the Worst-Case Shortest Vector Problem. In: The 41th STOC, pp. 333–342. ACM, New York (2009)
23. Regev, O.: New lattice-based cryptographic constructions. Journal of the ACM 51, 899–942 (2004)
24. Regev, O.: On lattices, learning with errors, random linear codes, and cryptography. In: The 37th STOC, pp. 84–93. ACM, New York (2005)
25. Rivest, R., Shamir, A., Adleman, L.: A Method for Obtaining Digital Signatures and Public-Key Cryptosystems. Communications of the ACM 21, 120–126 (1978)
26. Shor, P.: Algorithms for Quantum Computation: Discrete Logarithms and Factoring. In: The 35th Annual Symposium on Foundations of Computer Science, pp. 124–134. IEEE Computer Science Press, Santa Fe (1994)
27. Shoup, V.: NTL: A library for doing number theory, `http://www.shoup.net/ntl/`
28. Stehlé, D., Steinfeld, R.: Making NTRU as Secure as Worst-Case Problems over Ideal Lattices. In Paterson. In: Paterson, K.G. (ed.) EUROCRYPT 2011. LNCS, vol. 6632, pp. 27–47. Springer, Heidelberg (2011)

Achieving Short Ciphertexts or Short Secret-Keys for Adaptively Secure General Inner-Product Encryption

Tatsuaki Okamoto[1] and Katsuyuki Takashima[2]

[1] NTT
okamoto.tatsuaki@lab.ntt.co.jp
[2] Mitsubishi Electric
Takashima.Katsuyuki@aj.MitsubishiElectric.co.jp

Abstract. In this paper, we present two *non-zero inner-product* encryption (NIPE) schemes that are *adaptively secure* under a standard assumption, the decisional linear (DLIN) assumption, in the standard model. One of the proposed NIPE schemes features *constant-size ciphertexts* and the other features *constant-size secret-keys*. Our NIPE schemes imply an identity-based revocation (IBR) system with constant-size ciphertexts or constant-size secret-keys that is adaptively secure under the DLIN assumption. Any previous IBR scheme with constant-size ciphertexts or constant-size secret-keys was *not adaptively secure* in the standard model. This paper also presents two zero inner-product encryption (ZIPE) schemes each of which has constant-size ciphertexts or constant-size secret-keys and is adaptively secure under the DLIN assumption in the standard model. They imply an identity-based broadcast encryption (IBBE) system with constant-size ciphertexts or constant-size secret-keys that is adaptively secure under the DLIN assumption.

1 Introduction

1.1 Background

Functional encryption (FE) is an advanced concept of encryption or a generalization of public-key encryption (PKE) and identity-based encryption (IBE). In FE systems, a receiver can decrypt a ciphertext using a secret-key corresponding to a parameter x if x is suitably related to another parameter y specified for the ciphertext, or $R(x, y) = 1$ for some relation R (i.e., relation R holds for (x, y)) .

The first flavor of functional encryption traces back to the work of Sahai and Waters [18], which was subsequently extended in [4,9,13,17]. In their concept called attribute-based encryption (ABE), for example, parameter x for a secret-key is an access control policy, and parameter y for a ciphertext is a set of attributes. Decryption requires attribute set y to satisfy policy x, i.e., relation $R^{\mathsf{ABE}}(x, y) = 1$ iff y satisfies x. Identity-based broadcast encryption (IBBE) [1,5,6,8,19] and revocation (IBR) [12] schemes can also be thought of as

D. Lin, G. Tsudik, and X. Wang (Eds.): CANS 2011, LNCS 7092, pp. 138–159, 2011.

functional encryption systems where a ciphertext is encrypted for a set of identities $S = \{ID_1, \ldots, ID_n\}$ in IBBE (resp. IBR) systems, and to decrypt it by a secret-key associated with ID requires that $ID \in S$ (resp. $ID \notin S$), i.e., relation $R^{\mathsf{IBBE}}(ID, S) = 1$ (resp. $R^{\mathsf{IBR}}(ID, S) = 1$) iff $ID \in S$ (resp. $ID \notin S$).

Katz, Sahai and Waters [11] introduced a functional encryption scheme for zero inner products, zero inner product encryption (ZIPE) where a ciphertext encrypted with vector $\vec{x}$ can be decrypted by any key associated with vector $\vec{y}$ such that $\vec{x} \cdot \vec{y} = 0$, i.e., relation $R^{\mathsf{ZIPE}}(\vec{x}, \vec{y}) = 1$ iff $\vec{x} \cdot \vec{y} = 0$. Their scheme is *selectively secure* in the standard model and the ciphertext size is *linear* in the dimension of vectors, n, although it achieves an additional security property, *attribute-hiding*, in which $\vec{x}$ is hidden from the ciphertext. As shown in [11], ZIPE provides functional encryption for a wide class of relations corresponding to equalities, polynomials and CNF/DNF formulae.

Attrapadung and Libert [2] proposed a ZIPE scheme as well as a non-zero IPE (NIPE) scheme, where NIPE relation $R^{\mathsf{NIPE}}(\vec{x}, \vec{y}) = 1$ iff $\vec{x} \cdot \vec{y} \neq 0$. NIPE supports a wide class of relations corresponding to the complement of those for ZIPE. In their ZIPE and NIPE schemes, without retaining the attribute-hiding property, the ciphertext size reduces to a *constant* in n (the dimension of vectors, $\vec{x}$ and $\vec{y}$), as long as the description of the vector is not considered a part of the ciphertext, which is a common assumption in the broadcast encryption/revocation applications. Hereafter in this paper, "constant" will be used in this sense. In addition, the number of pairing operations for decryption in [2] is constant. Their ZIPE system is *adaptively secure* in the standard model, but the NIPE scheme is *not adaptively secure* (co-selectively secure) in the standard model.

The ZIPE system [2] implies an *adaptively secure* identity-based broadcast encryption (IBBE) scheme with constant-size ciphertexts in the standard model, while previous IBBE schemes with constant-size ciphertexts were either only selective-ID secure [1,5,6] or secure in a non-standard model [8,19]. Among IBBE systems with short ciphertexts (including selective-ID secure ones), the IBBE scheme [2] is the only one relying on standard assumptions, DBDH and DLIN assumptions. The NIPE scheme [2] implies a co-selectively secure (not adaptively secure) identity-based revocation (IBR) system [12] with constant-size ciphertexts in the standard model. Lewko, Sahai and Waters [12] presented IBR systems with constant-size public and secret keys that are not adaptively secure. Hence, the following problems are still remained.

1. No NIPE scheme with constant-size ciphertexts is *adaptively secure* in the standard model, and no IBR scheme with constant-size ciphertexts or constant-size secret-keys is *adaptively secure* in the standard model. No NIPE scheme with constant-size *secret-keys* has been presented.
2. No ZIPE (or no IBBE) scheme with constant-size ciphertexts is adaptively (or selectively) secure under a *single* standard assumption in the standard model. No ZIPE scheme with constant-size *secret-keys* has been presented.

1.2 Our Results

This paper presents the first NIPE scheme that has constant-size ciphertexts or constant-size secret-keys and that is *adaptively secure* in the standard model. The security assumption is a standard one, the decisional linear (DLIN) assumption. This implies the first IBR scheme with constant-size ciphertexts or constant-size secret-keys that is *adaptively secure* in the standard model.

This paper also presents the first ZIPE scheme that has constant-size ciphertexts or constant-size secret-keys and is adaptively secure solely under a *single* standard assumption, the DLIN assumption, in the standard model. This implies the first IBBE scheme with constant-size ciphertexts that is adaptively secure solely under a *single* standard assumption in the standard model. Our ZIPE scheme also implies a constant-size ciphertext *hierarchical* ZIPE (HIPE) scheme that is adaptively secure under DLIN, by employing delegation and re-randomization similar to those in [13,16]. It will be given in the full version.

In addition, the number of pairing operations for decryption is constant in all the proposed schemes. We summarize a comparison of our results with those of [2] in Table 1 in Section 8 (see the items of 'Security', 'Assump.', 'CT Size' and 'SK Size' in Table 1, for the features discussed in Sections 1.1 and 1.2).

1.3 Related Works

Several ABE schemes [3,7,10] with *constant-size ciphertexts* have been proposed. Among them, [7,10] only support limited classes of predicates that do not cover the classes supported by ZIPE and NIPE, while [3] supports a wider class of relations, non-monotone predicates, than those by ZIPE and NIPE. All of these ABE schemes, however, are only *selectively secure* in the standard model. Adaptively secure and *attribute-hiding* ZIPE scheme under the DLIN assumption has been presented [17], but the ciphertext-size is linear in n (*not constant*), while our ZIPE scheme has *constant-size* ciphertexts and is adaptively secure but *not attribute-hiding*.

1.4 Key Techniques

All of the proposed schemes in this paper are constructed on dual system encryption [20,14] and dual pairing vector spaces (DPVS) [16,13,17]. See Section 1.5 for some notations in this section. In DPVS, a pair of dual (or orthonormal) bases, $\mathbb{B}$ and $\mathbb{B}^*$, are randomly generated using a *fully* random linear transformation $X \xleftarrow{\mathsf{U}} GL(N, \mathbb{F}_q)$ (N: dimension of $\mathsf{span}\langle\mathbb{B}\rangle$ and $\mathsf{span}\langle\mathbb{B}^*\rangle$) such that $\mathbb{B}$ and $\mathbb{B}^*$ are transformed from canonical basis $\mathbb{A}$ by X and $(X^{-1})^{\mathrm{T}}$, respectively (see Section 2 and [16,13,17]). In a typical application of DPVS to cryptography, a part of $\mathbb{B}$ (say $\hat{\mathbb{B}}$) is used as a public key and the corresponding part of $\mathbb{B}^*$ (say $\hat{\mathbb{B}}^*$) is used as a secret key or trapdoor.

In this paper, we develop a novel technique on DPVS, where we employ a *special form* of random linear transformation $X \in GL(N, \mathbb{F}_q)$, or $X \in \mathcal{L}(N, \mathbb{F}_q)$ of Eq. (2) in Section 5.2, in place of *fully* random linear transformation $X \xleftarrow{\mathsf{U}} GL(N, \mathbb{F}_q)$. This form of X provides us a framework to achieve short ciphertexts or short secret-keys as well as a small number of pairing operations in decryption. It, however, is a challenging task to find such a special form of X like Eq. (2) that meet the several requirements for the dual system encryption method to prove the adaptive security of ZIPE and NIPE schemes under the DLIN assumption. Such requirements are given hereafter. To reduce the security of our schemes, especially Problems 1 and 2 in this paper, to the DLIN assumption, the form of X should be *consistent* with the distribution of the DLIN problem. The form of X should be *sparse* enough to achieve short ciphertexts or secret-keys. We should also have a *special* pairwise independence lemma, Lemma 3 in Section 5.4, that is due to the special form of X, where linear random transformations U and Z are more restricted (or specific) than those of previous results, e.g., [17], with fully random X. See Section 5.1 for more details.

1.5 Notations

When A is a random variable or distribution, $y \xleftarrow{\mathsf{R}} A$ denotes that y is randomly selected from A according to its distribution. When A is a set, $y \xleftarrow{\mathsf{U}} A$ denotes that y is uniformly selected from A. A vector symbol denotes a vector representation over $\mathbb{F}_q$, e.g., $\vec{x}$ denotes $(x_1, \ldots, x_n) \in \mathbb{F}_q^n$. For two vectors $\vec{x} = (x_1, \ldots, x_n)$ and $\vec{v} = (v_1, \ldots, v_n)$, $\vec{x} \cdot \vec{v}$ denotes the inner-product $\sum_{i=1}^{n} x_i v_i$. The vector $\vec{0}$ is abused as the zero vector in $\mathbb{F}_q^n$ for any n. X^{T} denotes the transpose of matrix X. I_ℓ denotes the $\ell \times \ell$ identity matrix. A bold face letter denotes an element of vector space $\mathbb{V}$, e.g., $\boldsymbol{x} \in \mathbb{V}$. When $\boldsymbol{b}_i \in \mathbb{V}$ $(i = 1, \ldots, \ell)$, $\mathsf{span}\langle \boldsymbol{b}_1, \ldots, \boldsymbol{b}_\ell \rangle \subseteq \mathbb{V}$ (resp. $\mathsf{span}\langle \vec{x}_1, \ldots, \vec{x}_\ell \rangle$) denotes the subspace generated by $\boldsymbol{b}_1, \ldots, \boldsymbol{b}_\ell$ (resp. $\vec{x}_1, \ldots, \vec{x}_\ell$). For bases $\mathbb{B} := (\boldsymbol{b}_1, \ldots, \boldsymbol{b}_N)$ and $\mathbb{B}^* := (\boldsymbol{b}_1^*, \ldots, \boldsymbol{b}_N^*)$, $(x_1, \ldots, x_N)_{\mathbb{B}} := \sum_{i=1}^{N} x_i \boldsymbol{b}_i$ and $(y_1, \ldots, y_N)_{\mathbb{B}^*} := \sum_{i=1}^{N} y_i \boldsymbol{b}_i^*$. For dimension n of vectors, $\vec{e}_j$ denotes the canonical basis vector $(\overbrace{0 \cdots 0}^{j-1}, 1, \overbrace{0 \cdots 0}^{n-j}) \in \mathbb{F}_q^n$ for $j = 1, \ldots, n$. $GL(n, \mathbb{F}_q)$ denotes the general linear group of degree n over $\mathbb{F}_q$. For a linear subspace $V \subset \mathbb{F}_q^n$, $V^\perp$ denotes the orthogonal complement, i.e., $V^\perp := \{\vec{w} \in \mathbb{F}_q^n | \vec{w} \cdot \vec{v} = 0 \text{ for all } \vec{v} \in V\}$.

2 Dual Pairing Vector Spaces by Direct Product of Symmetric Pairing Groups

Definition 1. *"Symmetric bilinear pairing groups" $(q, \mathbb{G}, \mathbb{G}_T, G, e)$ are a tuple of a prime q, cyclic additive group $\mathbb{G}$ and multiplicative group $\mathbb{G}_T$ of order q, $G \neq 0 \in \mathbb{G}$, and a polynomial-time computable nondegenerate bilinear pairing $e : \mathbb{G} \times \mathbb{G} \to \mathbb{G}_T$ i.e., $e(sG, tG) = e(G, G)^{st}$ and $e(G, G) \neq 1$.*

Let $\mathcal{G}_{\mathsf{bpg}}$ *be an algorithm that takes input* 1^λ *and outputs a description of bilinear pairing groups* $(q, \mathbb{G}, \mathbb{G}_T, G, e)$ *with security parameter* λ.

Definition 2. *"Dual pairing vector spaces (DPVS)"* $(q, \mathbb{V}, \mathbb{G}_T, \mathbb{A}, e)$ *by a direct product of symmetric pairing groups* $(q, \mathbb{G}, \mathbb{G}_T, G, e)$ *are a tuple of prime* q, *N-dimensional vector space* $\mathbb{V} := \overbrace{\mathbb{G} \times \cdots \times \mathbb{G}}^{N}$ *over* $\mathbb{F}_q$, *cyclic group* $\mathbb{G}_T$ *of order* q, *canonical basis* $\mathbb{A} := (\boldsymbol{a}_1, \ldots, \boldsymbol{a}_N)$ *of* $\mathbb{V}$, *where* $\boldsymbol{a}_i := (\overbrace{0, \ldots, 0}^{i-1}, G, \overbrace{0, \ldots, 0}^{N-i})$, *and pairing* $e : \mathbb{V} \times \mathbb{V} \to \mathbb{G}_T$. *The pairing is defined by* $e(\boldsymbol{x}, \boldsymbol{y}) := \prod_{i=1}^{N} e(G_i, H_i) \in \mathbb{G}_T$ *where* $\boldsymbol{x} := (G_1, \ldots, G_N) \in \mathbb{V}$ *and* $\boldsymbol{y} := (H_1, \ldots, H_N) \in \mathbb{V}$. *This is nondegenerate bilinear i.e.,* $e(s\boldsymbol{x}, t\boldsymbol{y}) = e(\boldsymbol{x}, \boldsymbol{y})^{st}$ *and if* $e(\boldsymbol{x}, \boldsymbol{y}) = 1$ *for all* $\boldsymbol{y} \in \mathbb{V}$, *then* $\boldsymbol{x} = \boldsymbol{0}$. *For all* i *and* j, $e(\boldsymbol{a}_i, \boldsymbol{a}_j) = e(G, G)^{\delta_{i,j}}$ *where* $\delta_{i,j} = 1$ *if* $i = j$, *and* 0 *otherwise, and* $e(G, G) \neq 1 \in \mathbb{G}_T$.

DPVS also has linear transformations $\phi_{i,j}$ *on* $\mathbb{V}$ *s.t.* $\phi_{i,j}(\boldsymbol{a}_j) = \boldsymbol{a}_i$ *and* $\phi_{i,j}(\boldsymbol{a}_k) = \boldsymbol{0}$ *if* $k \neq j$, *which can be easily achieved by* $\phi_{i,j}(\boldsymbol{x}) := (\overbrace{0, \ldots, 0}^{i-1}, G_j, \overbrace{0, \ldots, 0}^{N-i})$ *where* $\boldsymbol{x} := (G_1, \ldots, G_N)$. *We call* $\phi_{i,j}$ *"canonical maps".*

DPVS generation algorithm $\mathcal{G}_{\mathsf{dpvs}}$ *takes input* 1^λ $(\lambda \in \mathbb{N})$ *and* $N \in \mathbb{N}$, *and outputs a description of* $\mathsf{param}'_{\mathbb{V}} := (q, \mathbb{V}, \mathbb{G}_T, \mathbb{A}, e)$ *with security parameter* λ *and* N*-dimensional* $\mathbb{V}$. *It can be constructed by using* $\mathcal{G}_{\mathsf{bpg}}$.

For the asymmetric version of DPVS, $(q, \mathbb{V}, \mathbb{V}^*, \mathbb{G}_T, \mathbb{A}, \mathbb{A}^*, e)$, see Appendix A.2 in [17].

3 Definitions of Zero and Non-zero Inner-Product Encryption (ZIPE / NIPE)

This section defines zero and non-zero inner-product encryption (ZIPE / NIPE) and their security. The relations R^{ZIPE} of ZIPE and R^{NIPE} of NIPE are defined over vectors $\vec{x} \in \mathbb{F}_q^n \setminus \{\vec{0}\}$ and $\vec{v} \in \mathbb{F}_q^n \setminus \{\vec{0}\}$, where $R^{\mathsf{ZIPE}}(\vec{x}, \vec{v}) := 1$ iff $\vec{x} \cdot \vec{v} = 0$, and $R^{\mathsf{NIPE}}(\vec{x}, \vec{v}) := 1$ iff $\vec{x} \cdot \vec{v} \neq 0$, respectively

Definition 3 (Zero and Non-zero Inner-Product Encryption: ZIPE / NIPE). *Let a relation* R *be* R^{ZIPE} *or* R^{NIPE}. *A zero (resp. non-zero) inner-product encryption scheme consists of four algorithms with* $R := R^{\mathsf{ZIPE}}$ *(resp.* $R := R^{\mathsf{NIPE}}$*).*

Setup. *This is a randomized algorithm that takes as input security parameter. It outputs public parameters* pk *and master secret key* sk.

KeyGen. *This is a randomized algorithm that takes as input vector* $\vec{v}$, pk *and* sk. *It outputs a decryption key* $\mathsf{sk}_{\vec{v}}$.

Enc. *This is a randomized algorithm that takes as input message* m, *a vector,* $\vec{x}$, *and public parameters* pk. *It outputs a ciphertext* $\mathsf{ct}_{\vec{x}}$.

Dec. *This takes as input ciphertext* $\mathsf{ct}_{\vec{x}}$ *that was encrypted under a vector* $\vec{x}$, *decryption key* $\mathsf{sk}_{\vec{v}}$ *for vector* $\vec{v}$, *and public parameters* pk. *It outputs either plaintext* m *or the distinguished symbol* $\perp$.

A ZIPE (or NIPE) scheme should have the following correctness property: for all $(\mathsf{pk}, \mathsf{sk}) \xleftarrow{\mathsf{R}} \mathsf{Setup}(1^\lambda)$, *all vectors* $\vec{v}$, *all decryption keys* $\mathsf{sk}_{\vec{v}} \xleftarrow{\mathsf{R}} \mathsf{KeyGen}(\mathsf{pk}, \mathsf{sk}, \vec{v})$, *all messages* m, *all vectors* $\vec{x}$, *all ciphertexts* $\mathsf{ct}_{\vec{x}} \xleftarrow{\mathsf{R}} \mathsf{Enc}(\mathsf{pk}, m, \vec{x})$, *it holds that* $m = \mathsf{Dec}(\mathsf{pk}, \mathsf{sk}_{\vec{v}}, \mathsf{ct}_{\vec{x}})$ *with overwhelming probability, if* $R(\vec{x}, \vec{v}) = 1$.

Definition 4. *The model for proving the adaptively payload-hiding security of ZIPE (or NIPE) under chosen plaintext attacks is given hereafter.*

Setup. *The challenger runs the setup algorithm,* $(\mathsf{pk}, \mathsf{sk}) \xleftarrow{\mathsf{R}} \mathsf{Setup}(1^\lambda)$, *and gives public parameters* pk *to the adversary.*

Phase 1. *The adversary is allowed to adaptively issue a polynomial number of queries,* $\vec{v}$, *to the challenger or oracle* $\mathsf{KeyGen}(\mathsf{pk}, \mathsf{sk}, \cdot)$ *for private keys,* $\mathsf{sk}_{\vec{v}}$, *associated with* $\vec{v}$.

Challenge. *The adversary submits two messages,* $m^{(0)}$ *and* $m^{(1)}$, *and a vector,* $\vec{x}$, *provided that no* $\vec{v}$ *queried to the challenger in Phase 1 satisfies* $R(\vec{x}, \vec{v}) = 1$. *The challenger flips a coin* $b \xleftarrow{\mathsf{U}} \{0, 1\}$, *and computes* $\mathsf{ct}_{\vec{x}}^{(b)} \xleftarrow{\mathsf{R}} \mathsf{Enc}(\mathsf{pk}, m^{(b)}, \vec{x})$. *It gives* $\mathsf{ct}_{\vec{x}}^{(b)}$ *to the adversary.*

Phase 2. *The adversary is allowed to adaptively issue a polynomial number of queries,* $\vec{v}$, *to the challenger or oracle* $\mathsf{KeyGen}(\mathsf{pk}, \mathsf{sk}, \cdot)$ *for private keys,* $\mathsf{sk}_{\vec{v}}$, *associated with* $\vec{v}$, *provided that* $R(\vec{x}, \vec{v}) \neq 1$.

Guess. *The adversary outputs a guess* b' *of* b.

The advantage of adversary $\mathcal{A}$ *in the above game,* $\mathsf{Adv}_{\mathcal{A}}^{\mathsf{ZIPE,PH}}(\lambda)$ *(or* $\mathsf{Adv}_{\mathcal{A}}^{\mathsf{NIPE,PH}}(\lambda)$*), is defined by* $\Pr[b' = b] - 1/2$ *for any security parameter* λ. *A ZIPE (or NIPE) scheme is adaptively payload-hiding secure if all polynomial time adversaries have at most a negligible advantage in the game.*

4 Decisional Linear (DLIN) Assumption

Definition 5. *The DLIN problem is to guess* $\beta \in \{0, 1\}$, *given* $(\mathsf{param}_{\mathbb{G}}, G, \xi G, \kappa G, \delta\xi G, \sigma\kappa G, Y_\beta) \xleftarrow{\mathsf{R}} \mathcal{G}_\beta^{\mathsf{DLIN}}(1^\lambda)$, *where* $\mathcal{G}_\beta^{\mathsf{DLIN}}(1^\lambda)$: $\mathsf{param}_{\mathbb{G}} := (q, \mathbb{G}, \mathbb{G}_T, G, e) \xleftarrow{\mathsf{R}} \mathcal{G}_{\mathsf{bpg}}(1^\lambda)$, $\kappa, \delta, \xi, \sigma \xleftarrow{\mathsf{U}} \mathbb{F}_q$, $Y_0 := (\delta + \sigma)G$, $Y_1 \xleftarrow{\mathsf{U}} \mathbb{G}$, return $(\mathsf{param}_{\mathbb{G}}, G, \xi G, \kappa G, \delta\xi G, \sigma\kappa G, Y_\beta)$, *for* $\beta \xleftarrow{\mathsf{U}} \{0, 1\}$. *For a probabilistic machine* $\mathcal{E}$, *we define the advantage of* $\mathcal{E}$ *for the DLIN problem as:* $\mathsf{Adv}_{\mathcal{E}}^{\mathsf{DLIN}}(\lambda) := \left| \Pr\left[\mathcal{E}(1^\lambda, \varrho) \to 1 \middle| \varrho \xleftarrow{\mathsf{R}} \mathcal{G}_0^{\mathsf{DLIN}}(1^\lambda)\right] - \Pr\left[\mathcal{E}(1^\lambda, \varrho) \to 1 \middle| \varrho \xleftarrow{\mathsf{R}} \mathcal{G}_1^{\mathsf{DLIN}}(1^\lambda)\right] \right|$. *The DLIN assumption is: For any probabilistic polynomial-time adversary* $\mathcal{E}$, *the advantage* $\mathsf{Adv}_{\mathcal{E}}^{\mathsf{DLIN}}(\lambda)$ *is negligible in* λ.

5 Proposed NIPE Scheme with Constant-Size Ciphertexts

5.1 Key Ideas in Constructing the Proposed NIPE Scheme

In this section, we will explain key ideas of constructing and proving the security of the proposed NIPE scheme.

First, we will show how short ciphertexts and efficient decryption can be achieved in our scheme. Here, we will use a simplified (or toy) version of the proposed NIPE scheme, for which the security is no more ensured in the standard model under the DLIN assumption.

A ciphertext in the simplified NIPE scheme consists of two vector elements, $(\boldsymbol{c}_0, \boldsymbol{c}_1) \in \mathbb{G}^5 \times \mathbb{G}^n$, and $c_3 \in \mathbb{G}_T$. A secret-key consists of two vector elements, $(\boldsymbol{k}_0^*, \boldsymbol{k}_1^*) \in \mathbb{G}^5 \times \mathbb{G}^n$. Therefore, to achieve constant-size ciphertexts, we have to compress $\boldsymbol{c}_1 \in \mathbb{G}^n$ to a constant size in n. We now employ a special form of basis generation matrix, $X := \begin{pmatrix} \mu & & \mu'_1 \\ & \ddots & \vdots \\ & & \mu \ \mu'_{n-1} \\ & & \mu'_n \end{pmatrix} \in \mathcal{H}(n, \mathbb{F}_q)$ of Eq. (1) in Section 5.2, where $\mu, \mu'_1, \ldots, \mu'_n \xleftarrow{\mathsf{U}} \mathbb{F}_q$ and a blank in the matrix denotes $0 \in \mathbb{F}_q$. The system parameter or DPVS public basis is $\mathbb{B} := \begin{pmatrix} \boldsymbol{b}_1 \\ \vdots \\ \boldsymbol{b}_n \end{pmatrix} := \begin{pmatrix} \mu G & & \mu'_1 G \\ & \ddots & \vdots \\ & & \mu G \ \mu'_{n-1} G \\ & & \mu'_n G \end{pmatrix}$. Let a ciphertext associated with $\vec{x} := (x_1, \ldots, x_n)$ be $\boldsymbol{c}_1 := (\omega \vec{x})_{\mathbb{B}} = \omega(x_1 \boldsymbol{b}_1 + \cdots + x_n \boldsymbol{b}_n) = (x_1 \omega \mu G, \ldots, x_{n-1} \omega \mu G,\ \omega(\sum_{i=1}^n x_i \mu'_i) G)$, where $\omega \xleftarrow{\mathsf{U}} \mathbb{F}_q$. Then, $\boldsymbol{c}_1$ can be compressed to only *two* group elements $(C_1 := \omega \mu G,\ C_2 := \omega(\sum_{i=1}^n x_i \mu'_i) G)$ as well as $\vec{x}$, since $\boldsymbol{c}_1$ can be obtained by $(x_1 C_1, \ldots,\ x_{n-1} C_1, C_2)$ (note that $x_i C_1 = x_i \omega \mu G$ for $i = 1, \ldots, n-1$). That is, a ciphertext (excluding $\vec{x}$) can be just two group elements, or the size is constant in n.

Let $\mathbb{B}^* := (\boldsymbol{b}_i^*)$ be the dual orthonormal basis of $\mathbb{B} := (\boldsymbol{b}_i)$, and $\mathbb{B}^*$ be the master secret key in the simplified NIPE scheme. We specify $(\boldsymbol{c}_0, \boldsymbol{k}_0^*, c_3)$ such that $e(\boldsymbol{c}_0, \boldsymbol{k}_0^*) = g_T^{\zeta} \cdot g_T^{\omega\delta}$ and $c_3 := g_T^{\zeta} m \in \mathbb{G}_T$. We also set a secret-key for $\vec{v}$ as $\boldsymbol{k}_1^* := (\delta \vec{v})_{\mathbb{B}^*} = \delta(v_1 \boldsymbol{b}_1^* + \cdots + v_n \boldsymbol{b}_n^*)$. From the dual orthonormality of $\mathbb{B}$ and $\mathbb{B}^*$, it then holds that $e(\boldsymbol{c}_1, \boldsymbol{k}_1^*) = g_T^{\omega\delta(\vec{x}\cdot\vec{v})}$. Hence, a decryptor can compute $g_T^{\omega\delta}$ if and only if $\vec{x} \cdot \vec{v} \neq 0$, i.e., can obtain plaintext m by $c_3 \cdot e(\boldsymbol{c}_0, \boldsymbol{k}_0^*)^{-1} \cdot e(\boldsymbol{c}_1, \boldsymbol{k}_1^*)^{(\vec{x}\cdot\vec{v})^{-1}}$. Since $\boldsymbol{c}_1$ is expressed as $(x_1 C_1, \ldots, x_{n-1} C_1, C_2) \in \mathbb{G}^n$ and $\boldsymbol{k}_1^*$ is parsed as a n-tuple $(K_1, \ldots, K_n) \in \mathbb{G}^n$, the value of $e(\boldsymbol{c}_1, \boldsymbol{k}_1^*)$ is $\prod_{i=1}^{n-1} e(x_i C_1, K_i) \cdot e(C_2, K_n) = \prod_{i=1}^{n-1} e(C_1, x_i K_i) \cdot e(C_2, K_n) = e(C_1, \sum_{i=1}^{n-1} x_i K_i) \cdot e(C_2, K_n)$. That is, $n-1$ scalar multiplications in $\mathbb{G}$ and *two* pairing operations are enough for computing $e(\boldsymbol{c}_1, \boldsymbol{k}_1^*)$. Therefore, only a small (constant) number of pairing operations are required for decryption.

We then explain how our *full* NIPE scheme is constructed on the above-mentioned simplified NIPE scheme. The target of designing the full NIPE scheme is to achieve the adaptive security under the DLIN assumption. Here, we adopt a strategy similar to that of [17], in which the dual system encryption methodology is employed in a modular or hierarchical manner. That is, two top level assumptions, the security of Problems 1 and 2, are directly used in the dual system encryption methodology and these assumptions are reduced to a primitive assumption, the DLIN assumption.

To meet the requirements for applying to the dual system encryption methodology and reducing to the DLIN assumption, the underlying vector space as well as the basis generator matrix X is four times greater than that of the above-mentioned simplified scheme. For example, $\boldsymbol{k}_1^* := (\ \delta\vec{v},\ 0^n,\ \varphi_1\vec{v},\ 0^n\ )_{\mathbb{B}^*}$, $\boldsymbol{c}_1 = (\omega\vec{x}, 0^n, 0^n, \eta_1\vec{x})_{\mathbb{B}}$, and $X := \begin{pmatrix} X_{1,1} & \cdots & X_{1,4} \\ \vdots & & \vdots \\ X_{4,1} & \cdots & X_{4,4} \end{pmatrix} \in \mathcal{L}(4n, \mathbb{F}_q)$ of Eq. (2) in Section 5.2, where each $X_{i,j}$ is of the form of $X \in \mathcal{H}(n, \mathbb{F}_q)$ in the simplified scheme. The vector space consists of four orthogonal subspaces, i.e., real encoding part, hidden part, secret-key randomness part, and ciphertext randomness part. The simplified NIPE scheme corresponds to the first real encoding part.

A key fact in the security reduction is that $\mathcal{L}(4n, \mathbb{F}_q)$ is a *subgroup* in $GL(4n, \mathbb{F}_q)$, which enables a *random-self-reducibility* argument for reducing the DLIN problem to Problems 1 and 2 in this paper. The property that $\mathcal{H}(n, \mathbb{F}_q) \cap GL(n, \mathbb{F}_q)$ is a *subgroup* in $GL(n, \mathbb{F}_q)$ is also crucial for a special form of pairwise independence lemma in this paper (Lemma 3), where $\mathcal{H}(n, \mathbb{F}_q)$ is specified in $\mathcal{L}(4n, \mathbb{F}_q)$ or X. Our Problem 2, which is based on this lemma, employs special form matrices $U \xleftarrow{\mathsf{U}} \mathcal{H}(n, \mathbb{F}_q) \cap GL(n, \mathbb{F}_q)$ and $Z := (U^{-1})^{\mathrm{T}}$. Informally, our pairwise independence lemma implies that, for all $(\vec{x}, \vec{v})$, a pair, $(\vec{x}U, \vec{v}Z)$, are uniformly distributed over $(\mathsf{span}\langle\vec{x}, \vec{e}_n\rangle \setminus \mathsf{span}\langle\vec{e}_n\rangle) \times (\mathbb{F}_q^n \setminus \mathsf{span}\langle\vec{e}_n\rangle^{\perp})$ with preserving the inner-product value, $\vec{x} \cdot \vec{v}$, i.e., $(\vec{x}U, \vec{v}Z)$ reveal no information but $\vec{x}$ and $\vec{x} \cdot \vec{v}$.

A difference of matrix X with the ZIPE scheme will be noted in Remark 9.

5.2 Dual Orthonormal Basis Generator

Let $N := 4n$,

$$\mathcal{H}(n, \mathbb{F}_q) := \left\{ \begin{pmatrix} u & & & u_1' \\ & \ddots & & \vdots \\ & & u & u_{n-1}' \\ & & & u_n' \end{pmatrix} \middle| \begin{array}{l} u, u_l' \in \mathbb{F}_q \text{ for } l = 1, \ldots, n, \\ \text{a blank element in the matrix} \\ \text{denotes } 0 \in \mathbb{F}_q \end{array} \right\}, \tag{1}$$

$$\mathcal{L}(N, \mathbb{F}_q) := \left\{ X := \begin{pmatrix} X_{1,1} & \cdots & X_{1,4} \\ \vdots & & \vdots \\ X_{4,1} & \cdots & X_{4,4} \end{pmatrix} \middle| X_{i,j} := \begin{pmatrix} \mu_{i,j} & & & \mu_{i,j,1}' \\ & \ddots & & \vdots \\ & & \mu_{i,j} & \mu_{i,j,n-1}' \\ & & & \mu_{i,j,n}' \end{pmatrix} \begin{array}{c} \in \mathcal{H}(n, \mathbb{F}_q) \\ \text{for } i, j = \\ 1, \ldots, 4 \end{array} \right\} \bigcap GL(N, \mathbb{F}_q). \tag{2}$$

We note that $\mathcal{H}(n, \mathbb{F}_q) \cap GL(n, \mathbb{F}_q)$ is a subgroup of $GL(n, \mathbb{F}_q)$ and $\mathcal{L}(N, \mathbb{F}_q)$ is a subgroup of $GL(N, \mathbb{F}_q)$. We describe random dual orthonormal basis generator $\mathcal{G}_{\mathsf{ob}}^{(1)}$ below, which is used as a subroutine in the proposed NIPE scheme.

$$\begin{aligned}
&\mathcal{G}_{\mathsf{ob}}^{(1)}(1^\lambda,\ n): \ \ \mathsf{param}_{\mathbb{G}} := (q, \mathbb{G}, \mathbb{G}_T, G, e) \xleftarrow{\mathsf{R}} \mathcal{G}_{\mathsf{bpg}}(1^\lambda), \quad N_0 := 5,\ N_1 := 4n, \\
&\quad \mathsf{param}_{\mathbb{V}_t} := (q, \mathbb{V}_t, \mathbb{G}_T, \mathbb{A}_t, e) := \mathcal{G}_{\mathsf{dpvs}}(1^\lambda, N_t, \mathsf{param}_{\mathbb{G}}) \ \text{ for } t = 0, 1, \\
&\quad \psi \xleftarrow{\mathsf{U}} \mathbb{F}_q^\times,\ g_T := e(G, G)^\psi,\ \mathsf{param}_n := (\{\mathsf{param}_{\mathbb{V}_t}\}_{t=0,1},\ g_T), \\
&\quad X_0 := (\chi_{0,i,j})_{i,j=1,\ldots,5} \xleftarrow{\mathsf{U}} GL(N_0, \mathbb{F}_q),\ X_1 \xleftarrow{\mathsf{U}} \mathcal{L}(N_1, \mathbb{F}_q), \text{ hereafter}, \\
&\quad \{\mu_{i,j}, \mu'_{i,j,l}\}_{i,j=1,\ldots 4; l=1,\ldots,n} \text{ denotes non-zero entries of } X_1 \text{ as in Eq. (2)}, \\
&\quad \boldsymbol{b}_{0,i} := (\chi_{0,i,1}, .., \chi_{0,i,5})_{\mathbb{A}} = \textstyle\sum_{j=1}^{5} \chi_{0,i,j} \boldsymbol{a}_j \text{ for } i = 1, .., 5,\ \mathbb{B}_0 := (\boldsymbol{b}_{0,1}, .., \boldsymbol{b}_{0,5}), \\
&\quad B_{i,j} := \mu_{i,j} G,\ B'_{i,j,l} := \mu'_{i,j,l} G \ \text{ for } i, j = 1, \ldots, 4; l = 1, \ldots, n, \\
&\quad \text{for } t = 0, 1,\ (\vartheta_{t,i,j})_{i,j=1,\ldots,N_t} := \psi \cdot (X_t^{\mathrm{T}})^{-1}, \\
&\quad \boldsymbol{b}^*_{t,i} := (\vartheta_{t,i,1}, .., \vartheta_{t,i,N_t})_{\mathbb{A}} = \textstyle\sum_{j=1}^{N_t} \vartheta_{t,i,j} \boldsymbol{a}_j \text{ for } i = 1, .., N_t,\ \mathbb{B}^*_t := (\boldsymbol{b}^*_{t,1}, .., \boldsymbol{b}^*_{t,N_t}), \\
&\quad \text{return } \ (\mathsf{param}_n, \mathbb{B}_0, \mathbb{B}^*_0, \{B_{i,j}, B'_{i,j,l}\}_{i,j=1,\ldots,4; l=1,\ldots,n}, \mathbb{B}^*_1).
\end{aligned}$$

Remark 1 . Let

$$\left.\begin{aligned}
&\begin{pmatrix} \boldsymbol{b}_{1,(i-1)n+1} \\ \vdots \\ \boldsymbol{b}_{1,in} \end{pmatrix} := \begin{pmatrix} B_{i,1} & & & B'_{i,1,1} & & B_{i,4} & & & B'_{i,4,1} \\ & \ddots & & \vdots & \cdots & & \ddots & & \vdots \\ & & B_{i,1} & B'_{i,1,n-1} & & & & B_{i,4} & B'_{i,4,n-1} \\ & & & B'_{i,1,n} & & & & & B'_{i,4,n} \end{pmatrix} \\
&\qquad \text{for } i = 1, \ldots, 4, \\
&\mathbb{B}_1 := (\boldsymbol{b}_{1,1}, \ldots, \boldsymbol{b}_{1,4n}),
\end{aligned}\right\} \quad (3)$$

where a blank element in the matrix denotes $0 \in \mathbb{G}$. $\mathbb{B}_1$ is the dual orthonormal basis of $\mathbb{B}^*_1$, i.e., $e(\boldsymbol{b}_{1,i}, \boldsymbol{b}^*_{1,i}) = g_T$ and $e(\boldsymbol{b}_{1,i}, \boldsymbol{b}^*_{1,j}) = 1$ for $1 \le i \ne j \le 4n$.

5.3 Construction

In the description of the scheme, we assume that input vector, $\vec{x} := (x_1, \ldots, x_n)$, has an index l $(1 \le l \le n-1)$ with $x_l \ne 0$, and that input vector, $\vec{v} := (v_1, \ldots, v_n)$, satisfies $v_n \ne 0$.

$$\begin{aligned}
&\mathsf{Setup}(1^\lambda,\ n): \ (\mathsf{param}_n, \mathbb{B}_0, \mathbb{B}^*_0, \{B_{i,j}, B'_{i,j,l}\}_{i,j=1,\ldots,4; l=1,\ldots,n}, \mathbb{B}^*_1) \xleftarrow{\mathsf{R}} \mathcal{G}_{\mathsf{ob}}^{(1)}(1^\lambda,\ n), \\
&\widehat{\mathbb{B}}_0 := (\boldsymbol{b}_{0,1}, \boldsymbol{b}_{0,3}, \boldsymbol{b}_{0,5}), \widehat{\mathbb{B}}^*_0 := (\boldsymbol{b}^*_{0,1}, \boldsymbol{b}^*_{0,3}, \boldsymbol{b}^*_{0,4}), \widehat{\mathbb{B}}^*_1 := (\boldsymbol{b}^*_{1,1}, .., \boldsymbol{b}^*_{1,n}, \boldsymbol{b}^*_{1,2n+1}, .., \boldsymbol{b}^*_{1,3n}), \\
&\text{return } \ \mathsf{pk} := (1^\lambda, \mathsf{param}_n, \widehat{\mathbb{B}}_0, \{B_{i,j}, B'_{i,j,l}\}_{i=1,4; j=1,\ldots,4; l=1,\ldots,n}),\ \mathsf{sk} := \{\widehat{\mathbb{B}}^*_t\}_{t=0,1}.
\end{aligned}$$

$\mathsf{KeyGen}(\mathsf{pk},\ \mathsf{sk},\ \vec{v}):\quad \delta, \varphi_0, \varphi_1 \xleftarrow{\mathsf{U}} \mathbb{F}_q,\quad \boldsymbol{k}_0^* := (\delta,\ 0,\ 1,\ \varphi_0,\ 0)_{\mathbb{B}_0^*},$

$\boldsymbol{k}_1^* := (\ \overbrace{\delta\vec{v}}^{n},\ \overbrace{0^n}^{n},\ \overbrace{\varphi_1\vec{v}}^{n},\ \overbrace{0^n}^{n}\ )_{\mathbb{B}_1^*},\quad \text{return}\ \ \mathsf{sk}_{\vec{v}} := (\vec{v}, \boldsymbol{k}_0^*, \boldsymbol{k}_1^*).$

$\mathsf{Enc}(\mathsf{pk},\ m,\ \vec{x}):\quad \omega, \eta_0, \eta_1, \zeta \xleftarrow{\mathsf{U}} \mathbb{F}_q,\quad \boldsymbol{c}_0 := (-\omega,\ 0,\ \zeta,\ 0,\ \eta_0)_{\mathbb{B}_0},\quad c_3 := g_T^{\zeta} m,$

$C_{1,j} := \omega B_{1,j} + \eta_1 B_{4,j},\quad C_{2,j} := \sum_{l=1}^{n} x_l(\omega B'_{1,j,l} + \eta_1 B'_{4,j,l})\ \text{ for } j = 1, \ldots, 4,$

return $\mathsf{ct}_{\vec{x}} := (\vec{x}, \boldsymbol{c}_0, \{C_{1,j}, C_{2,j}\}_{j=1,\ldots,4}, c_3)$.

$\mathsf{Dec}(\mathsf{pk},\ \mathsf{sk}_{\vec{v}} := (\vec{v}, \boldsymbol{k}_0^*, \boldsymbol{k}_1^*),\ \mathsf{ct}_{\vec{x}} := (\vec{x}, \boldsymbol{c}_0, \{C_{1,j}, C_{2,j}\}_{j=1,\ldots,4}, c_3)):$

Parse $\boldsymbol{k}_1^*$ as a $4n$-tuple $(K_1^*, \ldots, K_{4n}^*) \in \mathbb{G}^{4n}$,

$D_j^* := \sum_{l=1}^{n-1}((\vec{x}\cdot\vec{v})^{-1}x_l)\, K^*_{(j-1)n+l}\ \text{ for } j = 1, .., 4,$

$F := e(\boldsymbol{c}_0, \boldsymbol{k}_0^*) \cdot \prod_{j=1}^{4}\left(e(C_{1,j}, D_j^*) \cdot e(C_{2,j}, K_{jn}^*)\right),\quad \text{return}\ \ m' := c_3/F.$

Remark 2 . A part of output of $\mathsf{Setup}(1^\lambda, n)$, $\{B_{i,j}, B'_{i,j,l}\}_{i=1,4; j=1,\ldots,4; l=1,\ldots,n}$, can be identified with $\widehat{\mathbb{B}}_1 := (\boldsymbol{b}_{1,1}, \ldots, \boldsymbol{b}_{1,n}, \boldsymbol{b}_{1,3n+1}, .., \boldsymbol{b}_{1,4n})$ through the form of Eq. (3), while $\mathbb{B}_1 := (\boldsymbol{b}_{1,1}, \ldots, \boldsymbol{b}_{1,4n})$ is identified with $\{B_{i,j}, B'_{i,j,l}\}_{i,j=1,..,4; l=1,..,n}$ by Eq. (3). Decryption Dec can be alternatively described as:

$\mathsf{Dec}'(\mathsf{pk},\ \mathsf{sk}_{\vec{v}} := (\vec{v}, \boldsymbol{k}_0^*, \boldsymbol{k}_1^*),\ \mathsf{ct}_{\vec{x}} := (\vec{x}, \boldsymbol{c}_0, \{C_{1,j}, C_{2,j}\}_{j=1,\ldots,4}, c_3)):$

$\boldsymbol{c}_1 := (\ \overbrace{x_1 C_{1,1}, .., x_{n-1} C_{1,1}, C_{2,1}}^{n},\ \ \ldots,\ \ \overbrace{x_1 C_{1,4}, .., x_{n-1} C_{1,4}, C_{2,4}}^{n}\ ),$

that is, $\boldsymbol{c}_1 = (\ \overbrace{\omega\vec{x},}^{n}\ \overbrace{0^n,}^{n}\ \overbrace{0^n,}^{n}\ \overbrace{\eta_1\vec{x}}^{n}\ )_{\mathbb{B}_1},\ F := e(\boldsymbol{c}_0, \boldsymbol{k}_0^*) \cdot e(\boldsymbol{c}_1, (\vec{x}\cdot\vec{v})^{-1}\boldsymbol{k}_1^*)$,
return $m' := c_3/F$.

[Correctness]. Using the alternate decryption Dec', $F = e(\boldsymbol{c}_0, \boldsymbol{k}_0^*) \cdot e(\boldsymbol{c}_1, (\vec{x}\cdot\vec{v})^{-1}\boldsymbol{k}_1^*) = g_T^{-\omega\delta+\zeta} g_T^{\omega\delta(\vec{x}\cdot\vec{v})/(\vec{x}\cdot\vec{v})} = g_T^{\zeta}$ if $\vec{x}\cdot\vec{v} \neq 0$.

5.4 Security

The proofs of Lemmas 1–9 are given in the full version of this paper.

Theorem 1. *The proposed NIPE scheme is adaptively payload-hiding against chosen plaintext attacks under the DLIN assumption.*

For any adversary $\mathcal{A}$, there exist probabilistic machines $\mathcal{E}_1, \mathcal{E}_{2\text{-}1}$ and $\mathcal{E}_{2\text{-}2}$ whose running times are essentially the same as that of $\mathcal{A}$, such that for any security parameter λ, $\mathsf{Adv}_{\mathcal{A}}^{\mathsf{NIPE,PH}}(\lambda) \leq \mathsf{Adv}_{\mathcal{E}_1}^{\mathsf{DLIN}}(\lambda) + \sum_{h=1}^{\nu}\left(\mathsf{Adv}_{\mathcal{E}_{2\text{-}h\text{-}1}}^{\mathsf{DLIN}}(\lambda) + \mathsf{Adv}_{\mathcal{E}_{2\text{-}h\text{-}2}}^{\mathsf{DLIN}}(\lambda)\right)$ $+\epsilon$, where $\mathcal{E}_{2\text{-}h\text{-}1}(\cdot) := \mathcal{E}_{2\text{-}1}(h, \cdot)$, $\mathcal{E}_{2\text{-}h\text{-}2}(\cdot) := \mathcal{E}_{2\text{-}2}(h, \cdot)$, ν is the maximum number of $\mathcal{A}$'s key queries and $\epsilon := (11\nu + 6)/q$.

Lemmas for the Proof of Theorem 1. We will show Lemmas 1–3 for the proof of Theorem 1.

Definition 6 (Problem 1). *Problem 1 is to guess* β, *given* $(\mathsf{param}_n, \mathbb{B}_0, \widehat{\mathbb{B}}_0^*, \boldsymbol{e}_{\beta,0}, \{B_{i,j}, B'_{i,j,l}\}_{i,j=1,..,4;l=1,..,n}, \widehat{\mathbb{B}}_1^*, \{E_{\beta,j}, E'_{\beta,j,l}\}_{j=1,..,4;l=1,..,n}) \xleftarrow{\mathsf{R}} \mathcal{G}_\beta^{\mathsf{P1}}(1^\lambda, n)$, *where*

$$\begin{aligned}
\mathcal{G}_\beta^{\mathsf{P1}}(1^\lambda, n): \quad & (\mathsf{param}_n, \mathbb{B}_0, \mathbb{B}_0^*, \{B_{i,j}, B'_{i,j,l}\}_{i,j=1,\ldots,4;l=1,\ldots,n}, \widehat{\mathbb{B}}_1^*) \xleftarrow{\mathsf{R}} \mathcal{G}_{\mathsf{ob}}^{(1)}(1^\lambda, n),\\
& \widehat{\mathbb{B}}_0^* := (\boldsymbol{b}_{0,1}^*, \boldsymbol{b}_{0,3}^*, .., \boldsymbol{b}_{0,5}^*),\ \widehat{\mathbb{B}}_1^* := (\boldsymbol{b}_{1,1}^*, \ldots, \boldsymbol{b}_{1,n}^*, \boldsymbol{b}_{t,2n+1}^*, \ldots, \boldsymbol{b}_{t,4n}^*),\\
& \omega, \tau, \eta_0, \eta_1 \xleftarrow{\mathsf{U}} \mathbb{F}_q,\ U \xleftarrow{\mathsf{U}} \mathcal{H}(n, \mathbb{F}_q) \cap GL(n, \mathbb{F}_q),\ \text{hereafter},\ u, u'_n \in \mathbb{F}_q^\times,\\
& \quad u'_1, \ldots, u'_{n-1} \in \mathbb{F}_q \text{ denote non-zero entries of } U, \text{ as in Eq. } (1),\\
& \boldsymbol{e}_{0,0} := (\omega, 0, 0, 0, \eta_0)_{\mathbb{B}_0}, \quad \boldsymbol{e}_{1,0} := (\omega, \tau, 0, 0, \eta_0)_{\mathbb{B}_0},\\
& \text{for } j = 1, \ldots, 4;\\
& \quad E_{0,j} := \omega B_{1,j} + \eta_1 B_{4,j},\ E'_{0,j,l} := \omega B'_{1,j,l} + \eta_1 B'_{4,j,l} \text{ for } l = 1, \ldots, n,\\
& \quad E_{1,j} := \omega B_{1,j} + \tau u B_{2,j} + \eta_1 B_{4,j},\\
& \quad E'_{1,j,l} := \omega B'_{1,j,l} + \tau u B'_{2,j,l} + \tau u'_l B'_{2,j,n} + \eta_1 B'_{4,j,l}\\
& \qquad \text{for } l = 1, \ldots, n-1, \text{ and } E'_{1,j,n} := \omega B'_{1,j,n} + \tau u'_n B'_{2,j,n} + \eta_1 B'_{4,j,n},\\
& \text{return } (\mathsf{param}_n, \mathbb{B}_0, \widehat{\mathbb{B}}_0^*, \boldsymbol{e}_{\beta,0}, \{B_{i,j}, B'_{i,j,l}\}_{i,j=1,\ldots,4;l=1,\ldots,n}, \widehat{\mathbb{B}}_1^*,\\
& \qquad \{E_{\beta,j}, E'_{\beta,j,l}\}_{j=1,\ldots,4;l=1,\ldots,n}),
\end{aligned}$$

for $\beta \xleftarrow{\mathsf{U}} \{0,1\}$. *For a probabilistic machine* $\mathcal{B}$, *we define the advantage of* $\mathcal{B}$ *as the quantity* $\mathsf{Adv}_{\mathcal{B}}^{\mathsf{P1}}(\lambda) := \left|\Pr\left[\mathcal{B}(1^\lambda, \varrho) \to 1 \middle| \varrho \xleftarrow{\mathsf{R}} \mathcal{G}_0^{\mathsf{P1}}(1^\lambda, n)\right] - \Pr\left[\mathcal{B}(1^\lambda, \varrho) \to 1 \middle| \varrho \xleftarrow{\mathsf{R}} \mathcal{G}_1^{\mathsf{P1}}(1^\lambda, n)\right]\right|$.

Remark 3 . A part of output of $\mathcal{G}_\beta^{\mathsf{P1}}(1^\lambda, n)$, $\{B_{i,j}, B'_{i,j,l}\}_{i,j=1,\ldots,4;l=1,\ldots,n}$, is identified with $\mathbb{B}_1 := (\boldsymbol{b}_{1,1}, \ldots, \boldsymbol{b}_{1,4n})$ (Eq. (3)). If we make $\boldsymbol{e}_{\beta,1,l} \in \mathbb{V}_1$ for $\beta = 0,1; l = 1, \ldots, n$ as:

$$\begin{aligned}
\boldsymbol{e}_{\beta,1,l} &:= (\overbrace{0^{l-1}, E_{\beta,1}, 0^{n-l-1}, E'_{\beta,1,l}}^{n}, \ldots, \overbrace{0^{l-1}, E_{\beta,4}, 0^{n-l-1}, E'_{\beta,4,l}}^{n}) \quad \text{for } l = 1, \ldots, n-1,\\
\boldsymbol{e}_{\beta,1,n} &:= (0^{n-1}, E'_{\beta,1,n}, \ldots, 0^{n-1}, E'_{\beta,4,n}),
\end{aligned}$$

they are expressed over $\mathbb{B}_1$ as:

$$\begin{aligned}
\boldsymbol{e}_{0,1,l} &:= (\overbrace{\omega\vec{e}_l}^{n}, \overbrace{0^n}^{n}, \overbrace{0^n}^{n}, \overbrace{\eta_1\vec{e}_l}^{n})_{\mathbb{B}_1} \quad \text{for } l = 1, \ldots, n,\\
\boldsymbol{e}_{1,1,l} &:= (\omega\vec{e}_l, \tau\vec{e}_l U, 0^n, \eta_1\vec{e}_l)_{\mathbb{B}_1} \quad \text{for } l = 1, \ldots, n.
\end{aligned}$$

Using these vector expressions, the output of $\mathcal{G}_\beta^{\mathsf{P1}}(1^\lambda, n)$ is expressed as $(\mathsf{param}_n, \mathbb{B}_0, \widehat{\mathbb{B}}_0^*, \boldsymbol{e}_{\beta,0}, \mathbb{B}_1, \widehat{\mathbb{B}}_1^*, \{\boldsymbol{e}_{\beta,1,l}\}_{l=1,\ldots,n})$.

Lemma 1. *For any adversary $\mathcal{B}$, there exists a probabilistic machine $\mathcal{E}$, whose running times are essentially the same as that of $\mathcal{B}$, such that for any security parameter λ, $\mathsf{Adv}^{\mathsf{P1}}_{\mathcal{B}}(\lambda) \leq \mathsf{Adv}^{\mathsf{DLIN}}_{\mathcal{E}}(\lambda) + 5/q$.*

Definition 7 (Problem 2). *Problem 2 is to guess β, given $(\mathsf{param}_n, \widehat{\mathbb{B}}_0, \mathbb{B}_0^*, \boldsymbol{h}^*_{\beta,0}, \boldsymbol{e}_0, \{B_{i,j}, B'_{i,j,l}\}_{i=1,3,4;j=1,..,4;l=1,..,n}, \mathbb{B}_1^*, \{\boldsymbol{h}^*_{\beta,1,l}, E_j, E'_{j,l}\}_{j=1,..,4;l=1,..,n}) \xleftarrow{\mathsf{R}} \mathcal{G}^{\mathsf{P2}}_{\beta}(1^\lambda, n)$, where*

$$
\begin{aligned}
&\mathcal{G}^{\mathsf{P2}}_{\beta}(1^\lambda, n): \quad (\mathsf{param}_n, \mathbb{B}_0, \widehat{\mathbb{B}}_0^*, \{B_{i,j}, B'_{i,j,l}\}_{i,j=1,\ldots,4;l=1,\ldots,n}, \widehat{\mathbb{B}}_1^*) \xleftarrow{\mathsf{R}} \mathcal{G}^{(1)}_{\mathsf{ob}}(1^\lambda, n),\\
&\quad \widehat{\mathbb{B}}_0 := (\boldsymbol{b}_{0,1}, \boldsymbol{b}_{0,3}, .., \boldsymbol{b}_{0,5}), \quad \delta, \rho, \varphi_0, \varphi_1, \omega, \tau \xleftarrow{\mathsf{U}} \mathbb{F}_q,\\
&\quad U \xleftarrow{\mathsf{U}} \mathcal{H}(n, \mathbb{F}_q) \cap GL(n, \mathbb{F}_q), \ Z := (U^{-1})^{\mathrm{T}},\\
&\quad \text{hereafter, } u, u'_n \in \mathbb{F}_q^{\times}, u'_1, \ldots, u'_{n-1} \in \mathbb{F}_q \text{ and } z, z'_n \in \mathbb{F}_q^{\times}, z'_1, \ldots, z'_{n-1} \in \mathbb{F}_q\\
&\quad\ \text{denote non-zero entries of } U \text{ and } Z^{\mathrm{T}}, \text{ as in Eq. (1), respectively},\\
&\quad \boldsymbol{h}^*_{0,0} := (\delta, 0, 0, \varphi_0, 0)_{\mathbb{B}_0^*}, \ \boldsymbol{h}^*_{1,0} := (\delta, \rho, 0, \varphi_0, 0)_{\mathbb{B}_0^*}, \ \boldsymbol{e}_0 := (\omega, \tau, 0, 0, 0)_{\mathbb{B}_0},\\
&\quad \vec{e}_l := (0^{l-1}, 1, 0^{n-l}) \in \mathbb{F}_q^n \ \text{ for } l = 1, \ldots, n;\\
&\quad \boldsymbol{h}^*_{0,1,l} := (\overbrace{\delta\vec{e}_l}^{n}, \overbrace{0^n}^{n}, \overbrace{\varphi_1\vec{e}_l}^{n}, \overbrace{0^n}^{n})_{\mathbb{B}_1^*} \ \text{ for } l = 1, \ldots, n,\\
&\quad \boldsymbol{h}^*_{1,1,l} := (\delta\vec{e}_l, \rho\vec{e}_l Z, \varphi_1\vec{e}_l, 0^n)_{\mathbb{B}_1^*} \ \text{ for } l = 1, \ldots, n,\\
&\quad \text{for } j = 1, \ldots, 4; \quad E_j := \omega B_{1,j} + \tau u B_{2,j},\\
&\qquad E'_{j,l} := \omega B'_{1,j,l} + \tau u B'_{2,j,l} + \tau u'_l B'_{2,j,n} \quad \text{for } l = 1, \ldots, n-1,\\
&\qquad E'_{j,n} := \omega B'_{1,j,n} + \tau u'_n B'_{2,j,n},\\
&\quad \text{return } (\mathsf{param}_n, \widehat{\mathbb{B}}_0, \mathbb{B}_0^*, \boldsymbol{h}^*_{\beta,0}, \boldsymbol{e}_0, \{B_{i,j}, B'_{i,j,l}\}_{i=1,3,4;j=1,\ldots,4;l=1,\ldots,n}, \mathbb{B}_1^*,\\
&\qquad\qquad \{\boldsymbol{h}^*_{\beta,1,l}, E_j, E'_{j,l}\}_{j=1,\ldots,4;l=1,\ldots,n}),
\end{aligned}
$$

for $\beta \xleftarrow{\mathsf{U}} \{0,1\}$. For a probabilistic adversary $\mathcal{B}$, the advantage of $\mathcal{B}$ for Problem 2, $\mathsf{Adv}^{\mathsf{P2}}_{\mathcal{B}}(\lambda)$, is similarly defined as in Definition 6.

Remark 4. A part of output of $\mathcal{G}^{\mathsf{P2}}_{\beta}(1^\lambda, n)$, $\{B_{i,j}, B'_{i,j,l}\}_{i=1,3,4;j=1,\ldots,4;l=1,\ldots,n}$, can be identified with $\widehat{\mathbb{B}}_1 := (\boldsymbol{b}_{1,1}, \ldots, \boldsymbol{b}_{1,n}, \boldsymbol{b}_{1,2n+1}, \ldots, \boldsymbol{b}_{1,4n})$ in the form of Eq. (3), while $\mathbb{B}_1 := (\boldsymbol{b}_{1,1}, .., \boldsymbol{b}_{1,4n})$ is identified with $\{B_{i,j}, B'_{i,j,l}\}_{i,j=1,\ldots,4;l=1,\ldots,n}$ by Eq. (3). If we make $\boldsymbol{e}_{1,l} \in \mathbb{V}_1$ for $l = 1, \ldots, n$ as:

$$
\begin{aligned}
\boldsymbol{e}_{1,l} &:= (\overbrace{0^{l-1}, E_1, 0^{n-l-1}, E'_{1,l}}^{n}, \ \ldots, \ \overbrace{0^{l-1}, E_4, 0^{n-l-1}, E'_{4,l}}^{n}) \quad \text{for } l = 1, \ldots, n-1,\\
\boldsymbol{e}_{1,n} &:= (0^{n-1}, E'_{1,n}, \ \ldots, \ 0^{n-1}, E'_{4,n}),
\end{aligned}
$$

they are expressed over $\mathbb{B}_1$ as:

$$\boldsymbol{e}_{1,l} := (\overbrace{\omega\vec{e}_l,}^{n} \quad \overbrace{\tau\vec{e}_l U,}^{n} \quad \overbrace{0^n,}^{n} \quad \overbrace{0^n}^{n} \quad)_{\mathbb{B}_1} \quad \text{for } l = 1,\ldots,n.$$

Using these vector expressions, the output of $\mathcal{G}_\beta^{\mathsf{P2}}(1^\lambda, n)$ is expressed as $(\mathsf{param}_n, \widehat{\mathbb{B}}_0, \mathbb{B}_0^*, \boldsymbol{h}_{\beta,0}^*, \boldsymbol{e}_0, \widehat{\mathbb{B}}_1, \mathbb{B}_1^*, \{\boldsymbol{h}_{\beta,1,l}^*, \boldsymbol{e}_{1,l}\}_{l=1,\ldots,n})$.

Lemma 2. *For any adversary $\mathcal{B}$, there exists a probabilistic machine $\mathcal{E}$, whose running time is essentially the same as that of $\mathcal{B}$, such that for any security parameter λ, $\mathsf{Adv}_{\mathcal{B}}^{\mathsf{P2}}(\lambda) \le \mathsf{Adv}_{\mathcal{E}}^{\mathsf{DLIN}}(\lambda) + 5/q$.*

Lemma 3. *Let $\vec{e}_n := (0,\ldots,0,1) \in \mathbb{F}_q^n$. For all $\vec{x} \in \mathbb{F}_q^n \setminus \mathsf{span}\langle\vec{e}_n\rangle$ and $\pi \in \mathbb{F}_q$, let $W_{\vec{x},\pi} := \{(\vec{r},\vec{w}) \in (\mathsf{span}\langle\vec{x},\vec{e}_n\rangle \setminus \mathsf{span}\langle\vec{e}_n\rangle) \times (\mathbb{F}_q^n \setminus \mathsf{span}\langle\vec{e}_n\rangle^\perp) \mid \vec{r}\cdot\vec{w} = \pi\}$.*

For all $(\vec{x},\vec{v}) \in (\mathbb{F}_q^n \setminus \mathsf{span}\langle\vec{e}_n\rangle) \times (\mathbb{F}_q^n \setminus \mathsf{span}\langle\vec{e}_n\rangle^\perp)$, for all $(\vec{r},\vec{w}) \in W_{\vec{x},(\vec{x}\cdot\vec{v})}$, $\Pr[\,\vec{x}U = \vec{r} \wedge \vec{v}Z = \vec{w}\,] = 1/\sharp W_{\vec{x},(\vec{x}\cdot\vec{v})}$, where $U \xleftarrow{\mathsf{U}} \mathcal{H}(n,\mathbb{F}_q) \cap GL(n,\mathbb{F}_q)$ and $Z := (U^{-1})^{\mathrm{T}}$.

Proof Outline : At the top level of strategy of the security proof, we follow the dual system encryption methodology proposed by Waters [20]. In the methodology, ciphertexts and secret keys have two forms, *normal* and *semi-functional*. In the proof herein, we also introduce other forms of secret keys called *1st-pre-semi-functional* and *2nd-pre-semi-functional*. The real system uses only normal ciphertexts and normal secret keys, and semi-functional ciphertexts and semi-functional/1st-pre-semi-functional/2nd-pre-semi-functional keys are used only in a sequence of security games for the security proof. To prove this theorem, we employ Game 0 (original adaptive-security game) through Game 3. In Game 1, the challenge ciphertext is changed to semi-functional. When at most ν secret key queries are issued by an adversary, there are 3ν game changes from Game 1 (Game 2-0-3), Game 2-1-1, Game 2-1-2, Game 2-1-3 through Game 2-ν-3.

In Game 2-h-1, the first $(h-1)$ keys are semi-functional and the h-th key is *1st-pre-semi-functional*, while the remaining keys are normal, and the challenge ciphertext is semi-functional. In Game 2-h-2, the first $(h-1)$ keys are semi-functional and the h-th key is *2nd-pre-semi-functional*, while the remaining keys are normal, and the challenge ciphertext is semi-functional. In Game 2-h-3, the first h keys are semi-functional (i.e., and the h-th key is *semi-functional*), while the remaining keys are normal, and the challenge ciphertext is semi-functional.

The final game (Game 3) with advantage 0 is conceptually changed from Game 2-ν-3. As usual, we prove that the advantage gaps between neighboring games are negligible.

When at most ν key queries are issued by an adversary, we set a sequence of $\mathsf{sk} := \mathsf{sk}_{\vec{v}}$'s, i.e., $(\mathsf{sk}^{(1)*},\ldots,\mathsf{sk}^{(\nu)*})$, in the order of the adversary's queries. Here we focus on $\vec{\boldsymbol{k}}_{\vec{v}}^{(h)*} := (\boldsymbol{k}_0^{(h)*}, \boldsymbol{k}_1^{(h)*})$, and $\vec{\boldsymbol{c}}_{\vec{x}} := (\boldsymbol{c}_0, \{C_{1,j}, C_{2,j}\}_{j=1,\ldots,4}, c_3)$, and ignore the other part of $\mathsf{sk}_{\vec{v}}$ (resp. $\mathsf{ct}_{\vec{x}}$), i.e., $\vec{v}$ (resp. i.e., $\vec{x}$), and call them secret key and ciphertext, respectively, in this proof outline. In addition, we ignore a negligible factor in the (informal) descriptions of this proof outline. For example, we say "A is bounded by B" when $A \leq B+\epsilon(\lambda)$ where $\epsilon(\lambda)$ is negligible in security parameter λ.

A *normal* secret key, $\vec{\boldsymbol{k}}_{\vec{v}}^{(h)*\mathsf{norm}}$, is the correct form of the secret key of the proposed NIPE scheme, and is expressed by Eq. (4). Similarly, a *normal* ciphertext $\vec{\boldsymbol{c}}_{\vec{x}}^{\,\mathsf{norm}}$, is expressed by Eq. (5). A *1st-pre-semi-functional* secret key, $\vec{\boldsymbol{k}}_{\vec{v}}^{(h)*\ \mathsf{1st\text{-}psemi}}$, is expressed by Eq. (7), a *2nd-pre-semi-functional* secret key, $\vec{\boldsymbol{k}}_{\vec{v}}^{(h)*\ \mathsf{2nd\text{-}psemi}}$, is expressed by Eq. (8), a *semi-functional* secret key, $\vec{\boldsymbol{k}}_{\vec{v}}^{(h)*\ \mathsf{semi}}$, is expressed by Eq. (9), and a *semi-functional* ciphertext, $\vec{\boldsymbol{c}}_{\vec{x}}^{\,\mathsf{semi}}$, is expressed by Eq. (6).

To prove that the advantage gap between Games 0 and 1 is bounded by the advantage of Problem 1 (to guess $\beta \in \{0,1\}$), we construct a simulator of the challenger of Game 0 (or 1) (against an adversary $\mathcal{A}$) by using an instance with $\beta \xleftarrow{\mathsf{U}} \{0,1\}$ of Problem 1. We then show that the distribution of the secret keys and challenge ciphertext replied by the simulator is equivalent to those of Game 0 when $\beta = 0$ and Game 1 when $\beta = 1$. That is, the advantage gap between Games 0 and 1 is bounded by the advantage of Problem 1 (Lemma 4). The advantage of Problem 1 is proven to be bounded by that of the DLIN assumption (Lemma 1). The advantage gap between Games 2-($h-1$)-3 and 2-h-1 is similarly shown to be bounded by the advantage of Problem 2 (i.e., advantage of the DLIN assumption) (Lemmas 5 and 2). The distributions of *1st-pre-semi-functional* secret key $\vec{\boldsymbol{k}}_{\vec{v}}^{(h)*\ \mathsf{1st\text{-}psemi}}$ (Eq. (7)) and *2nd-pre-semi-functional* secret key $\vec{\boldsymbol{k}}_{\vec{v}}^{(h)*\ \mathsf{2nd\text{-}psemi}}$ (Eq. (8)) are distinguishable by the simulator or challenger, but the joint distributions of $(\vec{\boldsymbol{k}}_{\vec{v}}^{(h)*\ \mathsf{1st\text{-}psemi}}, \vec{\boldsymbol{c}}_{\vec{x}}^{\,\mathsf{semi}})$ and $(\vec{\boldsymbol{k}}_{\vec{v}}^{(h)*\ \mathsf{2nd\text{-}psemi}}, \vec{\boldsymbol{c}}_{\vec{x}}^{\,\mathsf{semi}})$ along with the other keys are (information theoretically) equivalent for the adversary's view, when $\vec{x} \cdot \vec{v} = 0$, i.e., $R^{\mathsf{NIPE}}(\vec{x}, \vec{v}) \neq 1$. Therefore, as shown in Lemma 6, the advantages of Games 2-h-1 and 2-h-2 are equivalent. The advantage gap between Games 2-h-2 and 2-h-3 is similarly shown to be bounded by the advantage of Problem 2 (i.e., advantage of the DLIN assumption) (Lemmas 7 and 2). Finally we show that Game 2-ν-3 can be conceptually changed to Game 3 (Lemma 8) by using the fact that basis vectors $\boldsymbol{b}_{0,2}$ and $\boldsymbol{b}_{0,3}^{*}$ are unknown to the adversary.

Proof of Theorem 1. To prove Theorem 1, we consider the following $(3\nu + 3)$ games. In Game 0, a part framed by a box indicates coefficients to be changed in a subsequent game. In the other games, a part framed by a box indicates coefficients that were changed in a game from the previous game.

Game 0 : Original game. That is, the reply to a key query for $\vec{v}$ is

$$\boldsymbol{k}_0^* := (\delta,\ \boxed{0},\ 1,\ \varphi_0,\ 0)_{\mathbb{B}_0^*}, \qquad \boldsymbol{k}_1^* := (\delta\vec{v},\ \boxed{0^n},\ \varphi_1\vec{v},\ 0^n)_{\mathbb{B}_1^*}, \tag{4}$$

where $\delta, \varphi_0, \varphi_1 \xleftarrow{\mathsf{U}} \mathbb{F}_q$ and $\vec{v} := (v_1, \ldots, v_n) \in \mathbb{F}_q^n$ with $v_n \neq 0$. The challenge ciphertext for challenge plaintexts $(m^{(0)}, m^{(1)})$ and $\vec{x}$, $(\vec{x}, \boldsymbol{c}_0, \{C_{1,j}, C_{2,j}\}_{j=1,..,4}, c_3)$, which is identified with $(\vec{x}, \boldsymbol{c}_0, \boldsymbol{c}_1, c_3)$ in Remark 2, is

$$\boldsymbol{c}_0 := (-\omega,\ \boxed{0},\ \boxed{\zeta},\ 0,\ \eta_0)_{\mathbb{B}_0},\quad \boldsymbol{c}_1 := (\omega\vec{x},\ \boxed{0^n},\ 0^n,\ \eta_1\vec{x})_{\mathbb{B}_1},\ c_3 := g_T^{\zeta} m, \quad (5)$$

where $b \xleftarrow{\mathsf{U}} \{0,1\}; \omega, \zeta, \eta_0, \eta_1 \xleftarrow{\mathsf{U}} \mathbb{F}_q$ and $\vec{x} := (x_1, \ldots, x_n) \in \mathbb{F}_q^n$ with $x_l \neq 0$ for some $l \in \{1, .., n-1\}$.

Game 1 : Same as Game 0 except that the challenge ciphertext for challenge plaintexts $(m^{(0)}, m^{(1)})$ and $\vec{x}$ is

$$\boldsymbol{c}_0 := (-\omega,\ \boxed{-\tau},\ \zeta,\ 0,\ \eta_0)_{\mathbb{B}_0},\quad \boldsymbol{c}_1 := (\omega\vec{x},\ \boxed{\tau\vec{x}U},\ 0^n,\ \eta_1\vec{x})_{\mathbb{B}_1},\ c_3 := g_T^{\zeta} m, (6)$$

where $\tau \xleftarrow{\mathsf{U}} \mathbb{F}_q, U \xleftarrow{\mathsf{U}} \mathcal{H}(n, \mathbb{F}_q) \cap GL(n, \mathbb{F}_q)$, and all the other variables are generated as in Game 0.

Game 2-*h*-1 ($h = 1, \ldots, \nu$) : Game 2-0-3 is Game 1. Game 2-h-1 is the same as Game 2-$(h-1)$-3 except that the reply to the h-th key query for $\vec{v}$, $(\boldsymbol{k}_0^*, \boldsymbol{k}_1^*)$, is

$$\boldsymbol{k}_0^* := (\delta,\ \boxed{\rho},\ 1,\ \varphi_0,\ 0)_{\mathbb{B}_0^*},\qquad \boldsymbol{k}_1^* := (\delta\vec{v},\ \boxed{\rho\vec{v}Z},\ \varphi_1\vec{v},\ 0^n)_{\mathbb{B}_1^*}, \qquad (7)$$

where $\rho \xleftarrow{\mathsf{U}} \mathbb{F}_q, Z := (U^{-1})^{\mathrm{T}}$ for $U \xleftarrow{\mathsf{U}} \mathcal{H}(n, \mathbb{F}_q) \cap GL(n, \mathbb{F}_q)$ used in Eq. (6) and all the other variables are generated as in Game 2-$(h-1)$-3.

Game 2-*h*-2 ($h = 1, \ldots, \nu$) : Game 2-h-2 is the same as Game 2-h-1 except that a part of the reply to the h-th key query for $\vec{v}$, $(\boldsymbol{k}_0^*, \boldsymbol{k}_1^*)$, is

$$\boldsymbol{k}_0^* := (\delta,\ \boxed{w},\ 1,\ \varphi_0,\ 0)_{\mathbb{B}_0^*},\qquad \boldsymbol{k}_1^* := (\delta\vec{v},\ \rho\vec{v}Z,\ \varphi_1\vec{v},\ 0^n)_{\mathbb{B}_1^*}, \qquad (8)$$

where $w \xleftarrow{\mathsf{U}} \mathbb{F}_q$ and all the other variables are generated as in Game 2-h-1.

Game 2-*h*-3 ($h = 1, \ldots, \nu$) : Game 2-h-3 is the same as Game 2-h-2 except that the reply to the h-th key query for $\vec{v}$, $(\boldsymbol{k}_0^*, \boldsymbol{k}_1^*)$, is

$$\boldsymbol{k}_0^* := (\delta,\ w,\ 1,\ \varphi_0,\ 0)_{\mathbb{B}_0^*},\qquad \boldsymbol{k}_1^* := (\delta\vec{v},\ \boxed{0^n},\ \varphi_1\vec{v},\ 0^n)_{\mathbb{B}_1^*}, \qquad (9)$$

where all the variables are generated as in Game 2-h-2.

Game 3 : Same as Game 2-ν-3 except that $\boldsymbol{c}_0$ and c_3 of the challenge ciphertext are

$$\boldsymbol{c}_0 := (-\omega,\ -\tau,\ \boxed{\zeta'},\ 0,\ \eta_0)_{\mathbb{B}_0},\quad c_3 := g_T^{\zeta} m^{(b)},$$

where $\zeta' \xleftarrow{\mathsf{U}} \mathbb{F}_q$ (i.e., independent from $\zeta \xleftarrow{\mathsf{U}} \mathbb{F}_q$), and all the other variables are generated as in Game 2-ν-3.

Let $\mathsf{Adv}_{\mathcal{A}}^{(0)}(\lambda), \mathsf{Adv}_{\mathcal{A}}^{(1)}(\lambda), \mathsf{Adv}_{\mathcal{A}}^{(2\text{-}h\text{-}\iota)}(\lambda)$ ($h = 1, \ldots, \nu; \iota = 1, 2, 3$) and $\mathsf{Adv}_{\mathcal{A}}^{(3)}(\lambda)$ be the advantage of $\mathcal{A}$ in Game $0, 1, 2$-h-ι and 3, respectively. $\mathsf{Adv}_{\mathcal{A}}^{(0)}(\lambda)$ is equivalent to $\mathsf{Adv}_{\mathcal{A}}^{\mathsf{NIPE,PH}}(\lambda)$ and it is obtained that $\mathsf{Adv}_{\mathcal{A}}^{(3)}(\lambda) = 0$ by Lemma 9. We will show five lemmas (Lemmas 4-8) that evaluate the gaps between pairs of $\mathsf{Adv}_{\mathcal{A}}^{(0)}(\lambda)$, $\mathsf{Adv}_{\mathcal{A}}^{(1)}(\lambda), \mathsf{Adv}_{\mathcal{A}}^{(2\text{-}h\text{-}\iota)}(\lambda)$ for $h = 1, \ldots, \nu; \iota = 1, 2, 3$ and $\mathsf{Adv}_{\mathcal{A}}^{(3)}(\lambda)$. From these lemmas and Lemmas 1 and 2, we obtain Theorem 1. □

Lemma 4. *For any adversary $\mathcal{A}$, there exists a probabilistic machine $\mathcal{B}_1$, whose running time is essentially the same as that of $\mathcal{A}$, such that for any security parameter λ, $|\mathsf{Adv}_{\mathcal{A}}^{(0)}(\lambda) - \mathsf{Adv}_{\mathcal{A}}^{(1)}(\lambda)| \leq \mathsf{Adv}_{\mathcal{B}_1}^{\mathsf{P1}}(\lambda)$.*

Lemma 5. *For any adversary $\mathcal{A}$, there exists a probabilistic machine $\mathcal{B}_{2\text{-}1}$, whose running time is essentially the same as that of $\mathcal{A}$, such that for any security parameter λ, $|\mathsf{Adv}_{\mathcal{A}}^{(2\text{-}(h-1)\text{-}3)}(\lambda) - \mathsf{Adv}_{\mathcal{A}}^{(2\text{-}h\text{-}1)}(\lambda)| \leq \mathsf{Adv}_{\mathcal{B}_{2\text{-}h\text{-}1}}^{\mathsf{P2}}(\lambda)$, where $\mathcal{B}_{2\text{-}h\text{-}1}(\cdot) := \mathcal{B}_{2\text{-}1}(h, \cdot)$.*

Lemma 6. *For any adversary $\mathcal{A}$, for any security parameter λ, $|\mathsf{Adv}_{\mathcal{A}}^{(2\text{-}h\text{-}1)}(\lambda) - \mathsf{Adv}_{\mathcal{A}}^{(2\text{-}h\text{-}2)}(\lambda)| \leq 1/q$.*

Lemma 7. *For any adversary $\mathcal{A}$, there exists a probabilistic machine $\mathcal{B}_{2\text{-}2}$, whose running time is essentially the same as that of $\mathcal{A}$, such that for any security parameter λ, $|\mathsf{Adv}_{\mathcal{A}}^{(2\text{-}h\text{-}2)}(\lambda) - \mathsf{Adv}_{\mathcal{A}}^{(2\text{-}h\text{-}3)}(\lambda)| \leq \mathsf{Adv}_{\mathcal{B}_{2\text{-}h\text{-}2}}^{\mathsf{P2}}(\lambda)$, where $\mathcal{B}_{2\text{-}h\text{-}2}(\cdot) := \mathcal{B}_{2\text{-}2}(h, \cdot)$.*

Lemma 8. *For any adversary $\mathcal{A}$, for any security parameter λ, $|\mathsf{Adv}_{\mathcal{A}}^{(2\text{-}\nu\text{-}3)}(\lambda) - \mathsf{Adv}_{\mathcal{A}}^{(3)}(\lambda)| \leq 1/q$.*

Lemma 9. *For any adversary $\mathcal{A}$, for any security parameter λ, $\mathsf{Adv}_{\mathcal{A}}^{(3)}(\lambda) = 0$.*

6 Proposed NIPE Scheme with Constant-Size Secret-Keys

6.1 Dual Orthonormal Basis Generator

We describe random dual orthonormal basis generator $\mathcal{G}_{\mathsf{ob}}^{(2)}$ below, which is used as a subroutine in the proposed NIPE scheme, where $\mathcal{G}_{\mathsf{ob}}^{(1)}$ is given in Section 5.2.

$$
\begin{aligned}
&\mathcal{G}_{\mathsf{ob}}^{(2)}(1^\lambda,\ n) : (\mathsf{param}_n, \mathbb{D}_0, \mathbb{D}_0^*, \{D_{i,j}, D'_{i,j,l}\}_{i,j=1,\ldots,4;l=1,\ldots,n}, \mathbb{D}_1^*) \xleftarrow{\mathsf{R}} \mathcal{G}_{\mathsf{ob}}^{(1)}(1^\lambda,\ n),\\
&\quad \mathbb{B}_0 := \mathbb{D}_0^*,\ \mathbb{B}_0^* := \mathbb{D}_0,\ \mathbb{B}_1 := \mathbb{D}_1^*,\ B_{i,j}^* := D_{i,j},\ B'^*_{i,j,l} := D'_{i,j,l}\\
&\qquad\qquad\qquad\qquad\qquad\qquad \text{for } i,j = 1,\ldots,4; l = 1,\ldots,n,\\
&\quad \text{return } \ (\mathsf{param}_n, \mathbb{B}_0, \mathbb{B}_0^*, \mathbb{B}_1, \{B_{i,j}^*, B'^*_{i,j,l}\}_{i,j=1,\ldots,4;l=1,\ldots,n}).
\end{aligned}
$$

Remark 5 . From Remark 1, $\{B_{i,j}^*, B'^*_{i,j,l}\}_{i,j=1,\ldots,4;l=1,\ldots,n}$ is identified with basis $\mathbb{B}_1^* := (\boldsymbol{b}_{1,1}^*, \ldots, \boldsymbol{b}_{1,4n}^*)$ dual to $\mathbb{B}_1$.

6.2 Construction and Security

In the description of the scheme, we assume that input vector, $\vec{v} := (v_1, \ldots, v_n)$, has an index l $(1 \leq l \leq n-1)$ with $v_l \neq 0$, and that input vector, $\vec{x} := (x_1, \ldots, x_n)$, satisfies $x_n \neq 0$.

$\mathsf{Setup}(1^\lambda,\ n) : (\mathsf{param}_n, \mathbb{B}_0, \mathbb{B}_0^*, \mathbb{B}_1, \{B_{i,j}^*, B'^*_{i,j,l}\}_{i,j=1,..,4;\, l=1,..,n}) \xleftarrow{\mathsf{R}} \mathcal{G}_{\mathsf{ob}}^{(2)}(1^\lambda,\ n)$,
$\widehat{\mathbb{B}}_0 := (\boldsymbol{b}_{0,1}, \boldsymbol{b}_{0,3}, \boldsymbol{b}_{0,5}),\ \widehat{\mathbb{B}}_0^* := (\boldsymbol{b}_{0,1}^*, \boldsymbol{b}_{0,3}^*, \boldsymbol{b}_{0,4}^*)$,
$\widehat{\mathbb{B}}_1 := (\boldsymbol{b}_{1,1}, .., \boldsymbol{b}_{1,n}, \boldsymbol{b}_{1,3n+1}, .., \boldsymbol{b}_{1,4n})$,
return $\mathsf{pk} := (1^\lambda, \mathsf{param}_n, \{\widehat{\mathbb{B}}_t\}_{t=0,1}),\ \mathsf{sk} := (\widehat{\mathbb{B}}_0^*, \{B_{i,j}^*, B'^*_{i,j,l}\}_{i=1,3;\, j=1,..,4;\, l=1,..,n})$.

$\mathsf{KeyGen}(\mathsf{pk},\ \mathsf{sk},\ \vec{v}) : \quad \delta, \varphi_0, \varphi_1 \xleftarrow{\mathsf{U}} \mathbb{F}_q, \quad \boldsymbol{k}_0^* := (\delta,\ 0,\ 1,\ \varphi_0,\ 0)_{\mathbb{B}_0^*}$,
$K_{1,j}^* := \delta B_{1,j}^* + \varphi_1 B_{3,j}^*,\quad K_{2,j}^* := \sum_{l=1}^n v_l(\delta B'^*_{1,j,l} + \varphi_1 B'^*_{3,j,l})$ for $j = 1, .., 4$,
return $\mathsf{sk}_{\vec{v}} := (\vec{v}, \boldsymbol{k}_0^*, \{K_{1,j}^*, K_{2,j}^*\}_{j=1,...,4})$.

$\mathsf{Enc}(\mathsf{pk},\ m,\ \vec{x}) : \quad \omega, \eta_0, \eta_1, \zeta \xleftarrow{\mathsf{U}} \mathbb{F}_q, \quad \boldsymbol{c}_0 := (-\omega,\ 0,\ 1,\ 0,\ \eta_0)_{\mathbb{B}_0}$,
$\boldsymbol{c}_1 := (\omega\vec{x},\ 0^n,\ 0^n,\ \eta_1\vec{x})_{\mathbb{B}_1}, \quad c_3 := g_T^\zeta m, \quad$ return $\mathsf{ct}_{\vec{x}} := (\vec{x}, \boldsymbol{c}_0, \boldsymbol{c}_1, c_3)$.
$\mathsf{Dec}(\mathsf{pk},\ \mathsf{sk}_{\vec{v}} := (\vec{v}, \boldsymbol{k}_0^*, \{K_{1,j}^*, K_{2,j}^*\}_{j=1,...,4}),\ \mathsf{ct}_{\vec{x}} := (\vec{x}, \boldsymbol{c}_0, \boldsymbol{c}_1, c_3)) :$
Parse $\boldsymbol{c}_1$ as a $4n$-tuple $(C_1, \ldots, C_{4n}) \in \mathbb{G}^{4n}$,
$D_j := \sum_{l=1}^{n-1}((\vec{x}\cdot\vec{v})^{-1}v_l)\, C_{(j-1)n+l}$ for $j = 1, .., 4$,
$F := e(\boldsymbol{c}_0, \boldsymbol{k}_0^*) \cdot \prod_{j=1}^4 \left(e(D_j, K_{1,j}^*) \cdot e(C_{jn}, K_{2,j}^*)\right)$, return $m' := c_3/F$.

Remark 6. A part of output of $\mathsf{Setup}(1^\lambda, n)$, $\{B_{i,j}^*, B'^*_{i,j,l}\}_{i=1,3;j=1,...,4;l=1,...,n}$, can be identified with $\widehat{\mathbb{B}}_1^* := (\boldsymbol{b}_{1,1}^*, \ldots, \boldsymbol{b}_{1,n}^*, \boldsymbol{b}_{1,2n+1}^*, \ldots, \boldsymbol{b}_{1,3n}^*)$, while $\mathbb{B}_1^* := (\boldsymbol{b}_{1,1}^*, \ldots, \boldsymbol{b}_{1,4n}^*)$ is identified with $\{B_{i,j}^*,\ B'^*_{i,j,l}\}_{i,j=1,...,4;l=1,...,n}$ in Remark 5. Decryption Dec can be alternatively described as:

$\mathsf{Dec}'(\mathsf{pk},\ \mathsf{sk}_{\vec{v}} := (\vec{v}, \boldsymbol{k}_0^*, \{K_{1,j}^*, K_{2,j}^*\}_{j=1,...,4}),\ \mathsf{ct}_{\vec{x}} := (\vec{x}, \boldsymbol{c}_0, \boldsymbol{c}_1, c_3)) :$

$$\boldsymbol{k}_1^* := (\ \overbrace{v_1 K_{1,1}^*, .., v_{n-1}K_{1,1}^*, K_{2,1}^*}^{n},\ \ldots,\ \overbrace{v_1 K_{1,4}^*, .., v_{n-1}K_{1,4}^*, K_{2,4}^*}^{n}\ ),$$
that is, $\boldsymbol{k}_1^* = (\delta\vec{v},\ 0^n,\ 0^n,\ \varphi_1\vec{v})_{\mathbb{B}_1^*}$, $F := e(\boldsymbol{c}_0, \boldsymbol{k}_0^*) \cdot e((\vec{x}\cdot\vec{v})^{-1}\boldsymbol{c}_1, \boldsymbol{k}_1^*)$,
return $m' := c_3/F$.

Theorem 2. *The proposed NIPE scheme is adaptively payload-hiding against chosen plaintext attacks under the DLIN assumption.*

For any adversary $\mathcal{A}$, there exist probabilistic machines $\mathcal{E}_1, \mathcal{E}_{2\text{-}1}$ and $\mathcal{E}_{2\text{-}2}$ whose running times are essentially the same as that of $\mathcal{A}$, such that for any security parameter λ, $\mathsf{Adv}_{\mathcal{A}}^{\mathsf{NIPE,PH}}(\lambda) \le \mathsf{Adv}_{\mathcal{E}_1}^{\mathsf{DLIN}}(\lambda) + \sum_{h=1}^{\nu}\left(\mathsf{Adv}_{\mathcal{E}_{2\text{-}h\text{-}1}}^{\mathsf{DLIN}}(\lambda) + \mathsf{Adv}_{\mathcal{E}_{2\text{-}h\text{-}2}}^{\mathsf{DLIN}}(\lambda)\right)$ $+\epsilon$, where $\mathcal{E}_{2\text{-}h\text{-}1}(\cdot) := \mathcal{E}_{2\text{-}1}(h,\cdot)$, $\mathcal{E}_{2\text{-}h\text{-}2}(\cdot) := \mathcal{E}_{2\text{-}2}(h,\cdot)$, ν is the maximum number of $\mathcal{A}$'s key queries and $\epsilon := (11\nu + 6)/q$.

The proof of Theorem 2 is given in the full version of this paper.

7 Proposed ZIPE Scheme with Constant-Size Ciphertexts

7.1 Dual Orthonormal Basis Generator

Let $N := 4n + 1$ and

$$\mathcal{L}'(N, \mathbb{F}_q) := \left\{ X := \begin{pmatrix} \chi_{0,0} & \chi_{0,1}\vec{e}_n & \cdots & \chi_{0,4}\vec{e}_n \\ \vec{\chi}_{1,0}^{\mathrm{T}} & X_{1,1} & \cdots & X_{1,4} \\ \vdots & \vdots & & \vdots \\ \vec{\chi}_{4,0}^{\mathrm{T}} & X_{4,1} & \cdots & X_{4,4} \end{pmatrix} \middle| \begin{array}{l} X_{i,j} \in \mathcal{H}(n, \mathbb{F}_q), \\ \vec{\chi}_{i,0} := (\chi_{i,0,l})_{l=1,\ldots,n} \in \mathbb{F}_q^n, \\ \chi_{0,0}, \chi_{0,j} \in \mathbb{F}_q \\ \text{for } i,j = 1,\ldots,4 \end{array} \right\}$$
$$\bigcap GL(N, \mathbb{F}_q). \quad (10)$$

We describe random dual orthonormal basis generator $\mathcal{G}_{\mathsf{ob}}^{(3)}$ below, which is used as a subroutine in the proposed Zero IPE scheme.

$$\begin{aligned}
&\mathcal{G}_{\mathsf{ob}}^{(3)}(1^\lambda,\ n): \quad \mathsf{param}_{\mathbb{G}} := (q, \mathbb{G}, \mathbb{G}_T, G, e) \xleftarrow{\mathsf{R}} \mathcal{G}_{\mathsf{bpg}}(1^\lambda),\ N := 4n+1, \\
&\psi \xleftarrow{\mathsf{U}} \mathbb{F}_q^\times,\ g_T := e(G,G)^\psi,\ \mathsf{param}_{\mathbb{V}} := (q, \mathbb{V}, \mathbb{G}_T, \mathbb{A}, e) := \mathcal{G}_{\mathsf{dpvs}}(1^\lambda, N, \mathsf{param}_{\mathbb{G}}), \\
&\mathsf{param}_n := (\mathsf{param}_{\mathbb{V}},\ g_T),\ X \xleftarrow{\mathsf{U}} \mathcal{L}'(N, \mathbb{F}_q),\ \text{hereafter}, \\
&\{\chi_{0,0}, \chi_{0,j}, \chi_{i,0,l}, \mu_{i,j}, \mu'_{i,j,l}\}_{i,j=1,\ldots 4; l=1,\ldots,n} \text{ denotes non-zero entries of } X, \\
&\quad \text{as in Eq. (10) and Eq. (1)},\ \ (\vartheta_{i,j})_{i,j=0,\ldots,4n} := \psi \cdot (X^{\mathrm{T}})^{-1}, \\
&B_{0,0} := \chi_{0,0}G,\ B_{0,j} := \chi_{0,j}G,\ B_{i,0,l} := \chi_{i,0,l}G,\ B_{i,j} := \mu_{i,j}G,\ B'_{i,j,l} := \mu'_{i,j,l}G \\
&\qquad\qquad \text{for } i,j = 1,\ldots,4; l = 1,\ldots,n, \\
&\boldsymbol{b}_i^* := (\vartheta_{i,1},\ldots,\vartheta_{i,N})_{\mathbb{A}} = \textstyle\sum_{j=0}^{4n} \vartheta_{i,j}\boldsymbol{a}_j \text{ for } i = 0,\ldots,4n,\ \mathbb{B}^* := (\boldsymbol{b}_0^*,\ldots,\boldsymbol{b}_{4n}^*), \\
&\quad \text{return}\ \ (\mathsf{param}_n, \{B_{0,0}, B_{0,j}, B_{i,0,l}, B_{i,j}, B'_{i,j,l}\}_{i,j=1,\ldots,4; l=1,\ldots,n}, \mathbb{B}^*).
\end{aligned}$$

Remark 7 . $\{B_{0,0}, B_{0,j}, B_{i,0,l}, B_{i,j}, B'_{i,j,l}\}_{i=1,\ldots,4; j=1,\ldots,4; l=1,\ldots,n}$ is identified with basis $\mathbb{B} := (\boldsymbol{b}_0, \ldots, \boldsymbol{b}_{4n})$ dual to $\mathbb{B}^*$ as in Remark 1.

7.2 Construction and Security

In the description of the scheme, we assume that input vector, $\vec{x} := (x_1, \ldots, x_n)$, has an index l $(1 \leq l \leq n-1)$ with $x_l \neq 0$, and that input vector, $\vec{v} := (v_1, \ldots, v_n)$, satisfies $v_n \neq 0$.

$$\begin{aligned}
&\mathsf{Setup}(1^\lambda,\ n): \\
&\quad (\mathsf{param}_n, \{B_{0,0}, B_{0,j}, B_{i,0,l}, B_{i,j}, B'_{i,j,l}\}_{i,j=1,\ldots,4;\, l=1,\ldots,n}, \mathbb{B}^*) \xleftarrow{\mathsf{R}} \mathcal{G}_{\mathsf{ob}}^{(3)}(1^\lambda,\ n), \\
&\quad \widehat{\mathbb{B}}^* := (\boldsymbol{b}_0^*, \ldots, \boldsymbol{b}_n^*, \boldsymbol{b}_{2n+1}^*, \ldots, \boldsymbol{b}_{3n}^*), \\
&\quad \text{return}\ \ \mathsf{pk} := (1^\lambda, \mathsf{param}_n, \{B_{0,0}, B_{0,j}, B_{i,0,l}, B_{i,j}, B'_{i,j,l}\}_{i=1,4; j=1,\ldots,4; l=1,\ldots,n}), \\
&\qquad\qquad \mathsf{sk} := \widehat{\mathbb{B}}^*.
\end{aligned}$$

$$\mathsf{KeyGen}(\mathsf{pk},\ \mathsf{sk},\ \vec{v}): \quad \delta, \varphi \xleftarrow{\mathsf{U}} \mathbb{F}_q, \quad \boldsymbol{k}^* := (\ 1,\ \overbrace{\delta\vec{v},}^{n}\ \ \overbrace{0^n,}^{n}\ \ \overbrace{\varphi\vec{v},}^{n}\ \ \overbrace{0^n}^{n}\ )_{\mathbb{B}^*},$$

return $\mathsf{sk}_{\vec{v}} := \boldsymbol{k}^*$.

$\mathsf{Enc}(\mathsf{pk},\ m,\ \vec{x})$: $\omega, \eta, \zeta \xleftarrow{\mathsf{U}} \mathbb{F}_q$, $C_0 := \zeta B_{0,0} + \sum_{l=1}^{n} x_l(\omega B_{1,0,l} + \eta B_{4,0,l})$,

$c_3 := g_T^{\zeta} m$, $C_{1,j} := \omega B_{1,j} + \eta B_{4,j}$,

$C_{2,j} := \zeta B_{0,j} + \sum_{l=1}^{n} x_l(\omega B'_{1,j,l} + \eta B'_{4,j,l})$ for $j = 1, \ldots, 4$,

return $\mathsf{ct}_{\vec{x}} := (\vec{x}, C_0, \{C_{1,j}, C_{2,j}\}_{j=1,\ldots,4}, c_3)$.

$\mathsf{Dec}(\mathsf{pk},\ \mathsf{sk}_{\vec{v}} := \boldsymbol{k}^*,\ \mathsf{ct}_{\vec{x}} := (\vec{x}, C_0, \{C_{1,j}, C_{2,j}\}_{j=1,\ldots,4}, c_3))$:

Parse $\boldsymbol{k}^*$ as a $(4n+1)$-tuple $(K_0^*, \ldots, K_{4n}^*) \in \mathbb{G}^{4n+1}$,

$D_j^* := \sum_{l=1}^{n-1} x_l K^*_{(j-1)n+l}$ for $j = 1, \ldots, 4$,

$F := e(C_0, K_0^*) \cdot \prod_{j=1}^{4} \left(e(C_{1,j}, D_j^*) \cdot e(C_{2,j}, K_{jn}^*) \right)$, return $m' := c_3/F$.

Remark 8 . A part of output of $\mathsf{Setup}(1^\lambda, n)$, $\{B_{0,0}, B_{0,j}, B_{i,0,l}, B_{i,j}, B'_{i,j,l}\}_{i=1,4;j=1,..,4;l=1,..,n}$, can be identified with $\widehat{\mathbb{B}} := (\boldsymbol{b}_0, .., \boldsymbol{b}_n, \boldsymbol{b}_{3n+1}, .., \boldsymbol{b}_{4n})$, while $\mathbb{B} := (\boldsymbol{b}_0, \ldots, \boldsymbol{b}_{4n})$ is identified with $\{B_{0,0}, B_{0,j}, B_{i,0,l}, B_{i,j},\ B'_{i,j,l}\}_{i,j=1,..,4;l=1,..,n}$ in Remark 7. Decryption Dec can be alternatively described as:

$\mathsf{Dec}'(\mathsf{pk},\ \mathsf{sk}_{\vec{v}} := \boldsymbol{k}^*,\ \mathsf{ct}_{\vec{x}} := (\vec{x}, C_0, \{C_{1,j}, C_{2,j}\}_{j=1,\ldots,4}, c_3))$:

$$\boldsymbol{c} := (\ C_0,\ \overbrace{x_1 C_{1,1}, .., x_{n-1} C_{1,1}, C_{2,1}}^{n},\ \ldots,\ \overbrace{x_1 C_{1,4}, .., x_{n-1} C_{1,4}, C_{2,4}}^{n}\ ),$$

that is, $\boldsymbol{c} = (\zeta,\ \overbrace{\omega\vec{x}}^{n},\ \overbrace{0^n}^{n},\ \overbrace{0^n}^{n},\ \overbrace{\eta\vec{x}}^{n}\ )_{\mathbb{B}}$, $F := e(\boldsymbol{c}, \boldsymbol{k}^*)$, return $m' := c_3/F$.

[Correctness]. Using the alternate decryption Dec', $F = e(\boldsymbol{c}, \boldsymbol{k}) = g_T^{\zeta + \omega\delta\vec{x}\cdot\vec{v}} = g_T^{\zeta}$ if $\vec{x} \cdot \vec{v} = 0$.

Remark 9 . The proposed ZIPE in this section employs a single basis, $\mathbb{B}$, generated by $X \in GL(4n+1, \mathbb{F}_q)$ (or $X \in \mathcal{L}'(4n+1, \mathbb{F}_q)$ of Eq. (10)), and a ciphertext can be expressed as $(\boldsymbol{c}, g_T^{\zeta} m)$ with $\boldsymbol{c} = (\zeta,\ \omega\vec{x},\ 0^{2n},\ \eta_1\vec{x})_{\mathbb{B}}$ as shown in Remark 8. The proposed NIPE scheme in Sec. 5.3 employs two bases, $\mathbb{B}_0$ and $\mathbb{B}_1$, generated by $X_0 \in GL(5, \mathbb{F}_q)$ and $X_1 \in GL(4n, \mathbb{F}_q)$, and a ciphertext can be expressed as $(\boldsymbol{c}_0, \boldsymbol{c}_1, g_T^{\zeta} m)$ with $\boldsymbol{c}_0 := (-\omega,\ 0,\ \zeta,\ 0,\ \eta_0)_{\mathbb{B}_0}$ and $\boldsymbol{c}_1 = (\omega\vec{x},\ 0^{2n},\ \eta_1\vec{x})_{\mathbb{B}_1}$. Hence, the ciphertext and secret key of the ZIPE scheme are shorter than those of the NIPE scheme (see Table 1 in Sec. 8). It is due to the difference of the decryption tricks in the ZIPE and NIPE schemes. Similarly to the fact on $\mathcal{L}(4n, \mathbb{F}_q)$ (for the security of the NIPE scheme) shown in Sec. 5.1, it is crucial for the security of the ZIPE scheme that $\mathcal{L}'(4n+1, \mathbb{F}_q)$ is a subgroup in $GL(4n+1, \mathbb{F}_q)$, and its security proof is made in the essentially same manner as explained in Sec. 5.1.

Theorem 3. *The proposed ZIPE scheme is adaptively payload-hiding against chosen plaintext attacks under the DLIN assumption. For any adversary $\mathcal{A}$, there exist probabilistic machines $\mathcal{E}_1, \mathcal{E}_{2,h}$ ($h = 1, \ldots, \nu$), whose running times* *are essentially the same as that of $\mathcal{A}$, such that for any security parameter λ, $\mathsf{Adv}^{\mathsf{ZIPE,PH}}_{\mathcal{A}}(\lambda) \leq \mathsf{Adv}^{\mathsf{DLIN}}_{\mathcal{E}_1}(\lambda) + \sum_{h=1}^{\nu} \mathsf{Adv}^{\mathsf{DLIN}}_{\mathcal{E}_{2,h}}(\lambda) + \epsilon$, where ν is the maximum number of $\mathcal{A}$'s key queries and $\epsilon := (11\nu + 6)/q$.*

The proof of Theorem 3 is given in the full version of this paper.

8 Performance

Table 1 compares the proposed ZIPE and NIPE schemes (ZIPE with short ciphertexts in Sec. 7, NIPE with short ciphertexts in Sec. 5, ZIPE with short secret-keys given in the full version, and NIPE with short secret-keys in Sec. 6) with the ZIPE and NIPE schemes in [2] that are secure under standard assumptions.

Table 1. Comparison with IPE schemes in [2], where $|\mathbb{G}|$, $|\mathbb{G}_T|$, $|\mathbb{F}_q|$, P and M represent size of $\mathbb{G}$, size of $\mathbb{G}_T$, size of $\mathbb{F}_q$, pairing operation, and scalar multiplication on $\mathbb{G}$, respectively. CT, SK, IP and DBDH stand for ciphertexts, secret-keys, inner-product and decisional bilinear Diffie-Hellman, respectively.

	AL10 [2] ZIPE with Short CTs	AL10 [2] NIPE with Short CTs	Proposed ZIPE with Short CTs	Proposed NIPE with Short CTs	Proposed ZIPE with Short SKs	Proposed NIPE with Short SKs
Security	Adaptive	Co-selective	Adaptive	Adaptive	Adaptive	Adaptive
Assump.	DLIN & DBDH	DLIN & DBDH	DLIN	DLIN	DLIN	DLIN
IP Rel.	Zero	Non-zero	Zero	Non-zero	Zero	Non-zero
PK Size	$(n+11)\lvert\mathbb{G}\rvert + \lvert\mathbb{G}_T\rvert$	$(n+11)\lvert\mathbb{G}\rvert + \lvert\mathbb{G}_T\rvert$	$(10n+13)\lvert\mathbb{G}\rvert + \lvert\mathbb{G}_T\rvert$	$(8n+23)\lvert\mathbb{G}\rvert + \lvert\mathbb{G}_T\rvert$	$(10n+13)\lvert\mathbb{G}\rvert + \lvert\mathbb{G}_T\rvert$	$(8n+23)\lvert\mathbb{G}\rvert + \lvert\mathbb{G}_T\rvert$
SK Size	$(n+6)\lvert\mathbb{G}\rvert + (n-1)\lvert\mathbb{F}_q\rvert$	$(n+6)\lvert\mathbb{G}\rvert$	$(4n+1)\lvert\mathbb{G}\rvert$	$(4n+5)\lvert\mathbb{G}\rvert$	$9\lvert\mathbb{G}\rvert$	$13\lvert\mathbb{G}\rvert$
CT Size	$9\lvert\mathbb{G}\rvert + \lvert\mathbb{G}_T\rvert + \lvert\mathbb{F}_q\rvert$	$9\lvert\mathbb{G}\rvert + \lvert\mathbb{G}_T\rvert$	$9\lvert\mathbb{G}\rvert + \lvert\mathbb{G}_T\rvert$	$13\lvert\mathbb{G}\rvert + \lvert\mathbb{G}_T\rvert$	$(4n+1)\lvert\mathbb{G}\rvert + \lvert\mathbb{G}_T\rvert$	$(4n+5)\lvert\mathbb{G}\rvert + \lvert\mathbb{G}_T\rvert$
Dec Time	9P + nM	9P + nM	9P + $4(n-1)$M	13P + $4(n-1)$M	9P + $4(n-1)$M	13P + $4(n-1)$M

9 Concluding Remarks

The technique with using special type matrices shown in this paper can reduce the size of ciphertexts or secret-keys of adaptively secure FE schemes in [17] from $O(dn)$ to $O(d)$, where d is the number of sub-universes of attributes, and n is the maximal length of attribute vectors. A key-policy attribute-based encryption (ABE) system with constant-size ciphertext [3] is selectively secure in the standard model. Therefore, it is an interesting open problem to realize an *adaptively secure and constant-size ciphertext* ABE scheme.

References

1. Abdalla, M., Kiltz, E., Neven, G.: Generalized key Delegation for Hierarchical Identity-Based Encryption. In: Biskup, J., López, J. (eds.) ESORICS 2007. LNCS, vol. 4734, pp. 139–154. Springer, Heidelberg (2007)
2. Attrapadung, N., Libert, B.: Functional Encryption for Inner Product: Achieving Constant-Size Ciphertexts with Adaptive Security or Support for Negation. In: Nguyen, P.Q., Pointcheval, D. (eds.) PKC 2010. LNCS, vol. 6056, pp. 384–402. Springer, Heidelberg (2010)
3. Attrapadung, N., Libert, B., de Panafieu, E.: Expressive Key-Policy Attribute-Based Encryption with Constant-Size Ciphertexts. In: Catalano, D., Fazio, N., Gennaro, R., Nicolosi, A. (eds.) PKC 2011. LNCS, vol. 6571, pp. 90–108. Springer, Heidelberg (2011)
4. Bethencourt, J., Sahai, A., Waters, B.: Ciphertext-policy attribute-based encryption. In: 2007 IEEE Symposium on Security and Privacy, pp. 321–334. IEEE Press (2007)
5. Boneh, D., Hamburg, M.: Generalized Identity Based and Broadcast Encryption Scheme. In: Pieprzyk, J. (ed.) ASIACRYPT 2008. LNCS, vol. 5350, pp. 455–470. Springer, Heidelberg (2008)
6. Delerablée, C.: Identity-Based Broadcast Encryption with Constant Size Ciphertexts and Private Keys. In: Kurosawa, K. (ed.) ASIACRYPT 2007. LNCS, vol. 4833, pp. 200–215. Springer, Heidelberg (2007)
7. Emura, K., Miyaji, A., Nomura, A., Omote, K., Soshi, M.: A Ciphertext-Policy Attribute-Based Encryption Scheme with Constant Ciphertext Length. In: Bao, F., Li, H., Wang, G. (eds.) ISPEC 2009. LNCS, vol. 5451, pp. 13–23. Springer, Heidelberg (2009)
8. Gentry, C., Waters, B.: Adaptive Security in Broadcast Encryption Systems (with short ciphertexts). In: Joux, A. (ed.) EUROCRYPT 2009. LNCS, vol. 5479, pp. 171–188. Springer, Heidelberg (2009)
9. Goyal, V., Pandey, O., Sahai, A., Waters, B.: Attribute-based encryption for fine-grained access control of encrypted data. In: ACM Conference on Computer and Communication Security 2006, pp. 89–98. ACM (2006)
10. Herranz, J., Laguillaumie, F., Ràfols, C.: Constant Size Ciphertexts in Thereshold Attribute-Based Encryption. In: Nguyen, P.Q., Pointcheval, D. (eds.) PKC 2010. LNCS, vol. 6056, pp. 19–34. Springer, Heidelberg (2010)
11. Katz, J., Sahai, A., Waters, B.: Predicate Encryption Supporting Disjunctions, Polynomial Equations, and Inner Products. In: Smart, N.P. (ed.) EUROCRYPT 2008. LNCS, vol. 4965, pp. 146–162. Springer, Heidelberg (2008)
12. Lewko, A., Sahai, A., Waters, B.: Revocation systems with very small private keys. In: IEEE Symposium on Security and Privacy 2010 (2010)
13. Lewko, A., Okamoto, T., Sahai, A., Takashima, K., Waters, B.: Fully Secure Functional Encryption: Attribute-Based Encryption and (Hierarchical) Inner Product Encryption. In: Gilbert, H. (ed.) EUROCRYPT 2010. LNCS, vol. 6110, pp. 62–91. Springer, Heidelberg (2010), `http://eprint.iacr.org/2010/110`
14. Lewko, A.B., Waters, B.: New Techniques for Dual System Encryption and Fully Secure HIBE with Short Ciphertexts. In: Micciancio, D. (ed.) TCC 2010. LNCS, vol. 5978, pp. 455–479. Springer, Heidelberg (2010)
15. Okamoto, T., Takashima, K.: Homomorphic Encryption and Signatures from Vector Decomposition. In: Galbraith, S.D., Paterson, K.G. (eds.) Pairing 2008. LNCS, vol. 5209, pp. 57–74. Springer, Heidelberg (2008)

16. Okamoto, T., Takashima, K.: Hierarchical Predicate Encryption for Inner-Products. In: Matsui, M. (ed.) ASIACRYPT 2009. LNCS, vol. 5912, pp. 214–231. Springer, Heidelberg (2009)
17. Okamoto, T., Takashima, K.: Fully Secure Functional Encryption with General Relations from the Decisional Linear Assumption. In: Rabin, T. (ed.) CRYPTO 2010. LNCS, vol. 6223, pp. 191–208. Springer, Heidelberg (2010), http://eprint.iacr.org/2010/563
18. Sahai, A., Waters, B.: Fuzzy Identity-Based Encryption. In: Cramer, R. (ed.) EUROCRYPT 2005. LNCS, vol. 3494, pp. 457–473. Springer, Heidelberg (2005)
19. Sakai, R., Furukawa, J.: Identity-based broadcast encryption, IACR ePrint Archive: Report 2007/217 (2007), http://eprint.iacr.org/2007/217
20. Waters, B.: Dual System Encryption: Realizing Fully Secure IBE and HIBE under Simple Assumptions. In: Halevi, S. (ed.) CRYPTO 2009. LNCS, vol. 5677, pp. 619–636. Springer, Heidelberg (2009)

Comments on the SM2 Key Exchange Protocol

Jing Xu and Dengguo Feng

State Key Laboratory of Information Security,
Institute of Software, Chinese Academy of Sciences, Beijing, P.R.China
xujing@is.iscas.ac.cn

Abstract. SM2 key exchange protocol is one part of the public key cryptographic algorithm SM2 which has been standardized by Chinese state cryptography administration for commercial applications. It became publicly available in 2010 and since then it was neither attacked nor proved to be secure. In this paper, we show that the SM2 key exchange protocol is insecure by presenting realistic attacks in the Canetti-Krawczyk model. The demonstrated attack breaks session-key security against an adversary who can only reveal session states. We also propose a simple modification method to solve this problem.

Keywords: Key exchange protocol, SM2, security model, attack.

1 Introduction

Key exchange protocols are cryptographic primitives that specify how two or more parties communicating over a public network establish a common session key. This session key is then typically used to build a confidential or integrity-preserving communication channel among the involved parties. Therefore, key exchange protocols form a crucial component in many network protocols. The most famous example is the classic Diffie-Hellman (DH) key exchange protocol that marked the birth of modern cryptography [1]. However, the original DH protocol did not provide authentication of the communicating parties, suffering from active attacks such as a man-in-the-middle attack. Authenticated key exchange (AKE) not only allows parties to compute the shared key but also ensures authenticity of the parties. AKE protocols operate in a public key infrastructure and the parties use each other's public keys to construct a shared secret.

1.1 Security Attributes

For authenticated key exchange protocols, it is desirable to possess the following security attributes:

(1) (implicit) key authentication: an agreed-upon session key should be known only by identified parties;
(2) forward secrecy: an agreed-upon session key should remain secret, even if both parties' long-term secret key is compromised;

D. Lin, G. Tsudik, and X. Wang (Eds.): CANS 2011, LNCS 7092, pp. 160–171, 2011.

(3) key compromise impersonation resilience: An adversary who reveals a long-term secret key of some party A should be unable to impersonate other parties to A (still, an adversary can impersonate A to anyone else).

In addition to above basic properties, another desirable attribute is resistance to unknown key-share (UKS) attacks. In an unknown key-share attack, a party A is coerced into sharing a key with any party E when in fact she thinks that she is sharing the key with a party B. UKS attacks were first discussed by Diffie *et al.* [2] and have been found in a number of protocols including MTI/A0 [3], the STS-MAC variant of the Station-to-Station (STS) protocol [4], MQV [5] and KEA [6]. Consider the situation where B is a bank system and A is an account holder. If the UKS attack described above is successful, then the adversary E could impersonate B (a banking system) and obtain A's credit card number over the resulting private communication link. Therefore, it is very significant to design protocols secure against UKS attacks.

1.2 Related Works and Our Contribution

The design and analysis of secure key exchange protocols have been proved to be a non-trivial task, with a large body of work written on the topic, including [7-12] and many more. Of these protocols, the most famous, most efficient and most standardized is the MQV protocol. The MQV protocols [7] are a family of authenticated Diffie-Hellman protocols and have been widely standardized [13-15]. The HMQV protocol [12] is a hashed variant of the MQV key agreement protocol with a rigorous security proof, which is currently being standardized by IEEE P1363 standards group [16].

SM2 key exchange protocol [17] is one part of the public key cryptographic algorithm SM2, which has been standardized by Chinese state cryptography administration for commercial applications and has been released in December 2010. This standard aims to provide a reference of products and techniques for security manufacturers in China, promoting the credibility and interoperability of security products. Indeed, SM2 key exchange protocol appears to be a remarkable protocol which provides the same efficiency as the MQV protocol. However, one question that has not been settled so far is whether the protocol can be proven secure in a rigorous model of key exchange security. In order to provide an answer to this question we analyze the SM2 protocol in the Canetti-Krawczyk security model. Unfortunately, we show that SM2 protocol is vulnerable to unknown key-share attacks in this model. The demonstrated attack breaks *session-key* security against an adversary who can only reveal session states. Then we present a simple patch which fixes the security problem.

1.3 Organization

The rest of this paper is organized as follows. Section 2 reviews SM2 key exchange protocol. Section 3 provides an overview of the formal security model of key exchange protocols, on which all of our analysis work is based. Section 4 presents our attacks on SM2 key exchange protocol and offers a security patch for the protocol. Section 5 concludes this work.

For ease of presentation, the notations and definitions used in this paper are shown in Table 1.

Table 1. Notations and definitions

Notation	Definition
A, B	Clients
d_A, d_B	The long-term private key of client A and client B respectively
P_A, P_B	The long-term public key of client A and client B respectively
$\mathbb{F}_q$	The finite field containing q elements.
$E(\mathbb{F}_q)$	The set of all points on an elliptic curve E defined over $\mathbb{F}_q$ and including the point at infinity $\mathcal{O}$
G	A distinguished point on an elliptic curve called the base point or generating point
$\sharp E(\mathbb{F}_q)$	The number of points on the curve (including the point at infinity $\mathcal{O}$)
h	$h = \sharp E(\mathbb{F}_q)/n$, where n is the order of the base point G; h is called the co-factor
$KDF(Z, klen)$	A key derivation function whose output length is $klen$
$x\|y$	Concatenation of two strings x and y
$Hash(\cdot)$	a one-way hash function
$[a, b]$	The set of integers x such that $a \leq x \leq b$
$\lceil x \rceil$	The ceiling of x; the smallest integer $\geq x$
$x \bmod n$	The unique remainder r, $0 \leq r \leq n-1$, when x is divided by n
&	Bitwise AND operator
$[k]P$	Scalar multiplication of a point P, $[k]P = \underbrace{P + P + \cdots + P}_{ktimes}$

2 Review of SM2 Key Exchange Protocol

There are two entities involved in the protocol: clients A and B who wish to establish a session key between them. Let $E(\mathbb{F}_q)$ be an elliptic curve defined over a finite field $\mathbb{F}_q$ and let G be the base point in $E(\mathbb{F}_q)$ of order n. The client A chooses a random $d_A \in [1, n-1]$ as its long-term private key and computes its long-term public key P_A as $P_A = [d_A]G$. Similarly, client B's long-term key pair is (d_B, P_B), where $P_B = [d_B]G$. We also assume that the public parameters $\langle E(\mathbb{F}_q), G, n, P_A, P_B, Z_A, Z_B\rangle$ have been fixed in advance and are known to A and B, where Z_A and Z_B are hash values of the identities of A and B, respectively. A high-level depiction of the protocol is given in Fig. 1, and a more detailed description follows:

(1) To establish a session key with client B, client A (the initiator) performs the steps:
   (a) Select $r_A \in [1, n-1]$ randomly and compute $R_A = [r_A]G = (x_1, y_1)$.
   (b) Send R_A to B.
   (c) Compute $t_A = (d_A + \overline{x}_1 \cdot r_A) \bmod n$, where $\overline{x}_1 = 2^w + (x_1 \& (2^w - 1))$ and $w = \lceil(\lceil log_2(n)\rceil/2)\rceil - 1$.

(2) Client B (the responder) performs the steps:
  (a) Select $r_B \in [1, n-1]$ randomly and compute $R_B = [r_B]G = (x_2, y_2)$.
  (b) Compute $t_B = (d_B + \overline{x}_2 \cdot r_B) \bmod n$, where $\overline{x}_2 = 2^w + (x_2 \& (2^w - 1))$ and $w = \lceil (\lceil log_2(n) \rceil / 2) \rceil - 1$.
  (c) Verify that R_A satisfies the defining equation of E. If it holds, compute $V = [h \cdot t_B](P_A + [\overline{x}_1]R_A) = (x_V, y_V)$. If $V = \mathcal{O}$, then B terminates the protocol run with failure.
  (d) Compute $K_B = KDF(x_V \| y_V \| Z_A \| Z_B, klen)$ and (optional) $S_B = Hash(0x02 \| y_V \| Hash(x_V \| Z_A \| Z_B \| x_1 \| y_1 \| x_2 \| y_2))$.
  (e) Send R_B and (optional) S_B to A.

(3) Upon receiving the message from B, A performs the steps:
  (a) Verify that R_B satisfies the defining equation of E. If it holds, compute $U = [h \cdot t_A](P_B + [\overline{x}_2]R_B) = (x_U, y_U)$. If $U = \mathcal{O}$, then A terminates the protocol run with failure.
  (b) (optional) Compare S_B with $Hash(0x02 \| y_U \| Hash (x_U \| Z_A \| Z_B \| x_1 \| y_1 \| x_2 \| y_2))$. If they are equal, B is authenticated.
  (c) Compute $K_A = KDF(x_U \| y_U \| Z_A \| Z_B, klen)$ and (optional) $S_A = Hash(0x03 \| y_U \| Hash(x_U \| Z_A \| Z_B \| x_1 \| y_1 \| x_2 \| y_2))$.
  (d) (optional) Send S_A to B for session key confirmation.

(4) (optional) Upon receiving the message S_A from A, B compare S_A with $Hash(0x03 \| y_V \| Hash(x_V \| Z_A \| Z_B \| x_1 \| y_1 \| x_2 \| y_2))$. If they are equal, A is authenticated.

It is straightforward to verify that both parties compute the same shared session key $K = K_A = K_B$.

3 Formal Model for Key Exchange Protocols

This section presents an abridged description of the Canetti-Krawczyk (CK) security model for key exchange protocols [18-19] on which all the analysis work in this paper is based.

In the CK model there are n parties each modeled by a probabilistic Turing machine. Each party has a long-term public-private key pair together with a certificate that binds the public key to that party. The binding assurance is provided by a certification authority (CA) which is only trusted to correctly verify the identity of the registrant of a public key before issuing the certificate. We do not assume that the CA requires a proof-of-possession of the private key from a registrant of a public key. For simplicity, we will only describe the model for Diffie-Hellman protocols that exchange ephemeral and long-term public keys. In particular, in the case of SM2, two parties A, B exchange long-term public keys P_A, P_B and ephemeral public keys R_A, R_B; the session key is obtained by combining P_A, P_B, R_A, R_B and possibly the identities A, B.

A party can be activated to run an instance of the protocol called a *session*, and each party may have multiple sessions running concurrently. The communications network is controlled by an adversary $\mathcal{A}$, an interactive probabilistic polynomial-time (PPT) machine, which schedules and mediates all sessions between the parties. $\mathcal{A}$ may activate a party A in two ways:

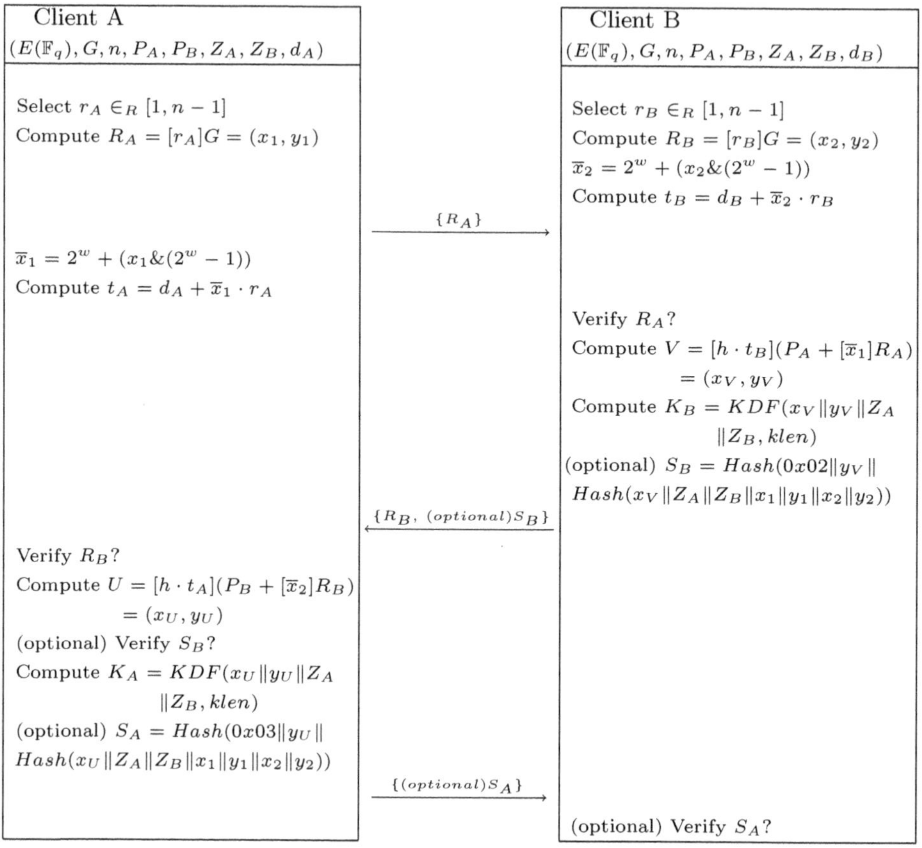

Fig. 1. SM2 key exchange protocol

1. By means of an **establish-session**(A, B) request, where B is another party with whom the key is to be established.
2. By means of a **send**(A, B, m) query, where B is a specified sender and m is an incoming message.

A session is associated with its *holder* or *owner* (the party at which the session exists), a *peer* (the party with which the session key is intended to be established), and a *session identifier*. The session identifier is a quadruple (A, B, Out, In) where A is the identity of the holder of the session, B the peer, Out the outgoing messages from A in the session, and In the incoming messages from B. identified via a session *identifier* s. In particular, in the case of SM2 this results in an identifier of the form (A, B, R_A, R_B) where R_A is the outgoing DH value and R_B the incoming DH value to the session. The peer that sends the first message in a session is called the *initiator* and the other the *responder*. The session (A, B, R_A, R_B) (if it exists) is said to be matching to the session (B, A, R_B, R_A).

In addition to the activation of parties, $\mathcal{A}$ can perform the following queries:

- **corrupt**(P): The adversary $\mathcal{A}$ learns the long term private key of a party P.
- **session-key-reveal**(s): The adversary $\mathcal{A}$ obtains the session key for a session s, provided that the session holds a session key.
- **session-state-reveal**(s): The adversary $\mathcal{A}$ learns the session state information of a particular session s, but does not include the long-term private key of the party associated with s. In particular, if $Hash(x_1, \cdots, x_n)$ is computed in the session s, $x_1, \cdots, x_n$ are parts of the local state and can therefore be revealed by **session-state-reveal**(s) query.
- **test-session**(s): To respond to this query, a random bit b is selected. If $b = 1$ then the session key is output. Otherwise, a random key is output chosen from the probability distribution of keys generated by the protocol. This query can only be issued to a session that has not been *exposed*. A session is exposed if the adversary performs any of the following actions:
  (1) A **session-state-reveal** or **session-key-reaveal** query to this session or to the matching session, or
  (2) A **corrupt** query to either partner associated with this session.

The security is defined based on a game played by the adversary $\mathcal{A}$. In this game $\mathcal{A}$ interacts with the protocol. In a first phase of the game, $\mathcal{A}$ is allowed to activate sessions and perform **corrupt**, **session-key-reveal** and **session-state-reveal** queries as described above. The adversary then performs a **test-session** query to a party and session of its choice. The adversary is not allowed to expose the test session. $\mathcal{A}$ may then continue with its regular actions with the exception that no more **test-session** queries can be issued. Eventually, $\mathcal{A}$ outputs a bit b' as its guess, then halts. $\mathcal{A}$ wins the game if $b = b'$. The definition of security follows.

Definition 1. A key agreement protocol π is *session-key* secure if for any adversary $\mathcal{A}$ the following conditions hold:

(1) If two honest parties complete matching sessions, then they both compute the same session key;
(2) The probability that $\mathcal{A}$ guesses correctly the bit b is no more than 1/2 plus a negligible function in the security parameter.

Krawczyk [12] provided a stronger version of the Canetti-Krawczyk model that captures additional security properties, including resistance to key-compromise impersonation (KCI) attacks, weak perfect forward secrecy, and resilience to the leakage of ephemeral private keys. Further details of the model can be found in the original papers [12].

4 Weaknesses of SM2 Key Exchange Protocol

This section presents unknown key-share (UKS) attacks on SM2 key exchange protocol and also offers a security patch for the protocol.

4.1 UKS Attack I

In this UKS attack, an adversary $\mathcal{E}$ selects $e \in [1, n-1]$ randomly, and registers its own public key as $P_{\mathcal{E}} = P_A + [e]G$. Note that $\mathcal{E}$ does not know the private key $d_A + e$ corresponding to the public key $P_{\mathcal{E}}$. In SM2 protocol, it is not required a **proof of possession** of the corresponding private key from the verification of a public key. The attack scenario is outlined in Fig. 2. A more detailed description of the attack is as follows:

(1) $\mathcal{E}$ intercepts the message $\{R_A\}$ from A to B and replaces the identity A with $\mathcal{E}$.
(2) B chooses his r_B, and computes R_B, S_B and the session key $K_B = KDF(x_V \| y_V \| Z_{\mathcal{E}} \| Z_B, klen)$. During the computation of K_B, the adversary $\mathcal{E}$ uses **session-state-reveal** query to learn the input to KDF. In particular, the adversary learns x_V and y_V.
(3) $\mathcal{E}$ receives B's message $\{R_B, S_B\}$. Since $\mathcal{E}$ knows $V = (x_V, y_V)$, $\mathcal{E}$ will successfully compute the session key K_A of A:

$$\begin{aligned} U &= [h \cdot t_A](P_B + [\overline{x}_2]R_B) \\ &= [h \cdot (d_A + \overline{x}_1 \cdot r_A)](P_B + [\overline{x}_2]R_B) \\ &= [h \cdot (d_B + \overline{x}_2 \cdot r_B)](P_A + [\overline{x}_1]R_A) \\ &= [h \cdot (d_B + \overline{x}_2 \cdot r_B)](P_{\mathcal{E}} - [e]G + [\overline{x}_1]R_A) \\ &= [h \cdot (d_B + \overline{x}_2 \cdot r_B)](P_{\mathcal{E}} + [\overline{x}_1]R_A) - [h \cdot (d_B + \overline{x}_2 \cdot r_B)][e]G \\ &= V - [e](P_B + [\overline{x}_2]R_B) \\ &= (x_U, y_U) \\ K_A &= KDF(x_U \| y_U \| Z_A \| Z_B, klen) \end{aligned}$$

In addition, $\mathcal{E}$ computes $S'_B = Hash(0x02 \| y_U \| Hash(x_U \| Z_A \| Z_B \| x_1 \| y_1 \| x_2 \| y_2))$ and $S'_A = Hash(0x03 \| y_V \| Hash(x_V \| Z_{\mathcal{E}} \| Z_B \| x_1 \| y_1 \| x_2 \| y_2))$.
(4) $\mathcal{E}$ sends the message $\{R_B, S'_B\}$ to A as coming from B, and sends the message $\{S'_A\}$ to B.

As a result, A believes that a session key K_A is shared with B, but all the while with B thinking it is sharing a key K_B with $\mathcal{E}$. Thus, the SM2 protocol cannot resist our unknown key-share attack.

4.2 UKS Attack II

Attack I can be prevented if certificates are only issued to users who have shown that they know the private key corresponding to their public key. However, even this is not sufficient to prevent UKS attacks in all cases. In our UKS attack II, an adversary $\mathcal{E}$ is a legitimate client. $\mathcal{E}$ selects its private key $d_{\mathcal{E}} \in [1, n-1]$ randomly, and registers its own public key as $P_{\mathcal{E}} = [d_{\mathcal{E}}]G$. The attack scenario is outlined in Fig. 3. A more detailed description of the attack is as follows:

A	$\mathcal{E}$	B
$r_A \in_R [1, n-1]$		
$R_A = [r_A]G$		
$\xrightarrow{\{R_A\}}$		
	$e \in [1, n-1]$	
	$P_{\mathcal{E}} = P_A + [e]G$	
	$\xrightarrow{\{R_A\}}$	
		$r_B \in_R [1, n-1]$
		$R_B = [r_B]G$
		$t_B = (d_B + \overline{x}_2 \cdot r_B) \bmod n$
		$V = [h \cdot t_B](P_{\mathcal{E}} + [\overline{x}_1]R_A) = (x_V, y_V)$
		$S_B = Hash(0x02 \| y_V \| Hash(x_V \| Z_{\mathcal{E}} \| Z_B \| x_1 \| y_1 \| x_2 \| y_2))$
	$\xleftarrow{\{R_B, S_B\}}$	
	session-state-reveal $\rightarrow V = (x_V, y_V)$	
	$U = V - [e](P_B + [\overline{x}_2]R_B) = (x_U, y_U)$	
	$S'_B = Hash(0x02 \| y_U \| Hash(x_U \| Z_A \| Z_B \| x_1 \| y_1 \| x_2 \| y_2))$	
	$S'_A = Hash(0x03 \| y_V \| Hash(x_V \| Z_{\mathcal{E}} \| Z_B \| x_1 \| y_1 \| x_2 \| y_2))$	
$\xleftarrow{\{R_B, S'_B\}}$		$\xrightarrow{\{S'_A\}}$
$K_A = KDF(x_U \| y_U \| Z_A \| Z_B, klen)$		$K_B = KDF(x_V \| y_V \| Z_{\mathcal{E}} \| Z_B, klen)$

Fig. 2. Attack I on SM2 protocol

(1) $\mathcal{E}$ intercepts the message $\{R_A\}$ from A to B and then computes $R_{\mathcal{E}} = P_A + [\overline{x}_1]R_A = (x_3, y_3)$, where $\overline{x}_1 = 2^w + (x_1 \& (2^w - 1))$ and $w = \lceil(\lceil log_2(n)\rceil/2)\rceil - 1$. Next, $\mathcal{E}$ sends $\{R_{\mathcal{E}}\}$ to B.

(2) Upon receiving the message, B thinks that the protocol run is initiated by $\mathcal{E}$. Then B responds it with the message $\{R_B, S_B\}$ by computing R_B, S_B and the session key $K_B = KDF(x_V \| y_V \| Z_{\mathcal{E}} \| Z_B, klen)$. During the computation of K_B, the adversary $\mathcal{E}$ uses **session-state-reveal** query to learn the input to KDF. In particular, the adversary learns x_V and y_V.

(3) $\mathcal{E}$ receives B's message $\{R_B, S_B\}$. Since

$$\begin{aligned} V &= [h \cdot t_B](P_{\mathcal{E}} + [\overline{x}_3]R_{\mathcal{E}}) \\ &= [h \cdot (d_B + \overline{x}_2 \cdot r_B)](P_{\mathcal{E}} + [\overline{x}_3](P_A + [\overline{x}_1]R_A)) \\ &= [h \cdot d_{\mathcal{E}}](P_B + [\overline{x}_2]R_B) + [\overline{x}_3][h \cdot (d_B + \overline{x}_2 \cdot r_B)](P_A + [\overline{x}_1]R_A) \\ &= [h \cdot d_{\mathcal{E}}](P_B + [\overline{x}_2]R_B) + [\overline{x}_3]U \end{aligned}$$

and $\mathcal{E}$ knows $V = (x_V, y_V)$, $\mathcal{E}$ will successfully compute the session key K_A of A:

$$U = [\overline{x}_3]^{-1}(V - [h \cdot d_{\mathcal{E}}](P_B + [\overline{x}_2]R_B)) = (x_U, y_U)$$
$$K_A = KDF(x_U \| y_U \| Z_A \| Z_B, klen),$$

where $\overline{x}_3 = 2^w + (x_3 \& (2^w - 1))$. In addition, $\mathcal{E}$ computes $S'_B = Hash(0x02 \| y_U \| Hash(x_U \| Z_A \| Z_B \| x_1 \| y_1 \| x_2 \| y_2))$ and $S'_A = Hash(0x03 \| y_V \| Hash(x_V \| Z_{\mathcal{E}} \| Z_B \| x_1 \| y_1 \| x_2 \| y_2))$.

(4) $\mathcal{E}$ sends the message $\{R_B, S'_B\}$ to A as coming from B and sends the message $\{S'_A\}$ to B.

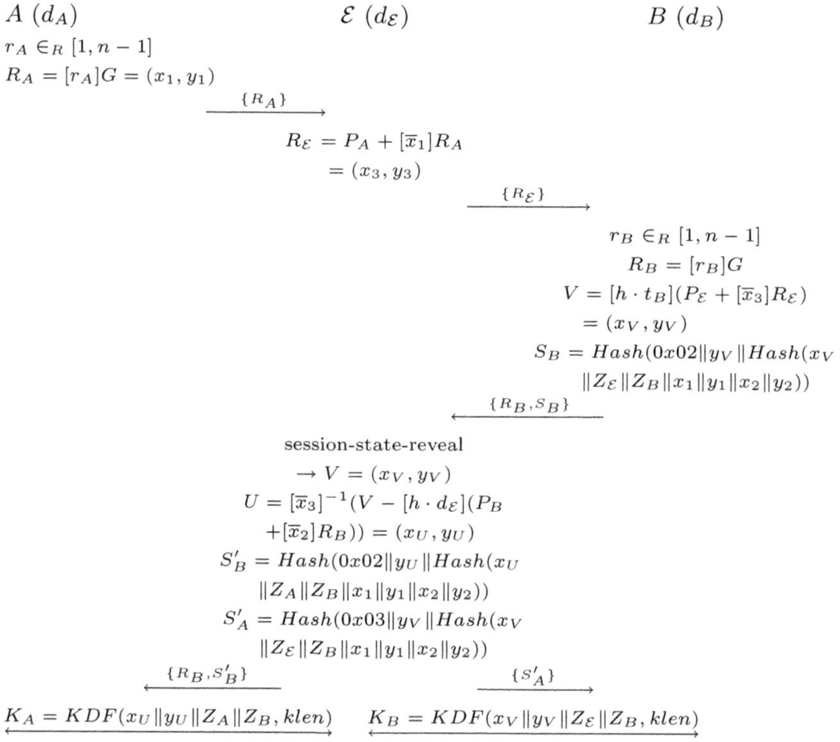

Fig. 3. Attack II on SM2 protocol

Clearly, A believes that a session key K_A is shared with B, but all the while with B thinking it is sharing a key K_B with $\mathcal{E}$. Compared to UKS attack I, this attack is much more effective. UKS attack I is readily prevented if the certificate authority requires proof of possession of the private key. The attack here succeeds despite such a requirement.

This attack is similar in spirit to Kaliski's attack [5] on the MQV protocol, however, the attack we present is more damaging. In Kaliski's attack, the adversary chooses its public and private keys after seeing A's initial value R_A. Meanwhile, in our attack, the public and private keys of the adversary are determined before the attack, which is *off-line*.

4.3 Formal Attack Description

We now interpret our UKS attack II in the context of the formal security model to show that the attack do indeed break the security of the protocol. The description of UKS attack I is omitted because of its similarity.

The UKS attack II on SM2 protocol is well captured in the Canetti-Krawczyk security model. Let again A and B denote two registered clients and $\mathcal{E}$ also be any registered client other than A and B. The goal of the adversary, denoted by $\mathcal{A}$, is to break the *session-key* security of the SM2 protocol. $\mathcal{A}$ begins by issuing **corrupt**($\mathcal{E}$) query to obtain the private key $d_{\mathcal{E}}$ of client $\mathcal{E}$. Then $\mathcal{A}$ issues **establish-session**(A, B) query to prompt instance $s_2 = (A, B, \times, \times)$ to initiate the protocol with client B. The rest of the oracle queries are straightforward from the attack scenario depicted in Fig.3. Table 2 summarizes the sequence of queries corresponding to our UKS attack II.

Clearly, $s_1 = (B, \mathcal{E}, R_B, R_{\mathcal{E}})$ and $s_2 = (A, B, R_A, R_B)$[1] are two non-matching sessions. Furthermore, the test session s_2 is not exposed because (1) no **corrupt** query has been asked for A or B, and (2) no **session-state-reveal** or **session-key-reveal** query has ever been made for the session s_2. Since $\mathcal{A}$ can compute the session key K_A of the session s_2, the probability that $\mathcal{A}$ guesses correctly the bit b used by the **test-session** query is 1. Therefore, $\mathcal{A}$ breaks the *session-key* security of the SM2 protocol.

4.4 Countermeasure

In the SM2 protocol, the client A cannot verify the real identity of the peer B. Therefore, the adversary $\mathcal{A}$ can corrupt any legitimate client to mount our described UKS attacks, and break the *session-key* security of the SM2 protocol. The weaknesses of the SM2 key exchange protocol are mainly due to the fact that the identities of the initiator and the responder are not appropriately integrated into the exchanged cryptographic messages. A simple improvement is to include the party's own identity and the peer's identity as input parameters for computing $\overline{x}_1$ and $\overline{x}_2$. Specifically, the values $\overline{x}_1 = 2^w + (x_1 \& (2^w - 1))$ and $\overline{x}_2 = 2^w + (x_2 \& (2^w - 1))$ in the SM2 protocol are replaced with $\overline{x}_1 = \overline{H}(x_1 \| Z_A \| Z_B)$ and $\overline{x}_2 = \overline{H}(x_2 \| Z_B \| Z_A)$, where $\overline{H}()$ is a one-way hash function whose output is $w = \lceil(\lceil log_2(n)\rceil/2)\rceil - 1$ bits.

Evidently, our improvement can effectively resist the UKS attack. For the adversary $\mathcal{E}$ who learns $V = (x_V, y_V)$, it is infeasible to compute $U = (x_U, y_U)$

[1] For simplicity, S_B, S'_A and S'_B are not included in the sessions s_1 and s_2—this is without loss of generality as they are optional parameters in SM2 protocol.

Table 2. The sequence of oracle queries corresponding to UKS attack II

	Query	Response
1	corrupt $(\mathcal{E})$	$d_{\mathcal{E}}$
2	establish-session (A, B)	$\{R_A\}$
3	send $(B, \mathcal{E}, \{R_{\mathcal{E}}\})$	$\{R_B, S_B\}$
4	session-state reveal $(s_1 = (B, \mathcal{E}, R_B, R_{\mathcal{E}}))$	$V = (x_V, y_V)$
5	send $(A, B, \{R_B, S'_B\})$	$\{S_A\}(accept)$
6	send $(B, \mathcal{E}, \{S'_A\})$	$(accept)$
7	test $(s_2 = (A, B, R_A, R_B))$	K_A

and then obtain the shared key $K_A = KDF(x_U \| y_U \| Z_A \| Z_B, klen)$ with A. This is due to the fact that the relation between U and V (as described in Section 4.2 and 4.3) does not exist.

5 Conclusion

In this paper, we have shown that the SM2 key exchange protocol is potentially vulnerable to an unknown key-share attack. The weakness is due to the fact that the identifiers of the communicants are not appropriately integrated into the exchanged cryptographic messages. However, the attack presented here should not discourage use of the SM2 protocol, as long as appropriate countermeasures are taken. We also have suggested a simple modification to resist our described attacks while the merits of the original protocol are left unchanged.

Acknowledgements. This work was supported by the National Grand Fundamental Research (973) Program of China under grant 2007CB311202, and the National Natural Science Foundation of China (NSFC) under grants 61170279 and 60873197. The authors would like to thanks the anonymous referees for their helpful comments.

References

1. Diffie, W., Hellman, H.: New directions in cryptography. IEEE Transactions of Information Theory 22(6), 644–654 (1976)
2. Diffie, W., van Oorschot, P., Wiener, M.: Authentication and authenticated key exchanges. Designs, Codes and Cryptography 2(2), 107–125 (1992)
3. Menezes, A., Qu, M., Vanstone, S.: Some new key agreement protocols providing mutual implicit authentication. In: Proceedings of the Second Workshop on Selected Areas in Cryptography (SAC 1995), pp. 22–32 (1995)
4. Blake-Wilson, S., Menezes, A.: Unknown Key-Share Attacks on the Station-to-Station (STS) Protocol. In: Imai, H., Zheng, Y. (eds.) PKC 1999. LNCS, vol. 1560, pp. 154–170. Springer, Heidelberg (1999)
5. Kaliski, B.: An unknown key-share attack on the MQV key agreement protocol. ACM Transactions on Information and System Security (TISSEC) 4(3), 275–288 (2001)

6. Lauter, K., Mityagin, A.: Security Analysis of KEA Authenticated Key Exchange Protocol. In: Yung, M., Dodis, Y., Kiayias, A., Malkin, T. (eds.) PKC 2006. LNCS, vol. 3958, pp. 378–394. Springer, Heidelberg (2006)
7. Law, L., Menezes, A., Qu, M., Solinas, J., Vanstone, S.: An efficient protocol for authenticated key agreement. Designs, Codes and Cryptography 28, 119–134 (2003)
8. Okamoto, T.: Authenticated Key Exchange and Key Encapsulation in the Standard Model. In: Kurosawa, K. (ed.) ASIACRYPT 2007. LNCS, vol. 4833, pp. 474–484. Springer, Heidelberg (2007)
9. LaMacchia, B.A., Lauter, K., Mityagin, A.: Stronger Security of Authenticated Key Exchange. In: Susilo, W., Liu, J.K., Mu, Y. (eds.) ProvSec 2007. LNCS, vol. 4784, pp. 1–16. Springer, Heidelberg (2007)
10. Ustaoglu, B.: Obtaining a secure and efficient key agreement protocol for (H)MQV and NAXOS. Designs, Codes and Cryptography 46(3), 329–342 (2008)
11. Cremers, C.J.F.: Session-State Reveal is Stronger than Ephemeral Key Reveal: Attacking the NAXOS Key Exchange Protocol. In: Abdalla, M., Pointcheval, D., Fouque, P.-A., Vergnaud, D. (eds.) ACNS 2009. LNCS, vol. 5536, pp. 20–33. Springer, Heidelberg (2009)
12. Krawczyk, H.: HMQV: A High-Performance Secure Diffie-Hellman Protocol. In: Shoup, V. (ed.) CRYPTO 2005. LNCS, vol. 3621, pp. 546–566. Springer, Heidelberg (2005)
13. ANSI X9.42, Public Key Cryptography for the Financial Services Industry: Agreement of Symmetric Keys Using Discrete Logarithm Cryptography. American National Standards Institute (2003)
14. ANSI X9.63, Public Key Cryptography for the Financial Services Industry: Key Agreement and Key Transport Using Elliptic Curve Cryptography. American National Standards Institute (2001)
15. SP 800-56A Special Publication 800-56A, Recommendation for Pair-Wise Key Establishment Schemes Using Discrete Logarithm Cryptography. National Institute of Standards and Technology (March 2006)
16. Krawczyk, H.: "HMQV in IEEE P1363", submission to the IEEE P1363 working group (July 7, 2006), `http://grouper.ieee.org/groups/1363/P1363-Reaffirm/submissions/krawczyk-hmqv-spec.pdf`
17. Public Key Cryptographic Algorithm SM2 Based on Elliptic Curves, Part 3: Key Exchange Protocol (in Chinese), `http://www.oscca.gov.cn/UpFile/2010122214822692.pdf`
18. Bellare, M., Canetti, R., Krawczyk, H.: A modular approach to the design and analysis of authentication and key exchange protocols. In: Proceedings of the Thirtieth Annual ACM Symposium on Theory of Computing, pp. 419–428 (1998)
19. Canetti, R., Krawczyk, H.: Analysis of Key-Exchange Protocols and their use for Building Secure Channels. In: Pfitzmann, B. (ed.) EUROCRYPT 2001. LNCS, vol. 2045, pp. 453–474. Springer, Heidelberg (2001)

Cryptanalysis of a Provably Secure Cross-Realm Client-to-Client Password-Authenticated Key Agreement Protocol of CANS '09*

Wei-Chuen Yau[1], Raphael C.-W. Phan[2,**], Bok-Min Goi[3], and Swee-Huay Heng[4]

[1] Faculty of Engineering,
Multimedia University,
63100 Cyberjaya, Malaysia
wcyau@mmu.edu.my

[2] Electronic, Electrical & Systems Engineering,
Loughborough University,
LE11 3TU Leicestershire, UK
r.phan@lboro.ac.uk

[3] Faculty of Engineering & Science,
Universiti Tunku Abdul Rahman,
53300 KL, Malaysia
goibm@utar.edu.my

[4] Faculty of Information Science & Technology,
Multimedia University,
75450 Melaka, Malaysia
shheng@mmu.edu.my

Abstract. In this paper, we cryptanalyze the recent smart card based client-to-client password-authenticated key agreement (C2C-PAKA-SC) protocol for cross-realm settings proposed at CANS '09. While client-to-client password-authenticated key exchange (C2C-PAKE) protocols exist in literature, what is interesting about this one is that it is the only such protocol claimed to offer security against password compromise impersonation without depending on public-key cryptography, and is one of the few C2C-PAKE protocols with provable security that has not been cryptanalyzed. We present three impersonation attacks on this protocol; the first two are easier to mount than the designer-considered password compromise impersonation. Our results are the first known cryptanalysis results on C2C-PAKA-SC.

Keywords: Client-to-client, password-authenticated key agreement, cross realm, impersonation, attack.

* Research partially funded by the Ministry of Science, Technology & Innovation (MOSTI) under grant no. MOSTI/BGM/R&D/500-2/8.

** Part of this work done while the author was visiting Multimedia University.

D. Lin, G. Tsudik, and X. Wang (Eds.): CANS 2011, LNCS 7092, pp. 172–184, 2011.

1 Introduction

The ease and low cost with which to access to networks through mobile devices has led to the current trend of users tending to stay connected while on the move. Mobile users (clients) roam freely within wireless networks from one realm (network) to another. For this case, cross-realm networks allow a client registered with a server in one realm to be still able to communicate securely with clients in a foreign realm that it roams into. In fact, this is a more realistic scenario than assuming all clients interacting with only one server, especially in current environments where different types of network infrastructures co-exist.

To establish a secure end-to-end communication between these clients of different realms, the basic approach is to allow them to share a secret via a key exchange protocol, and ideally to have them be able to authenticate each other despite being in different realms. Inherent to these kinds of networks involving human users, authenticated key exchange protocols use human-memorizable passwords.

A 2-party password-based authenticated key exchange (PAKE) protocol establishes a shared secret key between two parties. Authentication of parties is based on knowledge of a shared low-entropy password. The first known PAKE is due to Bellovin and Merritt [7]. This concept has also been extended to 3 parties, e.g. two clients and a trusted server or key distribution center (KDC).

While most existing literature consider PAKEs between a client and a server, Byun *et al.* [8] at ICICS '02 highlighted the need for PAKEs that allow to establish a secure end-to-end (client-to-client) channel between clients even in cross-realm networks. The basic idea is to use servers in the different realms as the go-between, i.e. to perform translation of encrypted or blinded secrets in one realm to the other under passwords shared between client and realm server and secret key shared between the realm servers. Such protocols are more popularly known as *cross-realm* C2C-PAKE protocols. For ease of notation, we will simply call these C2C-PAKEs for the rest of this paper.

Considering this cross realm setting, several additional security issues arise that would otherwise not be relevant in a single realm setting, e.g. protecting secrets of the client in one realm from a malicious server [13,20] or a malicious client [22,20] in the other realm. For more details of the variants and analyses, see [8,13,29,22,24,31,10,30,26,12,11,19,20].

1.1 Related Work and Motivation

PROVABLE SECURITY MODELS. The formal model for proving the security of 2-party PAKE protocols was proposed by Bellare *et al.* [6] so called the Bellare-Pointcheval-Rogaway (BPR2000) model, building on work by Bellare and Rogaway in [4,5]. Later, Abdalla *et al.* [2] extended this model to the 3-party case.

One informal approach to designing security protocols is to list all known attacks and argue why a protocol resists them. This list is not exhaustive, and sometimes fails to catch specific types of attacks. The main problem is that this heuristic approach assumes the particular behaviour of the adversary, i.e. he is assumed to attack in some way. History [6] has shown that this is not the right approach, because intuitively an adversary behaves in any way he prefers as long as he can break the system. Thus it is often that such a protocol is broken and a minor fix proposed, etc. This cycle continues resulting in many slightly different protocol variants because breaks and subsequent fixes are heuristically done.

In contrast the approach based on formal security models does not assume on any specific attack method an adversary may use. Instead a communication model is defined that describes how parties within the protocol, as well as an adversary, communicate with each other, and what sort of information formalized via the notion of oracle queries, is available to or may be under the control of the adversary. Then, security properties of a protocol are defined as one or more games each intended to capture a security property, played by the adversary within the pre-defined communication model. A protocol is secure with respect to the defined security properties if the adversary's advantage in winning the game(s) is negligible, and further that the task of an adversary winning is reduced to computationally intractable assumption(s). This approach is also known as *provable security* [28]. Once proven secure with respect to a particular defined security property, a protocol is guaranteed to resist attacks aimed to break the property by any adversary who works within the communication model regardless of what specific attacks are mounted, as long as the assumptions remain intractable.

However, defining an appropriate model is not a trivial task [27], because not including some types of queries e.g. the Corrupt query [14,15], or improperly defining the adversarial game [6] may result in a security proof that fails to capture valid attacks (see [6,14,15] for more details).

C2C-PAKE PROTOCOLS. The original C2C-PAKE protocol by Byun *et al.* [8] at ICICS '02 builds on the cross-realm extension [16] of the popular Kerberos network authentication protocol which is in turn based on the celebrated Needham-Shroeder protocol [23]. Then in 2006, Byun *et al.* [10], and Yin and Bao [30] independently proposed the first provably secure C2C-PAKE protocols, called EC2C-PAKE and C2C-PAKE-YB respectively. By a provably secure protocol, we mean one whose security is proven formally in a well-defined security model along the style discussed in the previous subsection. In [26], undetectable online dictionary attacks were mounted on both these protocols. Arguably, the practical significance of such types of attacks requires further investigation, as these online dictionary attacks require the adversary to be online during the attack to interact with a legitimate party in a protocol run. This contrasts with offline dictionary attacks that an adversary can just run offline without needing to be online in any protocol run. Subsequent to [10] and [30], Byun *et al.* [11] proposed a security model for C2C-PAKE protocols and presented the EC2C-PAKA scheme with security within that model. Feng and Xu [19] recently remarked that EC2C-PAKA does not achieve security against password compromise impersonation, and proceeded to propose a model that captures this kind of attack and then presented the C2C-PAKA scheme using public key cryptography that was proven secure within their model.

Jin and Xu [20] later observed that the Feng-Xu C2C-PAKA protocol is inefficient since it requires public key cryptography, while previous C2C-PAKE protocols only required symmetric cryptography. They thus proposed C2C-PAKA-SC that works without public key cryptography but rather utilizes smart cards in order to achieve security against password compromise impersonation attacks, and its security was proven within a model that captures password compromise impersonation. Jin and Xu also acknowledged [20] that it is desirable for C2C-PAKA-SC protocols to achieve unknown key-share resilience [17,21].

Our contributions. We analyze in detail the security of Jin and Xu's C2C-PAKA-SC. In doing so, we advocate that while provable security is the right approach to the analysis and design of C2C-PAKEs, and PAKEs in general, we nevertheless caution that proving such formal security is not an easy task. Already, some provably secure protocols have been shown [14,15] to exhibit flaws because of subtle points missed out in the security model used to conduct the proofs.

More precisely, we present three different attacks on C2C-PAKA-SC that work within its defined adversarial security model. These attacks are impersonation attacks and the first two imply unknown key-share attacks [17,21] since C2C-PAKA-SC is an authenticated key exchange protocol.

2 The C2C-PAKA-SC Protocol

In this section, we describe the C2C-PAKA-SC protocol of Jin and Xu [20] that we will analyze in Section 4. We will use the notations given in Table 1. Unless otherwise mentioned, all described operations are performed modulo p, except operations in the exponents, and all protocols are based on Diffie-Hellman (DH) type assumptions.

Basically the C2C-PAKE protocols operate in a setting where a KDC exists with many clients in each realm, and where each client shares a unique password with its realm KDC and each pair of KDCs of different realms share a secret key K. The original C2C-PAKE protocol by Byun *et al.* in [8] follows closely the design principle of Kerberos in which a client interacts with a KDC to obtain a ticket that leads to establishing a secret key for sharing between this client and another client. The subsequent provably-secure variants EC2C-PAKE [10], EC2C-PAKA [11], C2C-PAKA [19] and C2C-PAKA-SC [20] also follow this principle; while another i.e. C2C-PAKE-YB [30] uses a different paradigm because it is basically an extension of the 3-party PAKE in [3].

During the registration phase of Jin and Xu's C2C-PAKA-SC protocol, each client i registers with its realm KDC_i by inputting its identity ID_i and password pw_i, and obtains a smart card through a secure channel, within which are stored the values $\langle ID_i, R_i, H_1(\cdot), p\rangle$, where $R_i = H_1(ID_i)^x + H_1(pw_i)$, and x the secret key of KDC_i.

Table 1. Notations

A, B	The clients
ID_i	The identity of client i
KDC_i	Key distribution center which stores the ID_i and password (pw_i) of client i in its realm
pw_i	Client i's human-memorizable password shared with KDC_i
K	The symmetric secret key shared between different KDCs
x, y	The secret keys of KDC_A and KDC_B respectively
$E_K(\cdot)$	Symmetric encryption using the secret key, K
p, q	Sufficiently large primes such that $q \mid p-1$
g	The generator of a finite subgroup G of Z_p^* of order q
H_i	Cryptographic hash functions, $i = \{1, 2, 3\}$
$Ticket_i$	Ticket for receiving party i, equal to $E_K(k, ID_j, ID_i, L)$ where k is a random element of Z_q^*, L is the lifetime of $Ticket_i$ and ID_j the identity of the sender party
$MAC_K(\cdot)$	A message authentication code using the secret key, K
$\|\|$	Message concatenation
$a \in_\$ Z_q^*$	Randomly choosing an element a of Z_q^*

The main bulk of the protocol, i.e. its login and authentication phase is concisely shown in Figure 1, and consists of the following steps:

1. A attaches his smart card to a device reader and inputs his ID_A and password pw_A. The device obtains the stored value R_A from the smart card, and generates a random $\alpha \in Z_q^*$. It then computes $R'_A = (R_A - H_1(pw_A))^\alpha$, $W_A = H_1(ID_A)^\alpha$ and $C_A = H_1(T_1||R'_A||W_A||ID_A)$, where T_1 is a timestamp. The message $\langle ID_A, ID_B, T_1, C_A, W_A\rangle$ is then sent to KDC_A.
2. KDC_A checks the validity of T_1 and ID_A, and then computes its version of R'_A as $R''_A = W_A^x$. This computed R''_A is used together with the received values to compute $H_1(T_1||R''_A||W_A||ID_A)$ which is checked against the received C_A. If the check matches, then KDC_A proceeds to generate a random $k \in Z_q^*$ and computes $K_A = H_1(R''_A \oplus T_2)$ to be used as a password-based shared symmetric key between KDC_A and A. KDC_A then computes $V_A = E_{K_A}(k, ID_A, ID_B)$ and $Ticket_B = E_K(k, ID_A, ID_B, L)$ where L is the lifetime of $Ticket_B$. The message $\langle V_A, Ticket_B, T_2, L\rangle$ is sent to A.
3. A checks the validity of T_2, and computes $K'_A = H_1(R'_A \oplus T_2)$. It uses this to decrypt the received V_A in order to get k and be able to verify the validity of the decrypted ID_A, ID_B. It generates a random $a \in Z_q^*$ and computes $E_a = g^a||MAC_k(g^a)$. The message $\langle ID_A, E_a, Ticket_B\rangle$ is sent to B.
4. B attaches his smart card to a device reader and inputs his ID_B and password pw_B. The device reads the stored value R_B from the smart card and generates a random $\beta \in Z_q^*$; these are used to compute $R'_B = (R_B - H_2(pw_B))^\beta$, $W_B = H_2(ID_B)^\beta$ and $C_B = H_2(T_3||R'_B||W_B||ID_B)$. The message $\langle Ticket_B, T_3, C_B, W_B\rangle$ is sent to KDC_B.

5. On receipt of the message from B, then KDC_B checks the validity of T_3 and decrypts $Ticket_B$ to obtain k, and checks the decrypted IDs and lifetime L. It computes its version of R'_B as $R''_B = W_B^y$, and then uses this as well as other received values to compute $H_2(T_3||R''_B||W_B||ID_B)$ to check against the received C_B. KDC_B then computes $K_B = H_2(R''_B \oplus T_4)$ and uses this to compute $V_B = E_{K_B}(k, ID_A, ID_B)$. The message $\langle V_B, T_4 \rangle$ is sent to B.
6. B checks the validity of T_4 and computes $K'_B = H_2(R'_B \oplus T_4)$ to be used for decrypting V_B. This allows B to get k and check the validity of the decrypted identities ID_A and ID_B. B generates a random $b \in Z_q^*$ and computes $E_b = g^b||MAC_k(g^b)$; and $\langle E_b \rangle$ is sent from B to A.
7. Both A and B compute the session key as $sk = H_3(ID_A||ID_B||g^a||g^b||g^{ab})$.

3 Adversarial Capability in the C2C-PAKA-SC Security Model

Here we briefly review the adversarial capability of the security model of Jin-Xu [20] within which the C2C-PAKA-SC protocol was proven secure. The adversary is assumed to have oracle access to standard PAKE-style queries, and the Corrupt query includes consideration of the adversary obtaining access to information stored within a smart card.

- Execute(i, j, KDC_i, KDC_j, s): models passive eavesdropping on protocol session s involving clients i, j and their KDC_i, KDC_j.
- Reveal(i, s): models the ability of the adversary to obtain session keys established for some session s involving client i.
- Corrupt$(i, choice)$: models the ability of the adversary to obtain some long-term secret password or secret information stored in smart card of client i, where the index $choice$ denotes which secret is obtained, i.e. $choice = 1$ outputs the password pw_i to the adversary while $choice = 2$ outputs the information stored on the smart card of client i, i.e. $\langle ID_i, R_i, H_1(\cdot), p \rangle$.
- SendClient(i, s, m): models active attacks against the client i, i.e. a message m is sent to client i in some protocol session s.
- SendServer(KDC_i, s, m): models active attacks against the KDC_i, i.e. a message m is sent to KDC_i in some protocol session s.
- Test(i, s): defines the semantic security of the established key for session s involving client i. A coin b is flipped and this determines if the real session key (if $b = 1$) or a random key (if $b = 0$) is given to the adversary. This query is valid only on sessions for which no Reveal or Corrupt queries have been issued for client i or its session partner j both of which have accepted with partner IDs being $pid_{i,s} = pid_{j,s} = \langle i, j \rangle$, i.e. completed the protocol run without realizing anything is wrong and have established the key believed to be shared with the other partnering party.

KDC_A A B KDC_B

A:

$\alpha \in_{\$} Z_q^*$

$R'_A = (R_A - H_1(pw_A))^{\alpha}$

$W_A = H_1(ID_A)^{\alpha}$

$C_A = H_1(T_1 || R'_A || W_A || ID_A)$

$\xleftarrow{ID_A, ID_B, T_1, C_A, W_A}$

KDC_A:

Verify T_1, ID_A

$R''_A = W_A^x$

$C_A \stackrel{?}{=} H_1(T_1 || R''_A || W_A || ID_A)$

$k \in_{\$} Z_q^*$

$K_A = H_1(R''_A \oplus T_2)$

$V_A = E_{K_A}(k, ID_A, ID_B)$

$Ticket_B = E_K(k, ID_A, ID_B, L)$

$\xrightarrow{V_A, Ticket_B, T_2, L}$

A:

Verify T_2

$K'_A = H_1(R'_A \oplus T_2)$

Decrypt V_A

Verify ID_A, ID_B

$a \in_{\$} Z_q^*$

$E_a = g^a || MAC_k(g^a)$

$\xrightarrow{ID_A, E_a, Ticket_B}$

B:

$\beta \in_{\$} Z_q^*$

$R'_B = (R_B - H_2(pw_B))^{\beta}$

$W_B = H_2(ID_B)^{\beta}$

$C_B = H_2(T_3 || R'_B || W_B || ID_B)$

$\xrightarrow{Ticket_B, T_3, C_B, W_B}$

KDC_B:

Verify T_3

Decrypt $Ticket_B$

$R''_B = W_B^y$

$C_B \stackrel{?}{=} H_2(T_3 || R''_B || W_B || ID_B)$

$K_B = H_2(R''_B \oplus T_4)$

$V_B = E_{K_B}(k, ID_A, ID_B)$

$\xleftarrow{V_B, T_4}$

B:

Check T_4

$K'_B = H_2(R'_B \oplus T_4)$

Decrypt V_B

Check ID_A, ID_B

$b \in_{\$} Z_q^*$

$E_b = g^b || MAC_k(g^b)$

$\xleftarrow{E_b}$

A: $sk = H_3(ID_A || ID_B || g^a || g^b || g^{ab})$ **B:** $sk = H_3(ID_A || ID_B || g^a || g^b || g^{ab})$

Fig. 1. The C2C-PAKA-SC Protocol

4 Cryptanalysis of the C2C-PAKA-SC

In this section we describe the results of our cryptanalysis of the C2C-PAKA-SC. In more detail, we present three impersonation attacks.

4.1 By any Outsider C Impersonating A to B

This attack can be mounted by any outsider (denoted as C) and allows C to impersonate A to B. It proceeds as follows:

1. Adversary C calls the Execute query to eavesdrop on a valid protocol session between A and B, to obtain E_a and $Ticket_B$.
2. C initiates a new protocol run with B and calls the SendClient(B) query to replay the E_a and $Ticket_B$ obtained from Step 1 above.
3. The protocol steps continue as per normal protocol run.

In the end, B thinks it is sharing a key with A after the protocol run when A is in fact not present. This also corresponds to the goal of an unknown key-share attack; see [17,21] for more details, wherein some application scenarios are discussed.

In more detail, KDC_B cannot detect anything wrong and therefore believes all is normal because

- it checks T_3's freshness (but this is a timestamp generated by B so it is fresh)
- it can decrypt $Ticket_B$ properly because $Ticket_B$ is rightly generated by KDC_A
- C_B is verified because it was generated by B

Similarly, B cannot detect anything wrong because it has no ability to directly verify A's message initially, and then subsequently only checks KDC_B's message i.e.

- T_4 is fresh because it was generated by KDC_B
- V_B can be decrypted
- ID_A and ID_B are correct
- $MAC_k(g^a)$ resp. $MAC_k(g^b)$ are verified correctly since k is simply generated by KDC_A and contained in the replayed $Ticket_B$ that KDC_B decrypts and passes on to B.

The reason impersonation works is because during the protocol run, the only message required by B from A's realm is $\langle ID_A, E_a, Ticket_B\rangle$. E_a can be replayed because it is a function only of ephemeral values k and a. $Ticket_B$ can be replayed because it is a function only of the ephemeral value k and static values ID_A, ID_B, as long as it is within the lifetime L which cannot be too short since $Ticket_B$ needs to remain valid from the start of KDC_A's message transmission, through A and B to KDC_B.

The C2C-PAKA-2C scheme is both an authentication and key agreement scheme, thus the above attack demonstrates that its desired authentication and unknown key-share resilience requirement (the latter was explicitly listed as its desired requirement) are broken.

This attack is within the adversarial security model of C2C-PAKA-SC and in fact, does not even require any of the adversarial model's Corrupt or Reveal queries.

Another variant of this attack can be mounted by any outsider C allowing to not only impersonate A to B but also break the scheme in the sense of semantic security; semantic security of the scheme was proven in [20, Theorem 4.1]. This attack works as follows (see Figure 2):

1. Adversary C calls the Execute query on (A, B, s_1) to eavesdrop on a valid protocol session (of session ID s_1) between A and B, to obtain E_a and $Ticket_B$.
2. C impersonating A initiates a new protocol run with B and calls the Send-Client query on $(B, s_2, \cdot)$ to replay the E_a and $Ticket_B$ obtained from Step 1 above.
3. B replies to A with E_b, but it is intercepted by C.
4. B replies to C with $E_{b'}$; and completes protocol session s_2 in accepted state, i.e. it thinks it has completed a normal protocol session with A and computes the session key $sk_{s_2} = H(ID_A||ID_B||g^a||g^{b'}||g^{ab'})$.
5. C calls the SendClient query on $(A, s_1, E_{b'})$ to replay $E_{b'}$ to A. Then A completes protocol session s_1 in accepted state, i.e. it thinks it has completed a normal protocol session with B and computes the session key $sk_{s_1} = H_3(ID_A||ID_B||g^a||g^{b'}||g^{ab'})$.
6. C is able to issue a Reveal query on the session (A, s_1) and obtains the key $sk_{s_1} = H_3(ID_A||ID_B||g^a||g^{b'}||g^{ab'})$ for session s_1.
7. C issues the Test query to the session (B, s_2) and obtains the key sk for session s_2 which could be $sk_{s_2} = H(ID_A||ID_B||g^a||g^{b'}||g^{ab'}) = sk_{s_1}$ or a randomly generated one.
8. If sk equals sk_{s_1} obtained from step 6, then C outputs 1. Else, it outputs 0.

As can be seen, at the end of the game C is able to output the guess of b correctly with probability 1, thus breaking semantic security.

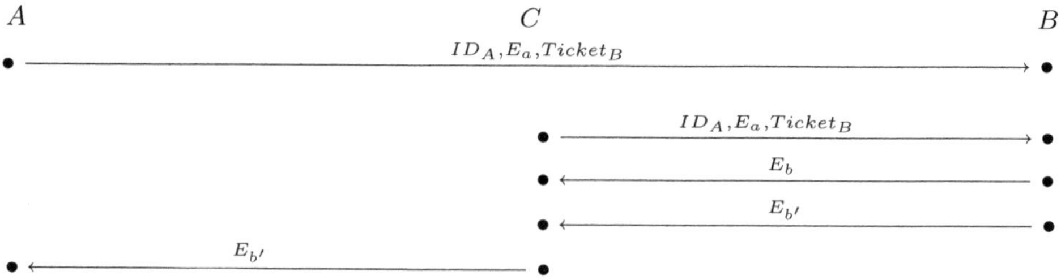

Fig. 2. Attack on Semantic Security

4.2 By Any Outsider C Impersonating B to A

An attack also exists allowing any outsider to impersonate B to A. The steps are as follows:

1. A initiates a protocol run, and the protocol flows normally.
2. When A sends a message $\langle ID_A, E_a, Ticket_B \rangle$ to B, the adversary C intercepts this and simply replies this back to A as $E_b = E_a$ via queries to the SendClient oracle.

The rest of the protocol proceeds normally, and C ends up authenticated to A as B with A thinking it has established a shared key with B.

The reason this simple attack works is because of the same structure between E_a and E_b and because the key k used within the MAC of E_a, E_b is not jointly established, thus even if B is absent, yet A's presence and thus its E_a is sufficient to be reused as E_b.

4.3 By Any Insider Client $B \neq A$ Impersonating A to KDC_A

The above two attacks considered violating the authentication security between clients A and B. In essence, A and B establish a shared session key not directly, but via their respective KDCs. More precisely, this is done by using the authentication security (via 2-party PAKE) between each client and its KDC based on the client's password being known to its KDC, and then by using the security between the KDCs based on their shared secret key K.

Going further, we show that even if there exists a 2-party PAKE type authentication security between a client A and its KDC based on the KDC knowing the client's password, yet it is still possible for any other client B of the same realm to impersonate A to its KDC without needing to know what the impersonated client password pw_A is.

Our attack works as follows, using adversarial capability as defined within the C2C-PAKA-SC security model:

1. B performs a Corrupt$(B, 2)$ query to access to stored content on his/her own smart card to get R_B.
2. B computes $R_B - H_1(pw_B) = H_1(ID_B)^x$, and can then compute $R'_A = H_1(ID_B)^x$, $W_A = H_1(ID_B)$ and $C_A = H_1(T_1|||R'_A||W_A||ID_A)$. Note that since B is an insider client, this implies it would obviously know its own password pw_B.
3. B issues a SendServer query to send the message $\langle ID_A, ID_B, T_1, C_A, W_A \rangle$ to KDC_A.
4. KDC_A will compute $R''_A = W_A^x = H_1(ID_B)^x = R'_A$, so its check of $C_A \stackrel{?}{=} H_1(T_1||R''_A||W_A||ID_A)$ will match and therefore it does not detect that anything is wrong.

Thus KDC_A ends up thinking A has authenticated itself when in fact it is another client B of the same realm; and B was able to impersonate A without needing to know A's password.

The problem here is that by right the authentication of A to KDC_A should only be possible with the knowledge of the password pw_A, yet there is a flaw in the way that KDC_A checks this password knowledge. More precisely, the checking is based on computing R''_A from the value W_A supplied by A (or someone claiming to be A) and checking this computed R''_A for equality against the received R'_A. While the computations of R'_A and R''_A differ, by design A can only compute R'_A via his knowledge of the password pw_A while KDC_A can only compute R''_A via knowledge of his own secret x; and irrespective of how the computation is done, the two values should match. Yet, our above attack shows that even without knowing the password pw_A, an adversary could generate the proper W_A and R'_A such that the check by KDC_A is satisfied, because essentially one only needs to be able to generate a pair of values W_A, R'_A such that the latter is the xth power of the former.

5 Concluding Remarks

Since 2006 [10,30], provable security models and protocols have been proposed in literature for C2C-PAKE, including some recent ones. These are nice results, since research has shown that ad hoc protocols fall to attacks over time. Nevertheless, provable security models and proofs for PAKE protocols should be done carefully to avoid miscatching known attacks. An example is the first provably secure n-party PAKE protocol in the DPWA setting [9] that was shown in [25] to fall to attacks that it was designed to resist, while other examples are in [14,15]. Although the responsibility rests on protocol designers to carefully define adequate security models and check the correctness of their security proofs, the community in particular protocol implementers should exercise caution when interpreting provable security models and proofs. Experience in the analysis and design of security protocols [1,6,14,15] has shown that even seemingly sound models, designs and proofs may exhibit problems, and though provable security is the right approach, years of public scrutiny should still complement the process before a protocol is deemed secure, or a model is deemed sufficient.

Our results also demonstrate that newer "improved" variants are not necessarily more securely designed. For instance, our attacks on C2C-PAKA-SC are more severe than previous attacks on its predecessor the Feng-Xu C2C-PAKA protocol; e.g. our first two attacks do not require knowledge of any secrets nor passwords, and counter-intuitively the original C2C-PAKA is not vulnerable to these attacks. Thus, we conclude that C2C-PAKA-SC should not be deemed secure for practical cross-realm applications.

Acknowledgements and Epilogue. We thank God for His many blessings. We thank the anonymous reviewers for their comments, notably those of Reviewers 1 and 3 which made this line of research much worth the while. Quoting herein: "mercifully does not attempt to go on to fix the now broken system", indeed we agree wholeheartedly. The literature is rich with many variants of Cross-Realm C2C-PAKA/E protocols, fortunately some of which are accompanied with proofs of security. Rather than propose fixes to broken systems especially when unbroken variants still exist, it is crucial to extract the gist of what lessons can be learnt from this experience. And beyond that, it is a good idea to move on to designing for other settings where less cryptographic protocols are known to exist.

References

1. Abadi, M.: Explicit Communication Revisited: Two New Attacks on Authentication Protocols. IEEE Transactions on Software Engineering 23(3), 185–186 (1997)
2. Abdalla, M., Fouque, P.-A., Pointcheval, D.: Password-Based Authenticated Key Exchange in the Three-Party Setting. In: Vaudenay, S. (ed.) PKC 2005. LNCS, vol. 3386, pp. 65–84. Springer, Heidelberg (2005)
3. Abdalla, M., Pointcheval, D.: Interactive Diffie-Hellman Assumptions with Applications to Password-Based Authentication. In: S. Patrick, A., Yung, M. (eds.) FC 2005. LNCS, vol. 3570, pp. 341–356. Springer, Heidelberg (2005)
4. Bellare, M., Rogaway, P.: Entity Authentication and Key Distribution. In: Stinson, D.R. (ed.) CRYPTO 1993. LNCS, vol. 773, pp. 232–249. Springer, Heidelberg (1994)
5. Bellare, M., Rogaway, P.: Provably Secure Session Key Distribution: the Three Party Case. In: Proc. ACM STOC 1995, pp. 57–66 (1995)
6. Bellare, M., Pointcheval, D., Rogaway, P.: Authenticated Key Exchange Secure against Dictionary Attacks. In: Preneel, B. (ed.) EUROCRYPT 2000. LNCS, vol. 1807, pp. 139–155. Springer, Heidelberg (2000)
7. Bellovin, S., Merritt, M.: Encrypted Key Exchange: Passwords based Protocols Secure against Dictionary Attacks. In: Proc. IEEE Symposium on Security & Privacy 1992, pp. 72–84 (1992)
8. Byun, J.W., Jeong, I.R., Lee, D.-H., Park, C.-S.: Password-Authenticated Key Exchange between Clients with Different Passwords. In: Deng, R.H., Qing, S., Bao, F., Zhou, J. (eds.) ICICS 2002. LNCS, vol. 2513, pp. 134–146. Springer, Heidelberg (2002)
9. Byun, J.W., Lee, D.-H.: N-Party Encrypted Diffie-Hellman Key Exchange Using Different Passwords. In: Ioannidis, J., Keromytis, A.D., Yung, M. (eds.) ACNS 2005. LNCS, vol. 3531, pp. 75–90. Springer, Heidelberg (2005)
10. Byun, J.W., Lee, D.-H., Lim, J.-I.: Efficient and Provably Secure Client-to-Client Password-Based Key Exchange Protocol. In: Zhou, X., Li, J., Shen, H.T., Kitsuregawa, M., Zhang, Y. (eds.) APWeb 2006. LNCS, vol. 3841, pp. 830–836. Springer, Heidelberg (2006)
11. Byun, J.W., Lee, D.H., Lim, J.I.: EC2C-PAKA: An Efficient Client-to-Client Password Authenticated Key Agreement. Information Sciences 177, 3995–4013 (2007)
12. Cao, T., Zhang, Y.: Cryptanalysis of Two Password-Authenticated Key Exchange Protocols between Clients with Different Passwords. International Mathematical Forum 2(11), 525–532 (2007)
13. Chen, L.: A Weakness of the Password-Authenticated Key Agreement between Clients with Different Passwords Scheme. Circulated for consideration at the 27th SC27/WG2 meeting in Paris, France, ISO/IEC JTC 1/SC27 N3716, 2003-10-20.24 (2003)
14. Choo, K.-K.R., Boyd, C., Hitchcock, Y.: Examining Indistinguishability-Based Proof Models for Key Establishment Protocols. In: Roy, B. (ed.) ASIACRYPT 2005. LNCS, vol. 3788, pp. 585–604. Springer, Heidelberg (2005)

15. Choo, K.-K.R., Boyd, C., Hitchcock, Y.: Errors in Computational Complexity Proofs for Protocols. In: Roy, B. (ed.) ASIACRYPT 2005. LNCS, vol. 3788, pp. 624–643. Springer, Heidelberg (2005)
16. Di Crescenzo, G., Kornievskaia, O.: Efficient Kerberized Multicast in a Practical Distributed Setting. In: Davida, G.I., Frankel, Y. (eds.) ISC 2001. LNCS, vol. 2200, pp. 27–45. Springer, Heidelberg (2001)
17. Diffie, W., van Oorschot, P.C., Wiener, M.J.: Authentication and Authenticated Key Exchanges. Design, Codes and Cryptography 2(2), 107–125 (1992)
18. Ding, Y., Horster, P.: Undetectable On-line Password Guessing Attacks. ACM Operating Systems Review 29(4), 77–86 (1995)
19. Feng, D.-G., Xu, J.: A New Client-to-Client Password-Authenticated Key Agreement Protocol. In: Chee, Y.M., Li, C., Ling, S., Wang, H., Xing, C. (eds.) IWCC 2009. LNCS, vol. 5557, pp. 63–76. Springer, Heidelberg (2009)
20. Jin, W., Xu, J.: An Efficient and Provably Secure Cross-Realm Client-to-Client Password-Authenticated Key Agreement Protocol with Smart Cards. In: Garay, J.A., Miyaji, A., Otsuka, A. (eds.) CANS 2009. LNCS, vol. 5888, pp. 299–314. Springer, Heidelberg (2009)
21. Kaliski Jr., B.S.: An Unknown Key-Share Attack on the MQV Key Agreement Protocol. ACM TISSEC 4(3), 275–288 (2001)
22. Kim, J., Kim, S., Kwak, J., Won, D.H.: Cryptanalysis and Improvement of Password Authenticated Key Exchange Scheme between Clients with Different Passwords. In: Laganá, A., Gavrilova, M.L., Kumar, V., Mun, Y., Tan, C.J.K., Gervasi, O. (eds.) ICCSA 2004. LNCS, vol. 3043, pp. 895–902. Springer, Heidelberg (2004)
23. Needham, R., Schroeder, M.: Using Encryption for Authentication in Large Networks of Computers. Communications of the ACM 21(12), 993–999 (1978)
24. Phan, R.C.-W., Goi, B.-M.: Cryptanalysis of an Improved Client-to-Client Password-Authenticated Key Exchange (C2C-PAKE) Scheme. In: Ioannidis, J., Keromytis, A.D., Yung, M. (eds.) ACNS 2005. LNCS, vol. 3531, pp. 33–39. Springer, Heidelberg (2005)
25. Phan, R.C.-W., Goi, B.-M.: Cryptanalysis of the N-Party Encrypted Diffie-Hellman Key Exchange Using Different Passwords. In: Zhou, J., Yung, M., Bao, F. (eds.) ACNS 2006. LNCS, vol. 3989, pp. 226–238. Springer, Heidelberg (2006)
26. Phan, R.C.-W., Goi, B.-M.: Cryptanalysis of Two Provably Secure Cross-Realm C2C-PAKE Protocols. In: Barua, R., Lange, T. (eds.) INDOCRYPT 2006. LNCS, vol. 4329, pp. 104–117. Springer, Heidelberg (2006)
27. Rogaway, P.: On the Role Definitions in and Beyond Cryptography. In: Maher, M.J. (ed.) ASIAN 2004. LNCS, vol. 3321, pp. 13–32. Springer, Heidelberg (2004)
28. Stern, J.: Why Provable Security Matters? In: Biham, E. (ed.) EUROCRYPT 2003. LNCS, vol. 2656, pp. 449–461. Springer, Heidelberg (2003)
29. Wang, S., Wang, J., Xu, M.: Weaknesses of a Password-Authenticated Key Exchange Protocol between Clients with Different Passwords. In: Jakobsson, M., Yung, M., Zhou, J. (eds.) ACNS 2004. LNCS, vol. 3089, pp. 414–425. Springer, Heidelberg (2004)
30. Yin, Y., Bao, L.: Secure Cross-Realm C2C-PAKE Protocol. In: Batten, L.M., Safavi-Naini, R. (eds.) ACISP 2006. LNCS, vol. 4058, pp. 395–406. Springer, Heidelberg (2006)
31. Yoon, E.-J., Yoo, K.-Y.: A Secure Password-Authenticated Key Exchange between Clients with Different Passwords. In: Zhou, X., Li, J., Shen, H.T., Kitsuregawa, M., Zhang, Y. (eds.) APWeb 2006. LNCS, vol. 3841, pp. 659–663. Springer, Heidelberg (2006)

Passive Attack on RFID LMAP++ Authentication Protocol

Shao-hui Wang[1,2,3] and Wei-wei Zhang[1]

[1] College of Computer, Nanjing University of Post and Telecommunication, Nanjing 210046, China
[2] Jiangsu High Technology Research Key Laboratory for Wireless Sensor Networks, Nanjing, Jiangsu 210003, China
[3] Network and Data Security Key Laboratory of Sichuan Province

Abstract. LMAP++ is an ultra-lightweight mutual authentication protocol designed for resource-constrained system such as RFID. The protocol is designed using only logical operator XOR and modular 2^{96} addition. In this paper, a passive attack on LMAP++ protocol is given after constructing the distinguisher for the random numbers used in the protocol. The attack shows the protocol cannot resist passive attack, and after eavesdropping about 480 authentication messages, the adversary can deduce the static identifier and two secrets with high probability.

Keywords: RFID, Authentication Protocol, Passive Attack, Distinguisher.

1 Introduction

Radio Frequency Identification (RFID) is a contactless technology used to identify and/or authenticate remote objects or persons, through a radio frequency channel. RFID is becoming more and more widespread in daily-life applications, from library management or pet identification, to anti counterfeiting, access control or biometric passports. An RFID system is usually divided into three components: reader, tag and backend database. Tags are usually micro-chips with constrained resources, and the unique identifier and some information related to the tag holder are stored in them. Readers can read and modify the messages stored in the tags, and will pass the messages to the backend database if needed. Usually the channel between the reader and the database is secure.

However, the ubiquity of RFID raises new concerns about privacy and security at the same time[1]. The attacks on the RFID system contain passive attacks and active attacks. In the passive attacks, the adversaries deduce some useful information only through eavesdropping, collecting and analyzing authentication messages between the reader and the tag, so the passive attacks are much more feasible and operational than active ones.

RFID tag constraints in processing power and memory make them tougher to deal with traditional algorithms, such as AES, RSA or hash functions[2]. These kinds of

D. Lin, G. Tsudik, and X. Wang (Eds.): CANS 2011, LNCS 7092, pp. 185–193, 2011.

constraints dictate a paradigm shift in security provision for RFID which is known as lightweight cryptography, and lightweight authentication protocol is a subset of lightweight cryptography[3,4,5]. In 2006, Peris et al. proposed a protocol named LMAP[6], an ultra-lightweight mutual authentication protocols, which is designed only using simple logical operations like XOR, AND, OR and Modular addition. Afterward, Li et.al.[7] proposed the SLMAP protocol in Chinacrypt'07 to simplify LMAP protocol. But these two protocols can not resist active attack and passive attack[8,9,10], because logical operations and modular addition do not have good diffusion effect. Logical operator XOR operates bit by bit, and the carry bit introduced by modular addition only affects its left bit position.

To resist the passive attack, Li presented the LMAP++ authentication protocol [11], and we can not use the passive attack method used in [9,10] directly to analyze LMAP++. Except for the research on the traceability of this protocol [12], there is not any attack outcome. In this paper, we give a passive attack on the LMAP++ protocol using the property of modular addition and XOR and the structure of the protocol. The attack shows that after eavesdropping about 480 authentication messages, the adversary can deduce the identifier and two secrets with high probability. The rest of the paper is organized as follows. In section 2, we review the LMAP++ protocol. The passive attack on LMAP++ is illustrated in section 3, and we state our conclusion in section 4.

2 LMAP++ Authentication Protocol

In 2008, based on LAMP architecture, Li [11] presented the LMAP++ authentication protocol for the low cost RFID tag to avoid the weakness of LMAP. The random numbers used are generated by the reader, and the protocol is designed using XOR($\oplus$) and modular 2^{96} addition. Because we can not deduce a formula between the random number and some known authentication messages, the passive attack method used for LMAP and SLMAP is not applicable directly to LMAP++.

In this section we introduce the LMAP++ protocol briefly. Every tag shares a fixed and unique identifier (ID) with the reader. At the n-th authentication, the tag and the reader share a pseudonym $IDS^{(n)}$ and two secrets ($K_1^{(n)}, K_2^{(n)}$), which will update to $IDS^{(n+1)}$, $K_1^{(n+1)}$ and $K_2^{(n+1)}$ if authentication is successful. We omit some superscript for convenience in the rest of the paper.

Every authentication contains three rounds: tag identification, mutual authentication and IDS, secrets updating. Here all the variables are 96 bit.

(I) Tag Identification. After receiving the "Hello" message from the reader, the tag sends the IDS to the reader, which will look up the tags in the database with the same pseudonym and get the corresponding information.

(II) Mutual Authentication. The communications between the reader and the tag are as follows:

a. Reader Authentication. Reader generates a random number r, computes and sends the tag the messages (A, B) as follow. The tag can deduce the random number

r through message A, and make sure whether the reader is valid via checking the correctness of message B:

$$A = (IDS \oplus K_1) + r \tag{1}$$

$$B = IDS + (K_2 \oplus r) \tag{2}$$

b. Tag Authentication. The tag sends back the answer message C using the random number r and static identifier ID:

$$C = (IDS + (ID \oplus r)) \oplus (K_1 + K_2 + r) \tag{3}$$

(III) IDS and Secrets Updating. After authenticating successfully, the reader and tag will update the pseudonym IDS and secrets as follow:

$$IDS^{(new)} = ((IDS + K_1) \oplus r) + ((ID + K_2) \oplus r) \tag{4}$$

$$K_1^{(new)} = (K_1 \oplus r) + (IDS + K_2 + ID) \tag{5}$$

$$K_2^{(new)} = (K_2 \oplus r) + (IDS + K_1 + ID) \tag{6}$$

3 Passive Attack on LMAP++

In this section, we illustrate in detail the mechanics to attack the LMAP++ passively. The passive adversaries can only eavesdrop the communication messages between the reader and the tag, i.e. the message (IDS, A, B, C). We first explain how to get the least significant bit of the identifier, then construct a distinguisher to obtain the least significant bit of the random number, thus give the algorithm to deduce the identifier and two secrets.

We denote by $[x]_i$ as the bit position i of the variable x, $[x]_{0...l-1}$ as the bit from position 0 to $l-1$ of x, so $[x]_0$ is the least significant bit of x.

3.1 The Least Significant Bit of Identifier $[ID]_0$

Obviously, as to the least significant bit, the modular 2^{96} addition and subtraction of a and b is just the same as logical operator XOR, i.e.:

$$[a+b]_0 = [a]_0 \oplus [b]_0 \text{ and } [a-b]_0 = [a]_0 \oplus [b]_0$$

Considering the least significant bit of the LMAP++ communication messages formulas (1), (2) and (3), we can get the following formulas:

$$[A]_0 = [IDS]_0 \oplus [K_1]_0 \oplus [r]_0 \tag{1'}$$

$$[B]_0 = [IDS]_0 \oplus [K_2]_0 \oplus [r]_0 \tag{2'}$$

$$[C]_0 = [IDS]_0 \oplus [ID]_0 \oplus [r]_0 \oplus [K_1]_0 \oplus [K_2]_0 \oplus [r]_0 = [IDS]_0 \oplus [ID]_0 \oplus [K_1]_0 \oplus [K_2]_0 \quad (3')$$

From (1') and (2'), it is easy to see:

$$[A]_0 \oplus [B]_0 = [K_1]_0 \oplus [K_2]_0$$

combining with the formula of $[C]_0$, we can get the least significant bit of ID:

$$[ID]_0 = [C]_0 \oplus [IDS]_0 \oplus [A]_0 \oplus [B]_0.$$

3.2 The Least Significant Bit of Random Number $[r]_0$

If we can write a formula between the random number r and the authentication messages and the identifier, we can obtain the least significant bit of random number r, thus to deduce the least significant bit of the secrets. But as to LMAP++ protocol, it is not feasible to get this kind of formula for the random number appears twice in formula (4) and (5). Next we show how to distinguish the different condition that $[r]_0$ equals to 0 or 1.

From the formula (1) and (2), we can get the equations of the secrets:

$$K_1 = (A - r) \oplus IDS \quad (7)$$

$$K_2 = (B - IDS) \oplus r \quad (8)$$

After substituting the formula (7) and (8) to the formula (3) and (4)， we can get the relationship of the message C and $IDS^{(new)}$ with the random number r and identifier ID:

$$C = (IDS + (ID \oplus r)) \oplus (((A - r) \oplus IDS) + ((B - IDS) \oplus r) + r) \quad (9)$$

$$IDS^{(new)} = ((IDS + ((A - r) \oplus IDS)) \oplus r) + ((ID + ((B - IDS) \oplus r)) \oplus r) \quad (10)$$

Now we consider the bit position 1 of the formula (9) and (10), and construct a distinguisher for $[r]_0 = 0$ and $[r]_0 = 1$ using $[ID]_1$. We must consider the carry and borrow when considering modular 2^{96} addition and subtraction. We use $s(a,b,i)$ to denote the carry at bit i of $a+b$, and $t(a,b,i)$ the borrow at the bit i of $a-b$. It is easy to have the following fact:

[Fact:] As to modular 2^m addition and subtraction, the following formulas must satisfy considering the bit position i ($i > 0$) :

$$[a+b]_i = [a]_i \oplus [b]_i \oplus s(a,b,i-1)$$
$$[a-b]_i = [a]_i \oplus [b]_i \oplus t(a,b,i-1)$$

If $[b]_0$=0, it is impossible to generate carry or borrow at the bit position 1. But if $[b]_0$ =1, the carry or borrow may appear or may not. Next we consider the bit position of 1 of formula (9) and (10) under the condition of $[r]_0$=0 or $[r]_0$=1.

1. The condition of $[r]_0$=0

Under this condition, there is no carry or borrow at bit position 1. We can get the following two formula (11) and (12) through formula (9) and (10):

$$\begin{aligned}[C]_1 &= [IDS]_1 \oplus [ID \oplus r]_1 \oplus s(IDS, ID, 0) \oplus [A]_1 \oplus [r]_1 \oplus [IDS]_1 \oplus \\ &\quad [B - IDS]_1 \oplus [r]_1 \oplus [r]_1 \oplus s(A \oplus IDS, B - IDS, 0) = \\ &\quad [ID]_1 \oplus s(IDS, ID, 0) \oplus [A]_1 \oplus [B - IDS]_1 \oplus s(A \oplus IDS, B - IDS, 0) \end{aligned} \tag{11}$$

$$\begin{aligned}[IDS^{(new)}]_1 &= [IDS]_1 \oplus [A]_1 \oplus [r]_1 \oplus [IDS]_1 \oplus s(IDS, A \oplus IDS, 0) \oplus [r]_1 \oplus [ID]_1 \oplus \\ &\quad [B - IDS]_1 \oplus [r]_1 \oplus s(ID, B - IDS, 0) \oplus [r]_1 \oplus s(A, ID + (B - IDS), 0) \\ &= [A]_1 \oplus s(IDS, A \oplus IDS, 0) \oplus [ID]_1 \oplus [B - IDS]_1 \oplus \\ &\quad s(ID, B - IDS, 0) \oplus s(A, ID + (B - IDS), 0) \end{aligned} \tag{12}$$

We can get two formulas about $[ID]_1$ through (11) and (12), and these two formulas must be equal, i.e. the necessary condition for $[r]_0$=0 is the following equation:

$$\begin{aligned}&[C]_1 \oplus s(IDS, ID, 0) \oplus s(A \oplus IDS, B - IDS, 0) = s(IDS, A \oplus IDS, 0) \oplus \\ &[IDS^{(new)}]_1 \oplus s(ID, B - IDS, 0) \oplus s(A, ID + (B - IDS), 0) \end{aligned} \tag{13}$$

2. the condition of $[r]_0$=1.

Here we use the notation $\overline{a}$ to denote a variable that is different from the variable a with the least significant bit, and the other bits is not change. Like the discussion above, we can get the equation (14) and (15) as follow:

$$\begin{aligned}[C]_1 &= [IDS]_1 \oplus [ID \oplus r]_1 \oplus s(IDS, \overline{ID}, 0) \oplus [A]_1 \oplus [r]_1 \oplus t(A, r, 0) \oplus [IDS]_1 \oplus \\ &\quad [B - IDS]_1 \oplus [r]_1 \oplus [r]_1 \oplus s(\overline{A} \oplus IDS, \overline{B - IDS}, r, 0) = [ID]_1 \oplus [A]_1 \oplus \\ &\quad s(IDS, \overline{ID}, 0) \oplus t(A, r, 0) \oplus [B - IDS]_1 \oplus s(\overline{A} \oplus IDS, \overline{B - IDS}, r, 0) \end{aligned} \tag{14}$$

$$\begin{aligned}[IDS^{(new)}]_1 &= [IDS]_1 \oplus [A]_1 \oplus [r]_1 \oplus t(A, r, 0) \oplus [IDS]_1 \oplus s(IDS, \overline{A} \oplus IDS, 0) \oplus \\ &\quad [r]_1 \oplus [ID]_1 \oplus [B - IDS]_1 \oplus [r]_1 \oplus s(ID, \overline{B - IDS}, 0) \oplus [r]_1 \oplus s(\overline{A}, ID + \\ &\quad (B - IDS), 0) = [A]_1 \oplus t(A, r, 0) \oplus s(IDS, \overline{A} \oplus IDS, 0) \oplus [ID]_1 \oplus [B - IDS]_1 \oplus \\ &\quad s(ID, \overline{B - IDS}, 0) \oplus s(A, ID + (B - IDS), 0) \end{aligned} \tag{15}$$

Similarly, we can get two formulas about $[ID]_1$ through (14) and (15), and these two formulas must be equal, i.e. the necessary condition for $[r]_0$ =1 is the following equation:

$$[C]_1 \oplus s(IDS,\overline{ID},0) \oplus s(\overline{A} \oplus IDS,\overline{B-IDS},r,0) = [IDS^{(\text{new})}]_1 \oplus$$
$$s(IDS,\overline{A} \oplus IDS,0) \oplus s(ID,\overline{B-IDS},0) \oplus s(A, ID+(B-IDS),0) \qquad (16)$$

Using the formulas (13) and (16), we can construct the distinguisher for $[r]_0 = 0$ or 1. Here we poll the least significant bit $([A]_0,[B]_0,[C]_0,[IDS]_0)$ from $(0,0,0,0)$ to $(1,1,1,1)$, and we can get $[ID]_0$ via subsection 3.1 and the left part and right part of formula (13) and (16) as shown in the following table:

Table 1. The left and right parts of (13) and (16) with different message values

$[A]_0$	$[B]_0$	$[C]_0$	$[IDS]_0$	$[ID]_0$	left part of (13)	right part of (13)	left part of (16)	right part of (16)
0	0	0	0	0	$[C]_1$	$[IDS^{(new)}]_1$	$[C]_1 \oplus 1$	$[IDS^{(new)}]_1$
0	0	0	1	1	$[C]_1$	$[IDS^{(new)}]_1$	$[C]_1$	$[IDS^{(new)}]_1$
0	0	1	0	1	$[C]_1$	$[IDS^{(new)}]_1$	$[C]_1 \oplus 1$	$[IDS^{(new)}]_1 + 1$
0	0	1	1	0	$[C]_1 \oplus 1$	$[IDS^{(new)}]_1 + 1$	$[C]_1 \oplus 1$	$[IDS^{(new)}]_1$
0	1	0	0	1	$[C]_1$	$[IDS^{(new)}]_1 + 1$	$[C]_1 \oplus 1$	$[IDS^{(new)}]_1$
0	1	0	1	0	$[C]_1$	$[IDS^{(new)}]_1 + 1$	$[C]_1$	$[IDS^{(new)}]_1$
0	1	1	0	0	$[C]_1$	$[IDS^{(new)}]_1$	$[C]_1 \oplus 1$	$[IDS^{(new)}]_1$
0	1	1	1	1	$[C]_1 \oplus 1$	$[IDS^{(new)}]_1 + 1$	$[C]_1 \oplus 1$	$[IDS^{(new)}]_1 + 1$
1	0	0	0	1	$[C]_1$	$[IDS^{(new)}]_1 + 1$	$[C]_1 \oplus 1$	$[IDS^{(new)}]_1$
1	0	0	1	0	$[C]_1$	$[IDS^{(new)}]_1 + 1$	$[C]_1$	$[IDS^{(new)}]_1$
1	0	1	0	0	$[C]_1$	$[IDS^{(new)}]_1$	$[C]_1 \oplus 1$	$[IDS^{(new)}]_1$
1	0	1	1	1	$[C]_1 \oplus 1$	$[IDS^{(new)}]_1 + 1$	$[C]_1 \oplus 1$	$[IDS^{(new)}]_1 + 1$
1	1	0	0	0	$[C]_1 \oplus 1$	$[IDS^{(new)}]_1 + 1$	$[C]_1$	$[IDS^{(new)}]_1 + 1$
1	1	0	1	1	$[C]_1 \oplus 1$	$[IDS^{(new)}]_1 + 1$	$[C]_1 \oplus 1$	$[IDS^{(new)}]_1 + 1$
1	1	1	0	1	$[C]_1 \oplus 1$	$[IDS^{(new)}]_1 + 1$	$[C]_1$	$[IDS^{(new)}]_1$
1	1	1	1	0	$[C]_1$	$[IDS^{(new)}]_1$	$[C]_1$	$[IDS^{(new)}]_1 + 1$

From the discussion above, we know that the sixth column equals to the seventh part is the necessary condition for $[r]_0$=0, and that the eighth column equals to the ninth part is the necessary condition for $[r]_0$=1. That is to say, if we observe the last significant bit of authentication messages $([A]_0,[B]_0,[C]_0,[IDS]_0)$ is (0,0,0,0), and the second to last bit of C and the pseudonym $IDS^{(new)}$ of the next authentication $[C]_1$ and $[IDS^{(new)}]_1$ are not equal, then it is impossible that $[r]_0$ is equal to 0. However, there exists some conditions that we can not determine whether $[r]_0$=0 or

$[r]_0 = 1$. For example, when $([A]_0,[B]_0,[C]_0,[IDS]_0)$ is (0,0,0,1), the necessary condition is the same for $[r]_0$=0 and 1. Suppose $([A]_0,[B]_0,[C]_0,[IDS]_0)$ can take the value from (0,0,0,0) to (1,1,1,1) with the same probability, we can get the conclusion that the probability to distinguish $[r]_0$ is 0.5.

3.3 Algorithm to Obtain the Identifier and Secrets

If the value of $[r]_0$ is known, we can get the least significant bit of two secrets from the formula (7) and (8):

$$[K_1]_0 = [IDS]_0 \oplus [A]_0 \oplus [r]_0 \quad \text{and} \quad [K_2]_0 = [IDS]_0 \oplus [B]_0 \oplus [r]_0$$

In addition, we can determine the value of $[ID]_1$ if $[r]_0$ is known using the equation (13) or (16). Using the method discuss in 3.2, for any given bit position i, we can construct the distinguisher for $[r]_i$ using $[ID]_{i+1}$, and the probability of distinguishing successfully is 0.5 also. We give the definition of distinguisher as follow:

Definition 3.1. If the messages (A,B,C,IDS) can distinguish successfully the bit position i of the random number r, i.e. $[r]_i$, we call the messages (A,B,C,IDS) is $[r]_i$-distinguisher.

Because the distinguisher is satisfied with the probability 0.5, if the i-th bit position $[r]_i$ can not be determined, then it is difficult to obtain the value of $[ID]_{i+1}$.

Before we give the algorithm to deduce all the bits of identifier and secrets, we first present the following theorem satisfied because of the structure of LMAP++.

Theorem 3.1. If in the $n-th$ authentication, we can get the values of identifier, random number, and secrets from bit position 0 to $l-1$, i.e. $[ID]_{0,l-1}$, $[r^{(n)}]_{0,l-1}$, $[K_1^{(n)}]_{0,l-1}$, $[K_2^{(n)}]_{0,l-1}$, then we can get the values of $[r^{(j)}]_{0,l-1}$, $[K_1^{(j)}]_{0,l-1}$, $[K_2^{(j)}]_{0,l-1}$ for any follow-up communication, i.e. $j > n$.

Proof: Here we only consider the $(n+1)-th$ authentication. From the equation (5) and (6), we can get values of the secrets $[K_1^{(n+1)}]_{0,l-1}$ and $[K_2^{(n+1)}]_{0,l-1}$ if $[ID]_{0,l-1}$, $[r^{(n)}]_{0,l-1}$, $[K_1^{(n)}]_{0,l-1}$ and $[K_2^{(n)}]_{0,l-1}$ are known. Thus the random number used in the $(n+1)-th$ communication $[r^{(n+1)}]_{0,l-1}$ can be determined using the formula (8) via the value of $B^{(n+1)}$ and $IDS^{(n+1)}$.

Using the theorem 3.1, we give the following algorithm to deduce the identifier and the two secrets.

Algorithm 3.1. Suppose the adversary has eavesdropped m authentication successively between the reader and the tag, he can get the messages $(A^{(n)},B^{(n)},C^{(n)},IDS^{(n)})$, $n=1,2,...,m$. After the adversary finds the $[r]_0$ -distinguisher(for example, the $j1-th$ communication) , he can calculate the least significant bit of the values of ID , r , K_1 and K_2 of $j1-th$ communication; If this $[r]_0$-distinguisher can still distinguish $[r]_1$, the adversary can go on computing $[ID]_1$, $[r^{(j1)}]_1$, $[K_1^{(j1)}]_1$ and $[K_2^{(j1)}]_1$. Otherwise the adversary will search for the $[r]_1$ -distinguisher after the $j1-th$ communication(for example, the $j2-th$ ($j2>j1$) communication). Using the theorem 3.1, the adversary can calculate the values of $[r^{(j2)}]_0$, $[K_1^{(j2)}]_0$ and $[K_2^{(j2)}]_0$ even if $(A^{(j2)},B^{(j2)},C^{(j2)},IDS^{(j2)})$ is not $[r]_0$-distinguisher, thus to obtain $[ID]_{0,1}$, $[r^{(j2)}]_{0,1}$, $[K_1^{(j2)}]_{0,1}$ and $[K_2^{(j2)}]_{0,1}$. The algorithm will finish when the adversary obtain all the 96 bit of these four variables.

Now we discuss the number of authentication (value of m) needed to eavesdrop. Suppose the message (A,B,C,IDS) take the value with the same probability, and the probability of $[r]_i$ - distinguisher appearing is 0.5, then in the successive 5 authentication, the probability that the $[r]_i$ -distinguisher will appear is $1-(0.5)^5$ $=0.96875$. We can get the conclusion the $[r]_i$ -distinguisher appear with high probability in 5 successive authentication, and the adversary can deduce the identifier and secrets after eavesdropping about 96*5=480 authentication messages. In fact, the actual number needed is smaller than 480, for a tuple of message (A,B,C,IDS) can be a distinguisher for several successive bits of r .

4 Conclusions

In this paper, through constructing the corresponding distinguisher for different bit position of random number, we give a passive attack on LMAP++, an ultra-lightweight mutual authentication protocol. The attack shows that after eavesdropping about 480 authentication messages, the adversary can deduce the identifier and two secrets with high probability. The bad diffusion effect of logical operator XOR and modular 2^m addition makes it hard to design lightweight protocol with high security. SASI protocol [13] and Gossamer protocol[14] introduce the circular shift operator, which improves the ability to resist passive attack.

Acknowledgements. This work is supported by the Priority Academic Program Development of Jiangsu Higher Education Institutions(PAPD), National Natural Science Funds (Grant No.60903181) and Nanjing University of Post and Telecommunication Funds (Grant No.NY208072).

References

1. Juels, A.: RFID Security and Privacy: A Research Survey. IEEE Journal on Selected Areas in Communications 24(2), 381–394 (2006)
2. Yuan, S.-g., Dai, H.-y., Lai, S.-l.: RFID authentication protocol based on Hash functions. Computer Engineering 34(12), 141–143
3. Sadighian, A., Jalili, R.: Afmap: Anonymous forward-secure mutual authentication protocols for RFID systems. In: The Third IEEE International Conference on Emerging Security Information, Systems and Technologies (SECURWARE 2009), pp. 31–36 (2009)
4. Vajda, I., Buttyan, L.: Light weight authentication protocols for low-cost RFID tags. In: Proceedings of Workshop on Security in Ubiquitous Computing (2003), http://www.hit.bme.hu/~buttyan/publications/VajdaB03suc.pdf
5. Juels, A.: Minimalist Cryptography for Low-Cost RFID Tags. In: Blundo, C., Cimato, S. (eds.) SCN 2004. LNCS, vol. 3352, pp. 149–164. Springer, Heidelberg (2005)
6. Peris-Lopez, P., Hernandez-Castro, J.C., Estevez- Tapiador, J., Ribagorda, A.: LMAP: A real lightweight mutual authentication protocol for low-cost RFID tags. In: Proceedings of the 2nd Workshop on RFID Security (2006), http://events.iaik.tugraz.at/RFIDSec06/Program/papers/013-LightweightMutualAuthentication.pdf
7. Li, T., Wang, G.: SLMAP - A secure ultra-lightweight RFID mutual authentication protocol. In: Chinacrypt 2007, pp. 19–22 (2007)
8. Li, T., Wang, G.: Security Analysis of Two ultra-Lightweight RFID Authentication Protocols. In: 22 IFIP International Information Security and Privacy, South Africa, pp. 65–78 (2007)
9. Barasz, M., Boros, B., Ligeti, P., et al.: Breaking LMAP. In: Proc. of RFIDSec 2007, pp. 11–16 (2007)
10. Wang, S., Zhang, W.-w.: Passive attack on SLMAP authentication protocol. Journal of Nanjing University of Posts and Telecommunications (submitted)
11. Li, T.: Employing lightweight primitives on low-cost rfid tags for authentication. In: VTC Fall, pp. 1–5 (2008)
12. Bagheri, N., Safkhani, M., et al.: Security Analysis of LMAP++, an RFID Authentication Protocol, http://www.eprint.iacr.org/2011/193.pdf
13. Chien, H.-Y.: SASI: A New Ultra-lightweight RFID Authentication Protocol Providing Strong Authentication and Strong Integrity. IEEE Transactions on Dependable and Secure Computing 4(4), 337–340 (2007)
14. Peris-Lopez, P., Hernandez-Castro, J.C., Estevez- Tapiador, J., Ribagorda, A.: Advances in Ultralightweight Cryptography for Low-Cost RFID Tags: Gossamer Protocol. In: Chung, K.-I., Sohn, K., Yung, M. (eds.) WISA 2008. LNCS, vol. 5379, pp. 56–68. Springer, Heidelberg (2009)

Multi-show Anonymous Credentials with Encrypted Attributes in the Standard Model

Sébastien Canard, Roch Lescuyer, and Jacques Traoré

Orange Labs, Applied Crypto Group, Caen, France
{sebastien.canard,roch.lescuyer,jacques.traore}@orange.com

Abstract. Anonymous credential systems allow users to obtain a certified credential (a driving license, a student card, etc.) from one organization and then later prove possession of this certified credential to another party, while minimizing the information given to the latter. At CANS 2010, Guajardo, Mennink and Schoenmakers have introduced the concept of anonymous credential schemes with encrypted attributes, where the attributes to be certified are encrypted and unknown to the user and/or issuing organization. Their construction is secure in the random oracle model and based on blind signatures, which, unfortunately, restrict the credentials to be used only once (one-show) to remain unlinkable. In their paper, Guajardo *et al.* left as an open problem to construct *multi-show* credential schemes with encrypted attributes, or to show the impossibility of such a construction. We here provide a positive answer to this problem: our multi-show anonymous credential scheme with encrypted attributes relies on the non-interactive Groth-Sahai proof system and the recent work on commuting signatures from Fuchsbauer (Eurocrypt 2011) and is proven secure in the standard model.

Keywords: Privacy, Anonymous credentials, Encrypted attributes.

1 Introduction

Anonymous credential systems, introduced by Chaum in [12], permit users to obtain the certification of their attributes by some authorized organizations. In this context, such a certification is called a "credential". For example, a university, as an organization, can deliver credentials on particular attributes (name, birth date, studies, etc.) to its students in order to certify their status. Such credential can next be used anonymously by users to prove to a third party the possession of the certified attributes, while minimizing the information given to this third party. For example, a legitimate student can prove that she is a student, namely that she owns a credential certified by a university, without revealing any other information. She can also prove that her credential attributes satisfy some properties, for example that she is under 25, without revealing her true age, nor her name or studies.

A lot of work has been done on anonymous credentials and, currently, there are mainly two kinds of constructions. The first one is based on the work from

D. Lin, G. Tsudik, and X. Wang (Eds.): CANS 2011, LNCS 7092, pp. 194–213, 2011.

Brands [6] and makes use of blind signatures [11]. Such constructions are secure in the random oracle model and very practical. Unfortunately, the resulting credentials are "one-show" as they become linkable if they are used several times. They are implemented by Microsoft in their U-Prove technology [24]. The second one is due to Camenisch and Lysyanskaya [8,10] and is based on the use of group signatures [13]. The resulting anonymous credential systems are less efficient than the Brands' based ones, but they are "multi-show" since they use the inherent unlinkability property of group signatures. This technology is implemented by IBM for their Idemix product [22]. Although the original scheme was secure in the random oracle model [8,10], recent variants such as [5] are secure in the standard model. More recent papers have also proposed several variants of anonymous credentials with additional features, such as the delegation of credentials [4,16], or revocation capabilities [9,7].

Anonymous Credential with Encrypted Attributes. At CANS 2010, Guajardo, Mennink and Schoenmakers [21] have introduced the concept of *anonymous credential schemes with encrypted attributes*. They argue that, in some practical scenarios, the user should not (or does not want to) learn the certified attributes. Anonymous credentials with encrypted attributes can also be used in the context of secure multi-party computation and in particular for the millionaires protocol (see [21] for details).

In [21], Guajardo *et al.* first give the security model for anonymous credential schemes with encrypted attributes. Such a scheme involves three kinds of participants: issuers (or organizations), users and verifiers. It is composed of three protocols: key generation, credential issuance and verification. The key generation protocol permits each party to compute their secret and public keys whereas the issuance protocol allows a user to obtain, from an issuer, credentials on some encrypted attributes. The idea here is that the user only has access to the attributes in encrypted form. Finally, a verification protocol is played between a user and a verifier, in which the user proves the possession of a credential on encrypted attributes (without obtaining them in clear) while the verifier may possess the decryption key to obtain the plain attributes. The authors also give some variations where the verifier (and sometimes the issuer) does not learn the attributes in the clear.

They next propose a practical construction of this new concept. Their scheme is based on the Brands anonymous credential scheme [6,24] and makes use of blind signatures. As a result, multiple uses (or multi-shows) of the same credential makes them, as with Brands' system, linkable: thus the resulting system is only "one-show" (*a.k.a.* "one-use"), as argued by the authors in [21].

The Multi-show Problem. In [21], the authors left as an open problem to construct *multi-show* credential schemes with encrypted attributes, or to show the impossibility of such a construction. In this paper, we provide a positive answer to this problem by giving a concrete construction.

The main difficulty is to obtain a system where, after one credential issuance with an issuer, the user can use the resulting credential several times in an unlinkable manner: in other words, nobody should be able to know whether two different verification protocols were played by the same user (using the same credential) or not. It is well-known [6,8,10] that blind signature based anonymous credential cannot reach such an unlinkability property and it seems, as argued in [21], that one may start from [8,10], which makes use of group signatures, that are by essence, unlinkable.

In a nutshell, a group signature based anonymous credential system works as follows. During the issuance protocol, the user obtains from the issuer a signature on her attributes. The verification protocol next consists in proving the possession of an issuer's signature on some attributes without revealing the signature (and thus reaching the unlinkability property) nor the private attributes. When trying to apply this technique to anonymous credentials with encrypted attributes, several solutions are conceivable.

- The issuer encrypts the attributes and next signs the resulting ciphertexts. The user therefore needs to prove the possession of an issuer's signature on the ciphertexts, without revealing the signature (which can be written, using classical notation for proofs of knowledge, $\text{PoK}(\sigma : \sigma = \text{Sign}_{\mathcal{I}}(c))$). However since the ciphertexts will remain unchanged, the unlinkability property collapses.
- One solution to the above problem would be to randomize the ciphertexts (provided that the underlying encryption scheme supports such randomization techniques). Unfortunately, the issuer's signature would not be valid on the resulting ciphertexts.
- One possibility to the above non validity of the given signature is for the issuer not to give the signature directly, but to prove the possession of such signature. It follows that the user next has to produce a proof of knowledge of such a proof of knowledge, which is known to be a meta proof [25]. However, the result would clearly be impractical.
- Another solution would be to keep the signature on the encrypted attributes. During a verification protocol, the user would first randomize the original ciphertext c to obtain $\tilde{c}$ and next prove that she knows a signature on a randomized version of $\tilde{c}$, without revealing the signature nor the ciphertext c (such a proof can be written $\text{PoK}(\sigma, c : \sigma = \text{Sign}_I(c) \wedge \tilde{c} = \text{ReRand}(c))$). The main problem is that, to the best of our knowledge, it does not exist a practical instantiation of such a proof, except by using some variants of commuting signatures [16], as we will do in our construction[1].

Our Solution. In this paper, we take a different approach which can be seen as a mix of the two last above solutions. We make use of the concept of commuting signature which has recently been introduced by Fuchsbauer [16]. Such signature schemes allow to use a ComSig procedure which on input one or several

[1] Even if the above proof is not exactly the one we will use.

(extractable) commitments on some messages plus a signing secret key, outputs a(n extractable) commitment on a signature on the committed messages along with a (Groth-Sahai) proof [20] that the signature on the committed messages can be recovered from the given commitment, without revealing the signature nor the committed messages. Fuchsbauer also gives a concrete and efficient construction of commuting signatures based on automorphic signatures [2,15]. We next associate a commuting signature to the randomization techniques on extractable commitments and Groth-Sahai proofs to obtain our multi-show anonymous credential scheme with encrypted attributes, which is secure in the standard model.

Organization of the Paper. The paper is organized as follows. In Section 2, we recall the concept and give the model for anonymous credential scheme with encrypted attributes in the multi-show setting. In Section 3, we give some useful tools, such as extractable commitments, Groth-Sahai proof systems and automorphic signatures. In particular, we describe an SXDH based Groth-Sahai proof of equality under different commitment keys (the DLIN version being given in [18]). Section 4 is devoted to commuting signatures. In this section, we introduce the way to produce a commuting signature on a vector of messages that are committed using different commitment keys. To the best of our knowledge, this tool is new and may be of independent interest. Finally, Section 5 describes our new anonymous credential scheme with encrypted attributes. The above extension on commuting signatures allows us to extend the work of Guajardo *et al.* to the case where the issuer certifies encrypted attributes to possibly different verifiers.

2 A Model for Anonymous Credential Systems with Encrypted Attributes

In this section, we recall the definition of encrypted credential schemes as given in [21]. We slightly differ from the Guajardo *et al.* model since we need to take into account the multi-show case.

2.1 Protocols

In an anonymous credential scheme with encrypted attributes, there is an issuer $\mathcal{I}$ who issues credentials on encrypted attributes, a user $\mathcal{U}$ who obtains credentials on some of her attributes she does not know, before anonymously proving the possession of such credentials, and a verifier $\mathcal{V}$ who is able to verify the validity of credentials and may obtain the plain attributes. The list of certified attributes are denoted $\mathbf{M}$ when they are in plain, and $\mathbf{C}$ when they are encrypted. An anonymous credential with encrypted attributes scheme Π is next composed of the following procedure, where λ is a security parameter.

– The key generation process is divided into three parts. The first one, denoted ParGen is played by any designated entity (possibly the issuer $\mathcal{I}$). It takes as input the security parameter 1^λ and outputs some parameters param for

the whole system. Next, the issuer executes IssGen which on inputs 1^λ and param, outputs $\mathsf{sk}_\mathcal{I}$. Finally, the verifier $\mathcal{V}$ uses VerGen to generate $\mathsf{sk}_\mathcal{V}$. This step finally publishes gpk as well as λ, param and the public keys related to $\mathsf{sk}_\mathcal{I}$ and $\mathsf{sk}_\mathcal{V}$.
- An issuance protocol Issue is played by $\mathcal{I}$ and $\mathcal{U}$. It takes as input gpk. The issuer additionally takes as input $\mathsf{sk}_\mathcal{I}$ and either the list $\mathbf{M}$ of plain attributes or the corresponding list $\mathbf{C}$ of encrypted attributes. The user always takes as input $\mathbf{C}$. This protocol outputs for the user a credential cred on the list $\mathbf{C}$ of encrypted attributes related to $\mathbf{M}$. The issuer outputs its view $\mathsf{view}^{\mathrm{Iss}}$ of the protocol.
- A verification protocol Verify is played by $\mathcal{U}$ and $\mathcal{V}$. It takes as input gpk. The verifier (resp. the user) additionally takes as input $\mathsf{sk}_\mathcal{V}$ (resp. cred and $\mathbf{C}$). The verifier outputs a bit representing either 1 and optionally a list $\mathbf{M}$ of plain attributes (for acceptance) or 0 (for rejection).

Completeness. Such a scheme should verify the *completeness* property which states that for any (gpk, $\mathsf{sk}_\mathcal{I}$, $\mathsf{sk}_\mathcal{V}$) output by the key generation procedures and related to honest issuer and verifier, the credential cred obtained by $\mathcal{U}$ during Issue will be accepted in the Verify protocol with overwhelming probability.

2.2 Security Properties

In [21], Guajardo *et al.* have given the security properties for the case of a one-show anonymous credential scheme with encrypted attributes. We thus need to modify their security model to reach the multi-show case. We moreover give more formal definitions for some properties. Let us consider an anonymous credential with encrypted attributes scheme denoted Π.

Used Oracles. Before giving the security experiments, we first describe the different oracles that will be used by the adversary. The security of our scheme is conducted in an adaptive corruption model, where the challenger $\mathcal{C}$ generates public keys for all entities and allows the adversary to get secret keys for some of them (the *corrupted* ones). We thus introduce a general key generation procedure, denoted KeyGen which corresponds to the above execution of ParGen, IssGen and VerGen. This procedure, executed by the challenger, takes as input 1^λ and outputs (gpk, $\mathsf{sk}_\mathcal{I}$, sk_V).

In the following experiments, the adversary can play either the issuer or the user during the Issue procedure. In the first case, the adversary requests the ObtainC oracle[2] on chosen attributes while in the second case, the requested oracle is denoted IssueC on attributes chosen by either the adversary or the challenger. When the role of the issuer is played by the challenger, the set of all issuer's views for the Issue protocol is denoted V. Each entry V_i of V includes the set $\mathbf{M}_i$ of certified plain attributes. We use similar notation for the Verify protocol, with a request to the ShowC when the adversary plays the role of the verifier and a request to VerifyC otherwise.

[2] By convention, the name of the oracle denotes the action executed by the challenger.

$\mathbf{Exp}^{\mathsf{unf}}_{\Pi,\mathcal{A}}(\lambda)$;
- $(\mathsf{gpk}, \mathsf{sk}_{\mathcal{I}}, \mathsf{sk}_{\mathcal{V}}) \leftarrow \textsc{KeyGen}(1^\lambda)$;
- $(st) \leftarrow \mathcal{A}_g^{\textsc{IssueC},\textsc{VerifyC},\textsc{ShowC}}(\mathsf{gpk}, \mathsf{sk}_{\mathcal{V}})$;
- $\textsc{Verify} : (\perp \leftarrow \mathcal{A}_c^{\mathcal{U}}(st)), \mathsf{out} \leftarrow \mathcal{C}^{\mathcal{V}}(\mathsf{sk}_{\mathcal{V}}))$;
- if $\mathsf{out} = 0$, then return 0;
- if $(\mathsf{out} = (1, \widetilde{\mathbf{M}}) \wedge \exists i : \mathbf{M}_i = \widetilde{\mathbf{M}}$, return 0;
- return 1.

$\mathbf{Exp}^{\mathsf{attmask}}_{\Pi,\mathcal{A}}(\lambda)$;
- $b \leftarrow \{0,1\}$;
- $(\mathsf{gpk}, \mathsf{sk}_{\mathcal{I}}, \mathsf{sk}_{\mathcal{V}}) \longleftarrow \textsc{KeyGen}(1^\lambda)$;
- $(\mathbf{M}_0, \mathbf{M}_1, st) \leftarrow \mathcal{A}_g^{\textsc{IssueC},\textsc{VerifyC}}(\mathsf{gpk})$;
- $\textsc{Issue} : (\perp \leftarrow \mathcal{C}^{\mathcal{I}}(\mathbf{M}_b)), (\tilde{st} \leftarrow \mathcal{A}_{ch}^{\mathcal{U}}(st))$;
- $b' \leftarrow \mathcal{A}_{gu}^{\textsc{IssueC},\textsc{VerifyC}}(\mathsf{gpk}, \tilde{st})$;
- return $(b = b')$.

$\mathbf{Exp}^{\mathsf{up}}_{\Pi,\mathcal{A}}(\lambda)$
- $b \leftarrow \{0,1\}$;
- $(\mathsf{gpk}, \mathsf{sk}_{\mathcal{I}}, \mathsf{sk}_{\mathcal{V}}) \longleftarrow \textsc{KeyGen}(1^\lambda)$;
- $(st, \mathbf{C}) \leftarrow \mathcal{A}_g^{\textsc{ObtainC},\textsc{ShowC}}(\mathsf{gpk}, \mathsf{sk}_{\mathcal{I}}, \mathsf{sk}_{\mathcal{V}})$;
- $\textsc{Issue} : (st_0 \leftarrow \mathcal{A}_{ch_1}^{\mathcal{I}}(st)), ((\mathsf{cred}_0) \leftarrow \mathcal{C}^{\mathcal{U}}(\mathbf{C}))$;
- $\textsc{Issue} : (st_1 \leftarrow \mathcal{A}_{ch_2}^{\mathcal{I}}(st_0)), ((\mathsf{cred}_1) \leftarrow \mathcal{C}^{\mathcal{U}}(\mathbf{C}))$;
- $\textsc{Verify} : (\perp \leftarrow \mathcal{C}^{\mathcal{U}}(\mathsf{cred}_b, \mathbf{C})), \tilde{st} \leftarrow \mathcal{A}_{ch_f}^{\mathcal{V}}(st_1))$;
- $b' \leftarrow \mathcal{A}_{gu}^{\textsc{ObtainC},\textsc{ShowC}}(\tilde{st})$;
- return $(b = b')$.

$\mathbf{Exp}^{\mathsf{hv\text{-}up}}_{\Pi,\mathcal{A}}(\lambda)$
- $b \leftarrow \{0,1\}$;
- $(\mathsf{gpk}, \mathsf{sk}_{\mathcal{I}}, \mathsf{sk}_{\mathcal{V}}) \longleftarrow \textsc{KeyGen}(1^\lambda)$;
- $(st, \mathbf{C}_0, \mathbf{C}_1) \leftarrow \mathcal{A}_g^{\textsc{ObtainC},\textsc{VerifyC}}(\mathsf{gpk}, \mathsf{sk}_{\mathcal{I}})$;
- if $|\mathbf{C}_0| \neq |\mathbf{C}_1|$, then return 0;
- $\textsc{Issue} : (st_0 \leftarrow \mathcal{A}_{ch_1}^{\mathcal{I}}(st)), ((\mathsf{cred}_0) \leftarrow \mathcal{C}^{\mathcal{U}}(\mathbf{C}_0))$;
- $\textsc{Issue} : (st_1 \leftarrow \mathcal{A}_{ch_2}^{\mathcal{I}}(st_0)), ((\mathsf{cred}_1) \leftarrow \mathcal{C}^{\mathcal{U}}(\mathbf{C}_1))$;
- $\textsc{Verify} : (\perp \leftarrow \mathcal{C}^{\mathcal{U}}(\mathsf{cred}_b, \mathbf{C}_b)), \tilde{st} \leftarrow \mathcal{A}_{ch_f}^{\mathcal{V}}(st_1))$;
- $b' \leftarrow \mathcal{A}_{gu}^{\textsc{ObtainC},\textsc{VerifyC}}(\tilde{st})$;
- return $(b = b')$.

Fig. 1. Security experiments

In the following, a protocol PROT between an entity $\mathcal{E}_0$, playing the role of $\mathcal{R}_0$ (of a user, a verifier or an issuer), taking on input i_0 and outputting o_0 and an entity $\mathcal{E}_1$, playing a role $\mathcal{R}_1$, taking on input i_1 and outputting o_1 is denoted $\textsc{Prot} : (o_0 \leftarrow \mathcal{E}_0^{\mathcal{R}_0}(i_0)), (o_1 \leftarrow \mathcal{E}_1^{\mathcal{R}_1}(i_1))$.

The different security experiments are next given in Figure 1 while the related security definitions are given as follows.

Unforgeability. As we are in the multi-show case and do not rely on blind signatures, we cannot use the same definition as Guajardo *et al.* [21] who ask the adversary to output more credentials than generated by the issuer. In fact, we give a single definition which embeds both the *one-more unforgeability* and the *blinding invariance unforgeability* properties introduced in the original model [21]. In particular, the blinding invariance unforgeability property states in [21] that for any attribute list output by the adversary, the number of credentials on this list does not exceed the number of times a credential has been issued on this list and the one-more unforgeability property prevents an adversary from outputting $K+1$ distinct credentials after having requested only K credentials. In our setting, this can be simplified by preventing an adversary from being accepted during a VERIFY protocol with a set of attributes which has never been certified by the issuer.

More precisely, our experiment asks the adversary to successfully play a VERIFY protocol such that the embedded attributes have never been certified by the issuer. The *unforgeability* experiment $\mathsf{Exp}^{\mathsf{unf}}_{\Pi,\mathcal{A}}(1^\lambda)$ for the adversary $\mathcal{A} = (\mathcal{A}_g, \mathcal{A}_c)$, with security parameter λ, is given in Figure 1 and an anonymous credential with encrypted attributes scheme satisfies the *unforgeability* property iff there exists a negligible function $\nu(\lambda)$ such that for any adversary $\mathcal{A}$, $\Pr(\mathsf{Exp}^{\mathsf{unf}}_{\Pi,\mathcal{A}}(1^\lambda) \longrightarrow 1) < \nu(\lambda)$.

Attribute Masking. This property says that no unauthorized party should learn the encrypted attributes. We here consider, as in [21], the case where only the user does not learn the plain attributes. Other cases (*e.g.* attributes not known by the issuer) can easily be adapted. Contrary to [21], we here provide a formal security definition, for which the experiment $\mathbf{Exp}^{\mathsf{attmask}}_{\Pi,\mathcal{A}}(\lambda)$ is given in Figure 1. It follows that an anonymous credential with encrypted attributes scheme satisfies the *attribute masking* property iff there exists a negligible function $\nu(\lambda)$ such that for any adversary $\mathcal{A} = (\mathcal{A}_g, \mathcal{A}_{ch}, \mathcal{A}_{gu})$, $\Pr\left(\mathsf{Exp}^{\mathsf{attmask}}_{\Pi,\mathcal{A}}(1^\lambda) \longrightarrow 1\right) < \frac{1}{2} + \nu(\lambda)$.

User Privacy. Contrary to the definition given in [21], we consider the case of a multi-show credential. Then, the user privacy should include the possibility for one single user to use several times the same credential, without being traced. In fact, there are two cases, depending on the possibility for the adversary to corrupt (*user privacy*) or not (*honest-verifier user privacy*) the verifier. In both cases, the adversary plays the role of the issuer $\mathcal{I}$ and executes two different ISSUE protocols with the challenger. Next, one of the two output credential is used by the challenger during a VERIFY protocol. If the verifier is corrupted, this experiment can easily be won by the adversary since the corrupted issuer can certify two different sets of attributes and the corrupted verifier can easily check which one is used during the VERIFY protocol by decrypting the encrypted attributes. Thus, when the verifier is corrupted, this experiment is only relevant when the plain attributes are similar in both ISSUE protocols. For this purpose (see $\mathbf{Exp}^{\mathsf{up}}$ in Figure 1), the adversary output one single set of encrypted attributes $\mathbf{C}$, which is used twice in both ISSUE protocols. We do not need this restriction for the case where the verifier is honest and the adversary thus output two different encrypted attributes $\mathbf{C}_0$ and $\mathbf{C}_1$ (see $\mathbf{Exp}^{\mathsf{hv\text{-}up}}$ in Figure 1).

Both experiments are given in Figure 1. Next, the scheme satisfies the *user privacy* (resp. *honest-verifier user privacy*) property iff there exists a negligible function $\nu(\lambda)$ such that for any adversary $\mathcal{A} = (\mathcal{A}_g, \mathcal{A}_{ch_1}, \mathcal{A}_{ch_2}, \mathcal{A}_{ch_f}, \mathcal{A}_{gu})$, $\Pr\left(\mathsf{Exp}^{\mathsf{up}}_{\Pi,\mathcal{A}}(1^\lambda) \longrightarrow 1\right) < \frac{1}{2} + \nu(\lambda)$ (resp. $\Pr\left(\mathsf{Exp}^{\mathsf{hv\text{-}up}}_{\Pi,\mathcal{A}}(1^\lambda) \longrightarrow 1\right) < \frac{1}{2} + \nu(\lambda)$).

3 Cryptographic Tools

We here introduce the cryptographic tools we need in the following. This includes extractable commitment schemes, Groth-Sahai (GS) proofs [20] and automorphic signatures [2,15].

In the following, a bilinear environment is given by the tuple (p, $\mathbb{G}_1$, $\mathbb{G}_2$, $\mathbb{G}_T$, e, g_1, g_2) where p is a prime number, $\mathbb{G}_1$, $\mathbb{G}_2$ and $\mathbb{G}_T$ are groups of order p, g_1 (resp. g_2) is a generator of $\mathbb{G}_1$ (resp. $\mathbb{G}_2$), and $e : \mathbb{G}_1 \times \mathbb{G}_2 \longrightarrow \mathbb{G}_T$ is a pairing with the non-degeneracy ($e(g_1, g_2) \neq 1$) and bilinearity (for all $u \in \mathbb{G}_1$, $v \in \mathbb{G}_2$ and $a, b \in \mathbb{Z}_p$, $e(u^a, v^b) = e(u, v)^{ab}$) properties. For vectors of group elements, "$\odot$" denotes the component-wise group operation.

As we use the Abe *et al.* proposal [3] for signing a vector of messages, we also need an injective mapping $\langle . \rangle : \{1, ..., n_{\max}\} \rightarrow \mathbb{G}_1 \times \mathbb{G}_2$ such that for all $n, n' \in \{1, ..., n_{\max}\}$, $\langle n \rangle \odot \langle n' \rangle \neq (1, 1)$, where $n_{max} \in \mathbb{N}$ is a fixed parameter. In the following, we consider $\langle . \rangle : n \mapsto ({g_1}^n, {g_2}^n)$ since for all reasonably small $n, n' \in \{1, ..., n_{\max}\}$ (we consider n_{max} as small in the following constructions), we have $({g_1}^{n+n'}, {g_2}^{n+n'}) \neq (1, 1)$.

3.1 Randomizable and Extractable Commitment Schemes

A *commitment scheme* permits one user to commit to a message, using some randomness, such that it is possible to further give the message and the randomness to prove that this was truly the committed message. The commitment becomes *extractable* when the commit process makes use of a public key which is related to a secret key allowing the owner of the latter to open any commitment and retrieve the initially committed message. Finally, the commitment scheme is said *randomizable* if it exists a public procedure which permits to randomize a given commitment, without obtaining any information about the initially committed message, and such that it is infeasible to know whether two given commitments are related to the same message or not. A formal definition of such a commitment can be found in [20].

SXDH Commitments. In the following, we will use SXDH randomizable and extractable commitment schemes [20], which can be described as follows.

Key generation. Given a bilinear environment, the extractable keys are $\alpha_1, \alpha_2 \in \mathbb{Z}_p$ and the public key ck of the commitment schemes is composed of $\mathbf{u} := (\mathbf{u}_1, \mathbf{u}_2), \mathbf{v} := (\mathbf{v}_1, \mathbf{v}_2)$ where (for $t_1, t_2 \in \mathbb{Z}_p$)

$$\mathbf{u}_1 := (g_1, {g_1}^{\alpha_1}), \mathbf{u}_2 := ({g_1}^{t_1}, {g_1}^{\alpha_1 t_1}), \mathbf{v}_1 := (g_2, {g_2}^{\alpha_2}), \mathbf{v}_2 := ({g_2}^{t_2}, {g_2}^{\alpha_2 t_2}).$$

Commitment. The commitment to a group element X, with randomness $\rho = (\rho_1, \rho_2) \in \mathbb{Z}_p^2$ is

$$\mathbf{c} := (c_1, c_2) = ({u_{11}}^{\rho_1} \cdot {u_{21}}^{\rho_2}, X \cdot {u_{12}}^{\rho_1} \cdot {u_{22}}^{\rho_2}) \text{ if } X \in \mathbb{G}_1 \text{ and}$$
$$\mathbf{c} := (c_1, c_2) = ({v_{11}}^{\rho_1} \cdot {v_{21}}^{\rho_2}, X \cdot {v_{12}}^{\rho_1} \cdot {v_{22}}^{\rho_2}) \text{ if } X \in \mathbb{G}_2.$$

Such a commitment is in the following denoted $\text{SXDHCom}(\mathsf{ck}, X, \rho)$.

Extraction. The extraction retrieves $X \in \mathbb{G}_1$ (resp. $X \in \mathbb{G}_2$) by computing $c_2 \cdot c_1^{-\alpha_1}$ (resp. $c_2 \cdot c_1^{-\alpha_2}$).

Randomization. Given a commitment $\mathbf{c}$ and some fresh randomness $\rho' = (\rho'_1, \rho'_2) \in \mathbb{Z}_p^2$, the randomization of $\mathbf{c}$ is done by computing

$$\mathbf{c}' := (c_1 \cdot {u_{11}}^{\rho'_1} \cdot {u_{21}}^{\rho'_2}, c_2 \cdot {u_{12}}^{\rho'_1} \cdot {u_{22}}^{\rho'_2}) \text{ if } \mathbf{c} \in \mathbb{G}_1^2 \text{ and}$$
$$\mathbf{c}' := (c_1 \cdot {v_{11}}^{\rho'_1} \cdot {v_{21}}^{\rho'_2}, c_2 \cdot {v_{12}}^{\rho'_1} \cdot {v_{22}}^{\rho'_2}) \text{ if } \mathbf{c} \in \mathbb{G}_2^2.$$

3.2 (SXDH) Groth-Sahai Proofs

Groth and Sahai have described in [20] a witness indistinguishable proof system, for a class of *pairing-product equations* (PPE for short) over variables $X_1, \ldots, X_m \in \mathbb{G}_1$ and $Y_1, \ldots, Y_n \in \mathbb{G}_2$ as

$$E(X_1, \ldots, X_m; Y_1, \ldots, Y_n) \ : \ \prod_{i=1}^{m} e(\underline{X_i}, B_i) \prod_{j=1}^{n} e(A_j, \underline{Y_j}) \prod_{i=1}^{m} \prod_{j=1}^{n} e(\underline{X_i}, \underline{Y_j})^{\gamma_{i,j}} = \mathbf{t}_\tau$$

defined by elements $A_j \in \mathbb{G}_1$, $B_i \in \mathbb{G}_2$, $\gamma_{i,j} \in \mathbb{Z}_p$ for $i \in [1, m]$, $j \in [1, n]$ and $\mathbf{t}_\tau \in \mathbb{G}_T$ and where the notation $\underline{X}$ means that the variable X is a secret value. In this paper, we use the SXDH version of GS proofs. Such a proof is denoted $\textsc{Prove}(\mathsf{ck}, E, (X_1, \ldots, X_m; Y_1, \ldots, Y_n), (r_1, \ldots, r_m; s_1, \ldots, s_m))$ where r_i, $s_j \in \mathbb{Z}_p$. We refer the reader to *e.g.* [20,16] for details.

In [4], it has been shown that such proofs can be publicly randomized, in such a way that it is infeasible to link the original proof to the randomized one. This procedure is in the following denoted $\textsc{RdProof}(\mathsf{ck}, E, (\mathbf{c}_i, r'_i)_{i=1}^m, (\mathbf{d}_j, s'_j)_{j=1}^n, \pi)$ where $\mathbf{c}_i$, for $i \in [1, m]$, denotes a commitment to X_i and $\mathbf{d}_j$, for $j \in [1, n]$, a commitment to Y_j, π is the proof to be randomized and the r'_i, s'_j correspond to the new randomness.

Diffie-Hellman Pairing-Product Equation. In the sequel, we will need several times to provide a GS proof with a *DH pairing-product equation.* This equation is denoted $E_{\mathcal{DH}}$ and, on the values $(M, N) \in \mathbb{G}_1 \times \mathbb{G}_2$ and where $g_1, g_2 \in \mathbb{G}_1 \times \mathbb{G}_2$, is given by $E_{\mathcal{DH}}^{(g_1, g_2)}(M, N) \ : \ e(\underline{M}, g_2) \cdot e({g_1}^{-1}, \underline{N}) = 1$.

3.3 GS Proof of Equality under Different Commitment Keys

In the following, we need to prove that two values X_1 and X_2, committed with two different keys, are equal. Such GS proof has already been given in [18] for the DLIN case but we here need it in the SXDH one. Such a proof is in the following denoted

$$\pi_{\text{eq}} \leftarrow d\textsc{Prove}^{\text{eq}}((\mathsf{ck}, \mathsf{ck}'), E_{\text{eq}}, (X_1, X_2), (r, s, r', s')).$$

When the values X_1 and X_2 are committed using the same key, the GS proof is related to the PPE

$$E_{\text{eq}}(X_1, X_2) \quad : \quad e(\underline{X_1}, g_2) \cdot e(\underline{X_2}, {g_2}^{-1}) = 1. \qquad (1)$$

Consider now two commitments $\mathbf{c}_1, \mathbf{c}_2$ of X_1, X_2 under *different* commitment keys $\mathsf{ck} := (\mathbf{u}, \mathbf{v})$ and $\mathsf{ck}' := (\mathbf{u}', \mathbf{v}')$ respectively. We want to construct a witness-indistinguishable proof system that X_1 and X_2 satisfy E_{eq} from $\mathbf{c}_1$ and $\mathbf{c}_2$. We have $\mathbf{c}_1 := ({u_{11}}^{r_1} \cdot {u_{21}}^{r_2}, X_1 \cdot {u_{12}}^{r_1} \cdot {u_{22}}^{r_2})$ for uniformly chosen $r_1, r_2 \xleftarrow{\$} \mathbb{Z}_p$. If we fix X_2, the proof that the committed value X_1 satisfies equation $e(\underline{X_1}, g_2) = e(X_2, g_2)$ can be reduced to $(\phi_{12} := {g_2}^{r_1}, \phi_{22} := {g_2}^{r_2})$ which can be checked[3] by

$$e(\mathbf{c}_{11}, g_2) = e(u_{11}, \phi_{12}) \cdot e(u_{21}, \phi_{22}) \text{ and} \tag{2a}$$
$$e(\mathbf{c}_{12}, g_2) = e(X_2, g_2) \cdot e(u_{12}, \phi_{12}) \cdot e(u_{22}, \phi_{22}), \tag{2b}$$

which gives us the first part of our GS proof.

Regarding now (2a) and (2b) as a set of equations over variables X_2, ϕ_{21} and ϕ_{22}, we use the GS proof system a second time by committing to these new variables under key $\mathbf{u}'$. Note that as we have already treated the case of $\mathbf{c}_{11}$ and $\mathbf{c}_{12}$, we consider them as fixed in the second part of the GS proof. This leads us to apply the proof algorithm on the following equations.

$$E_{\text{eq}'1}: \qquad e(u_{11}, \underline{{g_2}^{r_1}}) \cdot e(u_{21}, \underline{{g_2}^{r_2}}) = e(\mathbf{c}_{11}, g_2) \tag{3a}$$
$$E_{\text{eq}'2}: \qquad e(\underline{X_2}, g_2) \cdot e(u_{12}, \underline{{g_2}^{r_1}}) \cdot e(u_{22}, \underline{{g_2}^{r_2}}) = e(\mathbf{c}_{12}, g_2) \tag{3b}$$

The resulting complete GS proof will be given in the full version of the paper. Regarding the randomization of such a proof, one need to update commitments and proofs to the new randomness for the commitment on X_1.

3.4 Automorphic Signatures

Automorphic signatures have been introduced in [2,15] as a new signature scheme where (i) the verification keys lie in the message space, (ii) messages and signatures consist of elements of a bilinear group, and (iii) verification is done by evaluating a set of pairing-product equations. Automorphic signatures are used in [16] to construct commuting signatures, where a commuting signature is concretely a verifiably encrypted automorphic signature. We now described the instantiation given in [15].

Let $(p, \mathbb{G}_1, \mathbb{G}_2, \mathbb{G}_T, e, g_1, g_2)$ be a bilinear environment as defined above. We also need $h, k, u \in \mathbb{G}_1$. The message space is $\mathcal{DH} := \{({g_1}^m, {g_2}^m) \mid m \in \mathbb{Z}_p\}$. The secret key is $x \in \mathbb{Z}_p^*$ and the related public verification key is $\mathsf{vk} := (X, Y) = ({g_1}^x, {g_2}^x)$. The signature of a message $(M, N) \in \mathcal{DH}$ is done by picking $c, r \in \mathbb{Z}_p$ at random and computing

$$\sigma := \left(A := \left(h \cdot M \cdot k^r\right)^{\frac{1}{x+c}}, B := u^c, D := {g_2}^c, R := {g_1}^r, S := {g_2}^r\right).$$

A signature $\sigma = (A, B, D, R, S)$ on a message $(M, N) \in \mathcal{DH}$ is valid iff $e(A, Y \cdot D) = e(h \cdot M, g_2) \cdot e(k, S)$, $e(B, g_2) = e(u, D)$ and $e(R, g_2) = e(g_1, S)$. Details can be found in [15,16].

[3] As explained in Section 6.1 of the full version of [20], two group elements are enough for such a proof.

4 Commuting Signatures and Some New Extensions

We here introduce commuting signatures and some extensions which are of independent interest. In our main scheme, we need to sign a vector of messages while individual messages can be committed using different commitment extraction keys. To the best of our knowledge, this description is new and we here give a general way to treat such a case.

4.1 Additional Commitments

Fuchsbauer [16], when constructing commuting signatures, makes use of commitment on Diffie-Hellman (DH) tuples which are messages signed by the automorphic signature scheme introduced above.

Commitment on Diffie-Hellman Tuples. Let $k \in \mathbb{G}_1$. A commitment on a DH tuple (M, N) takes as input some randomness $(t, \mu, \nu, \rho, \sigma)$. It first computes $(P, Q) = ({g_1}^t, {g_2}^t)$ and $U = M \cdot k^t$. It next computes SXDH commitments $\mathbf{c}_M = \text{SXDHCom}(\mathsf{ck}, M, \mu)$, $\mathbf{c}_N = \text{SXDHCom}(\mathsf{ck}, N, \nu)$, $\mathbf{c}_P = \text{SXDHCom}(\mathsf{ck}, P, \rho)$ and $\mathbf{c}_Q = \text{SXDHCom}(\mathsf{ck}, Q, \sigma)$. Finally, it executes the SXDH GS proofs $\pi_M = \text{Prove}(ck, E_{\mathcal{DH}}^{(g_1,g_2)}, (M, N), (\mu, \nu))$, $\pi_P = \text{Prove}(ck, E_{\mathcal{DH}}^{(g_1,g_2)}, (P, Q), (\rho, \sigma))$ and $\pi_U = \text{Prove}(ck, E_U, (M, Q), (\mu, \sigma))$ where

$$E_U(M, Q) : e(k^{-1}, \underline{Q}) \cdot e(\underline{M}, {g_2}^{-1}) = e(U, g_2)^{-1}$$

The commitment on (M, N) is $C = \text{CommitDH}(\mathsf{ck}, (M, N), (t, \mu, \nu, \rho, \sigma)) = (\mathbf{c}_M, \mathbf{c}_N, \pi_M, \mathbf{c}_P, \mathbf{c}_Q, \pi_P, U, \pi_U)$. Such a commitment is randomizable [16] using the randomization techniques for SXDH commitments and GS proofs.

More Simple Commitment on DH Tuples. Before producing a commuting signature on a message (M, N), it is necessary to produce such a commitment on the message. In some other cases, we need to produce a commitment on a DH tuple but without any relation with the execution of a commuting signature on the underlying message. In this case, we do not need to make so a complicated commitment on a DH tuple. In fact, the values P and Q are in this case not necessary (see [16] for details). The above commitment, denoted $\text{CommitDH}^{-}(\mathsf{ck}, (M, N), (\mu, \nu))$ is thus reduced to the values $(\mathbf{c}_M, \mathbf{c}_N, \pi_M)$.

Commitment to an Automorphic Signature When the Message Is Known But Committed. An automorphic signature (see [15] and Section 3.4) can be committed as follows, when the signed message (M, N) is known[4] but committed as a DH tuple: $\text{CommitDH}(\mathsf{ck}, (M, N), (\mu, \nu))$.

Let $\sigma = (A, B, D, R, S)$ be an automorphic signature on the message (M, N). A commitment on a signature corresponds to $(\mathbf{c}_\sigma, \pi_\sigma) = ((\mathbf{c}_A, \mathbf{c}_B, \mathbf{c}_D, \mathbf{c}_R, \mathbf{c}_S), (\pi_A, \pi_B, \pi_R))$ where c_X corresponds to $\text{SXDHCom}(\mathsf{ck}, X, \alpha_X)$ (for $X \in \{A, B, D, R, S\}$) and where $\pi_A = \text{Prove}(\mathsf{ck}, E_A, (A, \alpha_A), (M, \mu), (D, \alpha_D), (S, \alpha_S))$,

[4] The case where the message is not known is different, see [16].

$\pi_B = \textsc{Prove}(\mathsf{ck}, E_{\mathcal{DH}}^{(u,g_2)}, (B, \alpha_B), (D, \alpha_D))$ and finally $\pi_R \leftarrow \textsc{Prove}(\mathsf{ck}, E_{\mathcal{DH}}^{(g_1,g_2)}, (R, \alpha_R), (S, \alpha_S))$ with

$$E_A(A, M, D, S) : e(k^{-1}, \underline{S}) \cdot e(\underline{A}, Y) \cdot e(\underline{M}, {g_2}^{-1}) \cdot e(\underline{A}, \underline{D}) = e(h, g_2).$$

In the following, such a procedure is denoted $\textsc{CommitSign}(\mathsf{ck}, \mathsf{vk}, (M, N), \sigma, (\mu, \nu, \alpha_A, \alpha_B, \alpha_D, \alpha_R, \alpha_S))$.

Commitment to an Automorphic Signature When the Message Is Known But not Committed. The case where the message (M, N) is publicly known (and thus not committed) can be simplified since the equation E_A becomes

$$E_A(A, M, D, S) : e(k^{-1}, \underline{S}) \cdot e(\underline{A}, Y) \cdot e(\underline{A}, \underline{D}) = e(h \cdot M, g_2).$$

The rest of the procedure is done similarly and the result is $\textsc{CommitSign}^{-}(\mathsf{ck}, \mathsf{vk}, (M, N), \sigma, (\mu, \nu, \alpha_A, \alpha_B, \alpha_D, \alpha_R, \alpha_S))$ in the following.

4.2 Simple Commuting Signature: One Committed Message and One Commitment Key

We now recall the $\textsc{SigCom}$ algorithms to sign a DH tuple committed using the above commitment scheme for DH tuples as described in [16]. The secret signing key is $\mathsf{sk} = x \in \mathbb{Z}_p^*$ and the corresponding public key is the DH tuple $\mathsf{vk} = (X = {g_1}^x, Y = {g_2}^x)$, as for an automorphic signature. The signature of a DH commitment $C = (\mathbf{c}_M, \mathbf{c}_N, \pi_M, \mathbf{c}_P, \mathbf{c}_Q, \pi_P, U, \pi_U)$ on (M, N) is done as follows (see [16] for details).

- The signer first picks fresh random values $c, r \xleftarrow{\$} \mathbb{Z}_p$ and $\alpha, \beta, \delta, \rho', \sigma' \xleftarrow{\$} \mathbb{Z}_p^2$.
- She next computes an automorphic signature (see [2,15] and Section 3.4) $A = (h \cdot U \cdot k^r)^{\frac{1}{x+c}}$, $B = u^c$, $D = {g_2}^c$, $R = {g_1}^r$ and $S = {g_2}^r$.
- She also computes the SXDH commitments $\mathbf{c}_A = \textsc{SXDHCom}(ck, A, \alpha)$, $\mathbf{c}_B = \textsc{SXDHCom}(\mathsf{ck}, B, \beta)$, $\mathbf{c}_D = \textsc{SXDHCom}(\mathsf{ck}, D, \delta)$, $\mathbf{c}_R = \mathbf{c}_P \odot \textsc{SXDHCom}(\mathsf{ck}, R, \rho')$ and $\mathbf{c}_S = \mathbf{c}_Q \odot \textsc{SXDHCom}(\mathsf{ck}, S, \sigma')$.
- She makes the GS proofs: $\pi'_A := \pi_U \odot \textsc{Prove}(\mathsf{ck}, E_{A^\dagger}, (A, D), (\alpha, \delta))$, $\pi_A := \textsc{RdProof}(\mathsf{ck}, E_A, (\mathbf{c}_A, 0), (\mathbf{c}_M, 0), (\mathbf{c}_D, 0), (\mathbf{c}_S, \sigma'), \pi'_A)$, $\pi_B = \textsc{Prove}(ck, E_{\mathcal{DH}}, (B, D), (\beta, \delta))$ and $\pi_R := \textsc{RdProof}(\mathsf{ck}, E_R, (\mathbf{c}_R, \rho'), (\mathbf{c}_S, \sigma'), \pi_P)$ where

$$\begin{aligned} E_A(A, M, D, S) &: e(k^{-1}, \underline{S}) \cdot e(\underline{A}, Y) \cdot e(\underline{M}, {g_2}^{-1}) \cdot e(\underline{A}, \underline{D}) = e(h, g_2) \\ E_{A^\dagger}(A, D) &: e(\underline{A}, Y) \cdot e(\underline{A}, \underline{D}) = 1 \\ E_B(B, D) &: e(u^{-1}, \underline{D}) \cdot e(\underline{B}, g_2) = 1 \\ E_R(R, S) &: e({g_1}^{-1}, \underline{S}) \cdot e(\underline{R}, g_2) = 1 \end{aligned}$$

The PPE $E_{A^\dagger}$ is not truly verified but is necessary to produce the GS proof on E_A (see [16]). The signature is $\Sigma := \big(\mathbf{c}_\Sigma = (\mathbf{c}_A, \mathbf{c}_B, \mathbf{c}_D, \mathbf{c}_R, \mathbf{c}_S), \pi_\Sigma = (\pi_A, \pi_B, \pi_R)\big)$.

4.3 Vector of Committed Messages and One Commitment Key

In our anonymous credential scheme with encrypted attributes, we need to sign several messages at the same time. For this purpose, we need to adapt the above commuting signature, which can easily be done by using the recent technique of [3] and [17]. On input a signing key x and a vector $(C_1, \ldots, C_n)$ of commitments to the DH tuples $(M_1, N_1), \ldots, (M_n, N_n)$, the whole procedure works as follows. Let $n_{\max} \in \mathbb{N}$ be the maximum number of messages we can sign together and let $\mathsf{sk} = x \in \mathbb{Z}_p^*$ be the used secret signing key, related to the public key $\mathsf{vk} = (X = {g_1}^x, Y = {g_2}^x)$.

1. Run the signing key generation $n+1$ times to get $(\mathsf{vk}_i = (X_i, Y_i), \mathsf{sk}_i = x_i)_{i=0}^n$, where $\mathsf{sk}_i \in \mathbb{Z}_p^*$.
2. Compute the following automorphic signatures (see Section 3.4): Γ_0 on vk_0 under sk and Δ_0 on $\langle n \rangle$ (as defined at the beginning of Section 3) under sk_0. The message vk_0 is next committed using CommitDH$^-$ to obtain $\mathbf{c}_{\mathsf{vk}_0}$. Finally, the signatures Γ_0 and Δ_0 are also committed using the procedures CommitSign (resp. CommitSign$^-$ since $\langle n \rangle$ is not committed), and obtain $(\mathbf{c}_{\Gamma_0}, \pi_{\Gamma_0})$ (resp. $(\mathbf{c}_{\Delta_0}, \pi_{\Delta_0})$).
3. Executes n times the following: produce an automorphic signature Γ_i on vk_i under sk_0 and a signature Δ_i on $\mathsf{vk}_i \odot \langle i \rangle$ under sk_0. In addition, commit to each message and each signature. Again, for $i \in [1, n]$, the message vk_i is committed using CommitDH$^-$ to obtain $\mathbf{c}_{\mathsf{vk}_i}$. The signatures Γ_i and Δ_i are also committed using CommitSign (this time, the message $\mathsf{vk}_i \odot \langle i \rangle$ related to Δ_i is not totally known), and obtain $(\mathbf{c}_{\Gamma_i}, \pi_{\Gamma_i})$ and $(\mathbf{c}_{\Delta_i}, \pi_{\Delta_i})$ respectively.
4. Executes n times the SigCom procedure for a single message (see Section 4.2 above). The i-th execution takes as inputs the commitment C_i and the related signing key sk_i and outputs a commuting signature $\Sigma_i = (\mathbf{c}_{\Sigma_i}, \pi_{\Sigma_i})$.
5. Commit to all public keys vk_i as $\mathbf{c}_{\mathsf{vk}_i} = (\mathbf{c}_{X_i} = \text{SXDHCom}(\mathsf{ck}, X_i, \zeta_i), \mathbf{c}_{Y_i} = \text{SXDHCom}(\mathsf{ck}, Y_i, \psi_i))$ and $\pi_{X_i} = \text{Prove}(\mathsf{ck}, E_{\mathcal{DH}}^{(g_1,g_2)}, (X_i, \zeta_i), (Y_i, \psi_i))$.
6. As the above signatures Σ_i are valid under the plain public keys vk_i, we need to use next the AdPrC$_K$ procedure[5] [16] to adapt Σ_i so that it is valid under the public keys which are committed in the $\mathbf{c}_{\mathsf{vk}_i}$'s. Given $\Sigma_i = (\mathbf{c}_{A_i}, \mathbf{c}_{B_i}, \mathbf{c}_{D_i}, \mathbf{c}_{R_i}, \mathbf{c}_{S_i}, \pi_{A_i}, \pi_{B_i}, \pi_{R_i})$, the resulting commuting signature is next $\Sigma_i' = (\mathbf{c}_{A_i}, \mathbf{c}_{B_i}, \mathbf{c}_{D_i}, \mathbf{c}_{R_i}, \mathbf{c}_{S_i}, \pi'_{A_i}, \pi_{B_i}, \pi_{R_i})$ where the proof π'_{A_i} = RdProof(ck, $E_{\hat{A}}$, $(\mathbf{c}_{A_i}, 0)$, $(\mathbf{c}_{M_i}, 0)$, $(\mathbf{c}_{S_i}, 0)$, (SXDHCom(ck, Y_i, 0), ψ_i), $(\mathbf{c}_{A_i}, 0)$, π_{A_i}), with

$$E_{\hat{A}}(A, M, S, Y, D) : e(k^{-1}, \underline{S}) \cdot e(\underline{A}, \underline{Y}) \cdot e(\underline{M}, {g_2}^{-1}) \cdot e(\underline{A}, \underline{D}) = e(h, g_2).$$

The whole commuting signature on the vector $\big((M_1, N_1), \ldots, (M_n, N_n)\big)$ is finally $\Sigma = \big((\mathbf{c}_{\mathsf{vk}_i}, \mathbf{c}_{\Gamma_i}, \mathbf{c}_{\Delta_i}, \pi_{\mathsf{vk}_i}, \pi_{\Gamma_i}, \pi_{\Delta_i})_{i=0}^n, (\Sigma_i')_{i=1}^n\big)$.

[5] The AdPrC$_K$ procedure (Adapt Proof when Committing to the Key) allows to adapt proofs when committing or decommitting to the verification key.

4.4 Vector of Committed Messages and Several Commitment Keys

We now introduce the way we will use such signatures in our anonymous credential scheme with encrypted attributes. To the best of our knowledge, this procedure is new and may be of independent interest.

We assume having the DH tuple commitments $(C_1, \ldots, C_n)$ on the messages $(M_1, N_1), \ldots, (M_n, N_n)$ under possibly several commitment keys $(\mathsf{ck}_1, \ldots, \mathsf{ck}_n)$, using the randomness $(\mu_1, \nu_1), \ldots, (\mu_n, \nu_n)$. We use a commuting signature with the commitment key denoted ck and the secret signing key $\mathsf{sk} = x$.

The idea is to use the above procedure. We commit to each message using the same commitment key and prove that the committed values are equals, using our new procedure given in Section 3.3. More precisely, we have the following.

1. For all $i \in [1, n]$, produce a DH tuple commitment on (M_i, N_i) using ck, which outputs $C_i^{(\mathsf{ck})} = \textsc{CommitDH}(\mathsf{ck}, (M_i, N_i), (\mu_i, \nu_i))$.
2. For all $i \in [1, n]$, produce a GS proof of equality under different commitment keys (see Section 3.3): $\pi_{\mathrm{eq}_i} = d\textsc{Prove}^{\mathrm{eq}}((\mathsf{ck}, \mathsf{ck}_i), E_{\mathrm{eq}}, (C_i^{(\mathsf{ck})}, C_i), (\mu_i^{(\mathsf{ck})}, \nu_i^{(\mathsf{ck})}, \mu_i, \nu_i))$.
3. Compute the commuting signature, using the secret key x, for the vector of messages $(C_1^{(\mathsf{ck})}, \ldots, C_n^{(\mathsf{ck})})$ on the single commitment key ck, as described in Section 4.3 just above. This procedure outputs Σ.

In this procedure, we remark that the two first steps are not necessarily executed by one single actor. If the messages are known by different parties, each one can execute these two first procedures and the owner of the signing key can next produce the commuting signature.

Note that not all the elements need to be hidden. In particular, thanks to the homomorphic properties of the SDXH commitment, there is no need to commit to $\mathsf{vk}_i \odot \langle i \rangle$ once vk_i is already committed. The proofs of validity are done *w.r.t.* the same commitment $\mathbf{c}_{\mathsf{vk}_i}$. Furthermore, $\langle n \rangle$ must stay in clear, since the length of the vector has to be checkable.

Security Considerations. Regarding the security of the above constructions, we first note that signatures on vector of committed messages have the same probability distribution as directly generated signatures on vector elements. Then the reduction from [3] is adapted to the unforgeability notion of commuting signatures. An adversary, given an access to an oracle which signs committed values, is not able to forge committed signatures on values that were not queried to the oracle in a committed form (thanks to the simulation algorithm, as in [16]). Thus, from a forgery on a vector of committed messages, we extract a forgery on the underlying commuting signature scheme with non-negligible probability.

4.5 Commuting Signatures in Privacy Enhancing Cryptography

In this paper, we mainly focus on anonymous credential systems as described and used in [12,8,21]. It exists in the literature several related cryptographic tools which also aims at preserving the privacy of consumers with close methods and problems. This is for example the case for group [13], blind [11] and traceable [23] signatures.

Commuting signatures have been introduced in [16] and the existing construction is based on automorphic signatures [2,15]. Automorphic signatures, as said in [2,15], permits to efficiently create blind signatures. It is also possible to give a more efficient variant of Groth's group signature scheme [19] by using automorphic signatures instead of certified signatures. For traceable signatures, recent papers [14,1] have also proposed variants of traceable signature, also using automorphic signatures. In particular, the signature confirmation/denial [1] and the efficient tracing [14] techniques can be incorporated into traditional anonymous credential (and related tools) systems.

In our case, this is slightly different since automorphic signatures are not enough. In fact, if commuting signatures can, in most cases, be used instead of automorphic signatures (even if the result is obviously less efficient), the contrary is false. As for delegatable anonymous credentials [4], we need at the same time (i) a signature process on a message which is not known by the signer (or the receiver in our case) and (ii) a process where a unique signature is used several times while being untraceable, even for the signer. This is exactly the aim of commuting signatures, as explained in [16], since signing and encrypting/commuting commute. The fact that a unique signature can be used several times, using the randomization techniques of commitment schemes and Groth-Sahai proofs, is what permits our scheme to be multi-show, while this was not the case for the Guajardo *et al.* construction [21] since they make use of blind signatures.

5 A Multi-show Anonymous Credential Scheme with Encrypted Attributes

We have now introduced all the elements we need to describe our multi-show anonymous credential scheme with encrypted attributes. We first sketch an overview before giving details and security arguments.

5.1 Overview of Our Solution

We want to design an anonymous credential scheme with encrypted attributes. We first take as a basis the work given on non-interactive anonymous credential schemes by Belenkiy *et al.* in [5], and later refined by Fuchsbauer [16] by using commuting signatures. To introduce the property of encrypted attributes, we make use of extractable commitments, arguing that a perfectly binding extractable committed value exactly corresponds to a ciphertext of a public key encryption scheme.

Our scheme next works as follows. In a nutshell, the issuance protocol consists for the issuer in producing a commuting signature on the attributes, using the commitment key of the verifier to encrypt/commit to the attributes. The resulting commuting signature is next given to the user who can play a verification protocol with the verifier by randomizing the commuting signature, which is composed of commitments and Groth-Sahai proofs. As two different randomizations of the same inputs are indistinguishable, we obtain the multi-show property, as expected.

Following [21], the user, during the issuance protocol, generates a random secret α for the issued credential cred. This secret (or the related public key) should be signed by the issuer in the above commuting signature. However, the verifier should not be able to obtain it and we thus need to use a commitment key which is different from the verifier's one for this particular message α. For this purpose, we use the mechanism proposed in Section 4.4, which permits to produce a commuting signature on messages committed with possibly different commitment keys. The other commitment extraction key is related to the verifier (who need to be known at the issuing process, as for the scheme in [21]) and the related commitment scheme is used by the issuer to commit to the attributes.

During the verification process, the user should prove the knowledge of the secret α, without revealing it, for obvious reasons. The GS proof system does not permit such a proof of knowledge and we need to do something more. In fact, we use the powerfulness of automorphic signatures [2,15] for which the verification keys lie in the message space. Thus, during the verification process, the user produces an automorphic signature on some plain message related to the context (or sent by the verifier), using the committed secret key α.

Finally, as we can use the general commuting signature scheme for a vector of committed messages and several commitment keys, we are also able to deal with several verifiers. Thus, the issuance protocol permits the issuer to create a credential with several encrypted attributes, for potentially several different verifiers, which is a new property not proposed in [21]. More formally, we give the following construction.

5.2 Algorithms and Protocols

Following [21], we outline the case where the issuer encrypts attributes for verifiers. Thus, the plain attributes remain hidden to the user. Our scheme is easily adaptable to other policies, giving the extraction key and making the extractable commitment accordingly.

Key Generation. Our scheme works on a bilinear environment (p, $\mathbb{G}_1$, $\mathbb{G}_2$, $\mathbb{G}_T$, e, g_1, g_2) as defined in Section 3. We also need randomly picked generators $h, k, u, v \in \mathbb{G}_1$. As shown above (see Section 4.4), we need a commitment key for the commuting signature. For this purpose, we generate a commitment public key $\mathsf{ck} := (\mathbf{u}, \mathbf{v})$, while the corresponding secret key is known by nobody[6]. We finally set $\mathsf{grp} = (p, \mathbb{G}_1, \mathbb{G}_2, \mathbb{G}_T, e, g_1, h, k, u, v, g_2, \mathsf{ck})$

[6] As we use Groth-Sahai proofs, we are in the common reference string model.

Issuer key generation. Each issuer generates her own keys. For this purpose, she picks at random $x_{\mathcal{I}} \in \mathbb{Z}_p^*$ as her secret key $\mathsf{sk}_{\mathcal{I}}$ and computes the public key as a DH tuple: $\mathsf{pk}_{\mathcal{I}} = (X_{\mathcal{I}}, Y_{\mathcal{I}}) = ({g_1}^{x_{\mathcal{I}}}, {g_2}^{x_{\mathcal{I}}})$.

Verifier key generation. Each verifier also generates her keys. As said before, these corresponds to the ones for an SXDH commitment scheme: the secret key is $(\alpha_{\mathcal{V}_1}, \alpha_{\mathcal{V}_2}) \in (\mathbb{Z}_p^*)^2$ and the corresponding public key is $\mathsf{ck}_{\mathcal{V}} := (\mathbf{u}_{\mathcal{V}}, \mathbf{v}_{\mathcal{V}})$ as defined in Section 3.1.

Issuance Protocol. This protocol is played by an issuer with keys $(\mathsf{sk}_{\mathcal{I}}, \mathsf{pk}_{\mathcal{I}})$ and is related to the attributes denoted $(m_1, \ldots, m_N)$, such that each $m_i = (M_i, N_i)$. We here consider that each attribute m_i is "encrypted" for the verifier i with public key $\mathsf{ck}_{\mathcal{V}_i}$ (with possibly several times the same verifier), which is not proposed in [21].

We first assume that some entity (possibly the issuer itself) first encrypts the plain attributes $(m_1, \ldots, m_N)$. For this purpose, it produces, for all $i \in [1, N]$, a commitment C_i, using $\mathsf{ck}_{\mathcal{V}_i}$, on each (M_i, N_i) and using the DH tuple commitment scheme described in Section 4.1: $c_i := \textsc{CommitDH}(\mathsf{ck}_{\mathcal{V}_i}, (M_i, N_i), (\mu_i, \nu_i))$. These commitments are next given on input to both the issuer (if necessary) and the user. The issuance protocol is next divided into several steps.

- <u>User</u> Generate an automorphic signing secret key $\alpha \in \mathbb{Z}_p^*$. The pair $(X = {g_1}^{\alpha}, Y = {g_2}^{\alpha})$ corresponds to the public verification key of an automorphic signature scheme related to α and is also used to commit to α. Thus, the user next commits to α, using the commitment key ck of the commuting signature scheme: $C_0 = \textsc{CommitDH}(\mathsf{ck}, (X, Y), (\xi, \xi'))$. The result is sent to the issuer.
- <u>Issuer</u> The issuer next produces a commuting signature on all the commitments $C_0, C_1, \cdots, C_N$, using the algorithm given in Section 4.4, the secret key $\mathsf{sk}_{\mathcal{I}}$ as the signing secret key, and the key ck for the related commitment scheme. As C_0 is already committed using ck, it is not necessary to commit again to it and produce a GS proof of equality. This is however necessary for the other C_i's.
  The resulting signature Σ is sent to the user, together with the C_i's.
- <u>User</u> Verify the commuting signature Σ (see [16]) and save the C_i's and the credential $((\alpha, \xi, \xi'), \mathsf{cred} := \Sigma)$.

Verification Protocol. Let $\mathcal{U}$ be a user having beforehand carried out an issuance protocol with an issuer, and thus having a credential $((\alpha, \xi, \xi'), \mathsf{cred})$ as defined above, and on some encrypted/committed attributes $(C_1, \ldots, C_N)$. She now interacts with a verifier having access to a decryption/extraction key $(\alpha_{\mathcal{V}_1}, \alpha_{\mathcal{V}_2})$ related to one commitment key $\mathsf{ck}_{\mathcal{V}}$ used to create cred. For the sake of simplicity, we assume that $\mathsf{ck}_{\mathcal{V}} = \mathsf{ck}_{\mathcal{V}_1}$ in the above issuance protocol.

- <u>User</u> Randomize her credential cred and commitments C_i's (to obtain the $\tilde{C}_i$'s) by using the randomization technique of commuting signatures [16].

The new commuting signature is $\tilde{\Sigma}$. Let $(\tilde{\xi}, \tilde{\xi}')$ be the new randomness associated to $\tilde{C}_0$ and α. User U produces[7] a signature σ on some fresh message related to the context[8] (or sent by $\mathcal{V}$), using the secret key α. Let us recall that α is related to $(X, Y) := ({g_1}^{\alpha}, {g_2}^{\alpha})$ and that (X, Y) is committed in $\tilde{C}_0$. The fresh message is hashed to $m \in \mathbb{Z}_p$ and mapped to $(M, N) := ({g_1}^{m}, {g_2}^{m})$. User produces an automorphic signature (see Section 3.4) $\sigma := (A, B, D, R, S) \leftarrow \textsc{Sign}(\alpha, (M, N))$ and proves that the signature σ is valid under the verification key committed in $\tilde{C}_0$ by computing $\pi \leftarrow \textsc{Prove}(\mathsf{ck}, E_Y, (X, Y), (\tilde{\xi}, \tilde{\xi}'))$ with

$$E_Y(Y) : e(A, \underline{Y} \cdot D) = e(h \cdot M, g_2) \cdot e(k, S).$$

She next sends to the verifier $\tilde{\Sigma}$, π and σ.

Verifier The verifier checks the commuting signature $\tilde{\Sigma}$, the proof π and the signature σ. She is next able to use her secret key $(\alpha_{\mathcal{V}_{1,1}}, \alpha_{\mathcal{V}_{1,2}})$ to extract the attribute (M_1, N_1) committed in C_1 (see Section 3.1).

Remark 1. The verifier retrieves a plain attribute as a DH tuple. The way for her to retrieve an *understandable* attribute can be treated by either considering bits (as done in [21]), or very small messages (to test all possibilities) or (for bigger messages) to publish a cross-reference table between understandable messages and corresponding DH tuples.

Regarding the security of our new construction, we give the following theorem, while the assumptions are given in Appendix A and the proof will be given in the full version of the paper.

Theorem 1. *Our anonymous credential scheme with encrypted attributes ensures the unforgeability, attribute masking and (honest-verifier) user privacy properties under the q-ADHSDH, the AWFCDH and the SXDH assumptions in* $(\mathbb{G}_1, \mathbb{G}_2)$.

Acknowledgments. This work has been supported by the French Agence Nationale de la Recherche under the PACE 07 TCOM Project, and by the European Commission through the ICT Program under Contract ICT-2007-216676 ECRYPT II. We are also grateful to Georg Fuchsbauer for his suggestions of improvement on Section 4.3, to Sherman Chow for his help on this final version and to anonymous referees for their valuable comments.

References

1. Abe, M., Chow, S.S.M., Haralambiev, K., Ohkubo, M.: Double-Trapdoor Anonymous Tags For traceable Signatures. In: Lopez, J., Tsudik, G. (eds.) ACNS 2011. LNCS, vol. 6715, pp. 183–200. Springer, Heidelberg (2011)

---

[7] This additional signature is added in order to prevent credentials sharing. This aspect is not taken into account in the model. To adopt an *all-or-nothing* policy, each α contained in each credential may be the same value, and this value is a user secret necessary to prove possession of a credential.

[8] Like the concatenation of the current time and the verifier public key.

2. Abe, M., Fuchsbauer, G., Groth, J., Haralambiev, K., Ohkubo, M.: Structure-Preserving Signatures and Commitments to Group Elements. In: Rabin, T. (ed.) CRYPTO 2010. LNCS, vol. 6223, pp. 209–236. Springer, Heidelberg (2010)
3. Abe, M., Haralambiev, K., Ohkubo, M.: Efficient Message Space Extension for Automorphic Signatures. In: Burmester, M., Tsudik, G., Magliveras, S., Ilić, I. (eds.) ISC 2010. LNCS, vol. 6531, pp. 319–330. Springer, Heidelberg (2011)
4. Belenkiy, M., Camenisch, J., Chase, M., Kohlweiss, M., Lysyanskaya, A., Shacham, H.: Randomizable Proofs and Delegatable Anonymous Credentials. In: Halevi, S. (ed.) CRYPTO 2009. LNCS, vol. 5677, pp. 108–125. Springer, Heidelberg (2009), http://eprint.iacr.org/2008/428
5. Belenkiy, M., Chase, M., Kohlweiss, M., Lysyanskaya, A.: P-Signatures and Noninteractive Anonymous Credentials. In: Canetti, R. (ed.) TCC 2008. LNCS, vol. 4948, pp. 356–374. Springer, Heidelberg (2008), http://eprint.iacr.org/2007/384
6. Brands, S.: Rethinking PKI and digital certificates - building in privacy. PhD thesis, Eindhoven Institute of Technology (1999)
7. Camenisch, J., Kohlweiss, M., Soriente, C.: An Accumulator Based on Bilinear Maps and Efficient Revocation for Anonymous Credentials. In: Jarecki, S., Tsudik, G. (eds.) PKC 2009. LNCS, vol. 5443, pp. 481–500. Springer, Heidelberg (2009)
8. Camenisch, J.L., Lysyanskaya, A.: An Efficient System for Non-Transferable Anonymous Credentials with Optional Anonymity Revocation. In: Pfitzmann, B. (ed.) EUROCRYPT 2001. LNCS, vol. 2045, pp. 93–118. Springer, Heidelberg (2001)
9. Camenisch, J.L., Lysyanskaya, A.: Dynamic Accumulators and Application to Efficient Revocation of Anonymous Credentials. In: Yung, M. (ed.) CRYPTO 2002. LNCS, vol. 2442, pp. 61–76. Springer, Heidelberg (2002)
10. Camenisch, J.L., Lysyanskaya, A.: Signature Schemes and Anonymous Credentials from Bilinear Maps. In: Franklin, M. (ed.) CRYPTO 2004. LNCS, vol. 3152, pp. 56–72. Springer, Heidelberg (2004)
11. Chaum, D.: Blind signatures for untraceable payments. In: CRYPTO 1982, pp. 199–203 (1983)
12. Chaum, D., Evertse, J.-H.: A Secure and Privacy-Protecting Protocol for Transmitting Personal Information Between Organizations. In: Odlyzko, A.M. (ed.) CRYPTO 1986. LNCS, vol. 263, pp. 118–167. Springer, Heidelberg (1987)
13. Chaum, D., van Heyst, E.: Group Signatures. In: Davies, D.W. (ed.) EUROCRYPT 1991. LNCS, vol. 547, pp. 257–265. Springer, Heidelberg (1991)
14. Chow, S.S.M.: Real Traceable Signatures. In: Jacobson Jr., M.J., Rijmen, V., Safavi-Naini, R. (eds.) SAC 2009. LNCS, vol. 5867, pp. 92–107. Springer, Heidelberg (2009)
15. Fuchsbauer, G.: Automorphic signatures in bilinear groups and an application to round-optimal blind signatures. Cryptology ePrint Archive, Report 2009/320 (2009), http://eprint.iacr.org/
16. Fuchsbauer, G.: Commuting Signatures and Verifiable Encryption. In: Paterson, K.G. (ed.) EUROCRYPT 2011. LNCS, vol. 6632, pp. 224–245. Springer, Heidelberg (2011)
17. Fuchsbauer, G.: Personal Communication (2011)
18. Fuchsbauer, G., Pointcheval, D., Vergnaud, D.: Transferable Constant-Size Fair E-Cash. In: Garay, J.A., Miyaji, A., Otsuka, A. (eds.) CANS 2009. LNCS, vol. 5888, pp. 226–247. Springer, Heidelberg (2009)
19. Groth, J.: Fully Anonymous Group Signatures without Random Oracles. In: Kurosawa, K. (ed.) ASIACRYPT 2007. LNCS, vol. 4833, pp. 164–180. Springer, Heidelberg (2007)

20. Groth, J., Sahai, A.: Efficient non-Interactive Proof Systems for Bilinear Groups. In: Smart, N.P. (ed.) EUROCRYPT 2008. LNCS, vol. 4965, pp. 415–432. Springer, Heidelberg (2008)
21. Guajardo, J., Mennink, B., Schoenmakers, B.: Anonymous Credential Schemes with Encrypted Attributes. In: Heng, S.-H., Wright, R.N., Goi, B.-M. (eds.) CANS 2010. LNCS, vol. 6467, pp. 314–333. Springer, Heidelberg (2010)
22. IBM. Identity mixer - Idemix, `http://www.zurich.ibm.com/security/idemix/`
23. Kiayias, A., Tsiounis, Y., Yung, M.: Traceable Signatures. In: Cachin, C., Camenisch, J.L. (eds.) EUROCRYPT 2004. LNCS, vol. 3027, pp. 571–589. Springer, Heidelberg (2004)
24. Microsoft. Microsoft U-Prove, `https://connect.microsoft.com/site1188`
25. De Santis, A., Yung, M.: Cryptographic Applications of the Non-Interactive Metaproof and many-Prover Systems. In: Menezes, A., Vanstone, S.A. (eds.) CRYPTO 1990. LNCS, vol. 537, pp. 366–377. Springer, Heidelberg (1991)

A Used Assumptions

[q-ADHSDH] The q-Asymmetrical Double Hidden Strong Diffie-Hell-Man Problem. Given $(g_1, X = g_1{}^x, h, u, g_2, Y = g_2{}^x) \in \mathbb{G}_1{}^4 \times \mathbb{G}_2{}^2$, $q-1$ tuples $\{(A_i = (h \cdot g_1{}^{v_i})^{\frac{1}{x+c_i}}, B_i = u^{c_i}, D_i = g_2{}^{c_i}, V_i = g_1{}^{v_i}, W_i = g_2{}^{v_i})\}_{i=1}^{q-1}$ for $c_i, v_i \xleftarrow{\$} \mathbb{Z}_p$ find a new tuple (A, B, D, V, W) such that $e(A, Y \cdot D) = e(h \cdot V, g_2)$, $e(B, g_2) = e(u, D)$ and $e(V, g_2) = e(g_1, W)$.

[AWFCDH] The Asymmetric Weak Flexible Computational Diffie-Hellman Problem. Given $(g_1, g_2) \in \mathbb{G}_1 \times \mathbb{G}_2$, $A = g_1{}^a$ for $a \xleftarrow{\$} \mathbb{Z}_p$, find a tuple $(R, S, M, N) \in (\mathbb{G}_1^*)^2 \times (\mathbb{G}_2^*)^2$ such that $e(A, S) = e(M, g_2) \quad e(M, g_2) = e(g_1, N) \quad e(R, g_2) = e(g_1, S)$, i.e. there exists $r \in \mathbb{Z}_p$ such that $(R, S, M, N) = (g_1{}^r, g_2{}^r, g_1{}^{ra}, g_2{}^{ra})$

[SXDH] The Symmetric External Diffie-Hellman Problem. Given $(g_1{}^r, g_1{}^s, g_1{}^t)$ (resp. $(g_2{}^r, g_2{}^s, g_2{}^t)$) for random $r, s \in \mathbb{Z}_p$ (resp. $r', s' \in \mathbb{Z}_p$), decide whether $t = rs \mod p$ or t is uniform in $\mathbb{Z}_p$.

For each problem given above, the corresponding assumption states that the problem is hard in $(\mathbb{G}_1, \mathbb{G}_2)$.

Group Signature with Constant Revocation Costs for Signers and Verifiers

Chun-I Fan[1], Ruei-Hau Hsu[1], and Mark Manulis[2]

[1] Computer Science Engineering National Sun Yat-sen University
Kaohsiung, Taiwan
cifan@faculty.cse.nsysu.edu.tw,
xyzhsu@gmail.com

[2] Cryptographic Protocols Group, Department of Computer Science
TU Darmstadt & CASED, Germany
mark@manulis.eu

Abstract. Membership revocation, being an important property for applications of group signatures, represents a bottleneck in today's schemes. Most revocation methods require linear amount of work to be performed by unrevoked signers or verifiers, who usually have to obtain fresh update information (sometimes of linear size) published by the group manager. We overcome these disadvantages by proposing a novel group signature scheme, where computation costs for unrevoked signers and potential verifiers remain constant, and so is the length of the update information that must be fetched by these parties from the data published by the group manager. We achieve this complexity by increasing the amount of work at the group manager's side, which growths quadratic with the total number of members. This increase is acceptable since algorithms of the group manager are typically executed on resourceful devices. Our scheme uses a slightly modified version of the pairing-based dynamic accumulator, introduced by Camenisch, Kohlweiss, and Soriente (PKC 2009), which we implicitly combine with the short (non-revocable) group signature scheme by Boneh, Boyen, and Shacham (CRYPTO 2004). We prove that our revocable scheme satisfies the desired security properties of anonymity, traceability, and non-frameability in the random oracle model, although for better efficiency we resort to a somewhat stronger hardness assumption.

1 Introduction

Revocable Group Signatures. Group Signatures (GS) [17] protect anonymity of signers, who are considered as members of the group, managed by a Group Manager (GM), and who can sign on behalf of the group, while remaining traceable (identifiable) only by the group manager. The tracing ability of the group manager is often used in case of dispute, e.g. if the signer misused his signing rights. In many situations, identification of the misbehaving signer should also lead to the revocation of his signing abilities. Group signatures, allowing the group manager to additionally revoke the signing rights of group members are called *revocable*. A revoked group member should no longer be able to produce valid group signatures. In traditional public key infrastructures revocation is typically handled by certification authorities that publish unique information

D. Lin, G. Tsudik, and X. Wang (Eds.): CANS 2011, LNCS 7092, pp. 214–233, 2011.

about the revoked certificates and which is then used by verifiers for checking the validity of certificates. In group signature schemes, however, revocation process must take into account the anonymity requirements offered by these schemes. Currently, there exist two main approaches for revocation: The first approach, originated by Camenisch and Lysyanskaya [16], uses so-called *dynamic accumulators*, where each secret signing key of an unrevoked group member contains a *witness* associated to the public accumulator value; upon revocation of some group member GM updates the accumulator value and publishes some update information, which in turn allows remaining group members to update their witnesses. The accumulator value is also used as input to the verification procedure. The second approach, termed *verifier-local revocation (VLR)*, originated by Boneh and Shacham [12], requires from the group manager to release a revocation token associated with the revoked member; all published revocation tokens are then used as input to the verification procedure, whereas unrevoked signers need not to update their secret signing keys.

Revocation and Security. Revocation of signing rights should not compromise the basic security properties of GS schemes. Modern GS schemes are proven secure in (variants of) the security model, introduced by Bellare, Micciancio, and Warinschi [4], that defined two main requirements, namely *full-anonymity* and *full-traceability*, capturing many previously stated (sometimes informally described) security properties, with regard to the anonymity of signers, unlinkability of their signatures, unforgeability of signatures, protection against framing attacks, in particular in the presence of malicious coalitions and possibly corrupted group managers. Security definitions from [4] were designed for static schemes and later refined in [5] to address caveats with full-traceability in dynamic schemes; in particular, full-traceability was relaxed to *traceability* and an additional requirement of *non-frameability* was used to address possible corruptions of the group manager in case of framing attacks. We observe that support for revocation introduces dynamic behavior, even for schemes that do not provide support for the dynamic admission of new group members. In schemes with VLR property an additional concern arises due to the implicit opening mechanism that is inherent to all these schemes, namely published revocation tokens also invalidate signatures that were produced by the revoked signer before the revocation took place, and by this introduce linkability amongst all signatures of that signer. More recent VLR schemes were enriched with the additional anonymity protection in form of *BU-anonymity* [25], where BU stands for "backward unlinkability", aiming to prevent linkability of signatures that were produced by the revoked signer while he was a legitimate member of the group. Note that BU-anonymity in VLR schemes is typically achieved by splitting the lifetime of the GS scheme in distinct time intervals and revoking a particular signer in all subsequent time intervals, starting with the interval in which his revocation took place for the first time.

Revocation Costs and Their Impact. Ideally, support for revocation should not introduce significant overhead with respect to the computational complexity, for at least the most frequent operations on the side of members and verifiers, namely signature generation and verification. Support for revocation should not significantly increase the size of main parameters, such as the length of group public keys and signatures. Note that many modern group signature schemes (without revocation) offer constant complexity for these parameters. However, all existing revocable GS schemes that either use dynamic accumulators or utilize VLR introduce linear costs in one or another way. For example, the length of (public) update information used by signers and/or verifiers is often linear in at least the number of revoked members: In schemes with dynamic accumulators public information is used by each (unrevoked) signer to update his secret signing key, resulting in the linear amount of computations on the signer's side. In schemes with VLR property public information contains revocation tokens of revoked signers, which are only used in the verification procedure, resulting in the linear amount of computations on the verifier's side (to perform the revocation check). Designing a revocable group signature scheme that would offer constant costs for unrevoked members to update their secret signing keys, constant costs for verifiers to perform the revocation check, and constant length of the update information, published by the group manager remains an open problem[1] so far.

2 Prior Work on Revocable Group Signatures

Revocation in group signature schemes was identified as a desirable property by Ateniese and Tsudik [1], who suggested that revoked signers should no longer be able to generate valid group signatures, while their earlier group signatures must remain anonymous. Thereafter, many revocable group signature schemes were built, e.g. [3, 11, 12, 13, 14, 15, 16, 18, 19, 20, 21, 23, 24, 25, 26, 31, 32]. The first popular approach for handling revocation in group signature schemes is based on dynamic accumulators, originally applied by Camenisch and Lysyanskaya [16], and later adopted to further constructions [20, 15, 30]. With this approach the group manager publishes an updated accumulator value together with some update information, which is processed by unrevoked signers prior or during the computation of subsequent group signatures.

An alternative revocation method is used in group signatures with VLR property, e.g. [12, 13, 19, 25, 26, 31, 32]. These schemes require from the group manager to publish a revocation list containing partial information about the secret keys of revoked signers. This list is then used as input to the verification algorithm, which performs the revocation check by processing all of its entries in the worst case. Unrevoked signers no longer need to update their signing keys. Many earlier VLR constructions were not able to offer anonymity with regard to group signatures, output in the past by the meanwhile revoked signers. This property, known as BU-anonymity, was introduced in [13] and further considered in [12, 19, 25, 26, 31, 32]. Many of existing revocable group signature schemes [3, 11, 12, 13, 14, 15, 16, 19, 25, 26, 31, 32] have linear computation complexity for the generation and/or verification of group signatures, either $O(N)$ with N being the number of group members, or $O(R)$ with R being the number of revoked members.

[1] Jin et al. [18] claim that their revocable group signature scheme has constant costs with regard to signing/verifying and lengths of signatures, group public key, and individual secret signing keys. A closer inspection of their scheme (which was also not proven secure) reveals, however, that one of the components published as revocation information is linear in the number of group members and that all these components must be fetched by unrevoked signers to perform the signing operation, resulting in its linear computation costs.

There exist, however, several more efficient constructions: The scheme by Nakanischi et al. [23], which improves upon [24], partitions the entire group into several subgroups, offering constant costs for signature generation and verification, yet requiring the signers to fetch public update information of size $O(N)$. The scheme by Nakanishi and Funabiki [20] offers revocation for larger groups by reducing this length to $O(\sqrt{N})$ and a more recent scheme by Nakanishi et al. [21] succeeded in reducing this size further to $O(R)$. Camenisch, Kohlweiss, and Soriente [15] introduced an accumulator-based revocation mechanism that can also be applied to achieve revocation for group signatures. Applying the specified version of their accumulator would increase the signing costs by $O(R)$ modular multiplications (for signers to update their keys), which is still more efficient than previous accumulator-based constructions, where the linear amount of work was dedicated to costlier operations, e.g. modular exponentiations. Furthermore, the length of the group public key would be increased to $O(N)$ as unrevoked members would have to download their witnesses to perform the update of secret signing keys. There is an informal discussion in [15] according to which the update procedure for witnesses can also be offloaded to GM (or some third party). This tweak would lead to the constant costs for signers to perform the update procedure and possibly result in constant-size group public keys.

3 Our Results and Organization

Revocable Group Signature. Our work aims at further improving revocation costs in group signature schemes. The main idea is to consider the accumulator-based approach and let the group manager, who is typically responsible for the update of publicly available revocation information, to invest more computational resources (in comparison to other schemes), and by this minimize the costs of other parties (signers and verifiers). In particular, our revocable group signature (RGS) scheme achieves constant computation costs for signature generation, verification, and update of individual secret signing keys. It offers constant lengths for the the group public key, the output group signatures, and the amount of public information that each unrevoked signer must fetch in order to update own secret signing key. Note that algorithms of the group manager are typically executed on devices with rich computational resources so that increasing the computation costs for those algorithms is a rather minor issue.

In Table 1 we emphasize our improvement through the comparison with revocable schemes from [11, 14, 15, 20, 21, 23, 24]. The table does not include VLR schemes, e.g. [12, 19, 22, 25, 26, 31, 32], which all have an intrinsic limitation of $O(R)$ work in the verification procedure, typically dedicated to pairing evaluations or modular exponentiations. We compare sizes of the group public key, signatures, and update information, which is fetched by unrevoked signers and verifiers to keep an up-to-date view over the current composition of the group. We further indicate computational costs for unrevoked signers to perform signature generation, for verifiers to perform signature verification with revocation check, and for group managers to compute the update information, which is newly published after any revocation event. Table 1 uses the following timing notations: T_e denotes the amount of time to perform one modular exponentiation (in a suitable group), T_p is the time of one pairing evaluation, T_m is the time for one modular multiplication, T_a measures one modular addition. Additionally, by N_s we denote the number of available subgroups (applies to [23, 20]) and l_n indicates the length of the RSA modulus (applies to [24], which is the predecessor of [23]). From the comparison we observe that, in general, our scheme performs by a magnitude better than the schemes from [11, 14, 15, 24]. Although computation costs for signature generation and verification of the schemes in [20, 21, 23] are comparable to ours, the size of the group public key in [21] and the length of the update information fetched by signers and verifiers in [20, 21, 23] are worse. In contrast to previous schemes, the group manager in our construction has quadratic costs of $O(NR)$. However, these costs refer to modular additions, which are known to be more efficient than modular multiplications.

Table 1. Comparison of Lengths and Computation Costs in Revocable Group Signature Schemes

	lengths			computation costs		
Schemes	GPK	GS	UI	Sign	Verify	GM Costs
[11]	$O(1)$	$O(1)$	$O(R)$	$O(R) \cdot T_e$	$O(1) \cdot (T_p + T_e)$	$O(R) \cdot T_e$
[14]	$O(1)$	$O(1)$	$O(R)$	$O(R) \cdot (T_e)$	$O(1) \cdot T_e$	$O(R) \cdot T_e$
[15]	$O(N)$	$O(1)$	$O(R)$	$O(R) \cdot T_m$	$O(1) \cdot (T_p + T_e)$	$O(R) \cdot T_m$
[20]	$O(1)$	$O(1)$	$O(N_s)$	$O(1) \cdot T_e$	$O(1) \cdot T_e$	$O(N_s)$
[21]	$O(\sqrt{N})$	$O(1)$	$O(R)$	$O(1) \cdot (T_p + T_e)$	$O(1) \cdot (T_p + T_e)$	$O(R) \cdot (T_p + T_e)$
[23]	$O(1)$	$O(1)$	$O(N_s)$	$O(1) \cdot T_e$	$O(1) \cdot T_e$	$O(N_s) \cdot T_e$
[24]	$O(1)$	$O(1)$	$O(N)$	$O(N/l_n) \cdot T_e$	$O(N/l_n) \cdot T_e$	$O(R) \cdot T_a$
Our RGS	$O(1)$	$O(1)$	$O(1)$	$O(1) \cdot (T_p + T_e)$	$O(1) \cdot (T_p + T_e)$	$O(NR) \cdot T_a$

GPK: group public key GS: group signature UI: update information

Our Techniques. Our RGS scheme implicitly applies a variant of the pairing-based dynamic accumulator by Camenisch, Kohlweiss, and Soriente [15] to the short pairing-based group signature scheme by Boneh, Boyen, and Shacham [11]. We stress that our revocable BBS scheme is different from the revocation mechanism that was discussed for the BBS scheme in [11] based on the ideas underlying the accumulator from [16], which is much less efficient than [15]. In our construction we slightly modify the process by which witnesses in the accumulator from [15] are updated and resort to a stronger hardness assumption for this purpose. Our modifications shift the computation costs for all updates from signers to the group manager and the new assumption, which we call Power Diffie-Hellman Exponent (PDHE) is stronger than the (non-standard) n-DHE assumption used in [15]. Under this assumption update of individual witnesses (which we call membership tokens in the scheme) requires quadratic amount $O(NR)$ of modular additions, performed by the group manager. At the same time we can obtain constant size for the group public key and for the amount of public update information, which an unrevoked signer must fetch prior to the generation of new group signatures. Upon revocation of some group member, the group manager will publish updated membership tokens of all unrevoked signers. In the signing phase an unrevoked signer will only use personal membership token (together with the secret signing key). In contrast, any verifier can check the validity of the group signature using the group public key and the up-to-date accumulator value. Taking into account the discussion in [15] we show how to achieve constant revocation costs for signers and constant lengths for the group public key. Additionally, we show how to reduce the amount of public update information that unrevoked signers must fetch prior to updating their secret signing keys from linear to constant. We observe that our work extends the initial ideas from [15] for offloading the computation of updated witnesses to GM towards a concrete realization and with some optimizations: In particular, we show that quadratic computation costs on the group manager's side to update witnesses for all members can further be optimized, i.e. we can replace quadratic amount of multiplications that would be necessary for the scheme in [15] with the quadratic amount of (considerably more efficient) additions, by slightly changing the computation of updated witnesses on the group manager's side. This modification, however, requires a slightly stronger hardness assumption to prove the non-frameability property of our scheme.

Organization. We proceed as follows. In Section 4, we recall the setting of bilinear groups and discuss several number-theoretic assumptions used in our work, including our PDHE assumption, which we introduce as a stronger variant of n-DHE from [15]. In Section 5, we describe definitions and security model for revocable group signature schemes. Section 6 provides high-level overview and full specification of our RGS scheme, whose security we prove in Section 7.

4 Preliminaries

4.1 Bilinear Groups

We will work in the pairing-based setting and thus recall the notion of bilinear groups:

1. $\mathbb{G}$, $\mathbb{G}'$, and $\mathbb{G}_T$ are cyclic groups, all of prime order p;
2. g_1 is a generator of $\mathbb{G}$; g_2 is a generator of $\mathbb{G}'$;
3. $e : \mathbb{G}\times\mathbb{G}' \to \mathbb{G}_T$ is an efficiently computable map with the following two properties:
   - Bilinear: for all $u \in \mathbb{G}$, $v \in \mathbb{G}'$, $a, b \in Z_p^*$: $e(u^a, v^b) = e(u, v)^{ab}$.
   - Non-degenerate: $e(g_1, g_2) \neq 1_{\mathbb{G}_T}$.

4.2 Hardness Assumptions

Here we first recall the two well-known hardness assumptions — q-SDH [7, 8] and DLIN [11]. We then introduce our Power Diffie-Hellman Exponent (PDHE) assumption as a stronger variant of the n-DHE assumption from [15].

Definition 1 (q-SDH Assumption). *For all probabilistic polynomial-time algorithms $\mathcal{A}$, the following success probability of $\mathcal{A}$ is assumed to be negligible:*

$$\Pr\left[\mathcal{A}(g_1, g_2, g_2^{\gamma}, ..., g_2^{(\gamma^q)}) = (g_1^{\frac{1}{\gamma+x}}, x) : g_1 \in \mathbb{G}, g_2 \in \mathbb{G}', (\gamma, x) \in \mathbb{Z}_p^2\right].$$

Definition 2 (Decision Linear (DLIN) Assumption). *For all probabilistic polynomial-time algorithms $\mathcal{A}$, the following advantage probability of $\mathcal{A}$ is assumed to be negligible:*

$$\Big| \Pr[\mathcal{A}(u, v, h, u^a, v^b, h^{a+b}) = 1 : (u, v, h) \in_R \mathbb{G}^3, (a, b) \in_R \mathbb{Z}_p^2] - \Pr[\mathcal{A}(u, v, h, u^a, v^b, \eta) = 1 : (u, v, h, \eta) \in_R \mathbb{G}^4, (a, b) \in_R \mathbb{Z}_p^2] \Big|.$$

Note that DLIN assumption serves as a basis for the well-known Linear Encryption scheme [11]. The following n-DHE assumption was introduced in [15].

Definition 3 (n-DHE Assumption [15]). *For all probabilistic polynomial-time algorithms $\mathcal{A}$, the following success probability of $\mathcal{A}$ is assumed to be negligible:*

$$\Pr\left[\begin{array}{l}\mathcal{A}(g, g_1, g_2, ..., g_n, g_{n+2}, ..., g_{2n}) = g_{n+1} : g \in \mathbb{G}', g_i = g^{\alpha^i}, \alpha \in_R \mathbb{Z}_p, \\ \hfill i = 1, ..., n, n+2, ..., 2n, n \in \mathbb{N}\end{array}\right].$$

We will rely on a stronger assumption, which we call PDHE and which can be seen as a variant of n-DHE. Note that in n-DHE assumption, the adversary must compute $g^{\alpha^{n+1}}$ and is only given a set of group elements. In contrast in PDHE assumption the adversary receives roughly twice as many group elements (from both groups) and an additional set of integers, each denoted by η_j.

Definition 4 (PDHE Assumption). *For all probabilistic polynomial-time algorithms $\mathcal{A}$, the following success probability of $\mathcal{A}$ is assumed to be negligible:*

$$\Pr\left[\begin{array}{l}\mathcal{A}\Big(g, \hat{g}, (\{g^{\alpha^{\psi+i}}, \hat{g}^{\alpha^{\psi+i}}, g^{\beta^{\psi+i}}, \hat{g}^{\beta^{\psi+i}}\}_i, \{g^{\alpha^{2\psi+j}} g^{\beta^{2\psi+j}}, \hat{g}^{\alpha^{2\psi+j}} \hat{g}^{\beta^{2\psi+j}}\}_j), \hat{g}^{\beta^{2\psi+n+1}}, \eta_j\Big) = \hat{g}^{\alpha^{2\psi+n+1}} \\ \quad : \quad \eta_j = \alpha^{2\psi+j} + \beta^{2\psi+j} \in \mathbb{Z}, (\alpha, \beta) \in \mathbb{Z}_p^{*2}, \psi \in \mathbb{Z}_q, \hat{g} \in \mathbb{G}, g \in \mathbb{G}', \\ \quad\quad i = 1, ..., n, j = 1, ..., n, n+2, ..., 2n, n \in \mathbb{N}\end{array}\right],$$

where α and β generate a subgroup of $\mathbb{Z}_p^*$ of prime order q, and p is a large prime such that $p = 2q + 1$.

Note that values α, β, and ψ remain unknown to the adversary. The main difference to n-DHE is that $\mathcal{A}$ learns integers $\eta_j = \alpha^{2\psi+j} + \beta^{2\psi+j}$ and must output $\hat{g}^{\alpha^{2\psi+n+1}}$. In the generic group model [29] security of PDHE could be argued as follows: A separate analysis can be performed to prove that the probability of $\mathcal{A}$ breaking the PDHE assumption using only set of elements $(g, \hat{g}, g^{\alpha^{\psi+i}}, \hat{g}^{\alpha^{\psi+i}}, g^{\beta^{\psi+i}}, \hat{g}^{\beta^{\psi+i}}, g^{\alpha^{2\psi+j}} g^{\beta^{2\psi+j}}, \hat{g}^{\alpha^{2\psi+j}} \hat{g}^{\beta^{2\psi+j}}, \hat{g}^{\beta^{2\psi+n+1}})$ from $\mathbb{G}$ and $\mathbb{G}'$ remains negligible. This proof is similar to that of the n-DHE assumption. One can then argue that integers η_j perfectly hide the additional values $\alpha^{2\psi+j}$ and $\beta^{2\psi+j}$ used in the PDHE assumption (note that ψ is unknown to the adversary).

5 Security Model and Definitions for Revocable Group Signatures

The security model of a revocable group signature scheme (RGS) defined in this section resembles the standard security model for static group signatures from [4], where we additionally consider the revocation algorithm and augment signing and verification operations of the group signature with revocation-relevant information. Therefore, this model has also partial connection to the security model from [12], where revocation information was handled within the verification procedure only.

Definition 5. *A Revocable Group Signature (RGS) scheme consists of the following algorithms:*

- **KeyGen**(λ, n): *This randomized algorithm takes as input a security parameter* $\lambda \in \mathbb{N}$, *and an integer* $n \in \mathbb{N}$ *(total number of group members). It outputs a* group public key gpk, *a* group manager's secret key gsk, *a* public membership information pmi, *an n-element vector of* membership tokens $\mathsf{mt} = \{\mathsf{mt}_1, ..., \mathsf{mt}_n\}$, *an n-vector of* secret signing keys $\mathsf{sk} = \{\mathsf{sk}_1, \ldots, \mathsf{sk}_n\}$, *and a set* S *of indices of unrevoked group members (initially set to contain all indices* $i \in [1, n]$*).*
- **Sign**$(\mathsf{gpk}, \mathsf{mt}_i, \mathsf{sk}_i, M)$: *On input* gpk, mt_i, *a secret signing key* sk_i *of user i, and a message* $M \in \{0, 1\}^*$, *this randomized algorithm outputs a* group signature σ.
- **Verify**$(\mathsf{gpk}, \mathsf{pmi}, \sigma, M)$: *On input* gpk, pmi, *a candidate group signature* σ, *and a message M, this deterministic algorithms outputs either "**true**" or "**false**". (The output of "**true**" indicates that* σ *is a valid signature on M, meaning also that its signer is not revoked.)*
- **Open**$(\mathsf{gpk}, \mathsf{gsk}, \sigma, M)$: *On input* gpk, gsk, *a candidate signature* σ, *and a message M, this algorithm outputs index i (meaning that i belongs to the signer of* σ*) or* $\perp$ *(meaning that* σ *is untraceable for* $i \notin [1, n]$*).*
- **Revoke**$(\mathsf{gsk}, \mathsf{S}, \mathsf{pmi}, \mathsf{mt}, i)$: *This deterministic algorithm takes as input* gsk, *the set* S *containing indices of unrevoked group member, the up-to-date* pmi, *the up-to-date n-element vector* mt, *and an index i (of the signer to be revoked). The algorithm updates* $\mathsf{S} = \mathsf{S} \backslash \{i\}$, pmi, *and* mt *(from which only* pmi *and* mt *will be published, see below for the explanation).*

An RGS scheme is correct *if: (1) for all* $(\mathsf{gpk}, \mathsf{gsk}, \mathsf{sk}, \mathsf{pmi}, \mathsf{mt}, \mathsf{S}) = $ **KeyGen**(λ, n), *all* $i \in \mathsf{S}$, *and any message* $M \in \{0, 1\}^*$*:*

$$\textbf{Verify}(\mathsf{gpk}, \mathsf{pmi}, \textbf{Sign}(\mathsf{gpk}, \mathsf{mt}_i, \mathsf{sk}_i, M), M) = \text{"}\textbf{true}\text{"}$$

and (2) for all $(\mathsf{gpk}, \mathsf{gsk}, \mathsf{sk}, \mathsf{pmi}, \mathsf{S}) = $ **KeyGen**(λ, n), *all* $i \in S$, *and any message* $M \in \{0, 1\}^*$*:*

$$\textbf{Open}(\mathsf{gpk}, \mathsf{gsk}, \textbf{Sign}(\mathsf{gpk}, \mathsf{mt}_i, \mathsf{sk}_i, M), M) = i.$$

In our description of the **Revoke** algorithm we implicitly assume that revocation is performed by the group manager (not necessarily the same party that also issues secret signing keys), who in order to revoke some group member $i \in [1, n]$ removes the corresponding index i from S, and updates pmi and mt according to the new set S. Note that the up-to-date set S is used by the group manager only to keep track of unrevoked members and to update pmi. In contrast, the public membership information pmi is distributed by the group manager and is used as input to the verification procedure. In our scheme pmi will correspond to the updated value of the accumulator, while individual membership tokens mt_i will correspond to the updated witnesses of unrevoked signers.

An RGS scheme should satisfy three main security requirements, discussed in the following. We start with the notion of full-traceability, which we define similar to [4], except that in order to account for the introduced dynamic behavior through revocation support, several modifications must be applied. This makes our definition somewhat related to the traceability definition from [5] for dynamic groups. In particular, the adversary must come up with a group signature which verifies successfully but for which the opening algorithm fails to output an index in $[1, n]$. The adversary is allowed to corrupt all members of the group.

Definition 6 (Full-traceability). *An RGS scheme is full-traceable if no probabilistic polynomial-time (PPT) adversary $\mathcal{A}$ can win the following game with non-negligible advantage by interacting with a challenger C.*

1. **Setup:** *C runs the algorithm* **KeyGen**(λ, n) *to generate a group public key* gpk*, a group master secret* gsk*, a public membership information* pmi*, an n-element vector of membership tokens* mt*, an n-element vector of secret signing keys* sk*, and a set* $\mathsf{S} = [1, n]$*. C also defines an initially empty set* U *of corrupted members. Then C invokes* $\mathcal{A}$ *on input* $(\mathsf{gpk}, \mathsf{gsk}, \mathsf{pmi}, \mathsf{mt}, \mathsf{S})$*, while keeping* sk *private.*
2. **Oracles:** *$\mathcal{A}$ can make a polynomial number of queries to the following oracles (which are answered by C):*
   - **Signing oracle:** *On input a message M and* $i \in \mathsf{S}$*, this oracle outputs* $\sigma =$ **Sign**$(\mathsf{gpk}, \mathsf{mt}_i, \mathsf{sk}, M)$.
   - **UCorruption oracle:** *On input* $i \in \mathsf{S}$*, this oracle outputs* sk_i *and updates* $\mathsf{U} = \mathsf{U} \cup \{i\}$.
   - **Revocation oracle:** *On input* $i \in [1, n]$*, this oracle responds with the output of* **Revoke**$(\mathsf{gsk}, \mathsf{S}, \mathsf{pmi}, \mathsf{mt}, i)$.
   - **Opening oracle:** *On input a signature σ and a message M, this oracle responds with the output of* **Open**$(\mathsf{gpk}, \mathsf{gsk}, \sigma, M)$.
3. **Output:** *Eventually, $\mathcal{A}$ stops and outputs a signature* σ^* *and a message* M^*.

$\mathcal{A}$ wins if all of the following holds:

- **Verify**$(\mathsf{gpk}, \mathsf{pmi}, \sigma^*, M^*) =$ **true**.
- **Open**$(\mathsf{gpk}, \mathsf{gsk}, \sigma^*, M^*) = \perp$.

The advantage of $\mathcal{A}$ in breaking full-traceability is defined as:

$$\mathbf{Adv}_{\mathcal{A}}^{trace}(\lambda) = \Pr[\mathcal{A} \text{ wins in the full-traceability game}],$$

where the probability is taken over the coin tosses of $\mathcal{A}$ and C.

Our next security requirement for RGS schemes is CPA-anonymity. Unlike [4, 5], by dealing with revocability we have to address anonymity of revoked signers. Our definition of CPA-anonymity comes close to the anonymity definition, that was used in the context of the BBS scheme in [11], where the adversary is not given access to the opening oracle.

Definition 7 (CPA-anonymity). *An RGS scheme is CPA-anonymous if no PPT adversary $\mathcal{A}$ can win the following game with non-negligible advantage by interacting with a challenger C.*

1. **Setup:** *C runs* **KeyGen**(λ, n) to generate a group public key gpk, a group master secret gsk, a public membership information pmi, an n-vector of membership token mt, group member's signing keys sk, and a set S of members' indices. C defines a set U of corrupt members' being empty initially. Then, C gives gpk, pmi, mt, and S to $\mathcal{A}$, while keeping gsk and sk private.
2. **Oracles:** *$\mathcal{A}$ can make a polynomial number of queries to the following oracles (which are answered by C):*

- **Signing oracle:** *On input a message M and $i \in \mathsf{S}$, this oracle outputs* $\sigma = \textbf{Sign}(\mathsf{gpk}, \mathsf{mt}_i, \mathsf{sk}_i, M)$.
- **UCorruption oracle:** *On input $i \in \mathsf{S}$, this oracle outputs* sk_i *and updates* $\mathsf{U} = \mathsf{U} \cup \{i\}$.
- **Revocation oracle:** *On input $i \in [1, n]$, this oracle responds with the output of* $\textbf{Revoke}(\mathsf{gsk}, \mathsf{S}, \mathsf{pmi}, i)$.

3. **Challenge:** *$\mathcal{A}$ selects a message M and two indices i_0 and i_1 with $i_0, i_1 \in \mathsf{S}$. C picks random bit $b \leftarrow \{0, 1\}$ and computes $\sigma^* = \textbf{Sign}(\mathsf{gpk}, \mathsf{mt}_i, \mathsf{sk}_{i_b}, M)$. Then, C sends σ^* to $\mathcal{A}$.*
4. **Output:** *$\mathcal{A}$ continues querying the oracles as above until it eventually stops and outputs a bit b' as its answer to the challenge.*

$\mathcal{A}$ wins the game if $b' = b$ and neither i_0 nor i_1 are revoked. The advantage of $\mathcal{A}$ in breaking anonymity is defined as:

$$\mathbf{Adv}_{\mathcal{A}}^{anon}(\lambda) = \Pr[\mathcal{A} \text{ wins in the CPA-anonymity game}],$$

where the probability is taken over the coin tosses of $\mathcal{A}$ and C.

The final security requirement is non-frameability [5], which accounts for framing attacks executed by a possibly corrupted group manager. Note that traceability only guaranties that any group signature remains traceable. However, it does not take into account potential attacks mounted by the group manager. As motivated in [5], such attacks cannot be captured in a meaningful way in the traceability definition for the dynamic setting (which we have here due to revocability) if the adversary learns the group manager's secret key. That is why non-frameability in the presence of corrupted group managers has to be defined separately.

Definition 8 (Non-frameability). *An RGS is non-frameable if no PPT adversary $\mathcal{A}$ can win the following game with non-negligible advantage by interacting with a challenger C.*

1. **Setup:** *C runs $\textbf{KeyGen}(\lambda, n)$ to obtain a group public key gpk, a group master secret gsk, a public membership information pmi, an n-vector of membership token mt, group members' signing keys sk, and a set S of members' indices. C also defines an empty set U as the set of corrupted members' indices. C then gives gpk, gsk, pmi, mt, and S to $\mathcal{A}$ while keeping sk private.*
2. **Oracles:** *$\mathcal{A}$ can make a polynomial number of queries to the following oracles (which are answered by C):*
   - **Signing oracle:** *On input a message M and $i \in \mathsf{S}$, this oracle outputs* $\sigma = \textbf{Sign}(\mathsf{gpk}, \mathsf{mt}_i, \mathsf{sk}_i, M)$.
   - **UCorruption oracle:** *On input $i \in \mathsf{S}$, this oracle responds with* sk_i *and updates* $\mathsf{U} = \mathsf{U} \cup \{i\}$.
   - **Revocation oracle:** *On input $i \in [1, n]$, this oracle responds with the output of* $\textbf{Revoke}(\mathsf{gsk}, \mathsf{S}, \mathsf{pmi}, i)$.
   - **Opening oracle:** *On input a signature σ and a message M, this oracle responds with the output of* $\textbf{Open}(\mathsf{gpk}, \mathsf{gsk}, \sigma, M)$.
3. **Output:** *Eventually, $\mathcal{A}$ stops and outputs a signature σ^* and a message M^*.*

$\mathcal{A}$ wins if all of the following holds

- **Verify**(gpk, pmi, σ^*, M^*) = **true**.
- **Open**(gpk, gsk, σ^*, M^*) = i^*, where $i^* \in [1, n]$, and **Sign**(gpk, mt_i, sk_{i^*}, M^*) has never been queried by $\mathcal{A}$.

The advantage of $\mathcal{A}$ in breaking non-frameability is defined as:

$$\mathbf{Adv}_{\mathcal{A}}^{non-frame}(\lambda) = \Pr[\mathcal{A} \text{ wins the non-frameability game}],$$

where the probability is taken over the coin tosses of $\mathcal{A}$ and C.

6 Our RGS Scheme with Constant Costs for Signers and Verifiers

In this section we provide specification of our RGS scheme. Prior to detailing its algorithms we give a high-level intuition for its construction.

6.1 High-Level Intuition

Our RGS scheme is mainly based on two previous techniques: The (non-revokable) group signature scheme by Boneh, Boyen, and Shacham (BBS) [11] and the dynamic accumulator by Camenisch, Kohlweiss, and Soriente (CKS) [15] (with slight modifications). The use of the BBS scheme in our RGS constructions helps to achieve constant size for group public keys and group signatures, while the efficient revocation is achieved due to the deployed CKS accumulator. The main technical problem in combining BBS scheme with CKS accumulator is as follows: The use of the CKS accumulator results in the linear length of the group public key and in the linear increase of computation costs for signers to update their witnesses with each revoked group member. Our main modification is to change the computation of witnesses in the CKS accumulator by shifting the significant amount of computation costs from the signers over to the group manager.

Combining Modified CKS Accumulator with BBS Group Signature Scheme. As our construction builds on the BBS scheme, we require that key generation is performed by a trusted key issuer, akin to [11]. The key issuer is responsible for the generation of: all secret signing keys, the group manager's secret key (which includes secrets to open signatures and to revoke members), the group public key, and the initial public membership information. In particular, the issuer picks a secret exponent x_i for each member i and computes a secret value $\alpha^{\psi+i}$ and a corresponding group element $\hat{g}^{x_i}\hat{g}^{\alpha^{\psi+i}}\tilde{g}$ in $\mathbb{G}$. It also computes the secret membership certificate $A_i = (\hat{g}^{x_i}\hat{g}^{\alpha^{\psi+i}}\tilde{g})^{\frac{1}{\mu_i+\omega}}$, which becomes part of the secret signing key sk_i. A group member i receives further a personal witness, containing $L_i = \sum_{j\in S \wedge j\neq i} \eta_{n+1-j+i}$, $g^{\alpha^{\psi+i}}$, and $g^{\beta^{\psi+i}}$, where L_i is the initial membership token mt_i, which will be publicly updated by the group manager on each revocation event, as long as i remains unrevoked. The public membership information pmi will contain up-to-date $L = \hat{g}^{\sum_{j\in S} \alpha^{\psi+n+1-j}}$ and $L' = \hat{g}^{\sum_{j\in S} \beta^{\psi+n+1-j}}$, which play the role of the public accumulator value. An unrevoked group member i can thus produce a signature to prove that it actually possesses a secret signing key, containing a witness accumulated in pmi. When a member i' is revoked, GM updates set S, values L and L' in pmi, and L_i of each unrevoked signer i. In order to generate a new signature i must first obtain up-to-date L, L', and its personal L_i (all of constant length). Note that major computation costs are in the update of corresponding L_i, which is performed by the group manager.

Reducing the Computation Cost of Witnesses of Accumulator. As soon as some member gets revoked remaining group members must implicitly update their secret signing keys upon the execution of the signing operation. For this purpose members use information, which is prepared for them by the group manager, who performs the revocation procedure. In our scheme for every signer there exists an individual public information (of constant length), which we call membership token, and which is updated by the group manager for all unrevoked signers. This information is represented by elements of bilinear groups that correspond to the group elements given to the adversary in the definition of the PDHE assumption (cf. Section 4.2). Using the CKS-like approach for witness updates, when applied in construction, each unrevoked signer would have to compute $\sum_{j\in \mathsf{S}\wedge j\neq i}\alpha^{2\psi+n+1-j+i}+\beta^{2\psi+n+1-j+i}$ in the exponent, where i is the index of the revoked member. This computation can be done by first requiring from the group manager to update j group elements for $j\in \mathsf{S}\wedge j\neq i$, whose discrete logarithms would correspond to each of the $\sum_{j\in \mathsf{S}\wedge j\neq i}\alpha^{2\psi+n+1-j+i}+\beta^{2\psi+n+1-j+i}$. Unrevoked signers would then compute a product of j different group elements. Instead, we let the group manager, who knows (α,β,ψ,i,j) anyway, compute and publish $\eta_j=\alpha^{2\psi+j}+\beta^{2\psi+j}$ for each j. In this case computation costs for witnesses would become constant on the signer's side. Note that each member would have to fetch only his updated witness.

6.2 Specification of RGS Algorithms

- **KeyGen(λ, n)** The key generation algorithm is executed by a trusted issuer (akin to [11, 12]), according to the following steps:
  1. Select bilinear groups $\mathbb{G}$, $\mathbb{G}'$, $\mathbb{G}_T$ of prime order $p<2^\lambda$ such that $q=(p-1)/2$ is a prime, and the bilinear map e. Pick a cryptographic hash function $H:\{0,1\}^*\to\mathbb{Z}_p^*$.
  2. Select $(\tilde{g},\hat{g},h,u,v)\in\mathbb{G}^5$, $g\in\mathbb{G}'$, $(\xi_1,\xi_2)\in_R\mathbb{Z}_p^{*2}$, $\omega\in\mathbb{Z}_p^*$, and $\psi\in\mathbb{Z}_q$, where $\hat{g}=u^{\xi_1}=v^{\xi_2}$. Note that $(\hat{g},u,v)$ represents a public key of the Linear Encryption scheme. Selects further two generators $(\alpha,\beta)\in\mathbb{Z}_p^{*2}$ of prime order q from $\mathbb{Z}_p$, and computes $\alpha^{\psi+i}$ and $\beta^{\psi+i}$, where $i=1,...,2n$.
  3. Compute $z=e(\hat{g}^{\alpha^{2\psi+n+1}},g)$, $z'=e(\hat{g}^{\beta^{2\psi+n+1}},g)$, $g_{\alpha,i}=g^{\alpha^{\psi+i}}$, $g_{\beta,i}=g^{\beta^{\psi+i}}$, $\eta_j=\alpha^{2\psi+j}+\beta^{2\psi+j}$, where $i=1,...,n$, and $j=1,...,n,n+2,...,2n$.
  4. Define the group master secret key $\mathsf{gsk}=(\xi_1,\xi_2,\alpha,\beta,\psi)$ and the group public key $\mathsf{gpk}=(H,p,\mathbb{G},\mathbb{G}',\mathbb{G}_T,e,\hat{g},\tilde{g},g,\Omega,z,z',h,u,v)$, where $\Omega=g^\omega$.
  5. Define $\mathsf{S}=\{1,...,n\}$ as the index set of group members. Compute $L=\hat{g}^{\sum_{j\in\mathsf{S}}\alpha^{\psi+n+1-j}}$, $L'=\hat{g}^{\sum_{j\in\mathsf{S}}\beta^{\psi+n+1-j}}$, and $L_i=\sum_{j\in\mathsf{S}\wedge j\neq i}(\alpha^{2\psi+n+1-j+i}+\beta^{2\psi+n+1-j+i})$ for all $i\in\mathsf{S}$. Define $\mathsf{pmi}=\{L,L'\}$ to be the public membership information, which will be updated by the group manager upon revocation of members, and set $\mathsf{mt}=\{L_1,...,L_n\}$ to be a vector of membership tokens; each unrevoked member will fetch its own membership token from this vector.

6. Compute secret signing keys $\mathsf{sk}_i = (A_i, g_{\alpha,i}, g_{\beta,i}, \mu_i, x_i)$ for each member $i \in \mathsf{S}$, where $A_i = (\hat{g}^{x_i}\hat{g}^{\alpha^{\psi+i}}\tilde{g})^{\frac{1}{\mu_i+\omega}}$.
7. Publish gpk, pmi, and mt. Output privately gsk, $\{A_i\}_{i\in\mathsf{S}}$, and S to the group manager (GM). Output privately sk_i to the corresponding member i. (Note that ω remains secret and should be ideally erased by the issuer.)

- **Sign**(gpk, mt_i, sk_i, M) Let gpk $= (H, p, \mathbb{G}, \mathbb{G}', \mathbb{G}_T, e, \hat{g}, \tilde{g}, g, \Omega, z, z', h, u, v)$, $\mathsf{mt}_i = L_i$, and $\mathsf{sk}_i = (A_i, g_{\alpha,i}, g_{\beta,i}, \mu_i, x_i)$. In order to sign some message $M \in \{0,1\}^*$ the signer i proceeds as follows:
  1. Select $\pi, \theta, \rho, \delta, r_\pi, r_\theta, r_{\mu_i}, r_\rho, r_{x_i}, r_{\pi\mu_i}, r_{\theta\mu_i}$, and r_δ at random from $\mathbb{Z}_p^*$.
  2. Compute $T_1 = A_i\hat{g}^{\pi+\theta}$, $T_2 = u^\pi$, $T_3 = v^\theta$, $T_4 = g^\rho$, $T_5 = \hat{g}^\rho$, $T_6 = g_{\alpha,i}^\rho$, $T_7 = (zz')^\rho$, $T_8 = e(h, T_4)^\delta$, $T_9 = \hat{g}^{L_i}h^\delta$, $T_{10} = g_{\beta,i}^\rho$, $R_1 = e(T_1, T_4)^{r_{\mu_i}} \cdot e(T_1, \Omega)^{r_\rho} / e(\hat{g}, T_4)^{r_{x_i}} \cdot e(\tilde{g}, g)^{r_\rho} \cdot e(\hat{g}, T_4)^{r_{\pi\mu_i}+r_{\theta\mu_i}} \cdot e(T_5, \Omega)^{r_\pi+r_\theta}$, $R_2 = u^{r_\pi}$, $R_3 = v^{r_\theta}$, $R_4 = g^{r_\rho}$, $R_5 = \hat{g}^{r_\rho}$, $R_6 = T_2^{r_{\mu_i}} \cdot u^{-r_{\pi\mu_i}}$, $R_7 = (zz')^{r_\rho}$, $R_8 = e(h, T_4)^{r_\delta}$, $R_9 = e(T_9, g)^{r_\rho}$, and $R_{10} = T_3^{r_{\mu_i}} \cdot v^{-r_{\theta\mu_i}}$, where (T_1, T_2, T_3) is a Linear Encryption ciphertext that encrypts A_i, and T_4 through T_{10} and R_1 through R_{10} are required to prove the knowledge of $\pi, \theta, \mu_i, \rho, x_i, \pi\mu_i, \theta\mu_i$ and of the accumulator $\{L, L'\}$.
  3. Compute challenge $c = H(\mathsf{gpk}, M, T_1, T_2, T_3, T_4, T_5, T_6, T_7, T_8, T_9, T_{10}, R_1, R_2, R_3, R_4, R_5, R_6, R_7, R_8, R_9, R_{10})$.
  4. Compute $s_\pi = r_\pi + c\pi$, $s_\theta = r_\theta + c\theta$, $s_\rho = r_\rho + c\rho$, $s_{x_i} = r_{x_i} + cx_i$, $s_{\pi\mu_i} = r_{\pi\mu_i} + c\pi\mu_i$, $s_{\theta\mu_i} = r_{\theta\mu_i} + c\theta\mu_i$, and $s_\delta = r_\delta + c\delta$.
  5. Output $\sigma = (c, T_1, T_2, T_3, T_4, T_5, T_6, T_7, T_8, T_9, T_{10}, s_\pi, s_\theta, s_\rho, s_{x_i}, s_{\pi\mu_i}, s_{\theta\mu_i}, s_\delta)$ on M.
- **Verify**(gpk, pmi, σ, M) Let gpk $= (H, p, \mathbb{G}, \mathbb{G}', \mathbb{G}_T, e, \hat{g}, \tilde{g}, g, \Omega, z, z', h, u, v)$, pmi $= (L, L')$, and $\sigma = (c, T_1, T_2, T_3, T_4, T_5, T_6, T_7, T_8, T_9, T_{10}, s_\pi, s_\theta, s_\rho, s_{x_i}, s_{\pi\mu_i}, s_{\theta\mu_i}, s_\delta)$. The validity of a candidate group signature σ on some message M can be checked according to the following procedure:
  1. Compute
  $R'_1 = e(T_1, T_4)^{s_{\mu_i}} \cdot e(T_1, \Omega)^{s_\rho} / e(\hat{g}, T_4)^{s_{x_i}} \cdot e(\hat{g}, T_6)^c \cdot e(\tilde{g}, g)^{s_\rho} \cdot e(\hat{g}, T_4)^{s_{\mu_i\pi}+s_{\mu_i\theta}} \cdot e(T_5, \Omega)^{s_\pi+s_\theta}$, $R'_2 = u^{s_\pi} \cdot (T_2)^{-c}$, $R'_3 = v^{s_\theta} \cdot (T_3)^{-c}$, $R'_4 = g^{s_\rho} \cdot (T_4)^{-c}$, $R'_5 = \hat{g}^{s_\rho} \cdot (T_5)^{-c}$, $R'_6 = T_2^{s_{\mu_i}} \cdot u^{-s_{\pi\mu_i}}$, $R'_7 = (zz')^{s_\rho} \cdot (T_7)^{-c}$, $R'_8 = e(h, T_4)^{s_\delta} \cdot T_8^{-c}$, and $R'_9 = (T_7^c \cdot e(T_9, g)^{s_\rho}) / (T_8 \cdot e(L, T_6) \cdot e(L', T_{10}))^c$, $R'_{10} = T_3^{s_{\mu_i}} \cdot v^{-s_{\theta\mu_i}}$
  2. Compute $c' = H(\mathsf{gpk}, M, T_1, T_2, T_3, T_4, T_5, T_6, T_7, T_8, T_9, T_{10}, R'_1, R'_2, R'_3, R'_4, R'_5, R'_6, R'_7, R'_8, R'_9, R'_{10})$.
  3. If $c' = c$ then output **true**; else output **false**.

  If the output is **true**, it means that σ on M is signed by valid signing key sk_i for $i \in \mathsf{S}$. Otherwise, sk_i is invalid or revoked (i.e., $i \notin \mathsf{S}$).
- **Open**(gpk, gsk, σ, M) Let gpk $= (H, p, \mathbb{G}, \mathbb{G}', \mathbb{G}_T, e, \hat{g}, \tilde{g}, g, \Omega, z, z', h, u, v)$, gsk $= (\xi_1, \xi_2, \alpha, \beta, \psi)$, and $\sigma = (c, T_1, T_2, T_3, T_4, T_5, T_6, T_7, T_8, T_9, T_{10}, s_\pi, s_\theta, s_\rho, s_{x_i}, s_{\pi\mu_i}, s_{\theta\mu_i}, s_\delta)$. In order to open some candidate group signature σ the group manager proceeds as follows:
  1. If **Verify**(gpk, pmi, σ, M)$\neq$**true** then return $\perp$.
  2. Otherwise, compute $A_i = T_1/(T_2^{\xi_1}T_3^{\xi_2})$ with ξ_1, ξ_2 extracted from gsk and T_2, T_3 extracted from σ. Returns i associated to A_i.
- **Revoke**(gsk, S, pmi, mt, i') In order to revoke some member i' the group manager proceeds as follows:

1. Update $\mathsf{S} = \mathsf{S} \backslash \{i'\}$, $L = L/(g^{\alpha^{\psi+n+1-i'}})$, and $L' = L'/(g^{\psi+\beta^{n+1-i'}})$.
2. Compute $L_i = L_i - \alpha^{2\psi+n+1-i'+i} - \beta^{2\psi+n+1-i'+i}$ in mt for each $i \in \mathsf{S}$.
3. Output updated S, pmi, and mt.

7 Security Analysis

Security of our RGS scheme with respect to definitions from Section 5 is established through the following theorems.

Theorem 1 (Full-traceability). *The proposed RGS scheme is fully-traceable in the random oracle model, based on the q-SDH assumption in bilinear groups $\mathbb{G}$ and $\mathbb{G}'$.*

Proof. The proof is given in Appendix A and follows some ideas from [11, 15].

Theorem 2 (CPA-anonymity). *The proposed RGS scheme is CPA-anonymous in the random oracle model, based on the DLIN assumption in $\mathbb{G}$.*

Proof. The proof is given in Appendix B.

Theorem 3 (Non-frameability). *The proposed RGS scheme is non-frameable in the random oracle model, based on the PDHE assumption and the hardness of computing discrete logarithms in $\mathbb{G}$.*

Proof. The proof is given in Appendix C.

8 Conclusion

In this work we made another step towards better efficiency in revocable group signatures. Our proposed RGS scheme achieves constant costs for signers and verifiers at the price of a higher amount of work for the group manager and a rather strong Power Diffie-Hellman Exponent (PDHE) assumption. In addition to constant computation costs, our scheme keeps group public keys, group signature, and the amount of public update information, to be fetched by either an unrevoked signer or a verifier, also constant. Our scheme, which is based on the combination of a modified CKS dynamic accumulator from [15] with the BBS group signature scheme from [11] preserves the original security properties of the BBS scheme, while also offering support for the revocation of the signing rights. An open problem would be to find a solution that achieves similar complexity under somewhat more standard (pairing-based) assumptions.

Acknowledgements. Ruei-Hau Hsu acknowledges the NSC-DAAD Sandwich Program of 2011 (Spring Season) sponsored by National Science Council (NSC) in Taiwan and German Academic Exchange Service (DAAD) for the received financial support during his research stay at TU Darmstadt. Mark Manulis was supported by grant MA 4957 of the German Research Foundation (DFG).

References

1. Ateniese, G., Tsudik, G.: Some Open Issues and New Directions in Group Signatures. In: Franklin, M.K. (ed.) FC 1999. LNCS, vol. 1648, pp. 196–211. Springer, Heidelberg (1999)
2. Ateniese, G., Camenisch, J.L., Joye, M., Tsudik, G.: A Practical and Provably Secure Coalition-Resistant Group Signature Scheme. In: Bellare, M. (ed.) CRYPTO 2000. LNCS, vol. 1880, pp. 255–270. Springer, Heidelberg (2000)
3. Ateniese, G., Song, D., Tsudik, G.: Quasi-Efficient Revocation of Group Signatures. In: Blaze, M. (ed.) FC 2002. LNCS, vol. 2357, pp. 183–197. Springer, Heidelberg (2003)
4. Bellare, M., Micciancio, D., Warinschi, B.: Foundations of Group Signatures: Formal Definitions, Simplified Requirements, and a Construction based on General Assumptions. In: Biham, E. (ed.) EUROCRYPT 2003. LNCS, vol. 2656, pp. 614–629. Springer, Heidelberg (2003)
5. Bellare, M., Shi, H., Zhang, C.: Foundations of Group Signatures: The Case of Dynamic Groups. In: Menezes, A. (ed.) CT-RSA 2005. LNCS, vol. 3376, pp. 136–153. Springer, Heidelberg (2005)
6. Biham, E., Shamir, A.: Differential Cryptanalysis of the Full 16-Round DES. In: Brickell, E.F. (ed.) CRYPTO 1992. LNCS, vol. 740, pp. 487–496. Springer, Heidelberg (1993)
7. Boneh, D., Boyen, X.: Short Signatures Without Random Oracles. In: Cachin, C., Camenisch, J.L. (eds.) EUROCRYPT 2004. LNCS, vol. 3027, pp. 56–73. Springer, Heidelberg (2004)
8. Boneh, D., Boyen, X.: Short Signatures without Random Oracles and the SDH Assumption in Bilinear Groups. Journal of Cryptology 21(2), 149–177 (2008)
9. Boneh, D., Boyen, X., Goh, E.-J.: Hierarchical Identity Based Encryption with Constant Size Ciphertext. In: Cramer, R. (ed.) EUROCRYPT 2005. LNCS, vol. 3494, pp. 440–456. Springer, Heidelberg (2005)
10. Boneh, D., Franklin, M.: Identity-Based Encryption from the Weil Pairing. In: Kilian, J. (ed.) CRYPTO 2001. LNCS, vol. 2139, pp. 213–229. Springer, Heidelberg (2001)
11. Boneh, D., Boyen, X., Shacham, H.: Short Group Signatures. In: Franklin, M. (ed.) CRYPTO 2004. LNCS, vol. 3152, pp. 41–55. Springer, Heidelberg (2004)
12. Boneh, D., Shacham, H.: Group Signatures with Verifier-Local Revocation. In: Proceedings of 11th ACM Conference on Computer and Communication Security: ACM-CCS 2004, pp. 168–177 (2004)
13. Bresson, E., Stern, J.: Efficient Revocation in Group Signatures. In: Kim, K.-c. (ed.) PKC 2001. LNCS, vol. 1992, pp. 190–206. Springer, Heidelberg (2001)
14. Camenisch, J.L., Groth, J.: Group Signatures: Better Efficiency and New Theoretical Aspects. In: Blundo, C., Cimato, S. (eds.) SCN 2004. LNCS, vol. 3352, pp. 120–133. Springer, Heidelberg (2005)
15. Camenisch, J., Kohlweiss, M., Soriente, C.: An Accumulator Based on Bilinear Maps and Efficient Revocation for Anonymous Credentials. In: Jarecki, S., Tsudik, G. (eds.) PKC 2009. LNCS, vol. 5443, pp. 481–500. Springer, Heidelberg (2009)
16. Camenisch, J.L., Lysyanskaya, A.: Dynamic Accumulators and Application to Efficient Revocation of Anonymous Credentials. In: Yung, M. (ed.) CRYPTO 2002. LNCS, vol. 2442, pp. 61–76. Springer, Heidelberg (2002)
17. Chaum, D., van Heyst, E.: Group Signatures. In: Davies, D.W. (ed.) EUROCRYPT 1991. LNCS, vol. 547, pp. 257–265. Springer, Heidelberg (1991)
18. Jin, H., Wong, D.S., Xu, Y.: Efficient Group Signature with Forward Secure Revocation. In: Ślęzak, D., Kim, T.-h., Fang, W.-C., Arnett, K.P. (eds.) SecTech 2009. CCIS, vol. 58, pp. 124–131. Springer, Heidelberg (2009); Proceedings of ANTS IV. LNCS 1838, pp.385–394. Springer (2000)

19. Libert, B., Vergnaud, D.: Group Signatures with Verifier-Local Revocation and Backward Unlinkability in the Standard Model. In: Garay, J.A., Miyaji, A., Otsuka, A. (eds.) CANS 2009. LNCS, vol. 5888, pp. 498–517. Springer, Heidelberg (2009)
20. Nakanishi, T., Funabiki, N.: Efficient Revocable Group Signature Schemes Using Primes. Journal of Information Processing 16, 110–121 (2008)
21. Nakanishi, T., Fujii, H., Hira, Y., Funabiki, N.: Revocable Group Signature Schemes with Constant Costs for Signing and Verifying. In: Jarecki, S., Tsudik, G. (eds.) PKC 2009. LNCS, vol. 5443, pp. 463–480. Springer, Heidelberg (2009)
22. Nakanishi, T., Funabiki, N.: "Verifier-Local Revocation Group Signature Schemes with Backward Unlinkability from Bilinear Maps. IEICE Transactions on Fundamentals of Electronics, Communications and Computer Sciences E90-A(1), 65–74 (2007)
23. Nakanishi, T., Kubooka, F., Hamada, N., Funabiki, N.: Group Signature Schemes with Membership Revocation for Large Groups. In: Boyd, C., González Nieto, J.M. (eds.) ACISP 2005. LNCS, vol. 3574, pp. 443–454. Springer, Heidelberg (2005)
24. Nakanishi, T., Sugiyama, Y.: A Group Signature Scheme with Efficient Membership Revocation for Reasonable Groups. In: Wang, H., Pieprzyk, J., Varadharajan, V. (eds.) ACISP 2004. LNCS, vol. 3108, pp. 336–347. Springer, Heidelberg (2004)
25. Nakanishi, T., Funabiki, N.: Verifier-Local Revocation Group Signature Schemes with Backward Unlinkability from Bilinear Maps. In: Roy, B. (ed.) ASIACRYPT 2005. LNCS, vol. 3788, pp. 533–548. Springer, Heidelberg (2005)
26. Nakanishi, T., Funabiki, N.: A Short Verifier-Local Revocation Group Signature Scheme with Backward Unlinkability. In: Yoshiura, H., Sakurai, K., Rannenberg, K., Murayama, Y., Kawamura, S.-i. (eds.) IWSEC 2006. LNCS, vol. 4266, pp. 17–32. Springer, Heidelberg (2006)
27. Pointcheval, D., Stern, J.: Security Proofs for Signature Schemes. In: Maurer, U.M. (ed.) EUROCRYPT 1996. LNCS, vol. 1070, pp. 387–398. Springer, Heidelberg (1996)
28. Pointcheval, D., Stern, J.: Security Arguments for Digital Signatures and Blind Signatures. Journal of Cryptology 13(3), 361–396 (2000)
29. Shoup, V.: Lower Bounds for Discrete Logarithms and Related Problems. In: Fumy, W. (ed.) EUROCRYPT 1997. LNCS, vol. 1233, pp. 256–266. Springer, Heidelberg (1997)
30. Tsudik, G., Xu, S.: Accumulating Composites and Improved Group Signing. In: Laih, C.-S. (ed.) ASIACRYPT 2003. LNCS, vol. 2894, pp. 269–286. Springer, Heidelberg (2003)
31. Zhou, S., Lin, D.: A Shorter Group Signature with Verifier-Local Revocation and Backward Unlinkability, Cryptology ePrint Archive: Report 2006/100 (2006), `http://eprint.iacr.org/2006/100`
32. Zhou, S., Lin, D.: Shorter Verifier-Local Revocation Group Signatures from Bilinear Maps. In: Pointcheval, D., Mu, Y., Chen, K. (eds.) CANS 2006. LNCS, vol. 4301, pp. 126–143. Springer, Heidelberg (2006)

A Proof of Theorem 1 (Full-Traceability)

Let $\mathcal{A}$ be an adversary that breaks the full-traceability of the proposed protocol by returning an untraceable signature σ_{i^*} with probability at least ϵ. We construct a PPT algorithm $\mathcal{B}$ that interacts with $\mathcal{A}$ and breaks the q-SDH assumption with probability at least ϵ'. The interaction of $\mathcal{B}$ with $\mathcal{A}$ proceeds as follows.

- **Setup:** $\mathcal{B}$ is given an n-SDH instance $(\tilde{g}, g, g^{\omega}, g^{\omega^2}, ..., g^{\omega^n})$ by a challenger C_s of n-SDH assumption, where $\tilde{g} \in \mathbb{G}$, $g \in \mathbb{G}'$, and n is the number of group members. $\mathcal{B}$ defines set $\mathsf{S} = [1, n]$ and an initially empty set U, and randomly selects an index

$i^* \in \mathsf{S}$ of a member to be attacked by $\mathcal{A}$. $\mathcal{B}$ then randomly selects $(\alpha, \beta, \tau, \xi_1, \xi_2) \in \mathbb{Z}_p^{*5}$, $\psi \in \mathbb{Z}_q$, and computes $\hat{g} = \tilde{g}^{\tau}$, $\{g_{\alpha,i} = g^{\alpha^{\psi+i}}, g_{\beta,i} = g^{\beta^{\psi+i}}\}_{1\le i\le n}$. $\mathcal{B}$ uses g^{ω} as Ω and computes gpk, gsk, pmi, and mt given to $\mathcal{A}$. After that, $\mathcal{B}$ turns an n-SDH instance into values $(\tilde{g}, g, \Omega = g^{\omega})$ and $n-1$ SDH pairs $(\tilde{g}^{\frac{1}{\mu_i+\omega}}, \mu_i)$ by Lemma 3.2 from [7] such that $e(g^{\frac{1}{\mu_i+\omega}}, \Omega g^{\mu_i}) = e(\tilde{g}, g)$. $\mathcal{B}$ then transforms the $n-1$ SDH pairs to $n-1$ members' signing key $\mathsf{sk}_i = (A_i (= (\tilde{g}^{\frac{1}{\mu_i+\omega}})^{\tau(x_i+\alpha^{\psi+i})+1} = (\hat{g}^{x_i}\hat{g}^{\alpha^{\psi+i}}\tilde{g})^{\frac{1}{\mu_i+\omega}}), g_{\alpha,i}, g_{\beta,i}, \mu_i, x_i)$, where $i \neq i^*$.
- **Oracles**: $\mathcal{B}$ simulates RGS by answering the following oracle queries.
  - **Hash oracle:** The hash oracle as a random oracle is simulated by $\mathcal{B}$. $\mathcal{B}$ randomly selects element in $\mathbb{Z}_p$ as the output of hash query and makes sure the responses are identical to the same queries by maintaining a hash list H**-list**.
  - **Signing oracle:** On input a pair (i, M), if $i \in [1, n]$, $\mathcal{B}$ can successfully responds with the corresponding σ by sk_i. If $i \in \mathsf{U}$, reject this request. If $i = i^*$, the simulation fails.
  - **UCorruption oracle:** On input i, if $i \in \mathsf{S} \wedge i \neq i^*$, $\mathcal{B}$ responds with $\mathsf{sk}_i = (A_i, g_{\alpha,i}, g_{\beta,i}, \mu_i, x_i)$ to $\mathcal{A}$ and appends i to U. If $i \in \mathsf{U}$, $\mathcal{B}$ returns sk_i without changing S and U. If $i = i^*$, the simulation fails.
  - **Revocation oracle:** On input$i$, if $i \in \mathsf{S}$, $\mathcal{B}$ updates $\mathsf{S} = \mathsf{S}\backslash\{i\}$, $\mathsf{pmi} = \{L, L'\}$, and $\mathsf{mt} = \{L_i\}_{i\in\mathsf{S}}$. $\mathcal{B}$ outputs S, pmi, and mt to $\mathcal{A}$.
  - **Opening oracle:** On inputσ, $\mathcal{A}$ decrypt (T_1, T_2, T_3) of σ to obtain A_i and returns the corresponding i of A_i to $\mathcal{A}$. Otherwise, return $\perp$.
- **Output:** Finally, $\mathcal{A}$ outputs a signature-message pair (σ^*, M^*).

$\mathcal{A}$ is the adversary to break the proposed scheme if σ^* is correct and belongs to some member $i \notin [1, n]$ with probability at least ϵ. Then $\mathcal{A}$ outputs a forged signature σ_{i^*} of the member i^* with probability at least ϵ/n. By Forking Lemma [27, 28], if $\mathcal{A}$ outputs a valid message-signature tuple $(M, \sigma_0 = (T_1, T_2, T_3, T_4, T_5, T_6, T_7, T_8, T_9, T_{10}, R_1, R_2, R_3, R_4, R_5, R_6, R_7, R_8, R_9, T_{10}), c, \sigma_1 = (s_\pi, s_\theta, s_\rho, s_{x_i}, s_{\pi\mu_i}, s_{\theta\mu_i}, s_\delta))$. We rewind the framework and $\mathcal{A}$ with the same random tape and the different random oracle H'. Then $\mathcal{A}$ still can output a forgery $(M, \sigma_0, c', \sigma_1')$ with probability at least $\epsilon/4$. Consequently, we obtain two valid tuples (M,σ_0,c,σ_1) and $(M,\sigma_0, c',\sigma_1')$ of the same member i^* with probability at least $\epsilon/4n$. $\mathcal{B}$ can extract the secrets $x_{i^*} = (s'_{x_{i^*}} - s_{x_{i^*}})/(c' - c)$, $\pi = (s_\pi - s'_\pi)/(c - c')$, and $\theta = (s_\theta - s'_\theta)/(c - c')$ from the above two valid signature tuples of the member i^*. After that, $\mathcal{B}$ computes $\tilde{g}^{\frac{1}{\mu_{i^*}+\omega}} = ((\hat{g}^{x_{i^*}}\hat{g}^{\alpha^{\psi}+i}\tilde{g})^{\frac{1}{\mu_{i^*}+\omega}})^{(\tau(x_{i^*}+\alpha^{\psi+i})+1)^{-1}}$ to obtain an SDH pair $(\tilde{g}^{\frac{1}{\mu_{i^*}+\omega}}, \mu_{i^*})$.

B Proof of Theorem 2 (CPA-Anonymity)

Suppose $\mathcal{A}$ is an adversary that breaks the CPA-anonymity of our RGS scheme with the advantage at least ϵ. We construct a PPT algorithm $\mathcal{B}$ that breaks Linear encryption (and by this the DLIN assumption) with the advantage at least ϵ by playing the anonymity game from Definition 7. The interaction between $\mathcal{B}$ and $\mathcal{A}$ proceeds as follows.

- **Setup:** First, $\mathcal{B}$ selects the groups $\mathbb{G}$, $\mathbb{G}'$, and $\mathbb{G}_T$ of prime order p. $\mathcal{B}$ is in possession of a public key $(u, v, \hat{g}) \in \mathbb{G}^3$ for the Linear encryption scheme (which it received from the Linear encryption challenger C_{le}). Recall that $\hat{g} = u^{\xi_1} = v^{\xi_2}$ for some unknown $(\xi_1, \xi_2) \in \mathbb{Z}_p^2$. $\mathcal{B}$ randomly picks $\tilde{g} \in \mathbb{G}$, $g \in \mathbb{G}'$, $\omega \in \mathbb{Z}_p^*$, $(\alpha,\beta) \in \mathbb{Z}_p^*$ of prime order q, and $\psi \in \mathbb{Z}_q$. $\mathcal{B}$ then initializes the sets $\mathsf{S} = \{1, ..., n\}$ and U (set of corrupted signers). Then, $\mathcal{B}$ generates the remaining parts of gpk, pmi, and mt according to the key generation procedure of the RGS scheme. $\mathcal{B}$ computes secret signing keys $\mathsf{sk}_i = \{A_i,\ g_{\alpha,j},\ g_{\beta,j},\ \mu_i,\ x_i\}$ for all $i \in \mathsf{S}$. $\mathcal{B}$ then provides $\mathcal{A}$ with $\mathsf{gpk} = (p, \mathbb{G}, \mathbb{G}', \mathbb{G}_T, e, \hat{g}, \tilde{g}, g, \Omega, z, z', h, u, v)$, $\mathsf{pmi} = \{L, L'\}$, $\mathsf{mt} = \{L_i\}_{i \in \mathsf{S}}$, S, U, and stores ω, α, β, and ψ for later use.

- **Oracles:** $\mathcal{B}$ answers oracle queries of $\mathcal{A}$ as follows.
  - **Hash oracle:** $\mathcal{B}$ simulates the random oracle H by maintaining a hash list H-**list**, and responds on new queries with random elements from $\mathbb{Z}_p$, while making sure that previous queries when asked again are answered consistently with H-**list**.
  - **Signing oracle:** Before the simulation, $\mathcal{B}$ also maintains a list E-**list** for storing the corresponding identity information of signatures. In the simulation of signing queries, $\mathcal{B}$ responds with σ_i for the oracle query with the corresponding input (i, M) of $\mathcal{A}$ as follows. $\mathcal{B}$ selects π and $\theta \in_R \mathbb{Z}_p^*$, and encrypts A_i as a ciphertext $(T_1 = A_i\hat{g}^{\pi+\theta}, T_2 = u^{\pi}, T_3 = v^{\theta})$. $\mathcal{B}$ then randomly selects ρ, δ, r_π, r_θ, r_{μ_i}, r_ρ, r_{x_i}, $r_{\pi\mu_i}$, $r_{\theta\mu_i}$, r_δ, and r_δ to generates $T_4, T_5, T_6, T_7, T_8, T_9, T_{10}, R_1, R_2, R_3, R_4, R_5, R_6, R_7, R_8, R_9$, and R_{10}. $\mathcal{B}$ also updates the output of hash list at (gpk, $M, T_1, T_2, T_3, T_4, T_5, T_6, T_7, T_8, T_9, T_{10}, R_1, R_2, R_3, R_4, R_5, R_6, R_7, R_8, R_9, R_{10}$) is equal to $c \in \mathbb{Z}_p$ and outputs $\sigma = (c, T_1, T_2, T_3, T_4, T_5, T_6, T_7, T_8, T_9, T_{10}, s_\pi, s_\theta, s_\rho, s_{x_i}, s_{\pi\mu_i}, s_{\theta\mu_i}, s_\delta)$. $\mathcal{B}$ also adds (σ, i) into E-**list**.
  - **UCorruption oracle:** On input $i \in \mathsf{S}$, $\mathcal{B}$ responds with the corresponding sk_i to $\mathcal{A}$ and appends i to U.
  - **Revocation oracle:** On input $i \in \mathsf{S}$, $\mathcal{B}$ removes i from S, re-computes $\mathsf{pmi} = \{L, L'\}$ and $\mathsf{mt} = \{L_i\}_{i \in \mathsf{S}}$, and responds with the updated S, pmi, and mt to $\mathcal{A}$.
  - **Signing oracle:** $\mathcal{B}$ also maintains a list E-**list** for keeping track of output signatures. In the simulation of signing queries, $\mathcal{B}$ responds with σ_i computed on input (i, M) from $\mathcal{A}$ as follows. $\mathcal{B}$ selects π and $\theta \in_R \mathbb{Z}_p^*$, and compute Linear encryption of A_i, i.e. ciphertext $(T_1 = A_i\hat{g}^{\pi+\theta}, T_2 = u^{\pi}, T_3 = v^{\theta})$. $\mathcal{B}$ then randomly selects ρ, δ, r_π, r_θ, r_{μ_i}, r_ρ, r_{x_i}, $r_{\pi\mu_i}$, $r_{\theta\mu_i}$, r_δ, and r_δ to generates $T_4, T_5, T_6, T_7, T_8, T_9, T_{10}, R_1, R_2, R_3, R_4, R_5, R_6, R_7, R_8, R_9$, and R_{10}. $\mathcal{B}$ defines the output of H(gpk, $M, T_1, T_2, T_3, T_4, T_5, T_6, T_7, T_8, T_9, T_{10}, R_1, R_2, R_3, R_4, R_5, R_6, R_7, R_8, R_9, R_{10}$) to be equal to $c \in \mathbb{Z}_p$ and outputs $\sigma = (c, T_1, T_2, T_3, T_4, T_5, T_6, T_7, T_8, T_9, T_{10}, s_\pi, s_\theta, s_\rho, s_{x_i}, s_{\pi\mu_i}, s_{\theta\mu_i}, s_\delta)$. Finally, $\mathcal{B}$ adds (σ, i) into E-**list**.
  - **UCorruption oracle:** If $\mathcal{A}$ corrupts $i \in \mathsf{S}$ then $\mathcal{B}$ hands the corresponding secret signing key sk_i over to $\mathcal{A}$ and includes i into U.
  - **Revocation oracle:** If $\mathcal{A}$ wishes to revoke some member $i \in \mathsf{S}$ then $\mathcal{B}$ removes i from S and updates $\mathsf{pmi} = \{L, L'\}$ and $\mathsf{mt} = \{L_i\}_{i \in \mathsf{S}}$ as specified in the RGS scheme. $\mathcal{B}$ then hand updated pmi and mt over to $\mathcal{A}$.
- **Challenge:** In the challenge phase, $\mathcal{A}$ selects a message M, two indices i_0 and i_1, and sends them to $\mathcal{B}$. If $(i_0, i_1) \in \mathsf{S}^2 \wedge (i_0, i_1) \notin \mathsf{U}^2$, $\mathcal{B}$ returns A_{i_0} and A_{i_1} as its challenge to C_{le}. C_{le} replies with Linear encryption ciphertext (T_1, T_2, T_3) of A_{i_b} according to some random (unknown) bit $b \in \{0, 1\}$. Then $\mathcal{B}$ randomly selects ρ and δ to compute $T_4, T_5, T_6, T_7, T_8, T_9$, and T_{10}. It further selects random s_π, s_θ, s_ρ, s_{x_i},

$s_{\pi\mu_i}$, $s_{\theta\mu_i}$, s_δ, and c from $\mathbb{Z}_p^*$, and computes $R_1 = e(T_1, T_4)^{s_{\mu_i}} \cdot e(T_1, \Omega)^{s_\rho} / e(\hat{g}, T_4)^{s_{x_i}} \cdot e(T_6, g)^c \cdot e(\tilde{g}, g)^{s_\rho} \cdot e(\hat{g}, T_4)^{s_{\mu_i\pi} + s_{\mu_i\theta}} \cdot e(T_5, \Omega)^{s_\pi + s_\theta}$, $R_2 = u^{s_\pi} \cdot (T_2)^{-c}$, $R_3' = v^{s_\theta} \cdot (T_3)^{-c}$, $R_4 = g^{s_\rho} \cdot (T_4)^{-c}$, $R_5 = \hat{g}^{s_\rho} \cdot (T_5)^{-c}$, $R_6 = T_2^{s_{\mu_i}} \cdot u^{-s_{\pi\mu_i}}$, $R_7 = (zz')^{s_\rho} \cdot (T_7)^{-c}$, $R_8 = e(h, T_4)^{s_\delta} \cdot T_8^{-c}$, $R_9 = (T_7^c \cdot e(T_9, g)^{s_\rho}) / (T_8 \cdot e(L, T_6) \cdot e(L', T_{10}))^c$, and $R_{10} = T_3^{s_{\mu_i}} \cdot v^{-s_{\theta\mu_i}}$. $\mathcal{B}$ then sends $\sigma_{i_b} = (c, T_1, T_2, T_3, T_4, T_5, T_6, T_7, T_8, T_9, T_{10}, s_\pi, s_\theta, s_\rho, s_{x_i}, s_{\pi\mu_i}, s_{\theta\mu_i}, s_\delta)$ to $\mathcal{A}$.

- $\mathcal{B}$ continues answering oracle queries of $\mathcal{A}$ as specified above, until $\mathcal{A}$ outputs a bit b'.

$\mathcal{B}$ forwards b' as its answer on the challenge of C_{le} in the Linear encryption game. Clearly, if $\mathcal{A}$ wins then $\mathcal{B}$ breaks the IND-CPA security of the Linear encryption scheme.

C Proof of Theorem 3 (Non-frameability)

We consider two types of adversaries to break non-frameability in the proposed scheme. Type I adversary can forge group signatures of the member i for $i \in \mathsf{S}$ and type II adversary for $i \in [1, n] \wedge i \notin S$.

Type I Adversary. Let $\mathcal{A}$ be the type I adversary of non-frameability of our RGS scheme with the probability at least ϵ by forging a group signature of member i for $i \in \mathsf{S}$. We then construct an algorithm $\mathcal{B}$ that can break the classical discrete logarithm (DL) assumption with probability at least ϵ' in polynomial time by playing the game in Definition 8. The algorithm $\mathcal{B}$ proceeds as follows.

- **Setup:** $\mathcal{B}$ is given a DL instance $(\hat{g}, \mathcal{U} = \hat{g}^x)$ by a challenger C_{DL} of DL assumption, where $\hat{g} \in \mathbb{G}$ and $x \in_R \mathbb{Z}_p$. $\mathcal{B}$ then prepares the sets S and U, and an index $i^* \in \mathsf{S}$ of the target member to be attacked by $\mathcal{A}$. After that, $\mathcal{B}$ generates gsk, gpk, pmi, mt, and $\{\mathsf{sk}_i\}_{i \in \mathsf{S}}$ as well as the proposed scheme except sk_{i^*}. Here $\mathcal{B}$ computes $A_{i^*} = (\mathcal{U}\tilde{g})^{\frac{1}{\mu_i + \omega}}$ and sets $x_{i^*} = x$, which is unknown.
- **Oracles:**
  - **Hash oracle:** The simulation of hash queries is the same as in the proof of Theorems 2 and 1.
  - **Signing oracle:** Here $\mathcal{B}$ also maintains a list E**-list** for storing the corresponding identity and signature pairs. On input a pair (i, M), if $i \in \mathsf{S}$ and $i \neq i^*$, $\mathcal{B}$ can successfully generate any signature of the member i. If $i = i^*$, $\mathcal{B}$ generates $T_1, T_2, T_3, T_4, T_5, T_6, T_7, T_8, T_9, T_{10}, R_1, R_2, R_3, R_4, R_6, R_7, R_8, R_9$, and T_{10} as in the RGS specification by using random s_π, s_θ, s_ρ, s_{x_i}, $s_{\pi\mu_i}$, $s_{\theta\mu_i}$, and s_δ from $\mathbb{Z}_p^*$. After that, $\mathcal{B}$ responds with $\sigma = (c, T_1, T_2, T_3, T_4, T_5, T_6, T_7, T_8, T_9, T_{10}, s_\pi, s_\theta, s_\rho, s_{x_i}, s_{\pi\mu_i}, s_{\theta\mu_i}, s_\delta)$.
  - **UCorruption oracle:**On input $i \in \mathsf{S}$, if $i \neq i^*$ then $\mathcal{B}$ returns sk_i. Otherwise, $\mathcal{B}$ aborts.
  - **Revocation and opening oracles:** The simulations of the revocation and opening oracles are the same as that of Theorem 2 and 1.
- **Output:** Finally, $\mathcal{B}$ outputs a signature-message pair (σ^*, M^*).

$\mathcal{A}$ breaks the proposed scheme if **Verify**(gpk, pmi, σ^*, M^*) = **true**. In addition, if $\mathcal{A}$ outputs a forged σ_i with probability at least ϵ, then $\mathcal{A}$ outputs a forged σ_{i^*} of the target member i^* with probability at least ϵ/n. $\mathcal{B}$ then can successfully obtain two forged signatures $\sigma_{i^*} = (\sigma_{i_0^*} = (T_1, T_2, T_3, T_4, T_5, T_6, T_7, T_8, T_9, T_{10}, R_1, R_2, R_3, R_4, R_5, R_6, R_7, R_8, R_9, R_{10}), c, \sigma_{i_1^*} = (s_\pi, s_\theta, s_\rho, s_{x_i}, s_{\pi\mu_i}, s_{\theta\mu_i}, s_\delta)$ and $\sigma'_{i^*} = (\sigma_{i_0^*} = (T_1, T_2, T_3, T_4, T_5, T_6, T_7, T_8, T_9, T_{10}, R_1, R_2, R_3, R_4, R_5, R_6, R_7, R_8, R_9, R_{10}), c', \sigma'_{i_1^*} = (s'_\pi, s'_\theta, s'_\rho, s'_{x_i}, s'_{\pi\mu_i}, s'_{\theta\mu_i}, s'_\delta))$ by applying Forking Lemma [27] and extract $x_{i^*} = (s'_{x_{i^*}} - s_{x_{i^*}})/(c' - c)$ as the solution of DL problem with probability at least $\epsilon/4n$. Hence, we have that $\epsilon/4n \leq \epsilon'$.

Type II Adversary. Let $\mathcal{A}$ be an adversary that breaks the non-frameability of the proposed scheme with the probability at least ϵ by forging a group signature of member i for $i \in [1, n] \wedge i \notin \mathsf{S}$. We then construct an algorithm $\mathcal{B}$ to break PDHE assumption with probability at least ϵ''. The construction of $\mathcal{B}$ and its interaction with $\mathcal{A}$ proceed as follows.

- **Setup:** $\mathcal{B}$ is given a PDHE instance $(g, \hat{g}, g^{\alpha^{\psi+i}}, \hat{g}^{\alpha^{\psi+i}}, g^{\beta^{\psi+i}}, \hat{g}^{\beta^{\psi+i}}, \eta_j\}_{1 \leq i,j \leq n \wedge i \neq j})$ by a challenger C_{PDHE} of PDHE assumption, where $i = 1, ..., n$ and $j = 1, ..., n, n + 2, ..., 2n$. $\mathcal{B}$ then prepares sets S and U being the same as that of the simulation for type I adversary and prepares gsk and gpk as well as that of the proposed scheme except that α, β, and ψ are unknown. Then $\mathcal{B}$ simulates $\mathsf{sk}_i = (A_i, g^{\alpha,i}, g^{\beta,i}, \mu_i, x_i)$ for the member i by using PDHE instance, gsk, and gpk.
- **Oracles:** $\mathcal{B}$ answers the following oracles to simulates RGS interacting with $\mathcal{A}$.
  - **Hash oracle:** The simulation of hash queries are the same as that of the simulation for type I adversary.
  - **Signing oracle:**On inputting a pair (i, M), if $i \in [1, n]$, $\mathcal{B}$ generates the corresponding σ by sk_i. Otherwise, reject this request.
  - **UCorruption, revocation, and opening oracles:** The simulations of ucorruption revocation and opening oracles are the same as that for type I adversary.
- **Output:** Finally, $\mathcal{A}$ output a signature-message pair (σ^*, M^*).

$\mathcal{A}$ is a type II adversary of non-frameability to break the proposed scheme if σ^* is correct and belongs to some $i \in [1, n] \wedge i \notin \mathsf{S}$ with probability at least ϵ. Then $\mathcal{B}$ can successfully break PDHE assumption as follows. $\mathcal{B}$ can apply Forking Lemma as well as the proof for type I adversary to extract secrets ρ and δ such that $(zz')^\rho = e(L, T_6) \cdot e(L', T_{11}) \cdot T_8/e(T_9, g)^\rho = (e(\hat{g}^{\alpha^{2\psi+n+1}}, g) \cdot e(\hat{g}^{\beta^{2\psi+n+1}}, g))^\rho$. This means that $\hat{g}^{\alpha^{2\psi+n+1}} \hat{g}^{\beta^{2\psi+n+1}} = \frac{\prod_{j \in \mathsf{S}} \hat{g}^{\alpha^{2\psi+n+1-j+i}} \prod_{j \in \mathsf{S}} \hat{g}^{\beta^{2\psi+n+1-j+i}}}{T_9 \cdot (h^\delta)^{-1}}$. $\mathcal{B}$ can directly compute $\prod_{j \in \mathsf{S}} \hat{g}^{\alpha^{2\psi+n+1-j+i}}$ and $\prod_{j \in \mathsf{S}} \hat{g}^{\beta^{2\psi+n+1-j+i}}$ since $i \notin \mathsf{S}$ such that $i \neq j$. Therefore, $\mathcal{B}$ can successfully break PDHE assumption by extracting $\hat{g}^{\alpha^{2\psi+n+1}}$ from the forged signature σ^* by using $\hat{g}^{\beta^{2\psi+n+1}}$.

Fast Computation on Encrypted Polynomials and Applications

Payman Mohassel

University of Calgary
pmohasse@cpsc.ucalgary.ca

Abstract. In this paper, we explore fast algorithms for computing on encrypted polynomials. More specifically, we describe efficient algorithms for computing the Discrete Fourier Transform, multiplication, division, and multipoint evaluation on encrypted polynomials. The encryption scheme we use needs to be *additively* homomorphic, with a plaintext domain that contains appropriate primitive roots of unity. We show that some modifications to the key generation setups and working with variants of the original hardness assumptions one can adapt the existing homomorphic encryption schemes to work in our algorithms.

The above set of algorithms on encrypted polynomials are useful building blocks for the design of secure computation protocols. We demonstrate their usefulness by utilizing them to solve two problems from the literature, namely the oblivious polynomial evaluation (OPE) and the private set intersection but expect the techniques to be applicable to other problems as well.

1 Introduction

Polynomials are powerful objects with numerous applications in computer science and cryptography in particular. It is easy to find their trace in a range of cryptographic primitives and protocols. A common approach is to use polynomials to represent the input data and then to perform the necessary computation on the new representation. Examples of this approach include well-studied problems such as private information retrieval [2], secret sharing [26], as well as more recent applications such as private set intersection [11], privacy-preserving set operations [20], and private keyword search [12].

In almost all of these constructions, the data is first represented using one or more polynomial and then various operations such as addition, multiplication, and point evaluation on the polynomials are performed in the relevant setting. Furthermore, due to the security requirements, in many cases the polynomials are encrypted, secret-shared, or committed to and the computation is performed on the encrypted versions. Currently, the *efficiency gains* in these protocols are often the result of focusing on the specific problem at hand and taking advantage of the unique properties of that problem.

D. Lin, G. Tsudik, and X. Wang (Eds.): CANS 2011, LNCS 7092, pp. 234–254, 2011.

However, faster algorithms for computing on encrypted polynomials, would potentially lead to more efficient protocols for such applications. *In fact, a natural question we study in this paper is whether the existing computer algebra techniques – that have been around for over half a century – can be used to improve the efficiency of the cryptographic protocols for applications that use polynomials to represent their data.* We are not aware of any previous work on this question. The advantage of such an approach compared to the customized improvements for specific problems is that the resulting techniques are more general and can be used in a range of applications. In essence, they provide a cryptographic toolkit that can be used by protocol designers in almost a black-box manner. For instance, as we will see in Section 4, our algorithms can be used (without much additional effort) to design simple and efficient protocols for private set intersection (PSI) and batch oblivious polynomial evaluation (OPE). The resulting PSI protocol is a logarithmic factor less efficient than the best existing constructions, while the Batch OPE construction is more efficient than the only other alternative (to the best of our knowledge) of repeating an OPE protocol multiple times. Nevertheless, we mostly present these protocols as simple demonstrations of applicability of our techniques to different privacy-preserving problems.

With the recent developments in designing fully-homomorphic encryption schemes [15], it is possible to run any plaintext algorithm (including fast computer algebra techniques) on encrypted data without the need for decryption (or interaction) and with efficiency that is asymptotically similar to the complexity of the original algorithms. However, the existing fully homomorphic schemes are not yet practical, and hence it is desirable to rely on more efficient schemes but with *limited* homomorphic properties.

When working with encryption schemes with limited homomorphic properties, we need to make sure that the computer algebra algorithms can be computed on encrypted data without the use of the decryption key. Also, since these algorithms work over rings with special properties, we need to ensure that the existing homomorphic encryption schemes can be adapted to work in the required plaintext domains.

In this paper, we demonstrate how given the right setup and the appropriate computational assumptions for the encryption schemes, one can implement some of the main computer algebra techniques for fast polynomial computation on encrypted data. We then show how to use these techniques to develop efficient protocols for several privacy-preserving problems studied in the literature. Next, we describe our contributions in more detail.

1.1 Our Contribution

Fast Computation on Encrypted Polynomials. We design efficient algorithms for performing different computational tasks on encrypted polynomials. Particularly, given polynomials that are encrypted using an *additively* homomorphic encryption scheme, we design algorithms for computing the encrypted Discrete Fourier Transform, polynomial multiplication, division[1], and multipoint evaluation all with computational complexities that are linear in the size of the polynomials (upto a logarithmic factor). The constant factors in the complexities are fairly small and specified in the body of the paper.

In designing our algorithms we rely heavily on computer algebra techniques designed for fast symbolic computation. We show that the additive homomorphic properties of the underlying encryption scheme are sufficient to implement these techniques efficiently and non-interactively.

The encrypted Fast Fourier Transform (FFT) algorithm which is the starting point of our work, operates over a commutative ring that contains primitive nth roots of unity for appropriate values of n that are powers of two. We take a closer look at two widely used additively homomorphic encryption schemes in the literature, i.e. the Paillier encryption [25], and the El Gamal encryption [13] with messages in the exponent. We show that modulo some modifications to the key generation setups and working with variants of the original hardness assumptions, namely the *fourier DDH* and *fourier DCRA* assumptions (see Section 2), one can adapt these schemes to work in our protocols.

A number of other papers in the literature investigate the connection between secure computation and polynomials. For example, the work of [10] and [7] looks at evaluating public multivariate polynomials on parties' private inputs. Similar to us (and independently), Cheon et al [4] also consider the use of fast computer algebra techniques with the aim of designing more efficient protocols (multiparty PSI in this case). Our main goal, and what separates our work from the related papers, is to introduce an efficient framework for fast computation on encrypted polynomials with the intention that they can be used in privacy preserving applications.

We demonstrate the usefulness of our encrypted polynomial toolkit by applying it to several cryptographic problems.

Batch Oblivious Polynomial Evaluation. We first look at the oblivious polynomial evaluation problem [24]. In the OPE problem a sender holds a polynomial f of degree n over some ring R. A receiver holding an input $u \in R$ wants to learn $f(u)$ without learning anything else about the polynomial f and without revealing to the sender any information about u. OPE can be seen as a generalization of the 1-out-of-n oblivious transfer, and the private keyword search problem.

One can envision a wide range of applications that can take advantage of oblivious evaluation of polynomials. It is common practice to approximate a function by using a finite number of terms of its Taylor series and evaluating the polynomial corresponding to it. Through such approximations, OPE can be used to obliviously evaluate a variety of functions. One concrete example is the Taylor series approximation of the $ln(x)$ function used in privacy-preserving

---

[1] We consider the variants of polynomial multiplication and division where one polynomial is encrypted and the other is in plaintext. This variant appears to be sufficient for all the applications we have in mind.

data mining protocols [21]. OPE has also been used to design privacy-preserving protocols for machine learning methods such as neural learning [3].

In most applications, one is interested in performing the polynomial evaluation many times and on different input values. The naive solution of rerunning the protocol for each instance, which to the best of our knowledge is the only existing solution, is costly since the computation cost grows multiplicatively with the number of points being evaluated.

In Section 4.1, we show how our techniques for computation on encrypted polynomials help with designing a protocol for batch evaluation of $k < n$ OPE instances with only $O(n)$ communication and $O(n \log n + k(\log k)^2)$ computation compared to the $O(kn)$ complexity of the naive solution. *Private Set Intersection.* We also use our algorithmic ideas to solve the private set intersection (PSI) problem. The PSI problem involves two or more parties each with their own private data sets who want to learn which data items they share without revealing anything more about their data. A number of organizations dealing with sensitive data such as healthcare providers, insurance companies, law enforcement agencies, and aviation security need to perform such an operation on their data. A large body of work has been studying the design of PSI protocols [11,20,16,18,6,17,9,19,8,1].

In Section 4, we provide two efficient protocols for the PSI problem. While the computational efficiency of our (more efficient) construction is a *logarithmic factor* worse than the best existing works, our protocols are obtained with little effort and are almost immediate applications of the algorithms we introduce. In fact, our main motivation is *not to design more efficient PSI protocols* but to demonstrate the usefulness of the techniques discussed in this paper, in the design of privacy-preserving protocols.

Our batch OPE protocol can be used to solve the private set intersection problem with almost linear computation. In order to find the intersection of two datasets A and B, it is sufficient to represent A via a polynomial f_A (where elements of A are roots of f_A), and then obliviously evaluate all elements of B at f_A. Those values evaluated to zero are in the intersection while the rest are not. The resulting PSI protocol requires $O(n)$ communication and $O(n(\log n)^2)$ computation.

Kissner and Song [20] designed a simple and elegant protocol for the private set intersection problem. Given two sets A and B of equal size n, the idea is to represent the sets using polynomials f_A and f_B. Then, one can show that roots of the polynomial $o = rf_A + sf_B$ for random polynomials r and s are either random or in the intersection of the two sets. This idea can be turned into a PSI protocol, and can easily be extended to work for the multiparty case. The main drawback of the construction, as pointed out in the literature, is that it requires computational complexity that is quadratic in the size of the datasets. However, in light of the new algorithms we designed for encrypted polynomial multiplication, the PSI protocol we derive has $O(n)$ communication and $O(n \log n)$ computation. The constant factors in the complexity are also small. In particular the computation involved consists of n encryptions, n decryptions, $2n \log n$ homomorphic additions/subtractions and $n \log n$ homomorphic multiplications.

2 Homomorphic Encryption and Hardness Assumptions

We use a semantically secure public-key encryption scheme that is also additively homomorphic. In particular, we call an encryption scheme E additively homomorphic if given two ciphertexts $E(m_1)$ and $E(m_2)$, we can efficiently compute an encryption of $m_1 + m_2$. We denote this by $E(m_1 + m_2) = E(m_1) +_h E(m_2)$. This implies that given an encryption $E(m)$ and a value c, we can efficiently compute a random encryption $E(cm)$; we denote this by $E(cm) = c \times_h E(m)$. For a vector $\boldsymbol{v}$ we denote by $E(\boldsymbol{v})$ an entry-wise encryption of the vector. We define the encryption of a polynomial by the encryption of the vector of its coefficients. We can add two encrypted vectors (polynomials) by adding each encrypted component individually (we use the same notation $+_h$ for this operation as well). When measuring efficiency of our algorithms, we often count the number of homomorphic additions/subtractions and multiplications separately, since homomorphic addition/subtraction tends to be significantly faster for all existing encryption schemes.

There are a number of homomorphic encryption schemes each with their own special properties. Generalized versions of our protocols would work with any encryption scheme that is additively homomorphic as long as the domain of the plaintexts is a *commutative ring*. However, the protocols become simpler and more efficient when we can guarantee that the underlying ring contains a *primitive nth root of unity* for an appropriate choice of n that is a power of two. Hence, it is important to review the existing homomorphic encryption schemes and to determine whether they can be instantiated with plaintext domains that satisfy our requirement. We take a closer look at two widely used additively homomorphic encryptions schemes in the literature, i.e. the Paillier encryption [25] and the El Gamal encryption [13] with messages in the exponent and introduce the two hardness assumptions *fourier DDH* and *fourier DCRA* which we need for our protocols. While these new assumptions are closely related to the standard DDH and DCRA assumptions, a better understanding of their hardness requires more careful analysis.

2.1 Additive Variant of El Gamal

Consider the additive variant of the El Gamal encryption scheme where messages are encoded in the exponent. In other words, a message m is encrypted by computing $E_{pk}(m, r) = (g^r, h^r g^m)$ where g is a generator for a cyclic group G_q over which the DDH assumption is hard. It is easy to see that multiplying two ciphertexts adds the two corresponding messages in the exponent. Decryption can be performed efficiently as long as the messages are small. For most of the applications we consider, it is sufficient to determine whether a ciphertext encrypts the message 0 or not, but otherwise there is no need to decrypt a ciphertext. The additive variant of the El Gamal encryption is sufficient for such applications. For example, in our private set intersection protocol (based on polynomial multiplication), it suffices for one party to test whether ciphertexts are encryptions of 0 without fully decrypting them. In applications such as private keyword search where one is required to decrypt larger plaintexts (i.e. the payload), it is still possible to get around this limitation at the cost of a less efficient construction. In particular, one can use a different polynomial for each bit (or a constant number of bits) of the payload, and invoke the OPE protocol once for each polynomial. Nevertheless, the additive variant of the El Gamal scheme is not a suitable choice for all applications, and should be used with this limitation in mind.

In most cases, the DDH assumption is considered over a finite field $\mathbb{F}_q$, where q is a prime. In order to make sure the underlying finite field contains the necessary primitive roots of unity, we take advantage of the following lemma:

Lemma 1. *[14] For a prime power q and $n \in \mathbb{N}$, a finite field $\mathbb{F}_q$ contains a primitive nth root of unity if and only if n divides $q-1$.*

Based on this Lemma, if we choose q carefully such that $q = r2^\ell + 1$ for positive integers r and ℓ, then the corresponding finite field $\mathbb{F}_q$ has nth roots of unity for any $n = 2^s$ for which $s \le \ell$. The primes of this form are sometimes referred to as *Fourier primes.*

Definition 1 (Fourier Primes). *A prime number p is called a fourier prime if $p = r2^\ell + 1$ for two positive integers r and ℓ.*

Given the above lemma, in order to use the additive variant of El Gamal in our protocols, we need to make sure that the prime we use is in fact a fourier prime. In other words, we need to modify the key generation step in order to sample from the space of primes of the form $r2^\ell + 1$. It is not hard to show that successively testing $2^\ell + 1, 2.2^\ell + 1, 3.2^\ell + 1, \ldots$, for primality is bound to find a fourier prime in a small number of iterations. See Chapter 18 of [14] for more details on efficiency of this step. In addition to generating a fourier prime, the key generation algorithm also needs to find a root of unity. The following lemma guarantees an efficient way of generating such roots of unity.

Lemma 2. *Let $p = r2^\ell + 1$ be a prime number where r and ℓ are positive integers. Let a be a nonsquare modulo p. $a^r \bmod p$ is a primitive 2^ℓth root of unity in $\mathbb{F}_p^*$.*

While we are not aware of any algorithms for solving the discrete log problem or any attacks against the DDH assumption that takes advantage of the fact that the prime number we work with is a fourier prime, the effect of this extra assumption requires further analysis. In this paper, we denote this new variant of the DDH assumption with the *fourier DDH assumption.*

Definition 2 (Fourier DDH Assumption). *We denote by fourier DDH, the DDH assumption over the group $\mathbb{F}_p$ where p is a fourier prime.*

2.2 Paillier's Encryption Scheme

Paillier's encryption [25] is another widely used additively homomorphic encryption scheme. In Paillier's encryption, plaintexts are in Z_N where $N = pq$ is product of two prime numbers. Once again, in order to take advantage of efficient FFT algorithms, we need to make sure that we can find nth roots of unity for all values of n we might be interested in for our applications. Consider the following lemma:

Lemma 3 ([14], Exercise 8.19). *Let p, q be two distinct odd primes and $N = pq$. Z_N^* contains a primitive kth root of unity if and only if $k|lcm(p-1, q-1)$.*

More specifically, in order to use Paillier's encryption scheme in our protocols, we need to make sure that at least one of the two primes p and q is a fourier prime. We can use the method discussed above to generate such a prime efficiently. Paillier's encryption is based on the hardness of a mathematical assumption called decisional composite residuosity assumption (DCRA). We denote the variant of the assumption where at least one of the primes is a fourier prime with *fourier DCRA*.

3 Non-interactive Computation on Encrypted Polynomials

In this section, we describe efficient and non-interactive algorithms for computing on encrypted polynomials. This collection of algorithms on encrypted data provides us with a useful toolkit for designing efficient privacy-preserving protocols for a number of problems.

We start by showing that the FFT algorithm and the interpolation on roots of unity (reviewed in Section A) can be efficiently extended to work on inputs that are encrypted using an *additively homomorphic encryption* scheme. More importantly, performing these computations on encrypted data can be done locally and without the use of the decryption algorithm. Then, we show how to use these two techniques to perform polynomial multiplication, division and multipoint evaluation on partially encrypted data (also non-interactively). The complexity of all the algorithms are linear, upto a logarithmic factor, in the degree of the polynomials we work with.

For simplicity, we assume that the plaintext domain of the encryption scheme is a commutative ring containing a primitive nth root of unity, for an appropriate value of n. As explained in Section 2, it is possible to modify both Paillier's and the El Gamal encryption schemes to meet this requirement.

Computing the DFT of Encrypted Polynomials. Let f be a polynomial of degree d with coefficients in a finite Ring R, and let $w \in R$ be an nth root of unity where $d < n$ and $n = 2^k$. Given the encryption of the polynomial f via an additively homomorphic encryption scheme (and without the knowledge of the decryption key), we want to compute an encryption of f's DFT, namely encryption of the vector $\langle f(w), f(w^2), \ldots, f(w^{n-1})\rangle$. In essence, DFT can be seen as a special case of multipoint evaluation, at the powers of nth root of unity w.

Using *Horner's rule* and the homomorphic properties of an additive encryption scheme, one can compute an encryption of each $f(w^i)$ via d homomorphic multiplications and d homomorphic additions. This leads to a total of $O(dn)$ homomorphic operations for computing the DFT. But this is not the best we can do. Using the *Fast Fourier Transform* we show how to reduce the total cost to $O(n \log n)$ homomorphic operations.

The Fast Fourier Transform, or FFT for short, can be used to compute the DFT quickly. The algorithm was (re)discovered by Cooley and Tukey [5], and is one of the most important algorithms in practice. The high level idea is to divide f by $x^{n/2} - 1$ and $x^{n/2} + 1$. Computing the DFT of f is then recursively reduced to computing the DFT of the remainder polynomials of the two divisions. We review the FFT algorithm in Appendix A and here only describe its adaptation to work on encrypted polynomials.

Our main observation is that the FFT algorithm lends itself quite nicely to (additive) homomorphic properties of the encryption scheme, and hence can be computed *non-interactively* and efficiently on encrypted data.

Encrypted FFT Algorithm
$\mathsf{EncFFT}_{w,n}(E_{pk}(\boldsymbol{f}))$

Input: $n = 2^k \in \mathbb{N}$ for $k \geq 1$; powers of a primitive nth root of unity $w \in R$; the public key pk for an additively homomorphic encryption scheme E with the plaintext domain R, and the encrypted vector $E_{pk}(\boldsymbol{f}) = \langle E_{pk}(f_0), E_{pk}(f_1), \ldots, E_{pk}(f_{n-1})\rangle$ where $f(x) = \sum_{0 \leq i < n} f_i x^i$.
Output: Encrypted vector $\langle E_{pk}(f(1)), E_{pk}(f(w)), \ldots, E_{pk}(f(w^{n-1}))\rangle$

1. If $n = 2$ then return $E_{pk}(f_0) + E_{pk}(f_1) \times_h w$.
2. For $0 \leq j < n/2$ compute $E_{pk}(r_{0,j}) = E_{pk}(f_j) +_h E_{pk}(f_{j+n/2})$.
3. For $0 \leq j < n/2$ compute $E_{pk}(r_{1,j}) = E_{pk}(f_j) -_h E_{pk}(f_{j+n/2})$.
4. For $0 \leq j < n/2$ compute $E_{pk}(r^*_{1,j}) = E_{pk}(r_{1,j}) \times_h w^j$.
5. Let $\boldsymbol{r_0} = \langle r_{0,0}, r_{0,1}, \ldots, r_{0,n/2-1}\rangle$ and $\boldsymbol{r^*_1} = \langle r^*_{1,0}, r^*_{1,1}, \ldots, r^*_{1,n/2-1}\rangle$.
6. Compute the two encrypted vectors $\boldsymbol{O_0} = \langle o_{0,0}, \ldots, o_{0,n/2-1}\rangle$ and $\boldsymbol{O_1} = \langle o'_{1,0}, \ldots, o'_{1,n/2-1}\rangle$ by letting $\boldsymbol{O_0} \leftarrow \mathsf{EncFFT}_{w^2,n/2}(E_{pk}(\boldsymbol{r_0}))$ and $\boldsymbol{O_1} \leftarrow \mathsf{EncFFT}_{w^2,n/2}(E_{pk}(\boldsymbol{r^*_1}))$.
7. Return $\langle o_{0,0}, o_{1,0}, o_{0,1}, o_{1,1}, \ldots, o_{0,n-1}, o_{1,n-1}\rangle$.

Efficiency. It is easy to verify that the above protocol requires $n \log n$ homomorphic additions/subtractions and $n/2 \log n$ homomorphic multiplications by w^i. As noted earlier, each homomorphic addition subtraction) translates to a group multiplication (inversion) while each homomorphic multiplication requires an exponentiation. On the other hand, if w is chosen to be small, some of these exponentiations (which are homomorphic multiplications with powers of w) become more efficient. In particular, based on Lemma 2, for a fourier prime $p = r2^\ell + 1$, we know that $w = a^r$ is primitive root of unity for any non-square a. Hence, as long as we make sure r is a small integer, w can also be chosen to be fairly small.

Encrypted Interpolation on Powers of Roots of Unity. In the interpolation problem, given the vector $\langle f(1), f(w), f(w^2), \ldots, f(w^{n-1})\rangle$, our goal is to compute the coefficients of the corresponding polynomial f. As discussed in detail in Appendix A.3, interpolation at powers of w can essentially be reduced to computing DFT of a polynomial of degree n, and hence be computed efficiently using the EncFFT algorithm. Next, we briefly describe this algorithm.

Encrypted Interpolation on Powers of nth Root of Unity
$\mathsf{EncInterpol}_{w,n}(\boldsymbol{v})$

Input: $n = 2^k \in \mathbb{N}$ with $k \geq 1$, a primitive nth root of unity $w \in R$, public key pk for an additively homomorphic encryption scheme E and an encrypted vector $E_{pk}(\boldsymbol{v}) = \langle E_{pk}(v_0), \cdots, E_{pk}(v_{n-1})\rangle$ where $w \in R^n$.

Output: The encrypted polynomial $E_{pk}(f)$ of degree n where $f(w^i) = v_i$ for $0 \leq i < n$.

1. Compute $w^{-1}, w^{-2}, \ldots, w^{-(n-1)}$.
2. Compute and return $1/n \times_h \mathsf{EncFFT}_{w^{-1},n}(E_{pk}(\boldsymbol{v}))$

The algorithm's efficiency is almost identical to that of the EncFFT algorithm as the bulk of the computation is one invocation of that algorithm.

Multiplying Encrypted Polynomials. Next we show how to use the EncFFT and EncInterpol algorithms described above to efficiently multiply encrypted polynomials. We are mostly interested in the variant of the multiplication where one polynomial is encrypted and the other polynomial is in plaintext. The idea for the fast polynomial multiplication is to compute the DFT of both polynomials, multiply the components of the DFT individually to obtain the DFT of the product polynomial and then interpolate to recover the product polynomial itself.

As we show next, given an additively homomorphic encryption scheme, this variant can be computed non-interactively and without the need for decrypting any ciphertexts. The algorithm only requires computation that is linear in the degree of the polynomials.

Encrypted Polynomial Multiplication Algorithm
$\mathsf{EncPolyMult}_{w,n}(E_{pk}(\boldsymbol{f}), g)$

Input: Encryption of the polynomial $f(x) = \sum_{0 \le i < d_1} f_i x^i$, i.e. $E_{pk}(\boldsymbol{f}) = \langle E_{pk}(f_0), \ldots, E_{pk}(f_{d_1})\rangle$ and the plaintext polynomial $g = \sum_{0 \le i < d_2} g_i x^i$, and a primitive nth root of unity w where $n = 2^k$ and $d_1 + d_2 < n$.
Output: Encryption of the product polynomial $h = fg$.

1. Compute $\langle E_{pk}(f(1)), \ldots, E_{pk}(f(w^{n-1}))\rangle \leftarrow \mathsf{EncFFT}_{w,n}(f)$.
2. Compute $\langle g(1), \ldots, g(w^{n-1})\rangle \leftarrow \mathsf{FFT}_{w,n}(g)$.
3. For $0 \le i < n$, compute $E_{pk}(h(i)) = g(i) \times_h E_{pk}(f(i))$.
4. Let $\boldsymbol{v_h} \leftarrow \langle h(1), \ldots, h(n-1)\rangle$.
5. Compute and return $\mathsf{EncInterpol}_{w,n}(E_{pk}(\boldsymbol{v_h}))$.

Efficiency. The algorithm requires $2n \log n$ homomorphic additions/subtractions and $n \log n$ homomorphic multiplications, since the EncFFT and the EncInterpol are each invoked exactly once.

Encrypted Polynomial Division. Here, we describe a protocol for performing the division with remainder on encrypted polynomials. We focus on the version of the division protocol where polynomial a of degree n is encrypted, a *monic* polynomial b of degree $m < n$ is in plaintext and we want to compute encryptions of two polynomials q and r such that $a = qb + r$ and r is of degree less than m. Next we review simple algebraic tricks that allow us to reduce the encrypted polynomial division algorithm to the encrypted polynomial multiplication algorithm we described earlier. We note that similar tricks were used in [22] in the context of secure computation but in a different setting and with different applications in mind.

We define *reversal* of a polynomial a as $rev_k(a) = x^k a(1/x)$. When $k = n$, this is the polynomial with the coefficients of a reversed, that is, if $a = a_n x^n + a_{n-1}x^{n-1} + \cdots + a_1 x + a_0$, then $rev(a) = rev_n(a) = a_0 x^n + \cdots + a_{n-1}x + a_n$.

We can now rewrite the division with remainder expression as

$$rev_n(a) = rev_{n-m}(q) rev_m(b) + x^{n-m+1} rev_{m-1}(r)$$

and therefore, $rev_n(a) = rev_{n-m}(q) rev_m(b) \bmod x^{n-m+1}$

Note that since we assume b is a monic polynomial, $rev_m(b)$ has the constant coefficient 1 and thus is invertible modulo x^{n-m+1}. Hence we have that

$$rev_{n-m}(q) \equiv rev_n(a) rev_m(b)^{-1} \bmod x^{n-m+1},$$

and can obtain $q = rev_{n-m}(rev_{n-m}(q))$ and $r = a - qb$.

In other words, performing the polynomial division with remainder is reduced to inverting the polynomial b modulo x^{n-m+1}, two polynomial multiplications and one polynomial subtraction. Since in our variant of the algorithm b is in plaintext, we can use standard computer algebra algorithms for inverting b which requires $O(n \log n)$ ring operations. Since in all our applications we are only interested in the remainder polynomial, we define the output of the protocol to be only r. The algorithm follows.

Encrypted Polynomial Division Algorithm
$\mathsf{EncPolyDiv}_{w,n'}(E_{pk}(\boldsymbol{a}), b)$

Input: Encryption of the polynomial $a(x) = \sum_{0 \leq i \leq n} a_i x^i$, i.e. $E_{pk}(\boldsymbol{a}) = \langle E_pk(a_0), \ldots, E_{pk}(a_n) \rangle$ and the plaintext polynomial $b(x) = \sum_{0 \leq i \leq m} b_i x^i$, and a primitive n'th root of unity w where $n' = 2^k$ and $n' > 2n - m + 1$.

Output: Encryption of the remainder polynomial r of degree less than m where $a = qb + r$.

1. Compute $w^2, \ldots, w^{n-1}$.
2. Let $a' = rev_n(a)$. By reversing the order of coefficients of $E_{pk}(a)$, we arrive at the encrypted version of a' denoted by $E_{pk}(\boldsymbol{a'})$.
3. Compute the polynomial $b' = rev_{n-m}(b)^{-1} \bmod x^{n-m+1}$ using standard computer algebra techniques.
4. Compute $E_{pk}(q_1) = \mathsf{EncPolyMult}_{w,n'}(b', E_{pk}(\boldsymbol{a'}))$.
5. Compute $E_{pk}(q_2) = E_{pk}(q_1) \bmod x^{n-m+1}$. This is a simple operation that can be performed non-interactively given the additive homomorphic property of the encryption scheme.
6. Compute $E_{pk}(q) = rev_{n-m}(E_{pk}(q_2))$ by reversing the coefficients.
7. Compute $E_{pk}(r) = E_{pk}(a) -_h \mathsf{EncPolyMult}_{w,n'}(E_{pk}(q), b)$.
8. Output $E_{pk}(r)$.

Efficiency. The protocol invokes the EncPolyMult protocol twice, and requires m and $n - m$ homomorphic additions/subtractions in steps 5 and 7, respectively. This leads to a total of $2n' \log n'$ homomorphic multiplication and $4n' \log n' + n'$ homomorphic additions.

Encrypted Multipoint Polynomial Evaluation. Given an encrypted polynomial f of degree n and n points $u_0, \cdots, u_{n-1} \in R$, our goal is to compute the encrypted vector $\langle f(u_0), f(u_1), \cdots, f(u_{n-1}) \rangle$. Through the use of the Horner's rule and the additive homomorphic properties of the encryption scheme, it is possible to perform this task with $O(n^2)$ homomorphic operations. However, we are interested in a significantly more efficient algorithm. We have already seen that in the special case when $u_i = w^i$ where w is a primitive nth root of unity, the EncFFT algorithm performs the same task with $O(n \log n)$ homomorphic operations. Our goal is to design an efficient algorithm for the general case of the problem.

Let $n = 2^k$ and $m_i = x - u_i$ for $0 \leq i < n$. We first compute the following sequence of polynomials:

$$M_{i,j} = m_{j2^i} m_{j2^i+1} \cdots m_{j2^i+(2^i-1)} = \prod_{0 \leq \ell < 2^i} m_{j2^i+\ell}$$

for $0 \leq i \leq k = \log n$ and $0 \leq j < 2^{k-i}$. In other words, each $M_{i,j}$ is a subproduct with 2^i factors from $M_{k,0} = \prod_{0 \leq \ell < n} m_\ell$. There exist a simple recursive algorithm for computing the polynomials $M_{i,j}$ for $0 \leq i \leq k$ and $0 \leq j < 2^{k-i}$, with $O(n(\log n)^2)$ ring operations. Note that since u_i's are in plaintext, the ring additions and multiplications are significantly cheaper than say encryption or homomorphic multiplication both of which require exponentiation.

The algorithm for multipoint evaluation uses these subproducts in a recursive way. The idea is to divide the polynomial we want to evaluate by two of these subproduct polynomials (see Step 3 of algorithm) and recursively run the multipoint evaluation algorithm on the remainder polynomials. Evaluating the remainder polynomials gives the same result as evaluating the original polynomial itself. We describe the detailed algorithm next:

Encrypted Multipoint Polynomial Evaluation Algorithm
$\mathsf{EncMultiEval}_n(E_{pk}(f), \boldsymbol{u})$

Input: Encryption of the polynomial $f(x) = \sum_{0 \leq i < n} f_i x^i$ over R, i.e. $E_{pk}(\boldsymbol{f}) = \langle E_{pk}(f_0), \ldots, E_{pk}(f_n) \rangle$, and the plaintext vector $\boldsymbol{u} = \langle u_0, u_1, \cdots, u_{n-1} \rangle$. Let $n = 2^k$, for $k \in \mathbb{N}$, and w the a primitive nth root of unity.
Output: The encrypted vector $\langle E_{pk}(f(u_1)), \cdots, E_{pk}(f(u_{n-1})) \rangle$.

1. Compute the subproduct polynomials $M_{i,j}$ for $0 \leq i \leq k$, and $0 \leq j < 2^{k-i}$, as described above.
2. If $n = 1$ then return f. f is a constant in this case.
3. Compute $E_{pk}(r_0) \leftarrow \mathsf{EncPolyDiv}_{w,n}(E_{pk}(f), M_{k-1,0})$, and $E_{pk}(r_1) \leftarrow \mathsf{EncPolyDiv}_{w,n}(E_{pk}(f), M_{k-1,1})$. Note that r_0 and r_1 are of degree less than $n/2$.
4. Let $\boldsymbol{u}^0 = \langle u_0, \cdots, u_{n/2-1} \rangle$ and $\boldsymbol{u}^1 = \langle u_{n/2}, \cdots, u_n \rangle$. Recursively call the algorithm twice
   (a) $\langle E_{pk}(r_0(u_0)), \cdots, E_{pk}(r_0(u_{n/2-1})) \rangle \leftarrow \mathsf{EncMultiEval}_{w,n/2}(E_{pk}(r_0), \boldsymbol{u}^0)$
   (b) $\langle E_{pk}(r_1(u_{n/2})), \cdots, E_{pk}(r_1(u_{n-1})) \rangle \leftarrow \mathsf{EncMultiEval}_{w,n/2}(E_{pk}(r_1), \boldsymbol{u}^1)$
5. Output $E_{pk}(r_0(u_0)), \cdots, E_{pk}(r_0(u_{n/2-1})), E_{pk}(r_1(u_{n/2})), \cdots, E_{pk}(r_1(u_{n-1}))$.

Efficiency. A careful calculation we omit here (See Chapter 10 of [14]) shows that the above algorithm requires at most $D(n) \log n$ operations where $D(n)$ is the number of operations needed for dividing a polynomial of degree less than $2n$ by a monic polynomial of degree n. Given the complexity of our division algorithm, this leads to at most $6n(\log n)^2$ homomorphic multiplications and $12n(\log n)^2 + 3n \log n$ homomorphic additions/subtractions.

4 Applications

4.1 Batch Oblivious Polynomial Evaluation

In the Oblivious Polynomial Evaluation (OPE) problem a sender holds a polynomial f of degree n over some ring R. A receiver holding an input $u \in R$ wants to learn $f(u)$ without learning anything else about the polynomial f and without revealing to the sender any information about u.

OPE was originally studied in [24], and can be seen as a generalization of a number of problems studied in the literature such as the 1-out-of-n oblivious transfer (e.g. see [23]), and private keyword search [12]. In case of 1-out-of-n oblivious transfer with a database $D = \{d_1, \ldots, d_n\}$, the sender can choose f such that $f(i) = d_i$. In case of private keyword (PKS) search where the database is $D = \{(w_1, d_1), \ldots, (w_n, d_n)\}$, the sender can choose a polynomial f where $f(w_i) = d_i$ and w_i is the keyword associated to d_i for $1 \leq i \leq n$.[2]

Given an additively homomorphic encryption scheme over the ring R, there exist a simple protocol for the OPE problem which requires $O(n)$ encryption/homomorphic operations when implemented using the Horner's rule. However, in most applications one is interested in evaluating the polynomial on many points. Given k evaluation points, the naive solution (and the only solution we are aware of) is to repeat the OPE protocol k times. This leads to $O(kn)$ encryption/homomorphic operations. For large values of k this is inefficient. Next, we use the techniques developed in previous sections to design a protocol for batch OPE that only requires $O(n(\log n) + k(\log k)^2)$ homomorphic operations. The protocol is a natural composition of the $\mathsf{EncPolyDiv}$ algorithm and the $\mathsf{EncMultiEval}$ algorithms we have designed. More specifically, to evaluate the polynomial f at points $u_1, \cdots, u_k$, we first divide f by the polynomial $(x - u_1) \cdots (x - u_k)$. Denote the resulting polynomial by r. It is easy to see that $r(u_i) = f(u_i)$ for $1 \leq i \leq k$. Therefore we can use the $\mathsf{EncMultiEval}$ protocol to evaluate r at $u_1, \ldots, u_k$. The protocol follows:

Batch Oblivious Polynomial Evaluation Protocol
$\mathsf{BatchOPE}(f, \boldsymbol{u})$

Sender's Input: A polynomial f of degree n with coefficients in R.
Receiver's Input: The vector $\boldsymbol{u} = \langle u_1, u_2, \cdots, u_k \rangle$ in the ring R^n.
Receiver's Output: $\langle f(u_1), \ldots, f(u_k) \rangle$

1. Sender generates a key pair (pk, sk) for the public key encryption scheme E, and sends pk to the receiver.
2. Sender sends $E_{pk}(f)$ to the receiver.
3. Receiver computes the polynomial $g = (x - u_1)(x - u_2) \cdots (x - u_k)$.
4. Receiver computes the encrypted polynomial $E_{pk}(r) \leftarrow \mathsf{EncPolyDiv}(E_{pk}(f), g)$ of degree k.
5. Receiver computes $E_{pk}(\boldsymbol{o}) \leftarrow \mathsf{EncMultiEval}_k(E_{pk}(r), \boldsymbol{u})$.
6. Receiver computes and sends $E_{pk}(\boldsymbol{o} + \boldsymbol{o_r}) = E_{pk}(\boldsymbol{o}) +_h E_{pk}(\boldsymbol{o_r})$ for a random vector $o_r \in R^k$ to the sender.
7. Sender decrypts the encrypted vector and sends $\boldsymbol{o} + \boldsymbol{o_r}$ back to the receiver.
8. Receiver computes and outputs $\boldsymbol{o} = \boldsymbol{o} + \boldsymbol{o_r} - \boldsymbol{o_r}$.

[2] A standard assumption made in PKS protocols is that the keywords are unique.

Efficiency: The protocol executes the EncPolyDiv protocol on a polynomial of degree n and the EncMultiEval protocol on a polynomial of degree k. This adds to a total of $O(n \log n + k(\log k)^2)$ homomorphic operations.

Claim. The BatchOPE protocol is secure against semihonest adversaries, if the encryption scheme E is semantically secure.

Proof Sketch: The proof of security of the protocol against semi-honest adversaries follows naturally from the semantic security of the encryption scheme, and the randomization steps that take place in the protocol. For completeness we include a sketch of the proof here. This also serves as a good example, since the proofs for all the other protocols follow the same pattern. The full proof of our protocol (deferred to the full version) follows the ideal/real simulation paradigm. In this extended abstract, however, we only give a proof sketch. In case of semi-honest adversaries, it is sufficient to simulate the view of the corrupted party given only his input/output. We have the following two cases:

Sender Is Corrupted. The Sender's view is only the message he receives in step 6 of the protocol. This message is an encryption of a uniformly random message vector. This is because the sender does not know the random vector $\boldsymbol{o_r}$ which masks the output vector $\boldsymbol{o}$. Hence the simulator can simulate the sender's view by computing an encryption of a randomly chosen message vector.

Receiver Is Corrupted. The simulator knows the receiver's input and randomness and the output vector $\boldsymbol{o}$ and wants to simulate the receiver's view in the real protocol. He first generates a dummy polynomial f' of appropriate degree (arbitrary coefficients) to use in place of the sender's input polynomial f. The simulator encrypts f' using the encryption scheme and performs all the computation on the encrypted f' instead to get an encrypted vector $E_{pk}(\boldsymbol{o'})$. Note that due to the semantic security of the encryption, the view generated using f' is computationally indistinguishable from the one generated using the real polynomial f.

To simulate the only remaining part of receiver's view (i.e. the message received in step 7), given $\boldsymbol{o_r}$ generated by the receiver, the simulator computes $\boldsymbol{o} + \boldsymbol{o_r}$. the generated vector is identical to the vector in receiver's view. This completes the proof sketch of security for the above scheme.

4.2 Private Set Intersection via OPE

The set intersection problem involves two or more parties each with their own private data sets who want to learn which data items they share without revealing anything more. As mentioned earlier, several recent works have focused on designing protocols with linear computation and communication complexity.

Interestingly, our batch oblivious polynomial evaluation protocol can be used to solve the private set intersection problem with linear complexity. In order to find the intersection of two datasets A and B, it is sufficient to represent A via a polynomial f_A (where elements of A are roots of f_A), and then obliviously evaluate all elements of B at f_A. Those values evaluated to zero are in the intersection while the rest are not. Simple randomization techniques can be added to avoid leaking any information about those elements that are not in the set. The protocol follows.

Private Set Intersection Protocol (via OPE)

Inputs: Alice holds the dataset A of size n_a, and Bob holds the dataset B of size n_b with elements in R. Without loss of generality we assume that $n_a > n_b$.
Output: Alice learns the intersection of A and B.

1. Alice computes the polynomial f_A of degree n_a by letting the roots of f_A be the elements in A.
2. Bob randomly permutes and arranges the elements of B in a vector $\boldsymbol{b} \in R^{n_b}$.
3. Alice and Bob run the Steps 1 to 5 of the $\mathsf{BatchOPE}(f_A, \boldsymbol{b})$ protocol. At this point Bob holds the encrypted vector $E_{pk}(\boldsymbol{o})$ which contains the evaluation of elements of B at polynomial f_A.
4. Bob generates two random vectors $\boldsymbol{r_1}, \boldsymbol{r_2} \in R^{n_b}$. He then computes and sends $E_{pk}(\boldsymbol{o_1}) = \boldsymbol{r_1} \times_h E_{pk}(\boldsymbol{o}) +_h \boldsymbol{b}$ and $E_{pk}(\boldsymbol{o_2}) = \boldsymbol{r_2} \times_h E_{pk}(\boldsymbol{o})$ where the vector multiplications are component-wise multiplications. Note that $\boldsymbol{o_2}$ is zero in components corresponding to the elements in the intersection, and random otherwise. For indices corresponding to elements in the intersection, $\boldsymbol{o_1}$ holds the actual values.
5. Alice decrypts $\boldsymbol{o_2}$ to learn the locations of the elements in the intersection (they are the ones with 0). She marks the indices for those locations, decrypts $\boldsymbol{o_1}$ and outputs the values in the marked indices as the final output.

The above protocol can be easily modified to compute the size of the intersection set instead. Particularly, if in the final stage we only compute the vector o_2 and count the number of zeros, we have a protocol that computes the size of the intersection.

Efficiency. The bulk of computation consists of running the $\mathsf{BatchOPE}$ protocol once, and hence the computational complexity of the scheme is $O(n_a \log n_a + n_b(\log n_b)^2)$ homomorphic operations.

Claim. the above protocol is secure against semihonest adversaries if the encryption scheme E is semantically secure.

Similar to proof of Claim 4.1, the security follows from the semantic security of the encryption scheme in a standard way. Details are deferred to the full version.

4.3 Private Set Intersection via Polynomial Multiplication

Kissner and Song [20] designed a simple and elegant protocol for the private set intersection problem. Given two sets A and B of equal size n, the idea is to represent the sets using polynomials f_A and f_B, respectively. Let r and s be two uniformly random polynomials of degree greater or equal to n over R. The polynomial $o = rf_A + sf_B = gcd(f_A, f_B)u$, where the polynomial u has coefficients uniformly distributed in R. Note that an element $a \in R$ is a root of $gcd(f_A, f_B)$ if and only if a appears in $A \cap B$. Furthermore, if R is large, the fact that u is uniformly distributed implies that with overwhelming probability, the roots of u do not represent any elements in A or B (see [20] for more detail). Hence, one can determine if an element is in the intersection set by testing whether the element evaluates to zero at polynomial o.

This construction easily extends to work for computing the intersection of many sets (held by many users). As discussed in the literature, the main drawback of the construction is that it requires computational complexity that is quadratic in the size of the datasets. However, in light of the algorithms we have designed for computing on encrypted polynomials, this can be improved.

Private Set Intersection Protocol (via EncPolyMult)

Inputs: Alice holds the dataset A of size n_a, and Bob holds the dataset B of size n_b with elements in R. Without loss of generality we assume that $n_a > n_b$.
Output: Bob learns the intersection of A and B.

1. Bob generates a key pair (pk, sk) for the public key encryption scheme E, and sends pk to Alice.
2. Alice and Bob represent their data sets using polynomials f_A and f_B of degree n_a and n_b respectively.
3. Bob encrypts his polynomial and sends $E_{pk}(f_B)$ to Alice.
4. Alice generates two uniformly random polynomials r and s over R of degree n_a. She then computes the encrypted polynomial $E_{pk}(o) = \mathsf{EncPolyMult}(r, E_{pk}(f_B)) +_h \mathsf{PolyMult}(s, f_A)$ and sends it to Bob.
5. Bob then evaluates the encrypted polynomial $E_{pk}(o)$ at his input set using the homomorphic properties of the encryption and for each ciphertext that is an encryption of 0, he outputs the corresponding input as part of the intersection.

Efficiency. Note that the protocol requires one invocation of the EncPolyMult algorithm. Hence the computation consists of n_b encryptions, $n_a + n_b$ decryptions, $2n_a \log n_a$ homomorphic additions and $n_a \log n_a$ homomorphic multiplications.

Claim. The above private set intersection protocol is secure against semi-honest adversaries if the encryption scheme E is semantically secure.

References

1. Ateniese, G., De Cristofaro, E., Tsudik, G.: (if) size matters: size-hiding private set intersection, pp. 156–173 (2011)
2. Beimel, A., Ishai, Y., Kushilevitz, E., Raymond, J.F.: Breaking the $O(n^{\frac{1}{2k-1}})$ barrier for information-theoretic private information retrieval. In: FOCS 2002, pp. 261–270 (2002)

3. Chang, Y.-C., Lu, C.-J.: Oblivious Polynomial Evaluation and Oblivious Neural Learning. In: Boyd, C. (ed.) ASIACRYPT 2001. LNCS, vol. 2248, pp. 369–384. Springer, Heidelberg (2001)
4. Cheon, J.H., Jarecki, S., Seo, J.H.: Multi-party privacy-preserving set intersection with quasi-linear complexity. Cryptology ePrint Archive, Report 2010/512 (2010)
5. Cooley, J.W., Tukey, J.W.: An algorithm for the machine calculation of complex Fourier series. Mathematics of Computation, 297–301 (1965)
6. Dachman-Soled, D., Malkin, T., Raykova, M., Yung, M.: Efficient Robust Private Set Intersection. In: Abdalla, M., Pointcheval, D., Fouque, P.-A., Vergnaud, D. (eds.) ACNS 2009. LNCS, vol. 5536, pp. 125–142. Springer, Heidelberg (2009)
7. Dachman-Soled, D., Malkin, T., Raykova, M., Yung, M.: Secure Efficient Multiparty Computing of Multivariate Polynomials and Applications. In: Lopez, J., Tsudik, G. (eds.) ACNS 2011. LNCS, vol. 6715, pp. 130–146. Springer, Heidelberg (2011)
8. De Cristofaro, E., Kim, J., Tsudik, G.: Linear-Complexity Private Set Intersection Protocols Secure in Malicious Model. In: Abe, M. (ed.) ASIACRYPT 2010. LNCS, vol. 6477, pp. 213–231. Springer, Heidelberg (2010)
9. De Cristofaro, E., Tsudik, G.: Practical Private Set Intersection Protocols with Linear Complexity. In: Sion, R. (ed.) FC 2010. LNCS, vol. 6052, pp. 143–159. Springer, Heidelberg (2010)
10. Franklin, M., Mohassel, P.: Efficient and Secure Evaluation of Multivariate Polynomials and Applications. In: Zhou, J., Yung, M. (eds.) ACNS 2010. LNCS, vol. 6123, pp. 236–254. Springer, Heidelberg (2010)
11. Freedman, M.J., Nissim, K., Pinkas, B.: Efficient Private Matching and Set Intersection. In: Cachin, C., Camenisch, J.L. (eds.) EUROCRYPT 2004. LNCS, vol. 3027, pp. 1–19. Springer, Heidelberg (2004)
12. Freedman, M.J., Ishai, Y., Pinkas, B., Reingold, O.: Keyword Search and Oblivious Pseudorandom Functions. In: Kilian, J. (ed.) TCC 2005. LNCS, vol. 3378, pp. 303–324. Springer, Heidelberg (2005)
13. El Gamal, T.: A Public Key Cryptosystem and a Signature Scheme Based on Discrete Logarithms. In: Blakely, G.R., Chaum, D. (eds.) CRYPTO 1984. LNCS, vol. 196, pp. 10–18. Springer, Heidelberg (1985)
14. von zur Gathen, J., Gerhard, J.: Modern computer algebra. Cambridge University Press, New York (1999)
15. Gentry, C.: Fully homomorphic encryption using ideal lattices. In: STOC, pp. 169–178 (2009)
16. Hazay, C., Lindell, Y.: Constructions of truly practical secure protocols using standard smartcards. In: ACM CCS, pp. 491–500 (2008)
17. Hazay, C., Nissim, K.: Efficient Set Operations in the Presence of Malicious Adversaries. In: Nguyen, P.Q., Pointcheval, D. (eds.) PKC 2010. LNCS, vol. 6056, pp. 312–331. Springer, Heidelberg (2010)
18. Jarecki, S., Liu, X.: Efficient Oblivious Pseudorandom Function with Applications to Adaptive ot and Secure Computation of Set Intersection. In: Reingold, O. (ed.) TCC 2009. LNCS, vol. 5444, pp. 577–594. Springer, Heidelberg (2009)
19. Jarecki, S., Liu, X.: Fast Secure Computation of Set Intersection. In: Garay, J.A., De Prisco, R. (eds.) SCN 2010. LNCS, vol. 6280, pp. 418–435. Springer, Heidelberg (2010)
20. Kissner, L., Song, D.: Privacy-Preserving Set Operations. In: Shoup, V. (ed.) CRYPTO 2005. LNCS, vol. 3621, pp. 241–257. Springer, Heidelberg (2005)
21. Lindell, Y., Pinkas, B.: Privacy Preserving Data Mining. In: Bellare, M. (ed.) CRYPTO 2000. LNCS, vol. 1880, pp. 36–54. Springer, Heidelberg (2000)

22. Mohassel, P., Franklin, M.K.: Efficient Polynomial Operations in the Shared-Coefficient Setting. In: Yung, M., Dodis, Y., Kiayias, A., Malkin, T. (eds.) PKC 2006. LNCS, vol. 3958, pp. 44–57. Springer, Heidelberg (2006)
23. Naor, M., Pinkas, B.: Oblivious transfer and polynomial evalutation. In: STOC, pp. 245–254 (1999)
24. Naor, M., Pinkas, B.: Oblivious polynomial evaluation. SIAM Journal on Computing, 12–54 (2006)
25. Paillier, P.: Public-Key Cryptosystems Based on Composite Degree Residuosity Classes. In: Stern, J. (ed.) EUROCRYPT 1999. LNCS, vol. 1592, pp. 223–238. Springer, Heidelberg (1999)
26. Shamir, A.: How to share a secret. Communications of the ACM, 612–613 (1979)

A Computer Algebra Techniques

A.1 Roots of Unity

Definition 3. *Let R be a ring, $n \in \mathbb{N}$, and $w \in R$.*

- *w is an nth root of unity if $w^n = 1$.*
- *w is a primitive nth root of unity, if it is an nth root of unity, $n \in R$ is a unit in R, and $w^{n/t} - 1$ is not a zero divisor for any prime divisor t of n.*

Here n has two meanings: in w it is an integer used as a counter to express the n-fold product of w with itself, and in $n \in R$ it stands for the ring element $n \cdot 1_R \in R$, the n-fold sum of 1_R with itself.

A.2 Computing the Discrete Fourier Transform

Definition 4. *Let $f \in R[x]$ be a polynomial of degree $d < n$. The Discrete Fourier Transform (DFT) mapping $DFT_w : R^n \to R^n$ denotes the evaluation of the polynomial f at the powers of w, i.e., $DFT_w(f) = \langle f(1), f(w), f(w^2), \ldots, f(w^{n-1})\rangle$.*

The Discrete Fourier Transform can be seen as a special case of multipoint evaluation, at the powers of nth root of unity w. Next, we introduce the Fast Fourier Transform, or FFT for short, that computes DFT quickly. The algorithm was (re)discovered by Cooley and Tukey [5], and is one of the most important algorithms in practice. Let $n \in \mathbb{N}$ be even, $w \in R$ a primitive nth root of unity, and $f \in R[x]$ of degree less than n. To evaluate f at the powers $1, w, w^2, \ldots, w^{n-1}$, we divide f by $x^{n/2} - 1$ and $x^{n/2} + 1$ with remainder:

$$f = q_0(x^{n/2} - 1) + r_0 = q_1(x^{n/2} + 1) + r_1$$

for some $q_0, r_0, q_1, r_1 \in R[x]$ of degree less than $n/2$. Due to the special form of the divisor polynomials, the computation of the remainders r_0 and r_1 can be done by adding the upper $n/2$ coefficients of f to, respectively subtracting them from, the lower $n/2$ coefficients. In other words, if $f = F_1 x^{n/2} + F_0$ with

$deg(F_0), deg(F_1) < n/2$, then $x^{n/2}-1$ divides $f-F_0-F_1$, and hence $r_0 = F_0+F_1$ and $r_1 = F_0 - F_1$. If we plug in a power of w for x we have:

$$f(w^{2\ell}) = q_0(w^{2\ell})(w^{n\ell} - 1) + r_0(w^{2\ell}) = r_0(w^{2\ell}) \quad (1)$$

$$f(w^{2\ell+1}) = q_1(w^{2\ell+1})(w^{n\ell}w^{n/2} + 1) + r_1(w^{2\ell+1}) = r_1(w^{2\ell+1}) \quad (2)$$

for all $0 \leq \ell < n/2$. In the above, we use the facts that $w^{n\ell} = 1$ and $w^{n/2} = -1$, since

$$0 = w^n - 1 = (w^{n/2} - 1)(w^{n/2} + 1)$$

and $w^{n/2}-1$ is not a zero divisor. It remains to evaluate r_0 at the even powers of w and r_1 at the odd powers. Now, w^2 is a primitive $(n/2)$th root of unity. It is easy ot see that the evaluation of r_0 reduces to a DFT of order $n/2$. The evaluation of $r_1(w^{2\ell+1}) = r_1^*(w^{2\ell})$ where $r_1^*(x) = r_1(wx)$, reduces to the computation of the coefficients of r_1^* which uses $n/2$ multiplications by powers of w, and a DFT of order $n/2$ for r_1^*. If n is a power of 2, we can proceed recursively to evaluate r_0 and r_1^* at the power of w^2, which leads to the following FFT algorithm:

Fast Fourier Transform
$\mathsf{FFT}_{w,n}(f)$

Input: $n = 2^k \in \mathbb{N}$ with $k \in \mathbb{N}$, $f = \sum_{0 \leq j \leq n} f_j x^j \in R[x]$, and the powers $w, w^2, \ldots, w^{n-1}$ of a primitive nth root of unity $w \in R$.

Output: $\mathrm{DFT}_{w,n}(f) = \langle f(1), f(w), \ldots, f(w^{n-1}) \rangle \in R^n$

1. If $n = 1$ then return f_0.
2. Compute $r_0 \leftarrow \sum_{0 \leq j < n/2}(f_j + f_{j+n/2})x^j$, and $r_1^* \leftarrow \sum_{0 \leq j < n/2}(f_j - f_{j+n/2})w^j x^j$
3. Call the algorithms $\mathsf{FFT}_{w^2,n/2}(r_0)$ and $\mathsf{FFT}_{w^2,n/2}(r_1^*)$ to compute r_0, r_1^* at the powers of w^2.
4. Return $r_0(1), r_1^*(1), r_0(w^2), r_1^*(w^2), \ldots, r_0(w^{n-2}), r_1^*(w^{n-2})$.

Efficiency. The above algorithm computes $\mathrm{DFT}_{w,n}(f)$ using $n \log n$ additions in R and $(n/2) \log n$ multiplications by powers of w.

A.3 Interpolation on Roots of Unity

It turns out that interpolation at powers of w is again essentially a Discrete Fourier Transform, and can be computed efficiently using the above algorithm. In the interpolation problem, given the vector $\langle f(1), f(w), f(w^2), \ldots, f(w^{n-1}) \rangle$, our goal is to compute the coefficients of f. Let V_w be the Vandermonde matrix:

$$V_w = \begin{pmatrix} 1 & 1 & 1 & \cdots & 1 \\ 1 & w & w^2 & \cdots & w^{n-1} \\ 1 & w^2 & w^4 & \cdots & w^{2(n-1)} \\ \vdots & \vdots & \vdots & \ddots & \vdots \\ 1 & w^{n-1} & w^{2(n-1)} & \cdots & w^{(n-1)^2} \end{pmatrix}$$

Then, we can compute the coefficients of f via the following matrix-vector multiplication:

$$\begin{pmatrix} f_0 \\ f_1 \\ \vdots \\ f_n \end{pmatrix} = (V_w)^{-1} \begin{pmatrix} f(1) \\ f(w) \\ \vdots \\ f(w^{n-1}) \end{pmatrix}$$

The following theorem shows how we can compute inverse of V_w via DFT computation.

Theorem 1 ([14]). *Let R be a ring (commutative, with 1), $n \in \mathbb{N}$, and $w \in R$ be a primitive nth root of unity. Then w^{-1} is a primitive nth root of unity and $(V_w)^{-1} = 1/nV_{w^{-1}}$.*

Based on above theorem we can interpolate on powers of an nth root of unity using the following algorithm:

Interpolation on powers of nth Root of Unity
$\mathsf{Interpol}_{w,n}(\boldsymbol{V})$

Input: $n = 2^k \in \mathbb{N}$ with $k \in \mathbb{N}$, a primitive nth root of unity $w \in R$ and a vector $\boldsymbol{V} = \langle v_0, v_2, \ldots, v_{n-1} \rangle \in R^n$.

Output: Polynomial f of degree n where $f(w^i) = v_i$ for $0 \leq i < n$.

1. Compute $w^{-1}, w^{-2}, \ldots, w^{-(n-1)}$.
2. Let the components of $\boldsymbol{V}$ represent the coefficients of a polynomial denoted by $v(x)$. Compute and return $1/n\ \mathsf{FFT}_{w^{-1},n}(v)$.

The efficiency of the algorithm is similar to that of the DFT algorithm. The only additional cost is to compute powers of w^{-1}. This requires n additional multiplications by w.

A.4 Polynomial Multiplication via FFT

The idea for the polynomial multiplication is to compute the DFT of both polynomials, multiply the components of the DFT individually to obtain the DFT of the product polynomial and then interpolate to recover the product polynomial itself.

Polynomial Multiplication via FFT
$\mathsf{PolyMult}(f, g)$

Input: $n = 2^k \in \mathbb{N}$ with $k \in \mathbb{N}$, $f = \sum_{0 \leq j \leq d_f} f_j x^j$ and $g = \sum_{0 \leq j \leq d_g} f_j x^j$ in $R[x]$ where $d_f + d_g < n$. The algorithm also takes input powers $w, w^2, \ldots, w^{n-1}$ of a primitive nth root of unity $w \in R$.

1. Compute $\langle f(1), \ldots, f(w^{n-1}) \rangle \leftarrow \mathsf{FFT}_{w,n}(f)$.
2. Compute $\langle g(1), \ldots, g(w^{n-1}) \rangle \leftarrow \mathsf{FFT}_{w,n}(g)$.
3. For $0 \leq i < n$ compute $S(i) = f(i)g(i)$.
4. Let $\boldsymbol{V_s} = \langle S(1), \ldots, S(n-1) \rangle$.
5. Computes $S \leftarrow \mathsf{Interpol}_{w,n}(\boldsymbol{V_s})$.

AniCAP: An Animated 3D CAPTCHA Scheme Based on Motion Parallax

Yang-Wai Chow[1] and Willy Susilo[2,*]

[1] Advanced Multimedia Research Laboratory
[2] Centre for Computer and Information Security Research
School of Computer Science and Software Engineering
University of Wollongong, Australia
{caseyc,wsusilo}@uow.edu.au

Abstract. CAPTCHAs are essentially challenge-response tests that are used to distinguish whether a user is a human or a computer. To date, numerous CAPTCHA schemes have been proposed and deployed on various websites to secure online services from abuse by automated programs. However, many of these CAPTCHAs have been found to suffer from design flaws that can be exploited to break the CAPTCHA. Hence, the development of a good CAPTCHA scheme that is both secure and human usable is an important research problem. This paper addresses this problem by presenting AniCAP, a new animated 3D CAPTCHA scheme that is designed to capitalize on the difference in ability between humans and computers at the task of perceiving depth through motion. In this paper, we present the design of AniCAP, along with a formal definition of its underlying Artificial Intelligence (AI) problem family. In addition, we analyze the security issues and considerations concerning AniCAP.

Keywords: CAPTCHA, animation, segmentation-resistant, motion parallax.

1 Introduction

CAPTCHAs (Completely Automated Public Turing test to tell Computers and Humans Apart) are automated tests that humans can pass but current computers programs cannot pass [23]. These days, CAPTCHAs are a ubiquitous part of the Internet and have been effective in deterring automated abuse of online services intended for humans [7]. A variety of different CAPTCHA schemes have emerged over the years, many of which have been deployed on numerous websites. Even major companies such as Google, Yahoo! and Microsoft, and social networks like Facebook, employ the use of CAPTCHAs to provide some level of security against online abuse.

However, many CAPTCHAs have been found to be insecure against automated attacks. Several researchers have demonstrated that certain design flaws in a number of CAPTCHAs can be exploited to break the CAPTCHA with a high degree of success [17,18,9,26,28,1,3]. This has given rise to the important research problem of how to develop CAPTCHAs that are secure against such attacks. The task of developing a good

* This work is supported by ARC Future Fellowship FT0991397.

D. Lin, G. Tsudik, and X. Wang (Eds.): CANS 2011, LNCS 7092, pp. 255–271, 2011.

CAPTCHA scheme is a challenging problem. This is because the resulting CAPTCHA must be secure against attacks, yet at the same time it must be usable by humans.

The tradeoff between CAPTCHA security and usability is a hard act to balance. In addition, it has been argued that the difficulty in creating robust CAPTCHAs is further compounded by the fact that the current collective understanding of CAPTCHAs is rather limited [28]. The design of a robust CAPTCHA must in someway capitalize on the difference in ability between humans and current computer programs [7]. This raises the question about whether or not it is possible to design a CAPTCHA that is easy for humans but difficult for computers [8].

This paper addresses the important problem of developing a robust CAPTCHA scheme. While there are three main categories of CAPTCHAs; namely, text-based CAPTCHAs, audio-based CAPTCHAs and image-based CAPTCHAs, this paper focuses on text-based CAPTCHAs. In this paper, we propose the design of a new animated 3D text-based CAPTCHA scheme that attempts to exploit the gap between natural human perception and the ability of computers to emulate perception.

Previous Work. This research was motivated by our previous work on stereoscopic 3D CAPTCHAs, or STE3D-CAP [22]. In previous work, we introduced a novel approach of presenting text-based CAPTCHA challenges in 3D by using stereoscopic images. The key idea behind STE3D-CAP is that humans can perceive depth from stereoscopic images. Thus, by adding random clutter to the scene, the resulting CAPTCHA would be hard for a computer to solve, whereas a human should easily be able solve the CAPTCHA as the text would appear to stand out from the rest of the scene when perceived in 3D. However, the limitation behind this approach was that it relied on the availability of specialized stereoscopic viewing devices, which may be the way of the future but are not ubiquitous at present. Nevertheless, our previous work gave rise to the notion of creating CAPTCHAs based on depth perception.

Our Contributions. This paper presents a new approach to creating animated 3D CAPTCHA challenges based on the concept of motion parallax. Motion parallax is a monocular cue that allows an observer to perceive depth information from the relative motion between objects in a scene. From the viewpoint of a moving observer, objects that are closer to the observer will appear to move by larger distances as compared to objects that are located further away from the observer. This apparent difference in the motion of objects is one of the means by which the human visual system can perceive depth.

We dubbed our novel animated 3D CAPTCHA scheme 'AniCAP', and its key features are listed as follows:

- Unlike other approaches that add random clutter to the CAPTCHA challenge in an attempt to deter automated attacks, AniCAP, which is a text-based CAPTCHA, uses text itself to increase the difficulty of the challenge. When viewed as a static image, AniCAP has the appearance of overlapping text-on-text, and with no distinct colors or borders around the characters it is not possible to solve the challenge in that manner. However, when AniCAP is viewed from the point of view of a moving camera, this gives rise to motion parallax. As such, it is not possible to use this unique text-on-text approach in the absence of depth perception.

- Hence, while most existing animated CAPTCHAs are 2D CAPTCHAs, AniCAP is actually a 3D CAPTCHA.
- In contrast to a number of other animated CAPTCHA schemes, where the challenge is only displayed for a certain period of time before the user has to wait for the next animation cycle, in AniCAP the challenge is present at all times throughout the animation cycle.
- In addition, the distortion in AniCAP constantly changes in successive frames, thus increasing security by making it difficult for a computer to compare pixel positions between frames.
- Furthermore, unlike the depth perception approach used in previous work [22], this approach does not rely on availability of specialized viewing devices. Instead, AniCAP can be implemented as an animated Graphics Interchange Format (GIF) file, or a video file, which can easily be included on webpages and viewed with standard computer equipment. We have provided an example of AniCAP (an actual animated version) that is available at

  http://www.uow.edu.au/~wsusilo/CAPTCHA/CAPTCHA.html.

  The correct solution to the challenge is 'SYAK'.

This paper presents the design principles and implementation details of AniCAP. We then formalized the notion of AniCAP and describe the hard Artificial Intelligence (AI) problem underlying this unique CAPTCHA scheme. Additionally, we present a discussion about the various security issues that had to be considered in relation to this novel CAPTCHA technique.

2 Background

2.1 Security and Usability

In order for a CAPTCHA scheme to have any practical value, humans must be able to correctly solve it with a high success rate, whilst the chances for a computer to solve it must be very small. While security considerations push designers to increase the difficulty of CAPTCHAs, usability issues force the designer to make the CAPTCHA only as hard as they need to be and still be effective at deterring abuse. These conflicting requirements have resulted in an ongoing arms race between CAPTCHA designers and those who try to break them [7].

With advances in research areas like computer vision, pattern recognition and machine learning, and enhancements in Optical Character Recognition (OCR) software, increasingly sophisticated attacks have been developed to break CAPTCHAs. On the other hand, humans have to rely on their inherent abilities and are unlikely to get better at solving CAPTCHAs. Hence, in order to exploit the gap in ability between human and computers it is vital to examine work by others, which highlight the security flaws and usability issues of various CAPTCHAs.

In terms of usability, text-based CAPTCHAs that are based on dictionary words are intuitive and easier for humans to solve. This is because humans find familiar text easier to read as opposed to unfamiliar text [24]. At the same time, CAPTCHAs based on language models are easier to break via dictionary attacks. Mori and Malik [17] were successful in breaking a number of CAPTCHAs that were based on the English language. Rather than attempting to identify individual characters, they used a holistic approach of recognizing entire words at once. Similar attacks exploiting language models have also been performed on a number of other CAPTCHAs [4,10].

To take advantage of text familiarity without using actual dictionary words, it is possible to use 'language-like' strings instead. Phonetic text or Markov dictionary strings are pronounceable strings that are not words of any language. Humans perform better at correctly identifying pronounceable strings in contrast to purely random character strings. Nevertheless, the disadvantage of using this approach is that certain characters (e.g. vowels) will appear at higher frequencies compared to other characters in pronounceable strings [24].

In an attempt to show that machine learning techniques could be used to break CAPTCHAs, Chellapilla and Simard [9] deliberately avoided exploiting language models and were still successful at breaking in a variety of CAPTCHAs. Solving text-based CAPTCHAs consists of a segmentation challenge, the identification of character locations in the right order, followed by recognition challenges, recognizing individual characters [7]. Their work demonstrated that computers could outperform humans at the task of character recognition. Hence, this led to the important principle that if a CAPTCHA could be segmented, it was essentially broken. As such, the state-of-the-art in robust text-based CAPTCHA design relies on the difference in ability between humans and computers when it comes to the task of segmentation [1].

In order to increase the difficulty of segmentation, techniques such as crowding or connecting characters together can be employed. In addition, the use of both local and global warping to distort characters can also make the task of segmentation harder [25,28]. It should also be noted that CAPTCHAs with fixed length strings, with characters that possibly appear at fixed locations, are easier to segment [26]. While color and/or textures can be used for aesthetic reasons, or for making it easier to distinguish text from background clutter, the inappropriate use of color and textures can have detrimental effects on both the security and usability of a CAPTCHA [27]. In general, if the use of color or textures does not contribute to the security strength of the CAPTCHA, it may be better not to use any.

2.2 Animated CAPTCHAs

Animated CAPTCHAs have been proposed to overcome the limitations of static CAPTCHAs. There are a number of existing animated text-based CAPTCHA schemes that are currently deployed on various websites. This section presents an overview of the main ideas behind the construction of a number of these animated CAPTCHAs.

The HELLOCAPTCHA [20] is an animated 2D CAPTCHA that is freely available via the developers' web service. The developers of HELLOCAPTCHA provide a number of different variants to their attractive CAPTCHA scheme. We select a characteristic subset of these, shown in Fig. 1, for discussion in relation to the security considerations presented in the preceding section. The examples depict frames taken at different times, where time increases from the left frame to the right frame. In most of the examples shown in Fig. 1, excluding Fig. 1(e), the challenge is not always on display. Therefore, if the users misses these specific frames, he/she will have to wait for the next animation cycle. The variant in Fig. 1(c), has multiple correct answers because of the changing sequence of characters. This will increase an attacker's chances of success. Background text in the variant shown in Fig. 1(d) can easily be filtered out as the challenge text is in a distinct color. In the examples shown in Fig. 1(a), Fig. 1(b), Fig. 1(c) and Fig. 1(d) the characters are located at fixed locations, thus making it easier to predict where the challenge will appear. All variants are fixed length character string challenges, thus a computer would have foreknowledge about the total number of required characters. In addition, none of the variants employ the segmentation-resistant principle or character warping, so an OCR program can easily recognize the characters. Hence, by correlating information between different frames, it is highly likely that a computer can break this CAPTCHA.

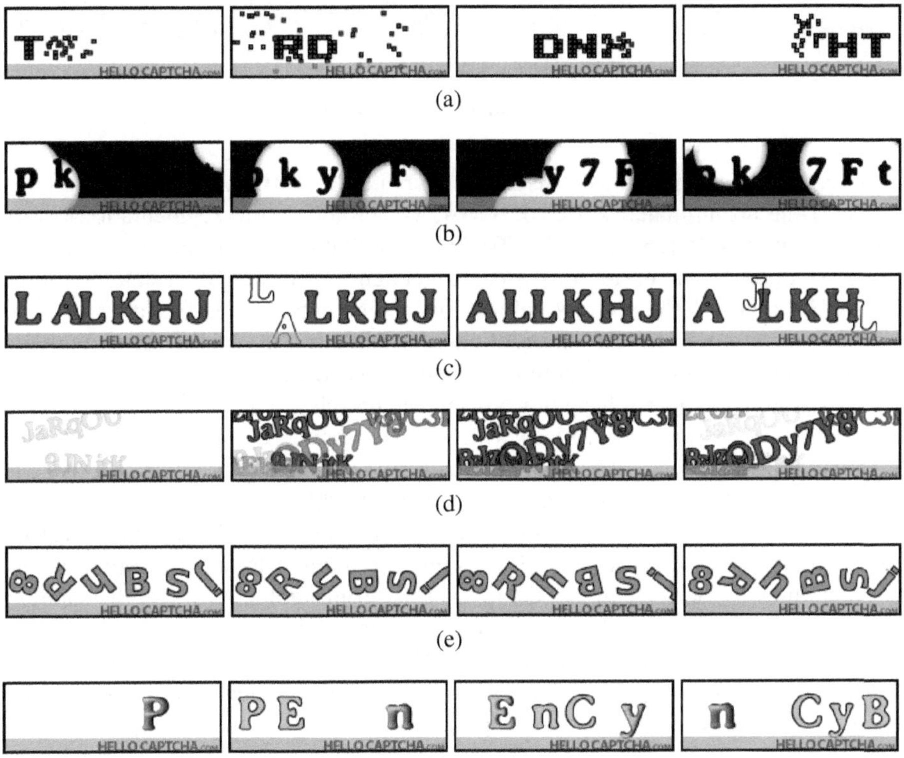

(a) (b) (c) (d) (e) (f)

Fig. 1. Examples showing different variants of HelloCaptcha [20]

The JkCaptcha, shown in Fig. 2, is an example of an animated 2D CAPTCHA that uses a text-on-text approach. In this CAPTCHA, what the user sees is a number of persistent characters over a continuously changing background. In the example shown in

Fig. 2(a), the challenge is displayed using a distinct color which can easily be separated from the background. From a usability point of view, the use of color in Fig. 2(b) is highly distracting as it changes from frame to frame. The user sees this as continuously flashing color. From a security standpoint, the persistent characters can easily be separated from the characters that change from frame to frame in the background. Furthermore, the characters in the foreground always occlude characters in the background. Additionally, the foreground characters are somewhat larger than the background characters so a simple pixel-count attack can easily be used to separate them [28]. Once separated, an OCR program can easily recognize the characters.

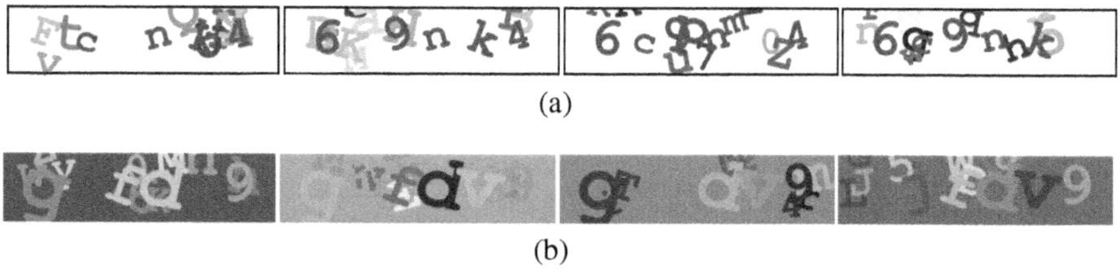

Fig. 2. Examples of JkCaptcha [15]

In contrast to the previously discussed animated CAPTCHAs, NuCaptcha [19] is a state-of-the-art animated 2D CAPTCHA, which adopts the segmentation-resistant principle. The developers of this CAPTCHA state that tests have shown that animated CAPTCHA puzzles are easier for humans to recognize and solve than static, scrambled CAPTCHA images. The concept behind NuCaptcha is that when the letters are moving, a human's mind sees the different parts and fills in the blanks; the parts that are moving together are grouped together, and a human can clearly differentiate the letters. Whereas computers do not have this advantage and see a smear of pixels. In addition, unlike CAPTCHAs created in Flash which are not secure, NuCaptcha is displayed as an H.264 MPEG-4 Video Stream that is rendered in the user's browser [19]. An easy-to-use example of NuCaptcha is shown in Fig. 3. The difficulty level of NuCaptcha can be augmented by increasing the number of characters in the challenge and by crowding the characters closer together. Fig. 3(a) to Fig. 3(c) demonstrate three frames taken at different times. It can be seen that the text scrolls from right to left, with the challenge, that is not always in the display, rendered in a distinct color.

In the research community, animated CAPTCHAs have been proposed by a number of researchers. Cui et al. [11,12] decribed a sketch of an animated CAPTCHA approach based on moving letters amidst a noisy background. However, this approach is hard for humans to use. An animated CAPTCHA based on the idea of presenting distorted text on the face of a deforming surface was proposed by Fischer and Herfet [14]. Another proposed animated CAPTCHA with images of moving objects was suggested by Athanasopoulos and Antonatos [2]. However, none of the above proposed methods are related to depth perception in animated images, nor do they analyze the security of their approaches.

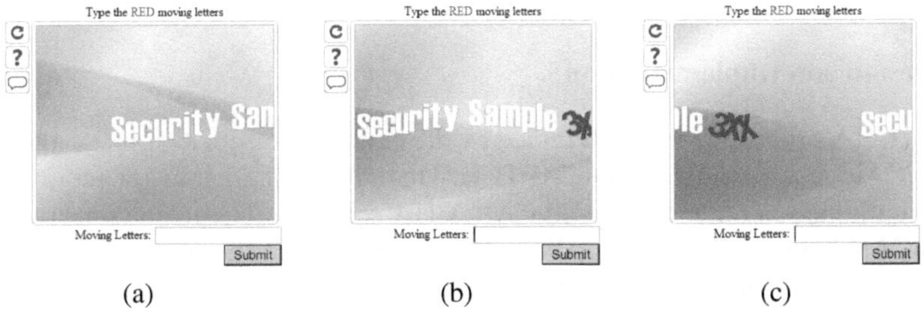

Fig. 3. Example of NuCaptcha [19]

2.3 CAPTCHA: Formal Definition and Notation

The term 'CAPTCHA' was introduced by von Ahn et al. [23]. In their seminal work, they describe CAPTCHAs as hard Artificial Intelligence (AI) problems that can be exploited for security purposes. The definitions and notation provided below are adapted and simplified from their work.

A CAPTCHA is a test V where most humans have success close to 1, while it is hard to write a computer program that has overwhelming probability of success over V. This means that any program that has a high probability of success over V can be used to solve a hard AI problem. Let $\mathcal{C}$ be a probability distribution. If $P(\cdot)$ is a probabilistic program, let $P_r(\cdot)$ denote the deterministic program that results when P uses random coins r.

Definition 1. [23] A test V is said to be (α, β)-human executable if at least an α portion of the human population has success probability greater than β over V.

Definition 2. [23] An *AI problem* is a triple $\mathcal{P} = (S, D, f)$ where S is a set of problem instances, D is a probability distribution over S and $f : S \to \{0,1\}^*$ answers the problem instances. Let $\delta \in (0, 1]$. For $\alpha > 0$ fraction of the humans H, we require $Pr_{x \leftarrow D}[H(x) = f(x)] > \delta$.

Definition 3. [23] An AI problem $\mathcal{P}$ is said to be (ψ, τ)-solved if there exists a program $\mathcal{A}$ that runs in time for at most τ on any input from S, such that

$$Pr_{x \leftarrow D,r}[\mathcal{A}_r(x) = f(x)] \geqslant \psi.$$

Definition 4. [23] An (α, β, η)-CAPTCHA is a test V that is (α, β)-human executable and if there exists $\mathcal{B}$ that has success probability greater than η over V to solve a (ψ, τ)-hard AI problem $\mathcal{P}$, then $\mathcal{B}$ is a (ψ, τ) solution to $\mathcal{P}$.

Definition 5. [23] An (α, β, η)-CAPTCHA is *secure* iff there exists no program $\mathcal{B}$ such that

$$Pr_{x \leftarrow D,r}[\mathcal{B}_r(x) = f(x)] \geqslant \eta$$

for the underlying AI problem $\mathcal{P}$.

3 AniCAP

3.1 Design and Implementation

A CAPTCHA's robustness is determined by the cumulative effects of its design choices [7]. AniCAP is an animated 3D CAPTCHA that was designed to overcome security flaws highlighted in other text-based CAPTCHA schemes. The main concept underlying AniCAP is that of motion parallax. This capitalizes on the inherent human ability to perceive depth from the apparent difference in motion of objects located at different distances from a moving viewpoint.

A number of approaches were employed to make AniCAP segmentation-resistant. Firstly, in AniCAP the main characters are rendered over background characters, and characters are overlapped and crowded together. To give rise to motion parallax, the foreground and background characters occupy different ranges of spatial depths. Secondly, some sections of the characters are rendered with a certain degree of translucency to prevent foreground characters from completely occluding background characters. This creates a somewhat 'see through' effect at certain places by blending the overlapping foreground and background characters together. Thirdly, all characters were deliberately rendered using the same font and color, with no distinct borders around the characters. Collectively, these factors make it difficult for a computer to segment the characters, whereas a human can distinguish the main characters in the foreground from the background characters due to motion parallax. This is because the foreground characters will appear to move at different rates compared to the background characters.

Since AniCAP is a 3D CAPTCHA, each 3D character can have random 3D transformations applied to it. For instance, rotations are in all three dimensions and are not merely restricted to the standard clockwise and counterclockwise rotations of 2D CAPTCHAs. When a 3D scene is viewed using perspective projection, objects that are closer to the viewpoint will appear larger than objects that are further away; a concept known as perspective foreshortening. This would mean that to ascertain the foreground characters, one simply had to identify the larger characters. To prevent this, we scale each character so that they all appear to have similar sizes, and AniCAP is made up of random character strings to prevent dictionary attacks.

In addition, characters are all rendered with local and global distortion to deter character recognition and pixel-count attacks. Local distortion refers to distortion applied to individual characters, whereas global distortion is distortion that is applied to the whole scene. To deter computer vision techniques like 3D scene reconstruction and optical flow (discussed in section 4), the global distortion appears to change from frame to frame. However, the change from frame to frame is not completely random, otherwise this would significantly impede human usability. Instead, the global distortion is based on the pixel's location in the frame. What the user sees is like a moving scene viewed through 'frosted glass'. Despite the distortion, when the moving scene is viewed as a whole, a human can perceive the characters because the human mind will group the fragments together as explained by the Gestalt principles of perceptual organization.

Examples of static AniCAP frames are shown in Fig. 4. Note that the same AniCAP challenge is used throughout this paper so that the reader can compare differences between the AniCAP images provided in this paper. A frame without any distortion is shown in Fig. 4(a), whereas a frame with local distortion only is provided in Fig. 4(b), Fig. 4(c) shows the same frame with both local and global distortion and Fig. 4(d) depicts the same frame with different distortion parameters. Fig. 5 shows a number of animation frames[1]. It can be seen from the static frames themselves that one cannot differentiate the foreground characters from the background characters.

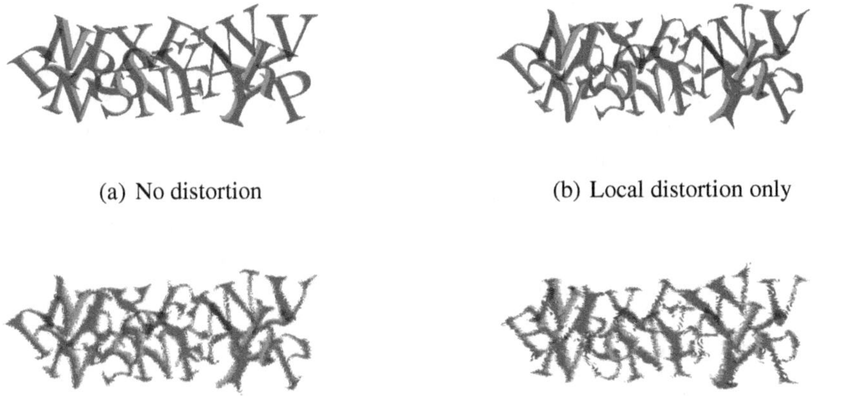

(a) No distortion

(b) Local distortion only

(c) Local and global distortion

(d) Different distortion parameters

Fig. 4. AniCAP distortion

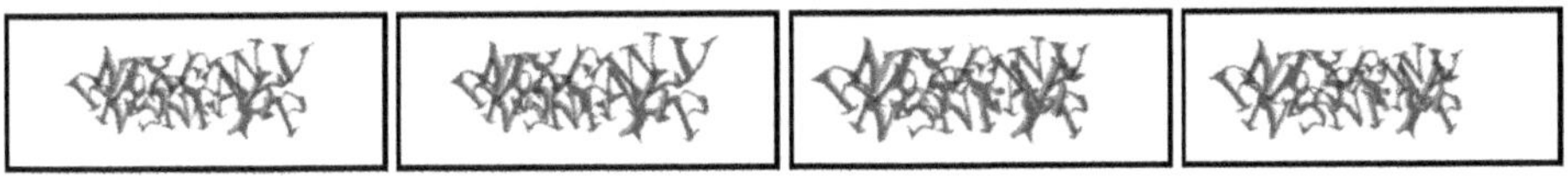

Fig. 5. Example AniCAP animation frames

The current implementation of AniCAP consists of 3 rows, with a variable number of characters per row. Characters in the rows are made to overlap in the vertical direction and the characters in the columns are crowded together in the horizontal direction, at times overlapping or joining together. The foreground characters consist of a certain number of characters, that are located in sequence somewhere in the middle row. The challenge is deliberately placed at random locations in the middle to as it is conceivable that a computer may be able to identify the shape of some non-overlapping characters at the edges. The reason why they are in sequence is to help human usability, because it is more difficult for a human to identify individual characters at random locations. We can adjust the difficulty level of AniCAP by varying the number of characters in the challenge, as well as the degree of character crowding and overlapping. The number

[1] Please refer to `http://www.uow.edu.au/~wsusilo/CAPTCHA/CAPTCHA.html` for the animated AniCAP example.

of background characters, crowding and overlapping will also have to vary proportionately. Additionally, the amount of distortion can also be varied.

In order to facilitate motion parallax, the choice of camera movement is important. The highest degree of motion parallax occurs when camera movement is perpendicular to the direction that the text is facing. Translating the camera closer or further away from the 3D text will also create motion parallax, as objects closer to the viewpoint will increase, or decrease, in size at higher rates as opposed to objects further away. However, motion parallax due to size changes are not as apparent compared to changes in horizontal and vertical movement. In addition, the camera can be rotated to view the 3D text from different angles, and it can also be made to focus on different sections of the text. In the current implementation, the camera's movement and rotation are randomized to incorporate all of the above movements at varying degrees. Depending of the camera's movement, the motion of the foreground characters can either be faster or slower than the background characters.

One of the drawbacks of AniCAP is that it may not immediately be obvious what the user is supposed to look for to solve the challenge. However, once this is described to the user, this should be obvious.

3.2 New AI Problem Family

Here we introduce a family of AI problems that is used to build AniCAP. Let us consider two layers of space, namely $\mathcal{P}_1$ and $\mathcal{P}_2$. Layer $\mathcal{P}_2$ consists of t sub-layers, namely $\{\mathcal{P}_{21}, \mathcal{P}_{22}, \cdots, \mathcal{P}_{2t}\}$. Each layer (or sub-layer) has a transparent background. The distance between $\mathcal{P}_1$ and $\mathcal{P}_2$ is denoted by δ_1, while the distance between $\mathcal{P}_{2i}$ and $\mathcal{P}_{2(i+1)}$ is δ_2. We require that δ_1 be sufficiently large to facilitate motion parallax, and typically $\delta_2 < \delta_1$.

Let $\mathcal{I}_{2d}$ be a distribution on characters, $\mathcal{I}_{3d}$ be a distribution on 3D characters. Let $\mathcal{I}_{mov}$ be a distribution of animation frames. Let $\Delta : \mathcal{I}_{2d} \rightarrow \mathcal{I}_{3d}$ be a lookup function that maps a character in $\mathcal{I}_{2d}$ and outputs a 3D character in $\mathcal{I}_{3d}$ with random 3D transformations. Let Ω_D be a distribution on local distortion factors. Let $\mathcal{D} : \mathcal{I}_{3d} \times \Omega_D \rightarrow \mathcal{I}_{3d}$ be a distribution of local distortion functions. Let $\mathcal{S} : \mathcal{I}_{3d} \rightarrow \mathcal{I}_{3d}$ be a distribution of scaling functions. Let $\Omega_{\tilde{D}}$ be a distribution on global distortion factors. Let $\widetilde{\mathcal{D}} : \mathcal{I}_{3d} \times \Omega_{\tilde{D}} \rightarrow \mathcal{I}_{3d}$ be a distribution of global distortion functions. The distortion function is a function that accepts a 3D image and a distortion factor $\in \Omega_D$ and outputs a distorted 3D image. Let $|A|$ denote the cardinality of A.

When a 3D character $\mathbf{i} \in \mathcal{I}_{3d}$ appears in $\mathcal{P}_1$ (or $\mathcal{P}_{2j}$, resp), we denote it as $\mathbf{i} \rhd \mathcal{P}_1$ (or $\mathbf{i} \rhd \mathcal{P}_{2j}$, resp). The camera $\mathfrak{C}$ views the stacks of layers of space from degree θ, where $\mathcal{P}_1$ is the top layer, followed by all the $\mathcal{P}_{2j}$, where $j \in \{1, \cdots, t\}$. This is denoted as $\mathfrak{C} \vdash_\theta \{\mathcal{P}_1, \mathcal{P}_{21}, \cdots, \mathcal{P}_{2j}\}$. The movement of the camera $\mathfrak{C}$ is recorded as $\mathtt{View}_{\mathfrak{C}} \in \mathcal{I}_{mov}$. Let $\mathcal{P}_1 || \mathcal{P}_{21} || \cdots || \mathcal{P}_{2t}$ be the stacks of layers of space that $\mathfrak{C}$ views. For clarify, for the rest of this paper, we will use **Roman boldface** characters to denote elements of $\mathcal{I}_{3d}$, while Sans Serif characters to denote elements of $\mathcal{I}_{2d}$.

Problem Family ($\mathcal{P}_{\mathsf{AniCAP}}$)
Assume that the CAPTCHA challenge length is ℓ. Let $\phi : \mathbb{Z} \rightarrow \{1, \cdots, \ell\}$ denote a function that maps any integer to the set $\{1, \cdots, \ell\}$. Let rand(c) be the pseudorandom generator function with the seed c. Consider the following experiment.

- **Stage 1. 3D Scene Generation**
  1. For i := 1 to ℓ do
     (a) Randomly select j $\in \mathcal{I}_{2d}$.
     (b) Compute **j** $\leftarrow \Delta$(j).
     (c) Select a local distortion function $d \leftarrow \mathcal{D}$.
     (d) Compute **k** $\leftarrow d(\mathbf{j}, \omega)$, where $\omega \in \Omega_D$ is selected randomly.
     (e) **k** $\rhd \mathcal{P}_1$.
  2. For i := 1 to t do
     (a) For k := 1 to $\phi(\text{rand}(time))$ do
        i. Randomly select j $\in \mathcal{I}_{2d}$.
        ii. Select the scaling function $s \in \mathcal{S}$.
        iii. Compute **j** $\leftarrow s(\Delta(\text{j}))$.
        iv. Select a local distortion function $d \leftarrow \mathcal{D}$.
        v. Compute **k** $\leftarrow d(\mathbf{j}, \omega)$, where $\omega \in \Omega_D$ is selected randomly.
        vi. **k** $\rhd \mathcal{P}_{2i}$.
- **Stage 2. Recording Animation**
  3. Select a global distortion function $d \leftarrow \widetilde{\mathcal{D}}$.
  4. For θ := start to end do
     (a) Compute $\psi \leftarrow d(\mathfrak{C} \vdash_\theta (\mathcal{P}_1 || \mathcal{P}_{21} || \cdots || \mathcal{P}_{2t}), \omega)$, where $\omega \in \Omega_{\tilde{D}}$ is selected randomly.
     (b) $\texttt{View}_{\mathfrak{C}} := \texttt{View}_{\mathfrak{C}} \cup \{\psi\}$.

The output of the experiment is $\texttt{View}_{\mathfrak{C}} \in \mathcal{I}_{mov}$, which is an animated CAPTCHA.

$\mathcal{P}_{\textsf{AniCAP}}$ is to write a program that takes $\texttt{View}_{\mathfrak{C}} \in \mathcal{I}_{mov}$, assuming the program has precise knowledge of $\mathfrak{C}$ and $\mathcal{I}_{2d}$, and outputs ℓ characters of j $\in \mathcal{I}_{2d}$.

Problem Description: $\mathcal{P}_{\textsf{AniCAP}}$
Essentially, a problem instance in $\mathcal{P}_{\textsf{AniCAP}}$ comprises two stages, namely 3D scene generation and recording the animation frames. The first stage, denoted as S_1, accepts the length of the CAPTCHA challenge, ℓ, and outputs a 3D scene, **im**. **im** consists of $t+1$ layers. Formally, this is defined as

$$\mathbf{im} \leftarrow \mathsf{S}_1(\ell).$$

The second stage, denoted as S_2, accepts a 3D scene, **im**, and a range of movements for the camera $\mathfrak{C}$, from $start$ to end, which defines the camera motion, and outputs the sequence of camera recordings $\texttt{View}_{\mathfrak{C}} \in \mathcal{I}_{mov}$. Formally, this is defined as

$$\texttt{View}_{\mathfrak{C}} \leftarrow \mathsf{S}_2(\mathbf{im}, start, end).$$

Hard Problem in $\mathcal{P}_{\textsf{AniCAP}}$
We believe that $\mathcal{P}_{\textsf{AniCAP}}$ contains a hard problem. Given the distribution of $\mathfrak{C}$ and $\mathcal{I}_{2d}$, for any program $\mathcal{B}$,

$$Pr_{\forall \mathcal{B}, \mathfrak{C}, \mathcal{I}_{2d}}(\text{j}^\ell \leftarrow ((\mathbf{im} \leftarrow \mathsf{S}_1(\ell)), (\texttt{View}_{\mathfrak{C}} \leftarrow \mathsf{S}_2(\mathbf{im}, start, end)))) < \eta,$$

where j $\in \mathcal{I}_{2d}$, and ℓ is the length of the CAPTCHA challenge. Based on this hard problem, we construct a secure (α, β, η)-CAPTCHA.

Theorem 1. *A secure (α, β, η)-CAPTCHA can be constructed from $\mathcal{P}_{\mathsf{AniCAP}}$ as defined above.*

Proof. Based on the problem family $\mathcal{P}_{\mathsf{AniCAP}}$, we construct a secure (α, β, η)-CAPTCHA. We show the proof of this statement in two stages, namely showing that the instance of $\mathcal{P}_{\mathsf{AniCAP}}$ is (α, β)-human executable. Then, we need to show that (α, β, η)-CAPTCHA is hard for a computer to solve.

Given $\mathcal{P}_{\mathsf{AniCAP}}$, humans receive an instance of $\mathtt{View}_{\mathfrak{C}} \leftarrow \mathsf{S}_2(\mathsf{S}_1(\ell), start, end)$. We note that the only viewable contents from this instance is $\mathtt{View}_{\mathfrak{C}}$. When the $start$ and end are selected to provide motion parallax, then humans can easily output j^{ℓ}, which is the ℓ characters in $\mathcal{P}_1$. Hence, the instance of $\mathcal{P}_{\mathsf{AniCAP}}$ is (α, β)-human executable.

However, given the instance of $\mathcal{P}_{\mathsf{AniCAP}}$, computers cannot output j^{ℓ}. Note that by only analyzing $\mathtt{View}_{\mathfrak{C}}$, computers need to use computer vision or other techniques to recognize the characters. Since machines cannot view the 3D contents, hence this problem is hard for computers. This justifies that the instance of $\mathcal{P}_{\mathsf{AniCAP}}$ is (α, β, η) hard, as claimed. □

4 Security Considerations for AniCAP

In this section, we analyze the security of AniCAP by considering several different attack scenarios.

4.1 Image Processing and Computer Vision Attacks

In image processing and computer vision attacks, the adversary $\mathcal{A}$ is provided with an AniCAP challenge, $\mathtt{View}_{\mathfrak{C}}$. The task of the adversary is to output the CAPTCHA solution, j^{ℓ}. In other words, $\mathcal{A}$ would like to extract j^{ℓ} from $\mathtt{View}_{\mathfrak{C}}$. In order to achieve this goal, $\mathcal{A}$ can launch attacks based on a number of different strategies. These are described as follows.

Edge Detection
The aim of the edge detection technique is to find the edges of the objects in the given image. To perform this attack, $\mathcal{A}$ will first have to decompose $\mathtt{View}_{\mathfrak{C}}$ to its constituting frames. For clarity, we denote the frames contained in $\mathtt{View}_{\mathfrak{C}}$ as

$$\mathtt{View}_{\mathfrak{C}} := \{\mathtt{View}_{\mathfrak{C}1}, \cdots, \mathtt{View}_{\mathfrak{C}n}\}$$

where without losing generality, we assume that there are n frames in $\mathtt{View}_{\mathfrak{C}}$. Note that $\mathtt{View}_{\mathfrak{C}i}$, $i \in \{1, \cdots, n\}$ is a 2D image. Then, $\mathcal{A}$ will conduct edge detection on these images which include all the foreground and background characters as well as the distortion embedded in the image. Since the global distortion, $d \in \widetilde{\mathcal{D}}$, changes from frame to frame, $\mathtt{View}_{\mathfrak{C}i}$, there is little correlation between the resulting edge detection images. Fig. 6 depicts an example of a resulting Canny edge detection image. As can be seen this does not help in solving the challenge.

Fig. 6. Example of an edge detection image

Image Difference

This attack can be conducted in a manner similar to that of the edge detection technique. First, the frames in $\texttt{View}_{\mathfrak{C}}$ will have to be decomposed. Hence, we obtain

$$\texttt{View}_{\mathfrak{C}} := \{\texttt{View}_{\mathfrak{C}1}, \cdots, \texttt{View}_{\mathfrak{C}n}\}$$

as defined earlier. Now, $\mathcal{A}$ will compute the difference between $\{\texttt{View}_{\mathfrak{C}i}\}$ and $\{\texttt{View}_{\mathfrak{C}i+j}\}$, where $i = \{1, \cdots, n-1\}$, $j = \{1, \cdots, n-i\}$. The results of difference images, obtained between two views, can be further analyzed by using other techniques, such as edge detection techniques. Nevertheless, this still does not yield much useful information for the task of segmentation, as there are too many overlapping characters. Fig. 7(a) shows an example of a resulting difference image between successive frames, and Fig. 7(b) shows the edge detection image after edge detection is performed on Fig. 7(a). It can be seen that no useful information can be obtained to solve the challenge.

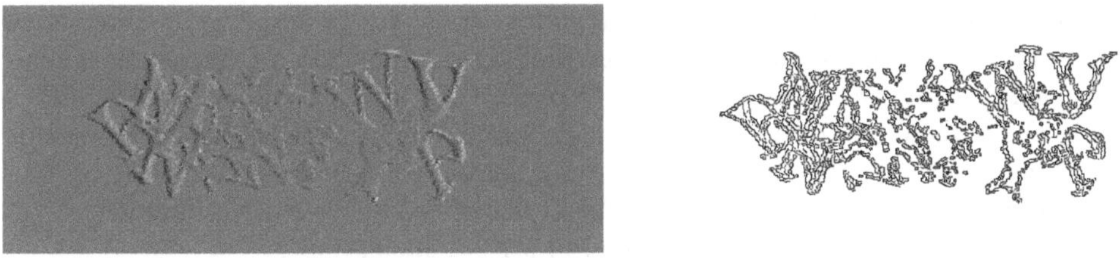

(a) (b)

Fig. 7. Difference image and edge detection image

3D Reconstruction

The purpose of this attack is to attempt to reconstruct an approximate 3D scene from $\texttt{View}_{\mathfrak{C}}$ in order to separate the foreground characters from the background characters in three dimensional space. Formally, $\mathcal{A}$ would like to solve

$$\mathsf{j}^{\ell} \leftarrow (\texttt{View}_{\mathfrak{C}} \leftarrow \mathsf{S}_2(\mathsf{S}_1(\ell), start, end))$$

A fundamental problem in 3D reconstruction is assigning correspondence between points in two or more images that are projections of the same point in three dimensional space. Most automated 3D reconstruction approaches use pixels or object silhouettes to specify correspondence [29]. Factors that impede correspondence between frames include noise, textureless regions, non-rigid objects, etc. as this creates ambiguity as to whether or not the selected point is actually the same point in other frames. Current 3D reconstruction algorithms are meant for images or image sequences typically captured from real world scenarios [16], which do no continuously distort from frame to frame. AniCAP is designed with global distortion that changes from frame to frame, in order to inhibit correspondence required for 3D reconstruction. In addition, the distortion, translucency and camera parameters are randomized, so $\mathcal{A}$ does not have prior knowledge about these.

Optical Flow
Optical flow in general refers to determining the apparent motion of objects in a scene based on the relative motion of the observer. Some definitions vary somewhat and differentiate between motion field estimation and apparent motion estimation [6]. Nevertheless, these are related techniques that may be used in an attempt to break AniCAP. The basic idea is similar to that of 3D reconstruction techniques in that certain points in $\texttt{View}_{\mathfrak{C}}$ have to be selected and tracked between successive frames. As with 3D reconstruction techniques, many current optical flow methods fail when it comes to handling hard problem that involve scenarios with noise, textureless regions, non-rigid objects, etc. [5]. This is again due to ambiguity in the selected points that have to be tracked from frame to frame [21]. AniCAP is designed to facilitate this ambiguity via randomized distortion, translucency and camera parameters, as well as textureless regions with no distinct colors or borders distinguishing the characters.

4.2 Brute Force Attacks

To attack AniCAP, $\mathcal{A}$ can conduct a straightforward attack by adopting the brute force strategy. In this type of attack, $\mathcal{A}$ will provide a random solution to the challenge until one succeeds. We note that the length of the CAPTCHA challenge in AniCAP is ℓ. Suppose there are 26 possible characters which comprise of uppercase alphabetic characters, then the chance of a successful brute force attack is $\frac{1}{26^{\ell}}$, which is negligible. Additionally, in practice CAPTCHAs are usually equipped with techniques such as token bucket algorithms to combat denial-of-service attacks [13].

4.3 Machine Learning Attacks

The aim of this attack is to provide *supervised* training data to the adversary, $\mathcal{A}$, in order to equip $\mathcal{A}$ with sufficient knowledge that can be used to attack the system. Intuitively, a training set of AniCAP challenges will have to be provided with their respective solutions, v's. Then, after the training is conducted, $\mathcal{A}$ will be given a fresh AniCAP challenge, in which $\mathcal{A}$ has to solve using the knowledge from its database. This attack is inspired by the supervised learning approach in machine learning and the notion of known plaintext attacks in cryptographic literature.

The outline of a practical situation adopting this attack is as follows. Consider a 'smart' attacker program being trained by a human. The human is presented with several AniCAP challenges, and the human can answer these challenges correctly. This information supplied to the attacker program as supervised training data and will be conducted during the *learning* stage. Once the learning stage is over, the program will be presented with a fresh AniCAP challenge. This time, the attacker program will need to answer the challenge itself, given the knowledge that it has gathered during the learning stage. The second stage is known as the *attacking* stage. The attack is considered successful if the attacker program can answer the fresh AniCAP challenge correctly. Formally, this attack is defined as a game among the challenger $\mathcal{C}$, an attacker $\mathcal{A}$ and a human $\mathcal{H}$ as follows.

Stage 1. Learning Stage

1. Define $\mathcal{L} := \emptyset$.
2. Repeat this process q times: For all CAPTCHA challenges given by $\mathcal{C}$ (i.e. $\texttt{View}_{\mathfrak{C}i}$), the human $\mathcal{H}$ will perform the following.
   (a) Output the correct answer υ_i.
   (b) Add this knowledge to $\mathcal{L}$, i.e. $\mathcal{L} := \mathcal{L} \cup \{\texttt{View}_{\mathfrak{C}i}, \upsilon_i\}$.
3. Output $\mathcal{L}$.

Stage 2. Attacking Stage
At this stage the attacker $\mathcal{A}$ is equipped with $\mathcal{L} = \forall_i(\texttt{View}_{\mathfrak{C}i}, \upsilon_i)$, where $|\mathcal{L}| = q$.

1. $\mathcal{C}$ outputs a fresh CAPTCHA challenge $\texttt{View}_{\mathfrak{C}}{}^* \not\subset \forall_i\{\texttt{View}_{\mathfrak{C}i}\}$, where $\forall_i\{\texttt{View}_{\mathfrak{C}i}\} \in \mathcal{L}$.
2. $\mathcal{A}$ needs to answer with the correct υ^*.

Note that the required $\texttt{View}_{\mathfrak{C}}{}^*$ in the attacking stage is $\texttt{View}_{\mathfrak{C}}{}^* \not\subset \forall_i\{\texttt{View}_{\mathfrak{C}i}\}$, where $\forall_i\{\texttt{View}_{\mathfrak{C}i}\} \in \mathcal{L}$.

Definition 1. *A CAPTCHA is secure against machine learning attacks if no adversary can win the above game with a probability that is non-negligibly greater than* $(\frac{1}{n})^\ell$, *where* ℓ *is the length of the CAPTCHA challenge, and* n *represents the number of characters used in the CAPTCHA challenge.*

Theorem 2. *AniCAP is secure against machine learning attacks.*

Proof (sketch). During the learning stage, $\mathcal{A}$ can form a data set $\mathcal{L} := \{\texttt{View}_{\mathfrak{C}i}, \upsilon_i\}$, for $i = 1, \cdots, n$. During the attacking stage, $\mathcal{A}$ will be provided with a AniCAP challenge $\texttt{View}_{\mathfrak{C}}{}^*$. Note that $\texttt{View}_{\mathfrak{C}}{}^* \not\subset \forall_i\{\texttt{View}_{\mathfrak{C}i}\}$, where $\forall_i\{\texttt{View}_{\mathfrak{C}i}\} \in \mathcal{L}$. Therefore,

$$Pr\,(\texttt{View}_{\mathfrak{C}}{}^*|\{\texttt{View}_{\mathfrak{C}i}, \upsilon_i\}, \text{where } \mathcal{L} := \{\texttt{View}_{\mathfrak{C}i}, \upsilon_i\}) = Pr(\texttt{View}_{\mathfrak{C}}{}^*).$$

Hence, the knowledge on $\mathcal{L}$ clearly does not help $\mathcal{A}$ to solve the fresh AniCAP challenge, $\texttt{View}_{\mathfrak{C}}{}^*$. □

5 Conclusion

This paper presents AniCAP, a novel text-based animated 3D CAPTCHA. AniCAP is built on the underlying concept that humans can perceive depth through motion parallax, thus capitalizing on the difference in ability between humans and computers at the task of perceiving depth through motion. Foreground characters and background characters in AniCAP are placed at different depths in the 3D scene. Thus, from the point of view of a moving camera, humans can distinguish the main characters in the foreground from the background characters, because the foreground characters will appear to move at different rates compared to the background characters.

AniCAP is designed to be segmentation-resistant by adopting a number of features such as the overlapping and crowding of characters together. Furthermore, by deliberately adopting a distortion approach that changes from frame to frame, this will prevent techniques that attempt to correlate or track points between frames, from succeeding. Other features employed in the design of AniCAP to deter automated attacks include randomized distortion, translucency and camera parameters, as well as textureless regions with no distinct colors or borders to distinguish between characters.

References

1. Ahmad, A.S.E., Yan, J., Marshall, L.: The Robustness of a New CAPTCHA. In: Costa, M., Kirda, E. (eds.) EUROSEC, pp. 36–41. ACM (2010)
2. Athanasopoulos, E., Antonatos, S.: Enhanced CAPTCHAS: Using Animation to Tell Humans and Computers Apart. In: Leitold, H., Markatos, E.P. (eds.) CMS 2006. LNCS, vol. 4237, pp. 97–108. Springer, Heidelberg (2006)
3. Baecher, P., Fischlin, M., Gordon, L., Langenberg, R., Lützow, M., Schröder, D.: Captchas: The good, the bad, and the ugly. In: Freiling, F.C. (ed.) Sicherheit. LNI, vol. 170, pp. 353–365. GI (2010)
4. Baird, H.S., Coates, A.L., Fateman, R.J.: PessimalPrint: a Reverse Turing Test. IJDAR 5(2-3), 158–163 (2003)
5. Baker, S., Scharstein, D., Lewis, J.P., Roth, S., Black, M.J., Szeliski, R.: A database and evaluation methodology for optical flow. In: ICCV, pp. 1–8. IEEE (2007)
6. Baker, S., Scharstein, D., Lewis, J.P., Roth, S., Black, M.J., Szeliski, R.: A database and evaluation methodology for optical flow. International Journal of Computer Vision 92(1), 1–31 (2011)
7. Chellapilla, K., Larson, K., Simard, P.Y., Czerwinski, M.: Building Segmentation Based Human-Friendly Human Interaction Proofs (HIPs). In: Baird, H.S., Lopresti, D.P. (eds.) HIP 2005. LNCS, vol. 3517, pp. 1–26. Springer, Heidelberg (2005)
8. Chellapilla, K., Larson, K., Simard, P.Y., Czerwinski, M.: Designing Human Friendly Human Interaction Proofs (HIPs). In: van der Veer, G.C., Gale, C. (eds.) CHI, pp. 711–720. ACM (2005)
9. Chellapilla, K., Simard, P.Y.: Using Machine Learning to Break Visual Human Interaction Proofs (HIPs). In: NIPS (2004)
10. Chew, M., Baird, H.S.: BaffleText: a Human Interactive Proof. In: Kanungo, T., Smith, E.H.B., Hu, J., Kantor, P.B. (eds.) DRR. SPIE Proceedings, vol. 5010, pp. 305–316. SPIE (2003)
11. Cui, J.-S., Mei, J.-T., Wang, X., Zhang, D., Zhang, W.-Z.: A captcha implementation based on 3d animation. In: Proceedings of the 2009 International Conference on Multimedia Information Networking and Security, MINES 2009, vol. 2, pp. 179–182. IEEE Computer Society, Washington, DC, USA (2009)
12. Cui, J.-S., Mei, J.-T., Zhang, W.-Z., Wang, X., Zhang, D.: A captcha implementation based on moving objects recognition problem. In: ICEE, pp. 1277–1280. IEEE (2010)

13. Elson, J., Douceur, J.R., Howell, J., Saul, J.: Asirra: a CAPTCHA that Exploits Interest-Aligned Manual Image Categorization. In: Ning, P., di Vimercati, S.D.C., Syverson, P.F. (eds.) ACM Conference on Computer and Communications Security, pp. 366–374. ACM (2007)
14. Fischer, I., Herfet, T.: Visual captchas for document authentication. In: 8th IEEE International Workshop on Multimedia Signal Processing (MMSP 2006), pp. 471–474 (2006)
15. Kessels, J.C.: JkCaptcha, `http://kessels.com/captcha/`
16. Lu, Y., Zhang, J.Z., Wu, Q.M.J., Li, Z.-N.: A survey of motion-parallax-based 3-d reconstruction algorithms. IEEE Transactions on Systems, Man, and Cybernetics 34, 532–548 (2004)
17. Mori, G., Malik, J.: Recognizing Objects in Adversarial Clutter: Breaking a Visual CAPTCHA. In: CVPR (1), pp. 134–144 (2003)
18. Moy, G., Jones, N., Harkless, C., Potter, R.: Distortion Estimation Techniques in Solving Visual CAPTCHAs. In: CVPR (2), pp. 23–28 (2004)
19. NuCaptcha Inc. NuCaptcha, `http://www.nucaptcha.com/`
20. Program Produkt. HELLOCAPTCHA, `http://www.hellocaptcha.com/`
21. Shi, J., Tomasi, C.: Good features to track. Technical report, Ithaca, NY, USA (1993)
22. Susilo, W., Chow, Y.-W., Zhou, H.-Y.: Ste3d-cap: Stereoscopic 3d Captcha. In: Heng, S.-H., Wright, R.N., Goi, B.-M. (eds.) CANS 2010. LNCS, vol. 6467, pp. 221–240. Springer, Heidelberg (2010)
23. von Ahn, L., Blum, M., Hopper, N.J., Langford, J.: CAPTCHA: Using Hard AI Problems for Security. In: Biham, E. (ed.) EUROCRYPT 2003. LNCS, vol. 2656, pp. 294–311. Springer, Heidelberg (2003)
24. Wang, S.-Y., Baird, H.S., Bentley, J.L.: CAPTCHA Challenge Tradeoffs: Familiarity of Strings versus Degradation of Images. In: ICPR (3), pp. 164–167. IEEE Computer Society (2006)
25. Yan, J., Ahmad, A.S.E.: Breaking Visual CAPTCHAs with Naive Pattern Recognition Algorithms. In: ACSAC, pp. 279–291. IEEE Computer Society (2007)
26. Yan, J., Ahmad, A.S.E.: A Low-Cost Attack on a Microsoft CAPTCHA. In: Ning, P., Syverson, P.F., Jha, S. (eds.) ACM Conference on Computer and Communications Security, pp. 543–554. ACM (2008)
27. Yan, J., Ahmad, A.S.E.: Usability of CAPTCHAs or Usability Issues in CAPTCHA Design. In: Cranor, L.F. (ed.) SOUPS. ACM International Conference Proceeding Series, pp. 44–52. ACM (2008)
28. Yan, J., Ahmad, A.S.E.: CAPTCHA Security: A Case Study. IEEE Security & Privacy 7(4), 22–28 (2009)
29. Ziegler, R., Matusik, W., Pfister, H., McMillan, L.: 3d reconstruction using labeled image regions. In: Proceedings of the 2003 Eurographics/ACM SIGGRAPH symposium on Geometry processing, SGP 2003, Aire-la-Ville, Switzerland, pp. 248–259. Eurographics Association (2003)

Towards Attribute Revocation in Key-Policy Attribute Based Encryption

Pengpian Wang[1,2], Dengguo Feng[1,2], and Liwu Zhang[1,2]

[1] State Key Laboratory of Information Security,
Institute of Software, The Chinese Academy of Sciences, Beijing, China
[2] National Engineering Research Center of Information Security, Beijing, China
{wangpengpian,feng,zlw}@is.iscas.ac.cn

Abstract. Attribute revocation is important to the attribute-based encryption (ABE). The existing ABE schemes supporting revocation mainly focus on the revocation of the user's identity, which could only revoke the user's whole attributes by revoking the user's identity. In some cases, we wish to revoke one attribute of a user instead of the whole attributes issued to him without affecting any other user's private key, such that the user still can use his private key to decrypt as long as the unrevoked attributes of him satisfy the decryption condition. In this paper, we propose two KP-ABE schemes realizing the attribute revocation under the direct revocation model.

Keywords: Attribute Revocation, Attribute Based Encryption, Key-Policy, Access Structure, Access Tree.

1 Introduction

As a generalization of IBE, the concept of attribute-based encryption (ABE) is first proposed by Sahai[1] in the context of fuzzy IBE. In ABE, each user possesses an attribute set which is associated with a private key issued by PKG (Private Key Generator), the sender should select an attribute set to encrypt the message, the receiver could decrypt the ciphertext correctly if and only if the private key he has satisfies the decryption condition which is referred to as policy. ABE schemes can be categorized by the place where the policy is embedded: whether the policy is embedded in the private key or in the ciphertext. These are called KP-ABE (Key Policy-Attribute Based Encryption) [3,4] and CP-ABE (Ciphertext Policy-Attribute Based Encryption) [5,6] scheme respectively. ABE schemes can also be categorized by the size of the attribute universe that such a scheme can deal with: whether it is of polynomial or super-polynomial size. These are called small and large universe scheme respectively[18].

ABE enables an access control mechanism over encrypted data by using the policy, which is very useful in secure cloud storage. In cloud storage, the resource owner stores the data in the cloud, and wants to control the accesses to the resource. Though the resource owner can give the access control policies to the cloud provider, a malicious cloud provider may return the resource to a request

D. Lin, G. Tsudik, and X. Wang (Eds.): CANS 2011, LNCS 7092, pp. 272–291, 2011.

without evaluating the access control policies, even more, the cloud provider may leak the sensitive data of the resource owner. By utilizing the ABE, we can resolve this problem, but another problem arises, that is, attribute revocation.

It is common to revoke an attribute of a user in traditional access control system which uses the reference monitor, but it is hard in ABE, the key point is how to revoke an attribute of a user without affecting any other user's private key. To the best of our knowledge, there exists no ABE scheme that meets this requirement. In this paper, we construct two KP-ABE schemes realizing attribute revocation to resolve this problem.

Related Work. Most of the ABE schemes don't consider the issue of attribute revocation specially, and use the same method for identity revocation of IBE schemes to realize the attribute revocation.

There are mainly two methods for identity revocation of IBE schemes: one is to use a trusted third party (TTP) that helps the user to decrypt, such as [7,8], the user's private key is divided into two parts, one is hold by the user, the other (not known by the user) is kept by TTP, if a user is revoked, the TTP will not help him to decrypt; The other method is to update the user's private key periodically, such as [9,10], the user's private key actually consists of two parts, one part contains the user's identity and is kept secretly by the user, the other part contains the time information and is published periodically by the PKG. When encrypting, The sender will use a user's identity and a time attribute to encrypt the message, if the receiver's identity is revoked at that time, since the PKG will not update the private key that contains the specific time attribute for the receiver, the receiver will not decrypt the ciphertext correctly. Though each ABE scheme could use these two methods to realize the attribute revocation, the TTP and PKG will be the bottleneck of the system.

In the literature [11,12], Attrapadung defines two revocation models explicitly: one is indirect revocation model, in this model, the user's identity (or attribute) is revoked by affecting the user's private key as mentioned above, when encrypting a message, the sender doesn't care the revocation list; the other is direct revocation model, in this model, the user's identity (or attribute) is revoked by embedding the revocation list in the ciphertext without affecting any user's private key. In some cases, for example the broadcast encryption [13], it is more efficient to use the direct revocation model, and Attrapadung [11] constructed four schemes supporting user's **identity revocation** under the direct revocation model(two KP-ABE schemes and two CP-ABE schemes). Actually, before the work of Attrapadung, there existed a few schemes that realized identity (or attribute) revocation under direct revocation model: Boneh[13] constructed a revocable broadcast IBE scheme; Staddon [14] gave a KP-ABE scheme supporting identity revocation, but it required that the size of attribute set used for encryption must be the half size of the attribute universe N. After Attrapadung's work, Lewko [15] proposed two IBE schemes supporting identity revocation under direct revocation model, one is very efficient with small private keys, the other is less efficient but the security is based on the standard DBDH assumption(See Appendix A). In addition, Yu [16] proposed a revocation system(in ABE setting) that is quite different from the existing ones, in their scheme, the attribute is revoked by changing the public parameters and any other user's private key simultaneously, so the cost of the attribute revocation is big.

We note that, for most of the existing ABE schemes supporting revocation, the issue of revocation is focused on the user's identity rather than the attribute. When the user's identity is revoked, each of the attributes he has will be revoked, too. In some cases, we wish to revoke one of the user's attributes instead of his identity(represents the whole attributes issued to him) under the direct revocation model, that is, the user can still use his private key to decrypt as long as the unrevoked attributes of him satisfy the decryption condition.

Our Contribution. We formalize the notion of attribute revocation for KP-ABE under the direct revocation model and present two concrete constructions for it, one is a small universe scheme and the other is a large universe scheme.

Our constructions is inspired by the schemes proposed by Attrapadung [11], which are based on the KP-ABE scheme proposed by Goyal [3] and broadcast encryption scheme proposed by Boneh [13]. In our schemes, each user has two same access trees, but each of these two trees is associated with a different private key. When giving a ciphertext encrypted under an attribute set with a revocation list of an attribute which belongs to the attribute set: if the user is not in the revocation list, he could use the first access tree to decrypt as long as the attribute set satisfies the access tree; if the user is in the revocation list, which means that one attribute of him is revoked, then he could only use the second access tree to decrypt as long as the attribute set excluding the revoked attribute satisfies the access tree. Though our schemes can only revoke one attribute per encryption, which is not desirable in practice, we believe it is the first step to realize the fine-grained attribute revocation, that is, the sender can revoke arbitrary number of attributes used for encryption.

Organization. The rest of the paper is organized as follows. In Section 2 we provide the relevant background on bilinear groups, access structure, access tree, lagrange coefficient and state our complexity assumption. Then we give the definition of KP-ABE that supports attribute revocation and the security model in Section 3. In Section 4 we give two concrete constructions and analysis the efficiency of those two schemes, then prove the security under the selective security model defined in Section 3. In Section 4, we briefly discuss how to achieve CCA-security and extend the Goyal's scheme [3] to support fine-grained attribute revocation. Finally, we conclude in Section 6.

2 Background

2.1 Bilinear Groups

We briefly review the necessary facts about bilinear map and bilinear map groups.

1. $\mathbb{G}_1$ and $\mathbb{G}_2$ are two (multiplicative) cyclic groups of prime order p;
2. g is a generator of $\mathbb{G}_1$;
3. $e : \mathbb{G}_1 \times \mathbb{G}_2 \rightarrow \mathbb{G}_2$ is a bilinear map.

Let $\mathbb{G}_1$ and $\mathbb{G}_2$ be two groups as above. A bilinear map is a map $e : \mathbb{G}_1 \times \mathbb{G}_2 \rightarrow \mathbb{G}_2$ with the following properties:

1. Bilinear: for all $u, v \in \mathbb{G}_1$ and $a, b \in \mathbb{Z}_p$, we have $e(u^a, v^b) = e(u, v)^{ab}$;
2. Non-degenerate: $e(g, g) \neq 1$.

We say that $\mathbb{G}_1$ is a bilinear group if the group action in $\mathbb{G}_1$ can be computed efficiently and there exists a group $\mathbb{G}_2$ and an efficiently computable bilinear map $e : \mathbb{G}_1 \times \mathbb{G}_2 \rightarrow \mathbb{G}_2$ as above. Note that, $e(,)$ is symmetric since $e(g^a, g^b) = e(g, g)^{ab} = e(g^b, g^a)$.

2.2 Access Structure and Access Tree

Definition 1. *(Access Structure [17]) Let $\{P_1, ..., P_n\}$ be a set of paries. A collection $\mathbb{A} \subseteq 2^{\{P_1,...,P_n\}}$ is monotone if $\forall B, C$:, if $B \in \mathbb{A}$ and $B \subseteq C$ then $C \in \mathbb{A}$. An access structure (respectively, monotonic access structure) is a collection (respectively, monotone collection) $\mathbb{A}$ of non-empty subsets of $\{P_1, P_2..., P_2\}$, ie., $\mathbb{A} \subseteq 2^{\{P_1,...,P_n\}} \backslash \{\emptyset\}$. The sets in $\mathbb{A}$ are called the authorized sets, and the sets not in $\mathbb{A}$ are called the unauthorized sets.*

Currently, there are mainly two methods to describe the access structure, one is the linear secret sharing schemes (LSSS)[17], the other is the access tree. We use the access tree to build our scheme for the reason that, in general, it is more practical to use access tree than to use LSSS. We briefly review the essential facts about access tree.

Access Tree. As mentioned above, an access tree represents an access structure. Each node represents a threshold, for the node x in the tree, denote n_x to be the number of its children, v_x to be the threshold value, we have $0 < v_x \leq n_x$, when $v_x = 1$, the threshold gate is an OR gate, when $v_x = n_x$, the threshold gate is an AND gate. In addition, each leaf-node also represents an attribute, and the threshold value of each leaf-node is 1, that is, each leaf-node represents an OR gate. We define four operations on the nodes of the access tree:

1. *parent*(x): For each node x except the root node, this denotes the parent node of x;
2. *children*(x): For each non-leaf node, this denotes the set of all the child nodes of x;
3. *index*(x): We assume that the access tree defines an ordering between the children of every node, so for each node except the root node, the *index*(x) returns such a number associated with node x, and we have that, $1 \leq index(x) \leq |children(parent(x))|$;
4. *attr*(x): For each leaf node, this denotes the attribute represented by the leaf node x.

We also define an function $\Gamma_x(\omega)$ on each node x of the access tree $\mathbb{A}$ under an attribute set ω, the function $\Gamma_x(\omega)$ is defined as below:

1. If x is a leaf node:

$$\Gamma_x(\omega) = \begin{cases} 0, & attr(x) \notin \omega \\ 1, & attr(x) \in \omega \end{cases}$$

2. If x is a non-leaf node:

$$\Gamma_x(\omega) = \begin{cases} 0, & \sum\limits_{z \in children(x)} \Gamma_z(\omega) < v_x \\ 1, & \sum\limits_{z \in children(x)} \Gamma_z(\omega) \geq v_x \end{cases}$$

Definition 2. *Let $\mathbb{A}$ be an access tree with the root node r, we say an attribute set ω **satisfies** the access tree $\mathbb{A}$ if and only if $\Gamma_r(\omega) = 1$.*

2.3 Lagrange Coefficient

By given $(n+1)$ different points (x_i, y_i) in $\mathbb{Z}_p$, we can use the polynomial interpolation to fix the polynomial $f(x)$, the degree of which is n, and:

$$f(x) = \prod_{i=1}^{n+1} (y_i \cdot \prod_{1 \leq j \leq n+1 \wedge j \neq i} \frac{x - x_j}{x_i - x_j})$$

Let $S \subseteq \mathbb{Z}_p$, for each $i \in S$, the lagrange coefficient is defined as:

$$\triangle_{i,S}(x) = \prod_{j \in S \wedge j \neq i} \frac{x - x_j}{x_i - x_j}$$

2.4 Decision q-BDHE Assumption

Security of our scheme is based on a complexity assumption called the Decision q-BDHE (Bilinear Diffie-Hellman Exponent) assumption. It is stated as follows:

Let $\mathbb{G}_1$ be a bilinear group of prime order p, given a vector of $2q+1$ elements:

$$(g, g^s, g^\alpha, g^{\alpha^2}, ..., g^{\alpha^q}, g^{\alpha^{q+2}}, ..., g^{\alpha^{2q}}) \in \mathbb{G}_1^{2q+1}$$

We say that the Decision q-BDHE assumption holds in $\mathbb{G}_1$ if no polynomial-time algorithm has a non-negligible advantage to distinguish $e(g,g)^{s\alpha^{q+1}}$ from a random element in $\mathbb{G}_2$.

3 Definition

Denote U={1,2,...n} to be the universe of all the users and N to be the universe of all the attributes. An attribute based encryption scheme that supports fine-grained attribute revocation under the direct revocation model consists of four algorithms ***Setup, Encryption, KeyGen, Decryption***, we describe each of these four algorithm below:

***Setup* $\rightarrow$ *(pk,msk)*:** This is a randomized algorithm that takes no input other than the implicit security parameters. It outputs a public key pk and a master secret key msk;

***Encryption*(ω, R_j, *M*, *pk*) $\rightarrow$ *C*:** This is a randomized algorithm that takes as input an attribute set $\omega \subseteq N$, a revocation list $R_j \subseteq U$ of attribute $j \in \omega$, a message M, and the public key pk. It outputs a ciphertext C.

***KeyGen*(*ID*, $\mathbb{A}$, *msk*, *pk*) $\rightarrow$ $SK_{ID,\mathbb{A}}$:** This is a randomized algorithm that takes as input a user index $ID \in U$, an access tree $\mathbb{A}$, the master secret key msk, and the public key pk. It outputs a user's private key $SK_{ID,\mathbb{A}}$.

***Decryption*(*C*, ω, R_j, $SK_{ID,\mathbb{A}}$, *pk*)$\rightarrow$ *M*:** This algorithm takes as input a ciphertext C that was encrypted under an attribute set ω with an attribute revocation list $R_j \subseteq U$ of attribute $j \in \omega$, the user's private key $SK_{ID,\mathbb{A}}$ for user $ID \in U$ with the access tree $\mathbb{A}$. Define the attribute set $\omega^{'}$ for the user ID: if $ID \in R_j$, let $\omega^{'} = \omega - \{j\}$; otherwise, let $\omega^{'} = \omega$. It outputs the message M if and only if the attribute set $\omega^{'}$ satisfies the access tree $\mathbb{A}$.

Selective Security Model. The selective security notion is defined in the following game:

***Init*:** The adversary declares an attribute set $\omega^* \subseteq N$ and an attribute revocation list $R_j^* \subseteq U$ of attribute $j \in \omega^*$ that it wishes to be challenged upon;

***Setup*:** The challenger runs the Setup algorithm of ABE and gives the public key pk to the adversary;

***Phase*1:** The adversary is allowed to issue queries for user private key $SK_{ID,\mathbb{A}}$ of the user $ID \in U$ with the access tree $\mathbb{A}$, such that $\omega^{'}$ (see the definition of *Decryption*) doesn't satisfy the access tree $\mathbb{A}$.

***Challenge*:** The adversary submits two equal-length messages M_0, M_1. The challenger chooses a random bit $b \in \{0, 1\}$, and encrypt the message M_b under the attribute set ω^* with the attribute revocation list R_j^*. Then the challenger gives the challenge ciphertext C^* to the adversary.

Phase2: Phase1 is repeated.

Guess: The adversary outputs a guess $b^{'}$ of b.

The advantage of the adversary in this game is defined as $\Pr[b^{'} = b]$-1/2. We note that the model can easily be extended to handle chosen-ciphertext attacks by allowing for decryption queries in *Phase*1 and *Phase*2.

Definition 3. *A KP-ABE scheme supporting attribute revocation under direct revocation model is secure in the selective security model if all polynomial time adversaries have at most a negligible advantage in the above game.*

4 Construction

In ours schemes, each attribute has two values, and each user has two same access trees: For the first access tree, PKG will use all the first values of the attributes and $\alpha^{ID}\gamma$ to generate the first part of the user's private key; For the second tree, all the second values of the attributes and $\alpha^{ID}\beta$ will be used to generate the second part of the user's private key. When encrypting a message under an attribute set ω with a revocation list R_j of the attribute $j \in \omega$, define $\omega^{'} = \omega - \{j\}$, $T_{i,0}$ as the first value of the attribute i, $T_{i,1}$ as the second value of the attribute i, then the sender will give all the values of $\{T_{i,0}^{s}\}_{i \in \omega}$ (used by the first part of the user's private key) and $\{T_{i,1}^{s}\}_{i \in \omega^{'}}$ (used by the second part of the user's private key). When a user with the private key $SK_{ID,\mathbb{A}}$ wants to decrypt the ciphertext:

1. if ω doesn't satisfy the access tree $\mathbb{A}$, we can know that, the $\omega^{'}$ doesn't satisfy the access tree $\mathbb{A}$ either. So neither of the two parts of the private key can be used to decrypt the ciphertext correctly;
2. if ω satisfies the access tree $\mathbb{A}$,
   (a) If *ID* is not in the attribute revocation list R_j, then he can use the first part of his private key to decrypt the ciphertext correctly. Note that,in such a case, $\omega^{'}$ may also satisfies the access tree, if so, the second part of the private key can decrypt the ciphertext correctly, too;
   (b) If *ID* is in the attribute revocation list R_j, which means the attribute j of the user is revoked. In such a case, the user could only use the second part of the private key to decrypt, and he can recover the message if and only if $\omega^{'}$ satisfies $\mathbb{A}$.

4.1 Small Universe Construction

Let $\mathbb{G}_1$ and $\mathbb{G}_2$ be groups of prime order p, and g is a generator of $\mathbb{G}_1$. In addition, define e: $\mathbb{G}_1 \times \mathbb{G}_1 \rightarrow \mathbb{G}_2$ to be a bilinear map.

Setup: Define the universe of users U={1,2,...,n} and attributes N={1,2,...,m}, let $U^{'} = U \cup \{n+2, n+3, ..., 2n\}$:

1. Choose a random element α from $\mathbb{G}_1$, for each $j \in U^{'}$, compute the $g_j = g^{\alpha^j}$;
2. For each attribute of $i \in N$, choose two random elements $t_{i,0}$ and $t_{i,1}$ from $\mathbb{Z}_p^*$, then set $T_{i,0} = g^{t_{i,0}}$ and $T_{i,1} = g^{t_{i,1}}$;

3. Choose two random elements γ and β;

The public key ***pk*** is:

$$\boldsymbol{pk} = \{\mathbb{G}_1, \mathbb{G}_2, g, \{(T_{i,0}, T_{i,1})\}_{i \in N}, \{g_j\}_{j \in U'}, g^\gamma, g^\beta\}$$

The master secret key ***msk*** is:

$$\boldsymbol{msk} = \{\{(t_{i,0}, t_{i,1})\}_{i \in N}, \alpha, \gamma, \beta\}$$

Encryption$(\omega,$ R_j, ***M***, ***pk***)**:** To encrypt a message $M \in \mathbb{G}_2$ under a set of attribute ω, with a user revocation list $R_j \subset U$ of attribute $j \in \omega$, choose a random value $s \in \mathbb{Z}_p$, let $S_0 = U - R_j, S_1 = U$, then set:

$$\begin{cases} C_0 = e(g_1, g_n)^s M, \\ C_1 = g^s, \\ C_{2,0} = \{T_{i,0}^s\}_{i \in \omega}, \\ C_{2,1} = \{T_{i,1}^s\}_{i \in \omega \wedge i \neq j}, \\ C_{3,0} = (g^\gamma \prod\limits_{i \in S_0} g_{n+1-i})^s, \\ C_{3,1} = (g^\beta \prod\limits_{i \in S_1} g_i)^s \end{cases}$$

The ciphertext ***C*** is published as:

$$\boldsymbol{C} = \{\omega, R_j, C_0, C_1, C_{2,0}, C_{2,1}, C_{3,0}, C_{3,1}\}$$

KeyGen(ID, $\mathbb{A}$, ***msk***, ***pk***)**:** To generate a secret key for user $ID \in U$ under an access tree $\mathbb{A}$, the algorithm proceeds in a top-down manner as follows:

1. For the root node r, choose two polynomials $q_{r,0}$ and $q_{r,1}$ of degree $v_r - 1$. For $q_{r,0}$, first let $q_{r,0}(0) = \alpha^{ID}\gamma$ and then randomly choose other $v_r - 1$ points to define $q_{r,0}$ completely. The way to choose $q_{r,1}$ is the same to $q_{r,0}$ except that, we let $q_{r,1}(0) = \alpha^{ID}\beta$
2. For any other node x in access tree $\mathbb{A}$, the way to choose polynomials $q_{x,0}$ and $q_{x,1}$ is the same to root node, except that, we let $q_{x,0}(0) = q_{parent(x),0}(index(x))$ and $q_{x,1}(0) = q_{parent(x),1}(index(x))$;
3. Once the polynomials have been decided, for each leaf node x, let $i{=}attr(x)$, we give the secret values to the user:

$$D_{x,0} = g^{\frac{q_{x,0}(0)}{t_{i,0}}}, D_{x,1} = g^{\frac{q_{x,1}(0)}{t_{i,1}}}$$

Define L to be the set of leaves of access tree $\mathbb{A}$, the user's secret key $SK_{ID,\mathbb{A}}$ is given as:

$$SK_{ID,\mathbb{A}} = \{(D_{x,0}, D_{x,1})\}_{x \in L}$$

Decryption$(\boldsymbol{C}, \omega, R_j, \boldsymbol{SK}_{\boldsymbol{ID},\mathbb{A}}$, ***pk***)**:** For a user with a secret key $SK_{ID,\mathbb{A}}$, if $ID \in R_j$, we let $\omega' = \omega - \{j\}$, if not, we let $\omega' = \omega$. The user could decrypt the ciphertext correctly if and only if $\Gamma_r(\omega') = 1$(r is the root node of $\mathbb{A}$), ie. the attribute set ω' **satisfies** the access tree $\mathbb{A}$. The decryption algorithm proceeds in a down-top manner as follows:

1. if $ID \notin R_j$:

   (a) For each leaf node x of access tree $\mathbb{A}$, let $i = attr(x)$: If $i \notin \omega'$, we let $F_x = \perp$; If $i \in \omega'$, we let $F_x = e(D_{x,0}, T^s_{i,0}) = e(g,g)^{s \cdot q_{x,0}(0)}$;
   (b) For any other node x(including the root node) of access tree $\mathbb{A}$, denote S_x to be the set of children of node x, such that for each node $z \in S_x, F_z \neq \perp$. If $|S_x| < v_x, F_x = \perp$;Otherwise, choose a subset $S'_x \subseteq S_x$, such that $|S'_x| = v_x$, then set $S'' = \{index(z)|z \in S'_x\}$ and use polynomial interpolation to compute:
   $$F_x = \prod_{z \in S'_x} F_z^{\triangle_{i,S''}(0)} = e(g,g)^{s \cdot q_{x,0}(0)}, i = index(z),$$
   We observe that, for the root node, we have:
   $$F_r = e(g,g)^{s \cdot q_{r,0}(0)} = e(g,g)^{s \cdot \alpha^{ID} \gamma}$$
   (c) Finally, we recover the Message M by computing:
   $$C_0 \cdot F_r \cdot \frac{e(\prod_{\substack{j \in S_0 \\ \wedge j \neq ID}} g_{n+1-j+ID}, C_1)}{e(g_{ID}, C_{3,0})}$$

Correctness: We can verify its correctness as(note that $ID \notin R_j$, so $ID \in S_0$):

$$C_0 \cdot F_r \cdot e(\prod_{j \in S_0 \wedge j \neq ID} g_{n+1-j+ID}, C_1) \cdot \frac{1}{e(g_{ID}, C_{3,0})}$$
$$= e(g_1, g_n)^s M \cdot e(g,g)^{s \cdot \alpha^{ID} \gamma} \cdot e(\prod_{j \in S_0 \wedge j \neq ID} g_{n+1-j+ID}, g)^s \cdot \frac{1}{e(g_{ID}, g^{\gamma} \prod_{i \in S_0} g_{n+1-i})^s}$$
$$= e(g_1, g_n)^s M \cdot e(g_{ID}, g^{\gamma})^s \cdot e(\prod_{j \in S_0 \wedge j \neq ID} g_{n+1-j+ID}, g)^s \cdot \frac{1}{e(g_{ID}, g^{\gamma})^s e(g, \prod_{i \in S_0} g_{n+1-i+ID})^s}$$
$$= M$$

2. if $ID \in R_j$

   (a) Same to the previous situation except that, If $i \in \omega'$, we let $F_x = e(D_{x,1}, T^s_{i,1}) = e(g,g)^{s \cdot q_{x,1}(0)}$;
   (b) Same to the previous situation, but for the root node, we have:
   $$F_r = e(g,g)^{s \cdot q_{r,1}(0)} = e(g,g)^{s \cdot \alpha^{ID} \beta}$$
   (c) Finally, we recover the Message M by computing:
   $$C_0 \cdot F_r \cdot \frac{e(\prod_{\substack{j \in S_1 \\ \wedge j \neq ID}} g_{n+1-j+ID}, C_1)}{e(g_{ID}, C_{3,1})}$$

4.2 Large Universe Construction

Let $\mathbb{G}_1$ and $\mathbb{G}_2$ be groups of prime order p, and g is a generator of $\mathbb{G}_1$. In addition, define e: $\mathbb{G}_1 \times \mathbb{G}_1 \rightarrow \mathbb{G}_2$ to be a bilinear map. We assume that the sender uses at most m attributes when encrypting (bounded ABE), denote N={1,2,...,m+1}.

***Setup*:** Define the universe of users U={1,2,...,n}, let $U' = U \cup \{n+2, n+3, ..., 2n\}$:

1. Choose a random element α from $\mathbb{G}_1$, for each $j \in U'$, compute the $g_j = g^{\alpha^j}$;
2. Choose two random elements γ and β;
3. For each $i \in N$, also choose two random elements $h_{i,0}$ and $h_{i,1}$ from $\mathbb{G}_1$, then define two functions;

$$T_0(x) = g^{\gamma x^m} \prod_{i=1}^{m+1} h_{i,0}^{\triangle_{i,N}(x)} \ , T_1(x) = g^{\beta x^m} \prod_{i=1}^{m+1} h_{i,1}^{\triangle_{i,N}(x)}$$

The public key ***pk*** is:

$$\boldsymbol{pk} = \{\mathbb{G}_1, \mathbb{G}_2, g, \{(h_{i,0}, h_{i,1})\}_{i \in N}, \{g_j\}_{j \in U'}, g^\gamma, g^\beta\}$$

The master secret key **msk** is:

$$\textbf{msk} = \{\alpha, \gamma, \beta\}$$

***Encryption*(ω, R_j, *M*, *pk*):** To encrypt a message $M \in \mathbb{G}_2$ under a set of attribute ω, with a user revocation list $R_j \subset U$ of attribute $j \in \omega$, choose a random value $s \in \mathbb{Z}_p$, let $S_0 = U - R_j, S_1 = U$, then set:

$$\begin{cases} C_0 = e(g_1, g_n)^s M, \\ C_1 = g^s, \\ C_{2,0} = \{T_0(i)^s\}_{i \in \omega}, \\ C_{2,1} = \{T_1(i)^s\}_{i \in \omega \wedge i \neq j}, \\ C_{3,0} = (g^\gamma \prod_{i \in S_0} g_{n+1-i})^s, \\ C_{3,1} = (g^\beta \prod_{i \in S_1} g_i)^s \end{cases}$$

The ciphertext ***C*** is published as:

$$\boldsymbol{C} = \{\omega, R_j, C_0, C_1, C_{2,0}, C_{2,1}, C_{3,0}, C_{3,1}\}$$

***KeyGen*(*ID*,$\mathbb{A}$, *msk*, *pk*):** To generate a secret key for user $ID \in U$ under an access tree $\mathbb{A}$, the algorithm proceeds exactly the same to our small universe construction except that, for each leaf node x, we choose two elements $r_{x,0}, r_{x,1}$ uniformly at random from $\mathbb{Z}_p^*$, let i=*attr*(x), we give the secret values to the user:

$$\begin{cases} D_{x,0} = g^{q_{x,0}(0)} \cdot T_0(i)^{r_{x,0}} \ , R_{x,0} = g^{r_{x,0}} \\ D_{x,1} = g^{q_{x,1}(0)} \cdot T_1(i)^{r_{x,1}} \ , R_{x,1} = g^{r_{x,1}} \end{cases}$$

Define L to be the set of leaves of access tree $\mathbb{A}$, the user secret key $SK_{ID,\mathbb{A}}$ is given as:

$$SK_{ID,\mathbb{A}} = \{(D_{x,0}, D_{x,1}, R_{x,0}, R_{x,1})\}_{x \in L}$$

Decryption$(C, \omega, R_j, SK_{ID,\mathbb{A}}, pk)$**:** The decryption algorithm proceeds exactly the same to our small universe construction except that, for each leaf node x, let $i = attr(x)$, If $i \notin \omega'$, we let $F_x = \perp$, otherwise:

1. if $ID \notin R_j$:

$$F_x = \frac{e(D_{x,0}, g^s)}{e(R_{x,0}, T_0(i)^s)} = \frac{e(g^{q_{x,0}(0)}, g^s)e(T_0(i)^{r_{x,0}}, g^s)}{e(g^{r_{x,0}}, T_0(i)^s)} = e(g,g)^{q_{x,0}(0)s}$$

2. if $ID \in R_j$:

$$F_x = \frac{e(D_{x,1}, g^s)}{e(R_{x,1}, T_1(i)^s)} = \frac{e(g^{q_{x,1}(0)}, g^s)e(T_1(i)^{r_{x,1}}, g^s)}{e(g^{r_{x,1}}, T_1(i)^s)} = e(g,g)^{q_{x,1}(0)s}$$

4.3 Efficiency

In this section, we analysis the efficiency of our two schemes. Denote n to be the number of users of the system, m to be the maximum number of attributes that can be used per encryption, L to be the set of the leaf nodes of access tree $\mathbb{A}$ (or to be the set of the rows in the LSSS access structure matrix), ω to be the set of attributes used for encryption, R to be the revocation list, we compare our schemes with Staddon's [14] and Attrapadung's [12] in Table 1.

Table 1. Comparisons of efficiency

Schemes	Staddon [14]	Attrapadung [12]		Our Schemes	
		1st scheme	2nd scheme	1st scheme	2nd scheme
size of PK	$m+n+4$	$m+n+1$	$m+6$	$2(m+n+1)$	$2(m+n+1)$
size of MSK	$3m+2n+2$	2	2	$2m+3$	3
size of SK	$2\|L\|$	$2\|L\|$	$2\|L\|+2$	$2\|L\|$	$4\|L\|$
size of Ciphertext	$m+\|R\|+2$	$\|\omega\|+3$	$\|\omega\|+\|R\|+2$	$2\|\omega\|+3$	$2\|\omega\|+3$
Attribute Revocation	×	×	×	√	√

Table 1 shows that, in each of our schemes, the size of the ciphertext is independent of the revocation list. Though our schemes could realize the attribute revocation, the sizes of the public keys are much larger than the schemes of Attrapadung's.

4.4 Security

Theorem 1. *If an adversary can break our small universe scheme with advantage ε in the selective security model, then a simulator can be constructed to solve the Decision n-BDHE problem.*

Proof: See Appendix B.

Theorem 2. *If an adversary can break our large universe scheme with advantage ε in the selective security model, then a simulator can be constructed to solve the Decision n-BDHE problem.*

Proof: See Appendix C.

5 Discussion

CCA Secure. Though our scheme is proven to be CPA-secure,by using the generic method presented by Amada [18], we can transform our CPA-secure schemes to be CCA-secure schemes under the same security model.

Fine-Grained Attribute Revocation. Note that, our schemes only support the revocation of one attribute per encryption. We can build a KP-ABE scheme supporting fine-grained attribute revocation by extending the Goyal's scheme [3], but the extension is somewhat trivial. The main idea is stated as follows:

1. ***Setup*:** For each attribute $i \in \{N\}$, choose n random elements $\{t_{i,j}\}_{1 \leq j \leq n}$ from $\mathbb{Z}_p^*$ as part of the master secret key, and publish the $\{T_{i,j}\}_{1 \leq i \leq m, 1 \leq j \leq n}$ as part of the parameters. The left parts of the public key and the master secret key are the same to that of the Goyal's scheme.
2. ***KeyGen*:** It is exactly the same to the Goyal's scheme except that, for user $ID \in U$, the PKG use $(y, \{t_{ID,j}\}_{1 \leq j \leq n})$ to generate the user's private key $SK_{ID,\mathbb{A}}$;
3. ***Encryption*:** To encrypt a message $M \in \mathbb{G}_2$ under an attribute set $\omega \subseteq N$, let $R_i \subseteq U$ to be the revocation list of attribute $i \in \omega$, define $S_i = U - R_i$, then for each attribute $i \in \omega$, compute $\{E_{i,j} = T_{i,j}^s\}_{j \in S_i}$ as part of the ciphertext.
4. ***Decryption*:** To decrypt a ciphertext encrypted under the attribute set $\omega \subseteq N$, with the revocation lists $\{R_i\}_{i \in \omega}$, we first define $\omega' = \{i | i \in \omega, ID \notin R_i\}$ to be the set of the unrevoked attribute for user ID. The user with private key $SK_{ID,\mathbb{A}}$ can decrypt the ciphertext correctly if and only if ω' satisfies the access tree $\mathbb{A}$.

The security of this KP-ABE scheme can be reduced to DBDH assumption(See Appendix A) as the Goyal's scheme, but we note that, the efficiency of this scheme is very low.

6 Conclusion and Future Work

In this paper, we presented two KP-ABE schemes supporting attribute revocation, which can revoke one attribute per encryption, and the security of these two schemes can be reduce to the q-BDHE assumption under the selective security model. Some possible future work includes: 1)realize a KP-ABE scheme with fine-grained attribute revocation supported which will be more efficient than the extended scheme given in Section 5; 2)construct a CP-ABE scheme supporting attribute revocation under the direct revocation model.

Acknowledgement. Thanks for the anonymous reviewers's comments and advices. This work was supported by the National Natural Science Foundation of China under Grant No.60803129, the Next Generation Internet Business and Equipment Industrialization Program under Grant No.CNGI-09-03-03 and the Opening project of Key Lab of Information Network Security of Ministry of Public Security (The Third Research Institute of Ministry of Public Security) under Grant No.C11604.

References

1. Sahai, A., Waters, B.: Fuzzy Identity-Based Encryption. In: Cramer, R. (ed.) EUROCRYPT 2005. LNCS, vol. 3494, pp. 457–473. Springer, Heidelberg (2005)
2. Boneh, D., Franklin, M.: Identity-Based Encryption from the Weil Pairing. In: Kilian, J. (ed.) CRYPTO 2001. LNCS, vol. 2139, pp. 213–229. Springer, Heidelberg (2001)
3. Goyal, V., Pandey, O., Sahai, A., Waters, B.: Attribute-based encryption for fine-grained access control of encrypted data. In: Proceedings of the 13th ACM Conference on Computer and Communications Security, pp. 89–98. ACM Press, New York (2006)
4. Ostrovsky, R., Sahai, A., Waters, B.: Attribute-based encryption with non-monotonic access structures. In: Proceedings of the 14th ACM Conference on Computer and Communications Security, pp. 195–203. ACM Press, New York (2007)
5. Bethencourt, J., Sahai, A., Waters, B.: Ciphertext-policy attribute-based encryption. In: Proceedings of the 2007 IEEE Symposium on Security and Privacy, pp. 321–334. IEEE Press, New York (2007)
6. Waters, B.: Ciphertext-Policy Attribute-Based Encryption: An Expressive, Efficient, and Provably Secure Realization. In: Catalano, D., Fazio, N., Gennaro, R., Nicolosi, A. (eds.) PKC 2011. LNCS, vol. 6571, pp. 53–70. Springer, Heidelberg (2011)
7. Hanaoka, Y., Hanaoka, G., Shikata, J., Imai, H.: Identity-Based Hierarchical Strongly Key-Insulated Encryption and Its Application. In: Roy, B. (ed.) ASIACRYPT 2005. LNCS, vol. 3788, pp. 495–514. Springer, Heidelberg (2005)
8. Libert, B., Quisquater, J.-J.: Efficient revocation and threshold pairing based cryptosystems. In: Proceedings of the Twenty-Second Annual Symposium on Principles of Distributed Computing, pp. 163–171. ACM Press, New York (2003)
9. Naor, M., Pinkas, B.: Efficient Trace and Revoke Schemes. In: Frankel, Y. (ed.) FC 2000. LNCS, vol. 1962, pp. 1–20. Springer, Heidelberg (2001)
10. Boldyreva, A., Goyal, V., Kumar, V.: Identity-based encryption with efficient revocation. In: Proceedings of the 15th ACM Conference on Computer and Communications Security, pp. 417–426. ACM Press, New York (2008)
11. Attrapadung, N., Imai, H.: Conjunctive Broadcast and Attribute-Based Encryption. In: Shacham, H., Waters, B. (eds.) Pairing 2009. LNCS, vol. 5671, pp. 248–265. Springer, Heidelberg (2009)
12. Attrapadung, N., Imai, H.: Attribute-Based Encryption Supporting Direct/Indirect Revocation Modes. In: Parker, M.G. (ed.) Cryptography and Coding 2009. LNCS, vol. 5921, pp. 278–300. Springer, Heidelberg (2009)
13. Boneh, D., Gentry, C., Waters, B.: Collusion Resistant Broadcast Encryption with Short Ciphertexts and Private Keys. In: Shoup, V. (ed.) CRYPTO 2005. LNCS, vol. 3621, pp. 258–275. Springer, Heidelberg (2005)

14. Staddon, A., Golle, P., Gagne, M., Rasmussen, P.: Content-driven access control system. In: Proceedings of the 7th symposium on Identity and trust on the Internet, pp. 26–35. ACM Press, New York (2008)
15. Lewko, A., Sahai, A., Waters, B.: Revocation systems with very small private keys. In: Proceedings of the 2010 IEEE Symposium on Security and Privacy, pp. 273–285. IEEE Press, New York (2010)
16. Yu, S.C., Wang, C., Ren, K., Lou, W.J.: Attribute based data sharing with attribute revocation. In: Proceedings of the 5th ACM Symposium on Information, Computer and Communications Security, pp. 261–270. ACM Press, New York (2010)
17. Beimel, A.: Secure Schemes for Secret Sharing and Key Distribution. PhD thesis, Israel Institute of Technology (1996)
18. Yamada, S., Attrapadung, N., Hanaoka, G., Kunihiro, N.: Generic Constructions for Chosen-Ciphertext Secure Attribute Based Encryption. In: Catalano, D., Fazio, N., Gennaro, R., Nicolosi, A. (eds.) PKC 2011. LNCS, vol. 6571, pp. 71–89. Springer, Heidelberg (2011)

Appendix A: DBDH Assumption

Decisional Bilinear Diffie-Hellman Assumption: Let $\mathbb{G}_1$ be a bilinear group of prime order p, given a vector of three random elements (g^a, g^b, g^c), We say that the DBDH assumption holds in $\mathbb{G}_1$ if no polynomial-time algorithm has a non-negligible advantage to distinguish $e(g,g)^{abc}$ from a random element in $\mathbb{G}_2$.

Appendix B: Security Proof of Theorem 1

Suppose there exists a polynomial-time adversary $\mathcal{A}$, that can attack our small universe scheme in the selective security model with advantage ε. We can build a simulator $\mathcal{B}$ that can solve the Decision n-BDHE problem with advantage $\varepsilon/2$. The simulation proceeds as follows:

The challenger chooses two groups $\mathbb{G}_1$ and $\mathbb{G}_2$ of prime order p, and randomly picks a generator g from $\mathbb{G}_1$, In addition, the challenger defines a bilinear map e:$\mathbb{G}_1 \times \mathbb{G}_1 \rightarrow \mathbb{G}_2$ and the user universe U=$\{1,2,...n\}$, the attribute universe N=$\{1,2,...,m\}$. The challenger sets:

$$Y = (g, g^s, g_1 = g^{\alpha}, g_2 = g^{\alpha^2}, ..., g_n = g^{\alpha^n}, g_{n+2} = g^{\alpha^{n+2}}, ..., g_{2n} = g^{\alpha^{2n}}).$$

Then the challenger flips a fair binary coin μ: If $\mu = 0$, the challenger set Z=$e(g_1, g_n)^s$; If $\mu = 1$, the challenger picks a random element Z from $\mathbb{G}_2$. Finally, the challenger gives (Y,Z) to the simulator $\mathcal{B}$. $\mathcal{B}$ proceeds as follows:

Init. The simulator $\mathcal{B}$ runs adversary $\mathcal{A}$. $\mathcal{A}$ selects an attribute set $\omega^* \subseteq N$ and a revocation list $R_j^* \subseteq U$ of attribute $j \in \omega^*$ that it wishes to be challenged upon.

Setup. The simulator $\mathcal{B}$ acts as follows:

1. Let $S_0 = U - R_j^*$, choose a random element $u_0 \in \mathbb{Z}_p^*$, then set:
$$g^\gamma = g^{u_0}(\prod_{i\in S_0} g_{n+1-i})^{-1};$$
2. Let $S_1 = U$, choose a random element $u_1 \in \mathbb{Z}_p^*$, then set:
$$g^\beta = g^{u_1}(\prod_{i\in S_1} g_i)^{-1};$$
3. For each attribute $i \in N$: if $i \in \omega^*$, it chooses a random value $r_{i,0} \in \mathbb{Z}_p^*$, and set $T_{i,0} = g^{r_{i,0}}$(thus,$t_{i,0} = r_{i,0}$); otherwise it chooses a random value $\eta_{i,0} \in \mathbb{Z}_p^*$, and sets $T_{i,0} = g^{\gamma\eta_{i,0}}$(thus, $t_{i,0} = \gamma\eta_{i,0}$);
4. Let $\omega' = \omega^* - \{j\}$, for each attribute $i \in N$: if $i \in \omega'$, it chooses a random value $r_{i,1} \in \mathbb{Z}_p^*$, and sets $T_{i,1} = g^{r_{i,1}}$(thus, $t_{i,1} = r_{i,1}$); otherwise it chooses a random value $\eta_{i,1} \in \mathbb{Z}_p^*$, and set $T_{i,1} = g^{\beta\eta_{i,1}}$(thus, $t_{i,1} = \beta\eta_{i,1}$);

Then $\mathcal{B}$ gives the public key $\boldsymbol{pk} = \{\mathbb{G}_1, \mathbb{G}_2, Y \setminus \{g^s\}, \{(T_{i,0}, T_{i,1})\}_{i\in N}, g^\gamma, g^\beta\}$ to the adversary $\mathcal{A}$. Note that, since g, α, u_0, u_1 are chosen uniformly at random, this public key has an identical distribution to that in the actual construction.

Phase1. At any time, the adversary $\mathcal{A}$ may make a private key extraction query of user $ID \in U$ with access tree $\mathbb{A}$, such that ω' doesn't satisfy the access tree $\mathbb{A}$, where $\omega' = \omega^*$ if $ID \notin R_j^*$, or $\omega' = \omega^* - \{j\}$ if $ID \in R_j^*$. The simulator $\mathcal{B}$ acts as follows to generate the private key $SK_{ID,\mathbb{A}}$:

1. When $ID \notin R_j^*$(in this case, we have $\omega' = \omega^*$):
   (a) For the root node r, choose two polynomials $q_{r,0}$ and $q_{r,1}$ of degree $v_r - 1$. The way to choose $q_{r,1}$ is the same to $q_{r,0}$, so we only describe how to choose $q_{r,0}$. First set $q_{r,0}(0) = \alpha^{ID}$. Denote $S_r = \{x\}_{x\in children(r)\wedge\Gamma_x(\omega')=1}$, For each node $x \in S_r$, select a random value k_x from $\mathbb{Z}_p^*$ and let $q_{r,0}(index(x)) = k_x$. Since ω' doesn't satisfy the access tree $\mathbb{A}$, we have $|S_r| < v_r$. Then select randomly other $v_r - |S_r| - 1$ points from $\mathbb{Z}_p^*$ to completely fix $q_{r,0}$; Note that, for the child node x of the root node r, if $\Gamma_x(\omega') = 1$, we know $q_{x,0}(0)$, if $\Gamma_x(\omega') = 0$, we can't get the value $q_{x,0}(0)$, but we can use the polynomial interpolation to compute $g^{q_{x,0}(0)}$.
   (b) Since the private key is generated in a top-down manner, according to the way of choosing $q_{r,0}$ and $q_{r,1}$, for every other node x in access tree $\mathbb{A}$: if $\Gamma_x(\omega') = 1$, we can get the value $q_{x,0}(0)$ and $q_{x,1}(0)$, then we can fix the $q_{x,0}, q_{x,1}$ as the actual construction; if $\Gamma_x(\omega') = 0$, we can get the value $g^{q_{x,0}}, g^{q_{x,1}}$(if $\Gamma_{parent(x)}(\omega') = 1$, we also know the value $q_{x,0}(0), q_{x,1}(0)$),then fix the $q_{x,0}, q_{x,1}$ as the root node.
   (c) For each node x of access tree $\mathbb{A}$, we define $Q_{x,0} = \gamma q_{x,0}, Q_{x,1} = \beta q_{x,1}$, thus $Q_{r,0} = \gamma\alpha^{ID}, Q_{r,1} = \beta\alpha^{ID}$. Then for each leaf node z, let $i = attr(z)$, we set:

$$D_{z,0} = \begin{cases} g^{\frac{Q_{z,0}(0)}{t_{i,0}}} = g^{\frac{\gamma q_{z,0}(0)}{r_{i,0}}} = (g^{\gamma})^{\frac{q_{z,0}(0)}{r_{i,0}}}, i \in \omega' \\ g^{\frac{Q_{z,0}(0)}{t_{i,0}}} = g^{\frac{\gamma q_{z,0}(0)}{\gamma \eta_{i,0}}} = (g^{q_{z,0}(0)})^{\frac{1}{\eta_{i,0}}}, i \notin \omega' \end{cases}$$

$$D_{z,1} = \begin{cases} g^{\frac{Q_{z,1}(0)}{t_{i,1}}} = g^{\frac{\beta q_{z,1}(0)}{r_{i,1}}} = (g^{\beta})^{\frac{q_{z,1}(0)}{r_{i,1}}}, i \in \omega' \wedge i \neq j \\ g^{\frac{Q_{z,1}(0)}{t_{i,1}}} = g^{\frac{\beta q_{z,1}(0)}{\beta \eta_{i,1}}} = (g^{q_{z,1}(0)})^{\frac{1}{\eta_{i,1}}}, i \notin \omega' \vee i = j \end{cases}$$

2. When $ID \in R_j^*$(in this case, we have $\omega' = \omega^* - \{j\}$), if ω^* doesn't satisfy the access tree $\mathbb{A}$, the simulation is exactly the same to the previous case, so we only concentrate on the simulation of the special case, in which ω^* satisfies the $\mathbb{A}$, but ω' doesn't satisfy the $\mathbb{A}$.
For $D_{z,1}$: Since ω' doesn't satisfy the $\mathbb{A}$, $\mathcal{B}$ simulates exactly the same with the previous case, and for each leaf node z, gives the $D_{z,1}$ (we assume that $i = attr(z)$):

$$D_{z,1} = \begin{cases} g^{\frac{Q_{z,1}(0)}{t_{i,1}}} = g^{\frac{\beta q_{z,1}(0)}{r_{i,1}}} = (g^{\beta})^{\frac{q_{z,1}(0)}{r_{i,1}}}, i \in \omega' \\ g^{\frac{Q_{z,1}(0)}{t_{i,1}}} = g^{\frac{\beta q_{z,1}(0)}{\beta \eta_{i,1}}} = (g^{q_{z,1}(0)})^{\frac{1}{\eta_{i,1}}}, i \notin \omega' \end{cases}$$

For $D_{z,0}$:
(a) For the root node r, choose a polynomial $q_{r,0}$ of degree $v_r - 1$. First set $q_{r,0}(0) = \alpha^{ID}$. Then select randomly other $v_r - 1$ points from $\mathbb{Z}_p^*$ to completely fix $q_{r,0}$; Note that, for the child x of the root r, we only know the value $g^{q_{x,0}(0)}$ by using the polynomial interpolation;
(b) For every other node x in access tree $\mathbb{A}$, the way to choose $q_{x,1}$ is the same to root node, and we know the value $g^{q_{x,0}(0)}$;
(c) For each node x in access tree $\mathbb{A}$, define $Q_{x,0} = \gamma q_{x,0}$. For root node r, we have:

$$g^{Q_{r,0}(0)} = (g^{u_0}(\prod_{i \in S_0} g_{n+1-i})^{-1})^{\alpha^{ID}} = (g_{ID})^{u_0}(\prod_{i \in S_0} g_{n+1-i+ID})^{-1})$$

Since $ID \in R_j^*$, we have $ID \notin S_0$, thus the term g_{n+1} will not exist in $(\prod_{i \in S_0} g_{n+1-i+ID})^{-1})$, and we can compute the value $g^{Q_{r,0}(0)}$ in polynomial time. Then for each node x in access tree $\mathbb{A}$, by using using the polynomial interpolation, we can also compute the value $g^{Q_{x,0}(0)}$;
(d) Then for each leaf node z, let $i = attr(z)$, we set:

$$D_{z,0} = \begin{cases} g^{\frac{Q_{z,0}(0)}{t_{i,0}}} = g^{\frac{Q_{z,0}(0)}{r_{i,0}}} = (g^{Q_{z,0}(0)})^{\frac{1}{r_{i,0}}}, i \in \omega^* \\ g^{\frac{Q_{z,0}(0)}{t_{i,0}}} = g^{\frac{\gamma q_{z,0}(0)}{\gamma \eta_{i,0}}} = (g^{q_{z,0}(0)})^{\frac{1}{\eta_{i,0}}}, i \notin \omega^* \end{cases}$$

Define L to be the set of leaves of access tree $\mathbb{A}$, $\mathcal{B}$ returns the user private key $SK_{ID,\mathbb{A}} = \{(D_{x,0}, D_{x,1})\}_{x \in L}$ to $\mathcal{A}$.

Challenge. The adversary $\mathcal{A}$ chooses two challenge messages $M_0, M_1 \in \mathbb{G}_2$ with equal length and sends to $\mathcal{B}$. $\mathcal{B}$ randomly chooses a bit $b \in \{0, 1\}$, and sets:

$$\begin{cases} C_0^* = ZM_b, \\ C_1^* = g^s, \\ C_{2,0}^* = \{T_{i,0}^s = g^{t_{i,0}s} = (g^s)^{r_{i,0}}\}_{i\in\omega^*}, \\ C_{2,1}^* = \{T_{i,0}^s = g^{t_{i,1}s} = (g^s)^{r_{i,1}}\}_{i\in\omega'}, \\ C_{3,0}^* = (g^\gamma \prod\limits_{i\in S_0} g_{n+1-i})^s = (g^s)^{u_0}, \\ C_{3,1}^* = (g^\beta \prod\limits_{i\in S_1} g_i)^s = (g^s)^{u_1} \end{cases}$$

The $\mathcal{B}$ sends the challenge ciphertext $\boldsymbol{C}^* = \{\omega^*, R_j^*, C_0^*, C_1^*, C_{2,0}^*, C_{2,1}^*, C_{3,0}^*, C_{3,1}^*\}$ to the adversary $\mathcal{A}$. If $\mu = 0$ then $Z = e(g_1, g_n)^s$, since s is chosen randomly from $\mathbb{G}_2$ by the challenger, the challenge ciphertext $\boldsymbol{C}^*$ is a valid random encryption of message M_b. If $\mu = 1$, then Z is a random element of $\mathbb{G}_2$, so C_0^* is also a random element of $\mathbb{G}_2$ from the adversary's view and contains no information of M_b.

Phase2. Phase1 is repeated.

Guess. The adversary outputs the guess $b^{'}$ of b.
Let $\mu^{'}$ be the guess of μ by the simulator $\mathcal{B}$:
If $b^{'} = b$, $\mathcal{B}$ outputs $\mu^{'} = 0$;
If $b^{'} \neq b$, $\mathcal{B}$ outputs $\mu^{'} = 1$;
Then we analysis the advantage of $\mathcal{B}$ to solve the Decision n-BDHE problem.

1. when $\mu = 0$: In this case, $Z = e(g_1, g_n)^s$, so the $\boldsymbol{C}^*$ is a valid ciphertext, then the adversary $\mathcal{A}$ sees an encryption of M_b. By definition, the advantage of $\mathcal{A}$ is ϵ, thus, we can conclude that $Pr[b^{'} = b|\mu = 0] = \epsilon$ - 1/2, then we have:
$$Pr[\mu^{'} = \mu|\mu = 0] = Pr[b^{'} = b|\mu = 0] = \epsilon - 1/2;$$
2. when $\mu = 1$: In this case, Z is a random element of $\mathbb{G}_2$, and the adversary gains no information of b. Therefore, we have $Pr[b^{'} = b|\mu = 1] = Pr[b^{'} \neq b]$ $= 1/2$, then we conclude that:
$$Pr[\mu^{'} = \mu|\mu = 1] = Pr[b^{'} \neq b|\mu = 1] = 1/2;$$
Finally, we have:
$$\begin{aligned} Pr[\mu^{'} = \mu] &= Pr[\mu^{'} = \mu|\mu = 0]Pr[\mu = 0] + Pr[\mu^{'} = \mu|\mu = 1]Pr[\mu = 1] \\ &= (\epsilon - 1/2) \cdot 1/2 + 1/2 \cdot 1/2 \\ &= \epsilon/2 \end{aligned}$$

This concludes the proof of Theorem 1.

Appendix C: Security Proof of Theorem 2

Suppose there exists a polynomial-time adversary $\mathcal{A}$, that can attack our large universe scheme in the selective security model with advantage ε. We can build a simulator $\mathcal{B}$ that solve the Decision n-BDHE problem with advantage $\varepsilon/2$. The simulation proceeds as follows:

The challenger chooses two groups $\mathbb{G}_1$ and $\mathbb{G}_2$ of prime order p, and randomly picks a generator g from $\mathbb{G}_1$, In addition, the challenger defines a bilinear map $e:\mathbb{G}_1\times\mathbb{G}_1\rightarrow\mathbb{G}_2$ and the user universe $U=\{1,2,...n\}$, The challenger sets:

$$Y=(g, g^s, g_1 = g^{\alpha}, g_2 = g^{\alpha^2}, ..., g_n = g^{\alpha^n}, g_{n+2} = g^{\alpha^{n+2}}, ..., g_{2n} = g^{\alpha^{2n}}).$$

Then the challenger flips a fair binary coin μ: If $\mu = 0$, the challenger sets $Z=e(g_1, g_n)^s$; If $\mu = 1$, the challenger picks a random element Z from $\mathbb{G}_2$. Finally, the challenger gives (Y,Z) to the simulator $\mathcal{B}$. $\mathcal{B}$ proceeds as follows:

Init. The simulator $\mathcal{B}$ runs adversary $\mathcal{A}$. $\mathcal{A}$ selects an attribute set $\omega^* \in \mathbb{Z}_p^m$ and a revocation list $R_j^* \subseteq U$ of attribute $j \in \omega^*$ that it wishes to be challenged upon.

Setup. The simulator $\mathcal{B}$ acts as follows:

1. Let $S_0 = U - R_j^*$, choose a random element $u_0 \in \mathbb{Z}_p^*$, then set:
$$g^{\gamma} = g^{u_0}(\prod_{i\in S_0} g_{n+1-i})^{-1};$$
2. Let $S_1 = U$, choose a random element $u_1 \in \mathbb{Z}_p^*$, then set:
$$g^{\beta} = g^{u_1}(\prod_{i\in S_1} g_i)^{-1};$$
3. Choose two random m degree polynomial $f_0(x), f_1(x)$, define $\omega' = \omega^* - \{j\}$, then calculate two polynomials k_0, k_1 as follows: set $k_0(x) = -x^m$ for all $x \in \omega^*$ and $k_0(x) \neq -x^m$ for some other x. set $k_1(x) = -x^m$ for all $x \in \omega'$ and $k_1(x) \neq -x^m$ for some other x;
4. For each $i \in \{1, 2, ..., m + 1\}$, calculate $h_{i,0} = g^{f_0(i)}(g^{\gamma})^{k_0(i)}, h_{i,1} = g^{f_1(i)}$ $(g^{\beta})^{k_1(i)}$, so the $T_0(x), T_1(x)$ are fixed.

Then $\mathcal{B}$ gives the public key $\boldsymbol{pk} = \{\mathbb{G}_1, \mathbb{G}_2, Y \setminus \{g^s\}, \{(h_{i,0}, h_{i,1})\}_{i\in\{1,2,...,m+1\}}, g^{\gamma}, g^{\beta}\}$ to the adversary $\mathcal{A}$. Note that, since $g, \alpha, u_0, u_1, f(x)$ are chosen uniformly at random, this public key has an identical distribution to that in the actual construction.

Phase1. At any time, the adversary $\mathcal{A}$ may make a private key extraction query of user $ID \in U$ with access tree $\mathbb{A}$, such that ω' doesn't satisfy the access tree $\mathbb{A}$, where $\omega' = \omega^*$ if $ID \notin R_j^*$, or $\omega' = \omega^* - \{j\}$ if $ID \in R_j^*$. The simulator $\mathcal{B}$ acts as follows to generate the private key $SK_{ID,\mathbb{A}}$:

1. When $ID \notin R_j^*$(in this case, we have $\omega' = \omega^*$):
   (a) The way to fix the polynomials $Q_{x,0} = \gamma q_{x,0}, Q_{x,1} = \beta q_{x,1}$ for each node x is same to that in the proof of Theorem 1 (See Appendix B).
   (b) For each leaf node z, let $i = attr(z)$:

i. If $i \in \omega'$, then we know the values of $q_{z,0}(0), q_{z,1}(0)$. Choose two random values $r_{z,0}, r_{z,1}$ from $\mathbb{Z}_p^*$ and set:

$$\begin{cases} R_{z,0} = g^{r_{z,0}} \\ R_{z,1} = g^{r_{z,1}} \\ D_{z,0} = g^{Q_{z,0}(0)} \cdot T_0(i)^{r_{z,0}} = (g^{\gamma})^{q_{z,0}(0)} \cdot T_0(i)^{r_{z,0}} \\ D_{z,1} = g^{Q_{z,1}(0)} \cdot T_1(i)^{r_{z,1}} = (g^{\beta})^{q_{z,1}(0)} \cdot T_0(i)^{r_{z,1}} \end{cases}$$

ii. If $i \notin \omega'$, then we only know the values of $g^{q_{z,0}(0)}, g^{q_{z,1}(0)}$. Choose two random values $r'_{z,0}, r'_{z,1}$ from $\mathbb{Z}_p^*$, let $r_{z,0} = \frac{-q_{z,0}(0)}{i^m + k_0(i)} + r'_{z,0}, r_{z,1} = \frac{-q_{z,1}(0)}{i^m + k_1(i)} + r'_{z,1}$, then set:

$$\begin{cases} R_{z,0} = g^{r_{z,0}} = g^{\frac{-q_{z,0}(0)}{i^m+k_0(i)} + r'_{z,0}} = (g^{q_{z,0}(0)})^{\frac{-1}{i^m+k_0(i)}} g^{r'_{z,0}} \\ R_{z,1} = g^{r_{z,1}} = g^{\frac{-q_{z,1}(0)}{i^m+k_1(i)} + r'_{z,1}} = (g^{q_{z,1}(0)})^{\frac{-1}{i^m+k_1(i)}} g^{r'_{z,1}} \\ D_{z,0} = g^{Q_{z,0}(0)} \cdot T_0(i)^{r_{z,0}} = (g^{q_{z,0}(0)})^{\frac{-f_0(i)}{i^m+k_0(i)}} T_0(i)^{r'_{z,0}} \\ D_{z,1} = g^{Q_{z,1}(0)} \cdot T_1(i)^{r_{z,1}} = (g^{q_{z,1}(0)})^{\frac{-f_1(i)}{i^m+k_1(i)}} T_1(i)^{r'_{z,1}} \end{cases}$$

2. When $ID \in R_j^*$(in this case, we have $\omega' = \omega^* - \{j\}$), if ω^* doesn't satisfy the access tree $\mathbb{A}$, the simulation is exactly the same to the previous case, so we only concentrate on the simulation of the special case, in which ω^* satisfies the $\mathbb{A}$, but ω' doesn't satisfy the $\mathbb{A}$.
For $D_{x,1}$: Since ω' doesn't satisfy the $\mathbb{A}$, $\mathcal{B}$ simulates exactly the same with the previous case, and for each leaf node z, gives the $D_{z,1}, R_{z,1}$;
For $D_{x,0}$: The simulation of this case is same to that in the proof of of Theorem 1 (See Appendix B), and for each node, we know the values of $g^{Q_{x,0}(0)}, g^{Q_{x,1}(0)}$. For each leaf node z, let $i = attr(z)$, then choose two random values $r_{z,0}, r_{z,1}$ from $\mathbb{Z}_p^*$ and set:

$$\begin{cases} R_{z,0} = g^{r_{z,0}} \\ R_{z,1} = g^{r_{z,1}} \\ D_{z,0} = g^{Q_{z,0}(0)} \cdot T_0(i)^{r_{z,0}} = (g^{\gamma})^{q_{z,0}(0)} \cdot T_0(i)^{r_{z,0}} \\ D_{z,1} = g^{Q_{z,1}(0)} \cdot T_1(i)^{r_{z,1}} = (g^{\beta})^{q_{z,1}(0)} \cdot T_0(i)^{r_{z,1}} \end{cases}$$

Define L to be the set of leaves of access tree $\mathbb{A}$, $\mathcal{B}$ returns the user private key $SK_{ID,\mathbb{A}} = \{(D_{x,0}, D_{x,1}, R_{x,0}, R_{x,1})\}_{x \in L}$ to $\mathcal{A}$.

Challenge. The adversary $\mathcal{A}$ chooses two challenge messages $M_0, M_1 \in \mathbb{G}_2$ with equal length and sends to $\mathcal{B}$. $\mathcal{B}$ randomly chooses a bit $b \in \{0,1\}$, and sets:

$$\begin{cases} C_0^* = ZM_b, \\ C_1^* = g^s, \\ C_{2,0}^* = \{T_0(i)^s = (g^{\gamma i^m} g^{f_0(i)+\gamma k_0(i)})^s = (g^s)^{f_0(i)}\}_{i \in \omega^*}, \\ C_{2,11}^* = \{T_1(i)^s = (g^{\beta i^m} g^{f_1(i)+\beta k_1(i)})^s = (g^s)^{f_1(i)}\}_{i \in \omega'}, \\ C_{3,0}^* = (g^{\gamma} \prod_{i \in S_0} g_{n+1-i})^s = (g^s)^{u_0}, \\ C_{3,1}^* = (g^{\beta} \prod_{i \in S_1} g_i)^s = (g^s)^{u_1} \end{cases}$$

The $\mathcal{B}$ sends the challenge ciphertext $\boldsymbol{C}^* = \{\omega^*, R_j^*, C_0^*, C_1^*, C_{2,0}^*, C_{2,1}^*, C_{3,0}^*, C_{3,1}^*\}$ to the adversary $\mathcal{A}$. If $\mu = 0$ then $Z = e(g_1, g_n)^s$, since s is chosen randomly from $\mathbb{G}_2$ by the challenger, the challenge ciphertext $\boldsymbol{C}^*$ is a valid random encryption of message M_b. If $\mu = 1$, then Z is a random element of $\mathbb{G}_2$, so C_0^* is also a random element of $\mathbb{G}_2$ from the adversary's view and contains no information of M_b.

Phase2. Phase1 is repeated.

Guess. The adversary outputs the guess b' of b.
Let μ' be the guess of μ by the simulator $\mathcal{B}$:
If $b' = b$, $\mathcal{B}$ outputs $\mu' = 0$;
If $b' \neq b$, $\mathcal{B}$ outputs $\mu' = 1$;
Then we analysis the advantage of $\mathcal{B}$ to solve the Decision n-BDHE problem.

1. when $\mu = 0$: In this case, $Z = e(g_1, g_n)^s$, so the $\boldsymbol{C}^*$ is a valid ciphertext, then the adversary $\mathcal{A}$ sees an encryption of M_b. By definition, the advantage of $\mathcal{A}$ is ϵ, thus, we can conclude that $Pr[b' = b|\mu = 0] = \epsilon$ - 1/2, then we have:
$$Pr[\mu' = \mu|\mu = 0] = Pr[b' = b|\mu = 0] = \epsilon - 1/2;$$
2. when $\mu = 1$: In this case, Z is a random element of $\mathbb{G}_2$, and the adversary gains no information of b. Therefore, we have $Pr[b' = b|\mu = 1] = Pr[b' \neq b]$ $= 1/2$, then we conclude that:
$$Pr[\mu' = \mu|\mu = 1] = Pr[b' \neq b|\mu = 1] = 1/2;$$
Finally, we have:
$$\begin{aligned} Pr[\mu' = \mu] &= Pr[\mu' = \mu|\mu = 0]Pr[\mu = 0] + Pr[\mu' = \mu|\mu = 1]Pr[\mu = 1] \\ &= (\epsilon - 1/2) \cdot 1/2 + 1/2 \cdot 1/2 \\ &= \epsilon/2 \end{aligned}$$

This concludes the proof of Theorem 2.

A Note on (Im)Possibilities of Obfuscating Programs of Zero-Knowledge Proofs of Knowledge

Ning Ding and Dawu Gu

Department of Computer Science and Engineering
Shanghai Jiao Tong University, China
{dingning,dwgu}@sjtu.edu.cn

Abstract. Program obfuscation seeks efficient methods to write programs in an incomprehensible way, while still preserving the functionalities of the programs. In this paper we continue this research w.r.t. zero-knowledge proofs of knowledge. Motivated by both theoretical and practical interests, we ask if the prover and verifier of a zero-knowledge proof of knowledge are obfuscatable. Our answer to this question is as follows. First we present two definitions of obfuscation for interactive probabilistic programs and then achieve the following results:

1. W.r.t. an average-case virtual black-box definition, we achieve some impossibilities of obfuscating provers of zero-knowledge and witness-indistinguishable proofs of knowledge. These results state that the honest prover with an instance and its witness hardwired of any zero-knowledge (or witness-indistinguishable) proof of knowledge of efficient prover's strategy is unobfuscatable if computing a witness (or a second witness) for this instance is hard. Moreover, we extend these results to t-composition setting and achieve similar results. These results imply that if an adversary obtains the prover's code (e.g. stealing a smartcard) he can indeed learn some knowledge from it beyond its functionality no matter what measures the card designer may use for resisting reverse-engineering.
2. W.r.t. a worst-case virtual black-box definition, we provide a possibility of obfuscating the honest verifier (with the public input hardwired) of Blum's 3-round zero-knowledge proof for Hamilton Cycle. Our investigation is motivated by an issue of *privacy protection* (e.g., if an adversary controls the verifier, he can obtain all provers' names and public inputs. Thus the provers' privacy may leak). We construct an obfuscator for the verifier, which implies that even if an adversary obtains the verifier's code, he cannot learn any knowledge, e.g. provers' names, from it. Thus we realize the anonymity of provers' accesses to the verifier and thus solve the issue of privacy protection.

Keywords: Program Obfuscation, Zero Knowledge, Witness Indistinguishability, Proofs of Knowledge.

D. Lin, G. Tsudik, and X. Wang (Eds.): CANS 2011, LNCS 7092, pp. 292–311, 2011.

1 Introduction

In recent years, cryptography community has focused on a fascinating research line of program obfuscation. Loosely speaking, obfuscating a program P is to construct a new program which can preserve the functionality of P, but its code is fully "unintelligent". Any adversary can only use the functionality of P and cannot learn anything more than this, i.e. cannot reverse-engineering nor understand it. In other words, an obfuscated program should not reveal anything useful beyond executing it.

[2] formalized the notion of obfuscation through a simulation-based definition called the virtual black-box property, which says that every adversary has a corresponding simulator that emulates the output of the adversary given oracle (i.e. black-box) access to the same functionality being obfuscated. Following [2], many works focused on how to obfuscate different cryptographic functionalities. Among them, there are some negative results, e.g. [2,19,22]. [2] showed there doesn't exist any general obfuscation method for all programs. [19,22] showed that this negative result holds even w.r.t. some other definitions of obfuscation. On the other hand, some obfuscation methods have been proposed for several cryptographic functionalities. [26,24] demonstrated how to securely obfuscate re-encryption and encrypted signature respectively. [13] showed that with fully homomorphic encryption in e.g. [17], a category of functionalities, which can be characterized as first secret operation and then public encryption, are obfuscatable. Some works e.g. [27,25,4,8,11,10,9] focused on obfuscation for a basic and simple primitive, i.e. (multiple-bit) point functions, traditionally used in some password based identification systems, and its applications in constructing encryption and signature schemes of strong security.

We continue this research line w.r.t. zero-knowledge [21] (as well as witness-indistinguishable [16]) protocols in this paper. Motivated by theoretical and practical interests, we ask whether or not the prover and verifier of a zero-knowledge (or witness-indistinguishable) proof or argument of knowledge are obfuscatable. Before demonstrating our motivation, we first present some works related to this issue. [23] first addressed this issue, which gave some definitions of obfuscators (which security allows the distinguisher to obtain the obfuscated program, like a stronger definition in [25], but only requires the indistinguishability holds w.r.t. an individual adversary). [23] showed that there exists a cheating verifier for a constant-round non-black-box zero-knowledge protocol which is unobfuscatable, while the first such protocol was later constructed by [1]. Actually the non-black-box simulator in [1] doesn't reverse-engineering verifiers' codes. [20] investigated obfuscation for protocols based on tamper-proof hardware, and showed that with the help of hardware, all programs including provers are one-time obfuscatable.

We now turn to our motivation. As we know, zero-knowledge proofs of knowledge are a secure way to implement identification as well as other tasks. We use identification as an example to illustrate the motivation. Briefly, an identification scheme consists of a trusted party and multi-clients and multi-servers. The trusted party generates a verifiable public input (a.k.a. statement or instance, e.g. a directed graph or a modular square) and a witness (e.g. a Hamilton cycle or a modular square root) for each client. A client's name and its public input constitute a record and the trusted party maintains all records. Each client stores his witness secretly and when he needs to identify himself to a server, he proves to the server the knowledge of the witness via the identification scheme.

We now first present the motivation of obfuscating provers (with public inputs and witnesses hardwired). Consider the applications of smartcards, in which a holder of a smartcard may use his card to identify himself to a server via some zero-knowledge proof of knowledge where the card runs as prover and the server runs as verifier. If an adversary obtains the card, he can perform side channel attacks, physical attacks and white-box cryptoanalysis etc. to recover the witness (i.e. secret) in the prover's program. Thus card designers need to develop many countermeasures in designing prover's programs against such attacks. Since these approaches are heuristic without rigorous security proofs, they are usually broken by new developed attacks. Thus it is very interesting to ask if there is a way to write the prover's program such that any adversary cannot break it even if he possesses the actual program. Namely, can we obfuscate the prover? We think that this question is of both theoretical and practical interests.

We then present the motivation of obfuscating verifiers. Note that the impossibility in [23] says that a cheating verifier with its random coins hardwired is unobfuscatable, while herein we investigate obfuscation for honest verifiers without random coins hardwired. Thus an obfuscated verifier in this paper, if possible, is still a PPT algorithm, which needs fresh coins in execution. Due to this motivation, a question immediately arises. That is, a honest verifier doesn't contain any secret and is publicly available, so why do we need to obfuscate it? Our answer to this question is that what we need to obfuscate is a honest verifier with a *public input* hardwired rather than the verifier's strategy only, and the obfuscation for such verifiers is meaningful for an issue of *privacy protection*. (Some works e.g. [12] discussed the input privacy of zero-knowledge protocols, which especially work as sub-protocols of large cryptographic protocols. We here only concern the privacy issue for verifiers of zero-knowledge protocols in the stand alone setting and obfuscation seems to be a conceptually direct solution.)

For instance, consider the case of ID cards, which have been wildly used in our society. However, some secret governmental organizations may need to read data from cards for some secret goals. Before their card readers can read data from cards, they usually need to pass identification. In this scenario the readers of the organizations act as prover and cards act as verifier. Thus cards should store all legal clients' (i.e. organizations) names and their public inputs, while on the contrary these organizations may not want their names stored in any card since they would like to access cards anonymously without being identified. If cards indeed store their names, a holder (or an adversary) can of course extract these names from the verifier's program in his own card, which frustrates the anonymity of their accesses. Thus, to realize the anonymity for them, it is very natural to ask if we can obfuscate the verifier's program (incorporated with clients' names and public inputs) before an ID card is issued officially such that any holder or adversary cannot find this information from the card. More discussion of this motivation can be referred to Section 5.1. In this paper we are interested in finding the answers to these questions, and correspondingly achieve the following results.

1.1 Our Results

As we see, most known works on obfuscation focused upon non-interactive programs, while the main objects in this paper, i.e. provers and verifiers, are interactive. Nevertheless, there are still some works focused upon interactive programs. For instance, [23] presented some definitions for interactive deterministic programs. In this paper we first discuss how to combine the known definitions with the probabilistic property of provers and verifiers to define obfuscation for interactive probabilistic programs. We then present two definitions. Both of these two definitions require a functionality property and a virtual black-box (VBB) property. They differs mainly in the characterization of the VBB property.

The first one adopts the average-case virtual black-box (ACVBB) requirement introduced by [25,26], which is proposed for investigating obfuscation for provers. Note that to use zero-knowledge proofs of knowledge, a general paradigm is that a trusted party first samples a random hard public input (i.e. instance) and a witness for each prover, which then keeps its witness secret, and later a prover proves to a verifier the knowledge of the witness. Since the public input and witness are chosen randomly, it is natural to require that the prover's program with the *random* public input and witness hardwired satisfies ACVBB only. For this consideration it is reasonable to adopt this definition when investigating the obfuscation for provers.

The second definition adopts the worst-case VBB requirement introduced by [7,8,29], which is proposed for investigating obfuscation for verifiers. In this case, our goal is to hide all provers' names and public inputs in the verifier's program. Although the public inputs are usually randomly chosen, provers' names are not random. Thus the worst-case VBB requirement is more suitable when investigating obfuscation for verifiers.

Based on the ACVBB definition, we achieve some impossibilities of obfuscating honest provers of zero-knowledge and witness-indistinguishable (WI) proofs of knowledge. These results state that the prover with a public input and its witness hardwired of any zero-knowledge (or witness-indistinguishable) proof of knowledge of efficient prover's strategy is unobfuscatable if computing a witness (or a second witness) for the public input is hard. Moreover, we extend these results to t-composition setting and achieve similar results. These results imply that if an adversary obtains the prover's code (e.g. stealing a smartcard) he can indeed learn some knowledge beyond its functionality from it no matter what measures the card designer may use for resisting reverse-engineering. Note that this result doesn't contradict the general obfuscation in [20] since [20] assumed the tamper-proof hardware.

Based on the worst-case VBB definition, we provide a possibility of obfuscating the honest verifier (with any public input hardwired) of Blum's 3-round zero-knowledge proof for Hamilton Cycle [5]. We construct an obfuscation for the verifier, which implies that even if an adversary obtains the verifier's code, he cannot learn any knowledge, e.g. a prover's name, from it. Thus we realize the anonymity of the prover's access to the verifier and thus solve the issue of privacy protection. We further present some second-level considerations, e.g. achieving negligible soundness error, extending the possibility to multi-prover setting etc. Lastly, we point out that the approach for obfuscating the verifier is not suitable for obfuscating verifiers of other known zero-knowledge protocols e.g. Goldreich et al.'s proof for Graph 3-colorability [18]. Note that this result doesn't contradict the impossibility in [23], which showed a deterministic cheating verifier (without any public input hardwired) of some non-black-box protocol is unobfuscatable, while our possibility says that the honest verifier of Blum's (black-box) proof with any public input hardwired is obfuscatable and the obfuscation is still a PPT algorithm.

1.2 Organizations

The rest of the paper is arranged as follows. In Section 2, we present the preliminaries for this paper. In Section 3 we present the two definitions of obfuscation for interactive probabilistic programs. In Section 4 we present the impossibilities of obfuscating honest provers of zero-knowledge and witness-indistinguishable protocols. In Section 5 we present an obfuscation for the honest verifier of Blum's proof for Hamilton Cycle.

2 Preliminaries

For space limitations we assume familiarity with the notions of negligible functions (we use $\mathsf{neg}(n)$ to denote an unspecified negligible function), computational indistinguishability, interactive proofs [21] and arguments [6], and show the definitions of the following notions.

2.1 Point Functions and Their Obfuscation

A point function, $I_x : \{0,1\}^{|x|} \to \{0,1\}$, outputs 1 if and only if its input matches x, i.e., $I_x(y) = 1$ iff $y = x$, and outputs 0 otherwise. Let $\mathcal{I}_n$ denote the family of all I_x where $|x| = n$. [7,8,29] showed some constructions of obfuscation for $\mathcal{I}_n$ based on some non-standard complexity assumptions w.r.t. the following definition.

Definition 1. *Let $\mathcal{F}$ be any family of functions. A PPT, O, is called an obfuscator of $\mathcal{F}$, if:*

Approximate Functionality. *For any $f \in \mathcal{F}$: $\Pr[\exists x, O(F)(x) \neq F(x)]$ is negligible, where the probability is taken over all choices of O's coins.*

Polynomial Slowdown. *There is a polynomial p such that, for any $f \in \mathcal{F}$, $O(F)$ runs in time at most $p(T_f)$, where T_f is the worst-case running time of f.*
VBB. *For any non-uniform PPT A and any polynomial p, there exists a non-uniform PPT S such that for any $f \in \mathcal{F}$ and sufficiently large n: $|\Pr[A(O(f)) = 1] - \Pr[S^f(1^n) = 1]| \leq 1/p(n)$.*

2.2 Zero-Knowledge

Zero-knowledge was introduced by [21]. Usually, it is defined as that the view of any verifier in a real interaction can be simulated by the simulator. Here we adopt an equivalent definition for this paper which says the verifier's output in a real interaction can be simulated by the simulator, as follows.

Definition 2. *Let $L = L(R)$ be some language and let (P, V) be an interactive proof or argument for L. We say (P, V) is zero-knowledge if there exists a probabilistic polynomial-time algorithm, called simulator, such that for every polynomial-sized circuit V^* and every $(x, w) \in R$, the following two probability distributions are computationally indistinguishable:*
1. V^*'s output in the real execution with $P(x, w)$.
2. The simulator's output on input (x, V^*).

Concurrent zero-knowledge [14] is a generalization of zero-knowledge, which says that for any verifier taking part in multiple sessions there exists a simulator which output is indistinguishable from the verifier's output in the multiple sessions. The detailed definition can be found in [14].

2.3 Witness Indistinguishability

Witness indistinguishability is a weaker property than zero-knowledge, introduced by [16]. In a witness indistinguishable proof system if both w_1 and w_2 are witnesses that $x \in L$, then it is infeasible for the verifier to distinguish whether the prover used w_1 or w_2 as auxiliary input. We adopt the following definition.

Definition 3. *Let $L = L(R)$ be some language and (P, V) be a proof or argument system for L. We say that (P, V) is witness indistinguishable if for any polynomial-sized circuit V^*, any x, w_1, w_2 where $(x, w_1) \in R$ and $(x, w_2) \in R$, it holds that V^*'s output in the interaction with $P(x, w_1)$ is computationally indistinguishable from V^*'s output in the interaction with $P(x, w_2)$.*

It is shown in [16] that witness indistinguishability is preserved under concurrent composition.

2.4 Proofs of Knowledge

We adopt the following definition of proofs of knowledge shown in [15,3].

Definition 4. *Let $L = L(R)$ and let (P, V) be a proof (argument) system for L. We say that (P, V) is a proof (argument) of knowledge for L if there exists a probabilistic polynomial-time algorithm E (called the knowledge extractor) such that for every polynomial-sized prover P^* and for every $x \in \{0, 1\}^n$, if we let p' denote the probability that V accepts x when interacting with P^*, then $\Pr[E(P^*, x) \in R(x)] \geq p'(n) - \mathsf{neg}(n)$.*

3 Definitions of Obfuscation for Interactive Probabilistic Programs

In this section we show the two definitions of obfuscation for interactive probabilistic programs. Basically, these definitions follow the paradigm in the known definitions by requiring the functionality and the (AC)VBB properties. In Section 3.1 we present the considerations in formalizing these two properties and in Section 3.2 we give the definitions.

3.1 Considerations

We first show our consideration in formalizing the functionality property. Recall that [23] presented some definitions for interactive deterministic programs, which require that an obfuscation of a program is of same functionality with it. First notice that an interactive program, on a message sequence, may generate a communicated message and a local output (if there is no such message or output, let it be empty). Thus the functionality property in [23] says that on input a same message sequence, the program and its obfuscation generate same communicated messages and same local outputs (if the message sequence is illegal, the communicated message and local output can be $\perp$).

For interactive probabilistic programs, we could also adopt a probabilistic variant of this functionality property. That is, we could require that on input a fixed same message sequence, a program and its obfuscation should generate identically distributed communicated messages and identically distributed local outputs. Actually, this is indeed the main idea in formalizing the functionality property in this work. But we relax it in the two definitions in different ways.

First, we can relax the exact identical distribution requirement to a computational indistinguishability requirement. Informally, let $f \in \mathcal{F}$ denote an interactive probabilistic program and assume O is a possible obfuscator for $\mathcal{F}$. Our relaxed requirement says that for any PPT machine M, in the executions of $M^{<f>}$ and $M^{<O(f)>}$, the outputs of M in the two executions are computationally indistinguishable and f's and $O(f)$'s local outputs (which M cannot access) are also computationally indistinguishable. We will adopt this relaxation in Definition 6 for investigating obfuscation for provers, which can make our impossibilities even stronger.

Here $< f >$ (resp. $< O(f) >$) denotes the (probabilistic) functionality of f (resp. $O(f)$) and any machine cannot access its internal state (including coins). The notation of $M^{\langle f \rangle}$ (resp. $M^{<O(f)>}$) denotes a joint computation between M and f (resp. $O(f)$). We introduce this new notation in order to distinguish it from some known notations in literature, e.g. A^I where A is a PPT machine and I is a point function, or S'^V where S' is a simulator for proving zero-knowledge and V is a verifier, or E^P where E is a knowledge extractor and P is a valid prover. The formal definition of this joint computation is shown in Definition 5.

Second, we can also adopt a similar relaxation in e.g. [8] by requiring that $O(f)$ and f satisfy the identical distribution requirement except for a negligible fraction of O's coins, i.e. the approximate functionality (for interactive probabilistic programs). Actually, we will adopt this relaxation in Definition 7 for investigating obfuscation for verifiers.

We now turn to our consideration in formalizing the (AC)VBB property. Informally, we require that for any adversary A, there is a PPT simulator S satisfying that $S^{<f>}$ can output a fake program such that A cannot tell it from $O(f)$ (w.r.t. ACVBB in Definition 6) or what A can learn from $O(f)$ is also computable by $S^{<f>}$ (w.r.t. VBB in Definition 7). Notice that here we also employ the new notation of $S^{<f>}$. Thus all that is left is to characterize the joint computation of $S^{<f>}, M^{<f>}, M^{<O(f)>}$, as follows.

We illustrate this w.r.t. $S^{<f>}$ and then $M^{<f>}, M^{<O(f)>}$ can be defined similarly. This is actually the issue how to characterize the black-box access to f. Notice that if f were deterministic, f's response to a same S's query, which can be either a message or a message sequence, would be always same. But the situation becomes complex if f is interactive and probabilistic. We think that a reasonable characterization that S can access oracle f which conforms to the intuition of black-box access to f is as follows. First, S can play a consecutive interaction with f, as a real interaction. In each step, f on receiving a new message, computes the response and the local output using fresh coins. S can either play with f throughout the whole interaction, or cancel the interaction by sending an invalid message to f which then aborts the current interaction. Notice that f is a stateful machine in a consecutive interaction. By stateful, we mean that f can preserve its current internal state for later computation. When this interaction is finished or aborted, f cleans its internal state and is reset to the initial state. Second, after one interaction, S may invoke a new interaction with f, which is also reset to the initial state and computes the message with the fresh coins for its first step.

(We remark that the access of S to f in $S^{<f>}$ should be distinguished from the access of a black-box simulator for proving zero-knowledge to a verifier. Notice that this simulator is allowed to obtain the verifier's code and the term of "black-box" only means that the simulator doesn't reverse-engineering the verifier, but it can access the verifier's internal state and accordingly the simulator can rewind the verifier etc. It can be seen that such kind of access requires the knowledge of the verifier's program instead of the knowledge of the verifier's input-output behavior only. In $S^{<f>}$, S cannot behave like this.)

Thus we can see in each S's access to f, S should specify the way it tries to access f. For simplicity of characterizing $S^{<f>}$, we assume that in each access S should additionally send a bit b indicating if this access is for a consecutive step of the current interaction or for the first step of a new interaction. More concretely, each S's access to f consists of a message m and a bit b. If m is for a consecutive step of the current interaction, S resets $b = 0$ and requests that f responds to m based on its current internal state. If m is for the first step of a new interaction (m can be arbitrary if it is f that sends the first message), S sets $b = 1$ and requests that f responds to m where f is reset to the initial state. Formally, our characterization of $S^{<f>}$ is shown as follows.

Definition 5. *(characterization of $S^{<f>}$) For any two interactive probabilistic machines S, f, $S^{<f>}$ refers to the following joint computation between S and f: S starts a local computation (with an input) and then during the computation it needs to access f for responses. Each S's access to f consists of a message m and a bit $b \in \{0, 1\}$. On receiving (m, b), f responds as follows:*

1. *if $b = 0$: f continues the interaction with it current internal state, computes the message corresponding to m and the local output where it tosses fresh coins if needed, and responds with the message. If this interaction is finished after this step, f is reset to the internal state.*
2. *if $b = 1$: f is reset to the internal state and computes the message corresponding to m and the local output where it tosses fresh coins if needed, and responds with the message. If this interaction is finished after this step, f is reset to the internal state.*
3. *if $b \notin \{0, 1\}$: f responds with $\perp$ and is reset to the internal state.*

When receiving f's response, S proceeds to its local computation until the next access to f.

3.2 Definitions

Now we present our definitions. Since our objects in this work are provers and verifiers which are programs with some inputs hardwired, it is convenient to use circuits to represent such programs. So we present the definitions w.r.t. circuits explicitly, in which the meaning of $M^{<f>}, M^{<O(f)>}, S^{<f>}$ conforms to Definition 5. First, we present the definition w.r.t. ACVBB which basically follows the known definitions in [25,26], as Definition 6 shows.

Definition 6. *Let $\mathcal{F}_n$ be a family of interactive probabilistic polynomial-size circuits. Let $\mathcal{O}$ be a PPT algorithm which maps (description of) each $f \in \mathcal{F}_n$ to a circuit $O(f)$. We say that $\mathcal{O}$ is an obfuscator for $\mathcal{F}_n$ w.r.t. distribution β_n iff the following holds:*

Computational Functionality: *for random $f \in \mathcal{F}_n$ chosen from distribution β_n, for a random sample of $O(f)$, for any PPT machine M, in the two executions of $M^{<f>}$ and $M^{<O(f)>}$, M's outputs are computationally indistinguishable and f's and $O(f)$'s local outputs are also computationally indistinguishable.*

ACVBB: *there is a PPT simulator S such that for each non-uniform PPT D and random $f \in \mathcal{F}_n$ chosen from distribution β_n, $|\Pr[D(O(f)) = 1] - \Pr[D(S^{<f>}(1^n)) = 1]| = \mathsf{neg}(n)$, where the probabilities are taken over all choices of f and O's, S's and (the oracle) f's coins.*

Definition 6 is proposed for establishing our impossibilities. We could further require that D is given the access to oracle f, as [25,26] adopted. This weaker handling makes the impossibilities stronger. On the other hand, to show the possibility of obfuscating verifiers, we follows [7,8,29] to adopt the following definition, which requires the VBB property holds for each f rather than for a random f.

Definition 7. *Let $\mathcal{F}_n, O$ be as Definition 6 shows. We say that O is an obfuscator for $\mathcal{F}_n$ iff the following holds:*

Approximate Functionality: *for each $f \in \mathcal{F}_n$, $O(f)$ and f are of equivalent functionality except for a negligible fraction of O's coins.*

VBB: *for each non-uniform PPT A and each polynomial p, there is an non-uniform PPT S such that for* each $f \in \mathcal{F}_n$, $|\Pr[A(O(f)) = 1] - \Pr[S^{<f>}(1^n) = 1]| < 1/p(n)$, *where the probabilities are taken over all choices of A's, S's and O's coins.*

As shown in [28], the above two definitions are incomparable. To use zero-knowledge proofs of knowledge to realize identification, a general paradigm is to first sample a random hard instance and a witness for each prover, which then keeps its witness secret. Thus when investigating the obfuscation for provers, it is natural to adopt Definition 6. On the other hand, when investigating the obfuscation for verifiers, our goal is to hide the instance and the client's name where the latter is not random. Thus Definition 7 is more suitable in this scenario.

4 Impossibilities of Obfuscating Provers

In this section we address the question whether or not the honest prover's algorithm incorporated with a public input and a witness is obfuscatable. The results in this section basically say that for a zero-knowledge (or WI) proof or argument of knowledge which uses an efficient prover's strategy, the honest prover is unobfuscatable w.r.t. Definition 6 under some hardness requirements. In Section 4.1 we present the negative results. In Section 4.2 we extend these negative results to t-composition setting.

4.1 Impossibilities for Zero Knowledge and Witness Indistinguishability

We now present the impossibilities for obfuscating provers of zero-knowledge and WI proofs of knowledge. We first present the impossibility for zero-knowledge. Assume (P, V) is a zero-knowledge proof or argument of knowledge for a relation R and P admits an efficient strategy if given $(x, w) \in R$. Moreover, we require that there is a hard-instance sampling algorithm that can randomly select a hard instance x and a witness w such that given x it is hard to compute a witness for x. More precisely, we have the following definition.

Definition 8. *(hard-instance sampler) A PPT machine Sam_R is called a hard-instance sampler for relation R, if for any (non-uniform) PPT A,* $\Pr[\mathsf{Sam}_R(1^n) \rightarrow (x, w), (x, w) \in R_n : A(x) \rightarrow w' \text{ s.t. } (x, w') \in R_n] = \mathsf{neg}(n)$, *where R_n denotes the subset of R in which all instances are of length n and $\mathsf{Sam}_R(1^n)$'s output is always in R_n, and the probability is taken over all choices of Sam_R's and A's coins.*

A probability distribution β_{R_n} *over* R_n *is called a distribution induced by* Sam_R*, if letting* ξ *denote the random variable conforming to* β_{R_n}*, for each* $(x, w) \in R_n$*,* $\Pr[\xi = (x, w)]$ *is defined as* $\Pr[\mathsf{Sam}_R(1^n) = (x, w)]$*, where the latter is taken over all choices of* Sam_R*'s coins.*

For instance, for an arbitrary one-way function f let R_f denote $\{(f(x), x)\}$ for all x. Then a hard-instance sampler for $\{(f(x), x)\}$ can be the one simply choosing $x_0 \in \{0,1\}^n$ uniformly and outputting $(f(x_0), x_0)$. Usually, it is unnecessary to calculate β_{R_n} explicitly. For instance, in the case of R_f, we don't need to calculate the induced distribution of $(f(x_0), x_0)$ generated by the sampler explicitly. Actually, we only need to be ensured that given $f(x_0)$ it is hard to compute a x' satisfying $f(x') = f(x_0)$.

For $(x, w) \in R_n$, let $P(x, w)$ denote P's program with (x, w) hardwired. Let $P(X, W)$ denote the family of all $P(x, w)$ for all $(x, w) \in R_n$. Our first impossibility says that $P(X, W)$ cannot be obfuscated, as the following claim states.

Claim 1. *Assume* (P, V) *is a zero-knowledge proof or argument of knowledge for relation* R*. Assume* Sam_R *is a hard-instance sampler for* R *and* β_{R_n} *is the probability distribution over* R_n *induced by* Sam_R*. Then* $P(X, W)$ *is unobfuscatable w.r.t.* β_{R_n} *under Definition 6.*

Proof. Suppose the claim is not true. This means that there exists an obfuscator O for $P(X, W)$. For $(x, w) \leftarrow_R \mathsf{Sam}_R(1^n)$ (i.e. choose (x, w) conforming to β_{R_n}), let Q denote $O(P(x, w))$. Then there exists a PPT simulator S satisfying that for any polynomial-sized distinguisher D, $|\Pr[D(Q) = 1] - \Pr[D(S^{<P(x,w)>}(1^n)) = 1]| = \mathsf{neg}(n)$. (Here we assume S can obtain x by querying $P(x, w)$ with a specific command. We do this for the consideration that we are trying to provide negative results. If S given x cannot satisfy ACVBB, then S without x by no means satisfies ACVBB.) Since Q is an obfuscation for $P(x, w)$, Q can convince V that $x \in L(R)$ with overwhelming probability for every true x or V rejects x with overwhelming probability for every false x. Since $|\Pr[D(Q) = 1] - \Pr[D(S^{<P(x,w)>}(1^n)) = 1]| = \mathsf{neg}(n)$, we have that S's output is also a valid prover's program which behaves indistinguishably from Q.

Now let us focus on $S^{<P(x,w)>}(1^n)$. First, according to Definition 5, we have that S can perform arbitrary interaction (including some half-baked executions) with $P(x, w)$. Since (P, V) is zero-knowledge, the interaction can be simulated by the simulator for proving zero-knowledge of (P, V). Let S' denote the simulator for proving zero-knowledge of (P, V). Then for S (now viewed as a verifier) and x, S' on input S's code and x can simulate S's output with the interaction with $P(x, w)$ (S' can send x to S). That is, S''s output is indistinguishable from S's output. Let Q' denote S''s output.

Thus, we can construct a PPT algorithm, denoted $D(S')$, which can compute a witness for $x \in L(R)$ with overwhelming probability. $D(S')$ works as follows. On input x generated by $\mathsf{Sam}_R(1^n)$, $D(S')$ runs $S'(x)$ (w.l.o.g. assume S' having S's code hardwired) to generate Q'. Then since (P, V) is a proof or argument of knowledge, D runs the knowledge extractor E of (P, V) on input Q' to output a witness. Since Q' is indistinguishable from S's output in the execution of $S^{<P(x,w)>}(1^n)$, which is indistinguishable from Q, Q' is also a valid prover program which can convince V $x \in L(R)$ with overwhelming probability. Thus the probability that E on input Q' succeeds in generating a witness for x is overwhelming. This contradicts the assumption on Sam_R. This absurdity is caused by the hypothesis that $P(X,W)$ is obfuscatable. So the claim follows.

□

Now let us consider the case that (P,V) is not zero-knowledge but witness-indistinguishable. Is $P(X,W)$ still unobfuscatable? Actually, we cannot show this fact. But if each x has at least two witnesses and assume that for any PPT A, given $(x,w) \leftarrow_R \mathsf{Sam}_R(1^n)$, $\Pr[\mathsf{Sam}_R(1^n) \to (x,w), (x,w) \in R_n : A(x,w) \to w' \text{ s.t. } (x,w') \in R_n, w' \neq w] = \mathsf{neg}(n)$, where the probability is taken over all choices of Sam_R's and A's coins, we have that $P(x,w)$ is indeed unobfuscatable. We say that Sam_R is now **second-witness resistant**. For instance, let f, g denote two one-way permutations, $R_{f,g}$ denote $\{((y_1,y_2),x)) : f(x) = y_1 \text{ or } g(x) = y_2.\}$. A second-witness resistant sampler for $R_{f,g}$ can simply choose $x_1, x_2 \in_R \{0,1\}^n$ and output $((f(x_1), g(x_2)), x_1))$. Any A on input $((f(x_1), g(x_2)), x_1))$ cannot output x_2 with non-negligible probability. Thus the impossibility for WI is shown as follows.

Claim 2. *Assume (P,V) is a WI proof or argument of knowledge for relation R and each x in $L(R)$ has at least two witnesses. Assume Sam_R is a second-witness resistant sampler for R and β_{R_n} is the probability distribution over R_n induced by Sam_R. Then $P(X,W)$ is unobfuscatable w.r.t. β_{R_n} under Definition 6.*

Proof. Suppose that $P(X,W)$ is obfuscatable. Then there exists an obfuscator O for $P(X,W)$. For $(x,w) \leftarrow_R \mathsf{Sam}_R(1^n)$ let Q denote $O(P(x,w))$. Then there exists a PPT simulator S such that for any polynomial-sized distinguisher D, $|\Pr[D(Q)=1] - \Pr[D(S^{<P(x,w)>}(1^n)) = 1]| = \mathsf{neg}(n)$.

Let Q' denote $S^{<P(x,w)>}(1^n)$'s output. Since Q is an obfuscation for $P(x,w)$, the extractor E of (P,V) on input Q's description can output a witness for $x \in L(R)$ with non-negligible probability. By the indistinguishability of Q and Q', we have that E on input Q''s description can also output a witness for $x \in L(R)$ with non-negligible probability.

Then we can show that there exists a verifier which can tell which witness P is using. The verifier runs as follows. It adopts S's strategy to interact with P and finally output a program, denoted Q'. Then it adopts E's strategy on input Q' to extract a witness and output it. Since Sam_R is second-witness resistant, we claim that this witness is w. Otherwise, we can construct an A which can violate the second-witness resistance. That is, on input $(x,w) \leftarrow_R \mathsf{Sam}_R(1^n)$, A constructs $P(x,w)$ and adopts S's strategy with $P(x,w)$ to generate Q' and then runs E with input Q' to output a witness. If this witness is not x, the second-witness resistance of Sam_R is violated.

Thus it can be seen that the verifier's outputs in different interactions where P uses different witnesses for $x \in L(R)$ are obviously distinguishable. However, notice that any sequential repetitions of (P,V) are still witness-indistinguishable. It is a contradiction. Thus the claim follows.

□

Note that the result in this subsection requires the proof of knowledge property for (P, V). Then what about those proof systems without the proof of knowledge property? We leave this an interesting question in future works.

4.2 Extending the Impossibilities to t-Composition Setting

Definition 6 defines that S given access to f can output a fake program which is indistinguishable from the true obfuscated program. What about the case that the distinguisher obtains multiple such programs? To address this question, [8,9,4] introduced the notion of t-composeability of obfuscated programs, which requires that ACVBB or VBB holds in t-composition setting. Basically, an obfuscation w.r.t. t-composition setting requires that even if the distinguisher can obtain t obfuscated programs, he cannot distinguish them from t fake programs generated by S given access to the t programs. We now follow [8,9,4] to extend Definition 6 to t-composition setting as follows.

Definition 9. *Let $\mathcal{F}_n, \mathcal{O}, V$ be those in Definition 6. We say that $\mathcal{O}$ is an obfuscator for $\mathcal{F}_n$ w.r.t. distribution β_n in t-composition setting for any polynomial $t(n)$ iff the same conditions appeared in Definition 6 hold except that the ACVBB property is now modified as follows:*

ACVBB: *there is a uniform PPT oracle simulator S such that for each non-uniform PPT D, and randomly chosen $f_1, \cdots, f_t$ conforming to distribution β_n, $Q_i = \mathcal{O}(f_i)$ for all i, $|\Pr[D(Q_1, \cdots, Q_t) = 1] - \Pr[D(S^{<f_1>, \cdots, <f_t>}(1^n, t)) = 1]| = \mathsf{neg}(n)$, where the probabilities are taken over all choices of S's coins and Q_i for all i, and the access of S to each f_i conforms to Definition 5.*

The result in this subsection says that if (P, V) is concurrent zero-knowledge or witness-indistinguishable, then the result in the previous subsection still holds.

Claim 3. *Assume (P, V) is a concurrent zero-knowledge proof or argument of knowledge for R. Assume Sam_R is a hard-instance sampler for R and β_{R_n} is the probability distribution over R_n induced by Sam_R. Then $P(X, W)$ is unobfuscatable w.r.t. β_n under Definition 9.*

Proof (Proof Sketch:). The proof is similar to that of Claim 1. The route for proving this claim is that if there exists a simulator S which can output t indistinguishable programs, then it is easy to compute a witness for $x_i \in L(R)$. The key point in the proof is that in the computation of $S^{<P(x_1,w_1)>, \cdots, <P(x_t,w_t)>}(1^n, t)$ S can interact with all $P(x_i, w_i)$ for all i and it can adopt any scheduling's strategy for all sessions, so the computation of $S^{<P(x_1,w_1)>, \cdots, <P(x_t,w_t)>}(1^n, t)$ is actually a concurrent interaction between S and these $P(x_i, w_i)$. Then using the similar proof and the simulator for proving concurrent zero-knowledge, we can also construct a PPT algorithm which computes a witness for x_i. It is a contradiction. The claim follows. □

Claim 4. *Given the same conditions in Claim 2, $P(X, W)$ is unobfuscatable w.r.t. β_{R_n} under Definition 9.*

Proof (Proof Sketch:). The proof is similar to that of Claim 2. The route for proving this claim is that if there exists a simulator S which can output t indistinguishable programs, then there exists a verifier interacting with $P(x_i, w_i)$ for all i which can tell the witnesses that t copies of P are using. Notice that witness indistinguishability can be preserved in concurrent setting. The construction of the verifier can be referred to the proof of Claim 2. The claim follows. □

5 Possibilities of Obfuscating Verifiers

In this section we turn to the possibility of obfuscating verifiers of zero-knowledge protocols. The result in this section is that we provide an obfuscation for the honest verifier of Blum's proof for HC w.r.t. Definition 7. In Section 5.1 we present the privacy issue and the motivation of obfuscating verifiers aroused by this issue in detail. In Section 5.2 we present an obfuscation for the honest verifier of Blum's proof for HC which can hide the privacy information (i.e. public inputs).

5.1 Motivation for Obfuscating Verifiers

To run a zero-knowledge protocol, the prover and verifier first need to know the public input. Then the prover proves to the verifier that the public input is true or he knows a witness for it. Usually, the public input is not a secret and both the parties as well as adversaries can obtain it. The security requirements of zero-knowledge protocols don't concern the secrecy of public inputs either. Thus, we ask *if public inputs to zero-knowledge protocols are really not secrets and thus we don't need to hide or protect them at all.*

Consider the scenario of identification. As we know, zero-knowledge proofs of knowledge are a secure way to implement identification. Briefly, an identification scheme consists of a trusted party and multi-clients (i.e. provers) and multi-servers (i.e. verifiers). The trusted party generates a verifiable public input (a.k.a. statement or instance, e.g. a directed graph or a modular square) and a witness (e.g. a Hamilton cycle or a modular square root) for each client. A client's name and its public input constitute a record and the trusted party maintains all records. Each client stores his witness secretly and when he needs to identify himself to a server, he proves to the server the knowledge of the witness via the identification scheme. It is obvious that the server needs to obtain his public input firstly. Actually, it has two options in obtaining this public input as follows.

One is that the server queries the trusted party online with a client's name for his public input each time the client needs to identify himself. But this online method is quite impractical or even impossible in some applications (consider the case that there are some secret governmental organizations that the trusted party cannot publish their names and public inputs, and the case that the trusted party cannot be available in many offline applications). The other one is that the server stores all clients' names and public inputs locally (incorporated in verifier's program) and when a client invokes an identification procedure by informing the server his name, the server can find the corresponding public input and proceeds with the procedure.

Although the second method is practical, there is a risk for leaking clients' privacy. That is, if an adversary can control a server absolutely, he can gain all clients' names from the verifier's program and thus the privacy of these clients leaks. First consider the case of medical systems. If an adversary controls some medical server, which clients are usually patients, then he may extract all patients' names in this server and thus these patients' privacy leaks.

Second, consider the case of ID cards. As we see, ID cards are wildly used in our society. Some secret governmental organizations may need to read data from cards for some secret goals. Before their card readers can read data from a card, they need to pass identification procedures. Thus, by adopting the second method, cards store all legal clients' names and their public inputs. However, these governmental organizations may not want their names stored in cards since they would like to access them anonymously without being identified. Thus a adversarial card-holder can of course extract their names from the verifier's program in his own card, which frustrates the anonymity of their accesses.

Thus, clients' names and public inputs should be protected in some applications. In both of the two cases above, even if a server encrypts clients' names and public inputs, adversaries can first extract the encryption key in the server when they control it and then perform decryption to obtain the desired data. This shows to us that it is important to obfuscate verifiers' programs to hide all information, especially, clients' names and public inputs. In the next subsection, we will provide a desired verifier's strategy of Blum's protocol which can solve this privacy issue.

5.2 Obfuscation for Verifiers

In this subsection we provide an obfuscation for the honest verifier of Blum's protocol [5] while keeping the prover's strategy unchanged. Before proceeding to the obfuscation, we first explain why to choose Blum's proof to demonstrate the possibility of obfuscating verifiers. Basically, when the question of obfuscating verifiers is proposed, we would like to first provide a (theoretical) solution and then propose some practical solutions (as the research community usually does for other problems). Currently we can only present such a possibility for the verifier of Blum's proof. Moreover, we remark in the end of this section that the approach in obfuscating the verifier of Blum's proof cannot be directly used for obfuscating verifiers of other protocols.

For each directed graph G, we use M to denote the entries of G's adjacency matrix. For each M, let I_M denote the point function that on input M outputs 1 or outputs 0 otherwise. Let O_{I_M} denote an obfuscation of I_M w.r.t. Definition 1.

We assume familiar with Blum's protocol. For convenience of constructing obfuscation, we require that the verifier of Blum's proof lastly sends its accept/reject decision to the prover. In the following we present the basic version of our modified protocol in detail and after that we present some second-level considerations such as reducing soundness error via parallel repetitions or

Prover's input: a directed graph $G = (V, E)$ with $n = |V|$; C: a directed Hamilton cycle, $C \subset E$, in G.
Verifier's input: O_{I_M}: an obfuscation of point function I_M where M is the entries of the adjacency matrix of G.

Step P1: The prover selects a random permutation π of the vertices V and computes a graph $G' = (V', E')$ where $V' = \pi(V)$ and E' consists of those edges $(\pi(i), \pi(j))$ for each $(i, j) \in E$. Then it commits to the entries of the adjacency matrix of G'. That is, it sends an n-by-n matrix of commitments such that the $(\pi(i), \pi(j))$ entry is a commitment to 1 if $(i, j) \in E$ and is a commitment to 0 otherwise.

Step V2: The verifier uniformly selects $\sigma \in \{0, 1\}$ and sends it to the prover.

Step P3: If $\sigma = 0$, then the prover sends π to the verifier along with the revealing (i.e., pre-images) of all commitments. Otherwise, the prover reveals to the verifier only the commitments to entries $(\pi(i), \pi(j))$, with $(i, j) \in C$.

Step V4: (This step differs from the original construction.) If $\sigma = 0$, then the verifier computes the inverse permutation of π, denoted π^{-1}, and computes $H = \pi^{-1}(G')$. Then it inputs the entries of the adjacency matrix of H to $O(I_M)$. If $O(I_M)$ returns 1, then the verifier accepts or else rejects. If $\sigma = 1$, the verifier simply checks that all revealed values are 1 and that the corresponding entries form a simple n-cycle. (Of course, in both cases, the verifier checks that the revealed values do fit the commitments.) The verifier accepts if and only if the corresponding condition holds. Lastly, the verifier sends its accept/reject decision to the prover.

Protocol 1 . The obfuscated verifier for Blum's proof for HC

sequential repetitions of the basic protocol, extending the protocol to multi-prover setting, adding provers' names in the verifier's program.

Our obfuscated verifier is shown in Protocol 1. It can be first seen that the verifier doesn't receive G as input and instead its input is O_{I_M}. Second, our verifier differs from the original verifier in [5] is in Step V4. Third, ensured by Claim 5, this protocol is also a zero-knowledge proof of knowledge. If it is used for identification, the prover needs to send its name to the verifier before invoking the identification procedure. But now we ignore the step of prover's sending its name and accordingly, we only need to refer to point function I_M. We will later show that when considering that step, we only need to slightly modify the point function by adding prover's name.

Let Ver denote the honest verifier of Blum's protocol, $Ver(G)$ denote Ver having G hardwired. Let Ver' denote the verifier in Protocol 1, $Ver'(O_{I_M})$ denote Ver' having O_{I_M} hardwired (satisfying that O_{I_M}'s description can be read from $Ver'(O_{I_M})$'s code). Then we have the following claim.

Claim 5. *Assume O_{I_M} is an obfuscation of point function I_M w.r.t. Definition 1. Then $Ver'(O_{I_M})$ is an obfuscation of $Ver(G)$ w.r.t. Definition 7.*

Proof. We now show that the two properties in Definition 7 can be satisfied.

Approximate Functionality. The difference between $Ver'(O_{I_M})$ and $Ver(G)$ lies in Step V4. So we only focus on the two verifiers' outputs in Step V4. By [5] and Protocol 1, in the case of $\sigma = 1$, the two verifiers behaves identically. Thus we only need to show that their outputs are also identical in the case of $\sigma = 0$ in the following. Due to [5], in this case $Ver(G)$ checks if G' and G are isomorphic via π. If they are isomorphic, $Ver(G)$ accepts or rejects otherwise. On the other hand, for $Ver'(O_{I_M})$, due to Protocol 1 if G' and G are isomorphic via π, then $H = \pi^{-1}(G')$ is indeed G. So by Definition 1, except for negligible probability (in generating O_{I_M}), O_{I_M} on input the adjacency matrix of H should output 1. Thus $Ver'(O_{I_M})$ accepts. If the two graphs are not isomorphic, both the two verifiers reject, except for negligible probability. Thus, the approximate functionality property is satisfied.

VBB. To establish the VBB property, we need to show that for each A and each polynomial $p(n)$, there exists a simulator S satisfying that $|\Pr[A(Ver'(O_{I_M})) = 1] - \Pr[S^{<Ver(G)>}(1^n) = 1]| < 1/p(n)$ where $n = |V|$. We can see that for A, there exists a PPT A' such that $A(Ver'(O_{I_M}))$'s output is identical to A''s output on input $O(I_M)$. Thus $\Pr[A(Ver'(O_{I_M})) = 1] = \Pr[A'(O_{I_M}) = 1]$.

Since O_{I_M} is an obfuscation of I_M w.r.t. Definition 1, for A' and $2p(n)$, there exists a simulator S' satisfying that $|\Pr[A'(O_{I_M}) = 1] - \Pr[S'^{I_M}(1^n) = 1]| < 1/2p(n)$. We now turn to construct the desired S such that $|\Pr[S'^{I_M}(1^n) = 1] - \Pr[S^{<Ver(G)>}(1^n) = 1]| < 1/2p$. S works as follows. It emulates S''s computation and is responsible for answering all S' queries. During the emulation, for each S''s query S adopts the following strategy. For an arbitrary S''s query M', Let $G_{M'}$ denote the graph corresponding to M' (if M' is an invalid encoding of the adjacency matrix of a graph of n vertexes, S simply responds with 0). S adopts the prover's strategy of Blum's protocol to commit to the entries of the adjacency matrix of a random isomorphism to $G_{M'}$ and sends the commitments to its oracle $Ver(G)$. Then $Ver(G)$ responds with a random bit σ. If $\sigma = 0$, then S sends the permutation and opens all the commitments to $Ver(G)$. If $Ver(G)$ accepts, S knows that $G_{M'}$ is identical to G and then responds to S' with 1. Otherwise, S responds to S' with 0. If $\sigma = 1$, S aborts the current interaction and re-interacts with $Ver(G)$ from the beginning. We allow S to perform the re-interactions at most n times.

It can be seen that S runs in polynomial-time and the probability that σ is always 1 in n interactions is 2^{-n} (notice that Ver is the honest verifier). On the occurring that there exists an interaction in which $\sigma = 0$, S can respond to S with the correct answer. Thus $|\Pr[S'^{I_M}(1^n) = 1] - \Pr[S^{<Ver(G)>}(1^n) = 1]| = 2^{-n} < 1/2p$. Thus, $|\Pr[A'(O_{I_M}) = 1] - \Pr[S^{<Ver(G)>}(1^n) = 1]| < 1/p$. Namely, $|\Pr[A(Ver'(O_{I_M})) = 1] - \Pr[S^{<Ver(G)>}(1^n) = 1]| < 1/p$. Thus the VBB property is satisfied. Taking the above two conclusions, we have the claim follows.

□

Claim 5 ensures that even if an adversary obtains $Ver'(O_{I_M})$'s description totally, he cannot learn anything from this program beyond its functionality, e.g. the prover's public input. We remark that to protect some provers' privacy in practice, the trusted party should keep these provers' names and public inputs secret, or else an adversary may obtain these records easily and input each of them to O_{I_M} one by one to locate the prover. Now we present several second-level considerations for Protocol 1 as follows.

Achieving Negligible Soundness Error. Protocol 1 admits almost 1/2 soundness error. We now may adopt $\omega(\log n)$ repetitions of Protocol 1 or $\omega(1)$ repetitions of the $\log n$ parallel composition of Protocol 1 to reduce the soundness error to negligible.

Extension to Multi-prover Setting. In practice there are multi-provers who need to prove to Ver' via Protocol 1. In this case, Ver' needs to store the public inputs of these provers securely such that an adversary cannot extract these inputs even he can obtain the verifier's program totally. Based on Protocol 1, we can easily extend the obfuscation of the verifier to the multi-prover setting.

Assume there exist m provers with public inputs $G_1, \cdots, G_m$ which are pair-wise different. Let $M_1, \ldots, M_m$ denote the the entries of the adjacency matrix of $G_1 \cdots, G_m$ respectively. The solution for hiding these public inputs in Ver' is to replace O_{I_M} in Protocol 1 by $O_{I_{M_1,\cdots,M_m}}$, which denotes an obfuscation for the multi-point function $I_{M_1,\cdots,M_m}$ shown in [8]. In Step V4 of each session, Ver' sends H to $O_{I_{M_1,\cdots,M_m}}$ and if $O_{I_{M_1,\cdots,M_m}}$ returns 1 then it accepts or else rejects in the case of $\sigma = 0$. It can be seen that Ver' with $O_{I_{M_1,\cdots,M_m}}$ hardwired is also an obfuscation of $Ver(G_1, \cdots, G_m)$.

Adding Provers' Names in the Point Function. As mentioned previously, a prover may send its name before invoking the identification procedure. We consider the case that there is only a prover for simplicity. Assume that the prover's name is $\alpha \in \{0,1\}^*$ and its public input is G. We construct an obfuscated point program $O_{I_{\alpha \circ M}}$, which now becomes Ver''s input. In Step V4, in the case of $\sigma = 0$, Ver' computes H and sends $\alpha \circ H$ to $O_{I_{\alpha \circ M}}$, where $\circ$ denotes concatenation. Notice that Ver' receives α from the prover before running the identification procedure. Using the similar analysis, we have the result in this subsection still holds.

Remark 1. It can be seen that our approach for obfuscating the verifier is to let the verifier receive an obfuscated whole information of G as input and then compare the information in Step P3 to the whole information. While, for some other constructions e.g. Goldreich et al.'s protocol [18], their verifiers don't behave in such way. Thus our approach doesn't suit for obfuscating these verifiers.

6 Conclusions

In this paper we address the question of obfuscating provers and verifiers of zero-knowledge proofs of knowledge, motivated by theoretical and practical interests. First, we achieve some impossibilities of obfuscating provers w.r.t. the ACVBB definition, i.e., provers are unobfuscatable. This negative result means that if someone obtains a prover's program, he can indeed learn some knowledge.

Namely we can say the prover's program contains knowledge beyond its functionality. Second, we present an obfuscation for the verifier of Blum's protocol w.r.t. the VBB definition. This positive result means that for anyone who possesses the verifier's program, he cannot learn any knowledge beyond its functionality, which solves the privacy issue of hiding provers' names and public inputs.

Acknowledgments. The authors show their deep thanks to the reviewers of CANS 2011 for their detailed and useful comments, with the help of which the presentation of this work is significantly improved. This work is supported by the National Natural Science Foundation of China (61100209) and Shanghai Postdoctoral Scientific Program (11R21414500) and Major Program of Shanghai Science and Technology Commission (10DZ1500200).

References

1. Barak, B.: How to go beyond the black-box simulation barrier. In: FOCS, pp. 106–115 (2001)
2. Barak, B., Goldreich, O., Impagliazzo, R., Rudich, S., Sahai, A., Vadhan, S.P., Yang, K.: On the (im)possibility of Obfuscating Programs. In: Kilian, J. (ed.) CRYPTO 2001. LNCS, vol. 2139, pp. 1–18. Springer, Heidelberg (2001)
3. Bellare, M., Goldreich, O.: On Defining Proofs of Knowledge. In: Brickell, E.F. (ed.) CRYPTO 1992. LNCS, vol. 740, pp. 390–420. Springer, Heidelberg (1993)
4. Bitansky, N., Canetti, R.: On Strong Simulation and Composable Point Obfuscation. In: Rabin, T. (ed.) CRYPTO 2010. LNCS, vol. 6223, pp. 520–537. Springer, Heidelberg (2010)
5. Blum, M.: How to prove a theorem so no one else can claim it. In: The International Congress of Mathematicians, pp. 1441–1451 (1986)
6. Brassard, G., Chaum, D., Crépeau, C.: Minimum disclosure proofs of knowledge. J. Comput. Syst. Sci. 37(2), 156–189 (1988)
7. Canetti, R.: Towards Realizing Random Oracles: Hash Functions that Hide all Partial Information. In: Kaliski Jr., B.S. (ed.) CRYPTO 1997. LNCS, vol. 1294, pp. 455–469. Springer, Heidelberg (1997)
8. Canetti, R., Dakdouk, R.R.: Obfuscating Point Functions with Multibit Output. In: Smart, N.P. (ed.) EUROCRYPT 2008. LNCS, vol. 4965, pp. 489–508. Springer, Heidelberg (2008)
9. Canetti, R., Kalai, Y.T., Varia, M., Wichs, D.: On Symmetric Encryption and Point Obfuscation. In: Micciancio, D. (ed.) TCC 2010. LNCS, vol. 5978, pp. 52–71. Springer, Heidelberg (2010)
10. Canetti, R., Rothblum, G.N., Varia, M.: Obfuscation of Hyperplane Membership. In: Micciancio, D. (ed.) TCC 2010. LNCS, vol. 5978, pp. 72–89. Springer, Heidelberg (2010)
11. Canetti, R., Varia, M.: Non-malleable Obfuscation. In: Reingold, O. (ed.) TCC 2009. LNCS, vol. 5444, pp. 73–90. Springer, Heidelberg (2009)
12. Crescenzo, G.D.: You Can Prove so Many Things in Zero-Knowledge. In: Feng, D., Lin, D., Yung, M. (eds.) CISC 2005. LNCS, vol. 3822, pp. 10–27. Springer, Heidelberg (2005)
13. Ding, N., Gu, D.: A Note on Obfuscation for Cryptographic Functionalities of Secret-Operation then Public-Encryption. In: Ogihara, M., Tarui, J. (eds.) TAMC 2011. LNCS, vol. 6648, pp. 377–389. Springer, Heidelberg (2011)

14. Dwork, C., Naor, M., Sahai, A.: Concurrent zero-knowledge. In: STOC, pp. 409–418 (1998)
15. Feige, U., Fiat, A., Shamir, A.: Zero-knowledge proofs of identity. J. Cryptology 1(2), 77–94 (1988)
16. Feige, U., Shamir, A.: Witness indistinguishable and witness hiding protocols. In: STOC, pp. 416–426. ACM (1990)
17. Gentry, C.: Fully homomorphic encryption using ideal lattices. In: STOC, pp. 169–178. ACM (2009)
18. Goldreich, O., Micali, S., Wigderson, A.: Proofs that yield nothing but their validity and a methodology of cryptographic protocol design (extended abstract). In: FOCS, pp. 174–187. IEEE (1986)
19. Goldwasser, S., Kalai, Y.T.: On the impossibility of obfuscation with auxiliary input. In: FOCS, pp. 553–562. IEEE Computer Society (2005)
20. Goldwasser, S., Kalai, Y.T., Rothblum, G.N.: One-Time Programs. In: Wagner, D. (ed.) CRYPTO 2008. LNCS, vol. 5157, pp. 39–56. Springer, Heidelberg (2008)
21. Goldwasser, S., Micali, S., Rackoff, C.: The knowledge complexity of interactive proof-systems (extended abstract). In: STOC, pp. 291–304. ACM (1985)
22. Goldwasser, S., Rothblum, G.N.: On Best-Possible Obfuscation. In: Vadhan, S.P. (ed.) TCC 2007. LNCS, vol. 4392, pp. 194–213. Springer, Heidelberg (2007)
23. Hada, S.: Zero-Knowledge and Code Obfuscation. In: Okamoto, T. (ed.) ASIACRYPT 2000. LNCS, vol. 1976, pp. 443–457. Springer, Heidelberg (2000)
24. Hada, S.: Secure Obfuscation for Encrypted Signatures. In: Gilbert, H. (ed.) EUROCRYPT 2010. LNCS, vol. 6110, pp. 92–112. Springer, Heidelberg (2010)
25. Hofheinz, D., Malone-Lee, J., Stam, M.: Obfuscation for cryptographic purposes. J. Cryptology 23(1), 121–168 (2010)
26. Hohenberger, S., Rothblum, G.N., Shelat, A., Vaikuntanathan, V.: Securely Obfuscating Re-Encryption. In: Vadhan, S.P. (ed.) TCC 2007. LNCS, vol. 4392, pp. 233–252. Springer, Heidelberg (2007)
27. Lynn, B., Prabhakaran, M., Sahai, A.: Positive Results and Techniques for Obfuscation. In: Cachin, C., Camenisch, J.L. (eds.) EUROCRYPT 2004. LNCS, vol. 3027, pp. 20–39. Springer, Heidelberg (2004)
28. Varia, M.: Studies in Program Obfuscation. Ph.D. thesis, Massachusetts Institute of Technology (2010)
29. Wee, H.: On obfuscating point functions. In: Gabow, H.N., Fagin, R. (eds.) STOC, pp. 523–532. ACM (2005)

Author Index

GPSR Compliance

The European Union's (EU) General Product Safety Regulation (GPSR) is a set of rules that requires consumer products to be safe and our obligations to ensure this.

If you have any concerns about our products, you can contact us on ProductSafety@springernature.com

In case Publisher is established outside the EU, the EU authorized representative is:

Springer Nature Customer Service Center GmbH
Europaplatz 3
69115 Heidelberg, Germany

Batch number: 09474011

Printed by Printforce, the Netherlands